AF214010

Commendations from Book Reviewers

As a cryptographer and cryptanalyst with decades of experience, I am impressed by the fact that this accessible and comprehensive treatment of blockchain technology emphasizes its burgeoning value across industrial and government sectors, rather than focusing only on the more popularly discussed use cases of cryptocurrencies and nonfungible tokens. While it is commonplace to dismiss the need for advanced identity and access management techniques, since, after all, effectively countering double-spending does not require knowledge of true identity or other attributes of entities that transact via blockchain, the authors take a holistic approach in order to maximize the value of the book for system designers and implementers who are rightfully concerned about security and privacy.

David W. Kravitz, Technical Advisor, Spring Labs, FIPS-186 Digital Signature Algorithm Inventor and Hyperledger Fabric Identity Management Architect

A comprehensive and cutting-edge method for securing blockchain applications—pushing the boundaries of the current state of the art in several key areas.

David Cervigni, Senior Security Research Engineer at R3

Filling a critical void in the blockchain landscape, this comprehensive guide reveals the secrets to designing secure and attack-resistant applications. By bridging the gap between theory and practice with realistic examples, "Blockchain Application Security" empowers you with the knowledge and expertise gathered by the authors, needed to safeguard your blockchain applications. Join the ranks of the most security-conscious architects and developers and learn how to build.

Steven Wierckx, Product Security Team Lead and Threatmodel Trainer at Toreon

"Blockchain Application Security" is a valuable resource for specialists that are involved in development and deployment of blockchain applications. As a Cloud Security specialist, I would like to highlight a hands-on example that describes deployment of a decentralized application (dApp) that integrates blockchain technology with AWS services. The authors provide a detailed overview of the infrastructure components and discuss security risks that are related to them, as well as provide guidance on how to secure the implementation.

Ihor Sasovets, Lead Security Engineer and Penetration Tester at TechMagic, Cloud Security Specialist, AWS Community Builder, Contributor of OWASP API Top 10 2019 Edition

"Blockchain Application Security" is not merely a theoretical guide; it is a practical roadmap for developers to navigate the complexities of blockchain security. The depth and breadth of your exploration into the intricate security aspects of blockchain application development are truly remarkable. We will be recommending this book to our clients and incorporating its principles into our training programs. It is a testament to your expertise and dedication to fostering a more secure blockchain landscape.

Vijay Dhanasekaran, Founder and Chief Blockchain Officer,
Consultant at Blocknetics

The book masterfully guides readers from blockchain fundamentals to advanced implementation with security as its cornerstone. The author balances accessible explanations of blockchain, distributed ledgers and wallet integration with sophisticated dApp vulnerability analysis across multiple blockchain ecosystems. This comprehensive resource stands out by providing both conceptual frameworks and practical implementation patterns that protect users and developers alike. For anyone creating decentralized applications, from beginners to experienced developers, this book delivers essential knowledge in a remarkably clear progression from theory to implementation, making it an indispensable addition to any blockchain developer's library.

Mohd Mehdi, Head of DevOps, DevSecOps and Infrastructure at InfStones

Blockchain Application Security

How to Design Secure and Attack Resilient Blockchain Applications

Marco Morana
Harpreet Singh
Francesco Piccoli

WILEY

Copyright © 2026 by John Wiley & Sons, Inc. All rights reserved, including rights for text and data mining and training of artificial intelligence technologies or similar technologies.

Published by John Wiley & Sons, Inc., Hoboken, New Jersey.
Published simultaneously in Canada.

No part of this publication may be reproduced, stored in a retrieval system, or transmitted in any form or by any means, electronic, mechanical, photocopying, recording, scanning, or otherwise, except as permitted under Section 107 or 108 of the 1976 United States Copyright Act, without either the prior written permission of the Publisher, or authorization through payment of the appropriate per-copy fee to the Copyright Clearance Center, Inc., 222 Rosewood Drive, Danvers, MA 01923, (978) 750-8400, fax (978) 750-4470, or on the web at www.copyright.com. Requests to the Publisher for permission should be addressed to the Permissions Department, John Wiley & Sons, Inc., 111 River Street, Hoboken, NJ 07030, (201) 748-6011, fax (201) 748-6008, or online at http://www.wiley.com/go/permission.

The manufacturer's authorized representative according to the EU General Product Safety Regulation is Wiley-VCH GmbH, Boschstr. 12, 69469 Weinheim, Germany, e-mail: Product_Safety@wiley.com.

Trademarks: Wiley and the Wiley logo are trademarks or registered trademarks of John Wiley & Sons, Inc. and/or its affiliates in the United States and other countries and may not be used without written permission. All other trademarks are the property of their respective owners. John Wiley & Sons, Inc. is not associated with any product or vendor mentioned in this book.

Limit of Liability/Disclaimer of Warranty: While the publisher and the authors have used their best efforts in preparing this work, including a review of the content of the work, neither the publisher nor the authors make any representations or warranties with respect to the accuracy or completeness of the contents of this work and specifically disclaim all warranties, including without limitation any implied warranties of merchantability or fitness for a particular purpose. Certain AI systems have been used in the creation of this work. No warranty may be created or extended by sales representatives, written sales materials or promotional statements for this work. The fact that an organization, website, or product is referred to in this work as a citation and/or potential source of further information does not mean that the publisher and authors endorse the information or services the organization, website, or product may provide or recommendations it may make. This work is sold with the understanding that the publisher is not engaged in rendering professional services. The advice and strategies contained herein may not be suitable for your situation. You should consult with a specialist where appropriate. Further, readers should be aware that websites listed in this work may have changed or disappeared between when this work was written and when it is read. Neither the publisher nor authors shall be liable for any loss of profit or any other commercial damages, including but not limited to special, incidental, consequential, or other damages.

For general information on our other products and services or for technical support, please contact our Customer Care Department within the United States at (800) 762-2974, outside the United States at (317) 572-3993 or fax (317) 572-4002.

Wiley also publishes its books in a variety of electronic formats. Some content that appears in print may not be available in electronic formats. For more information about Wiley products, visit our web site at www.wiley.com.

Library of Congress Cataloging-in-Publication Data

Names: Morana, Marco M. author | Singh, Harpreet (Engineering manager) author | Piccoli, Francesco (Chief executive officer) author
Title: Blockchain application security : how to design secure and attack resilient blockchain applications / Marco Morana, Harpreet Singh, Francesco Piccoli.
Description: Hoboken, New Jersey : Wiley, [2026] | Includes bibliographical references and index.
Identifiers: LCCN 2025030392 | ISBN 9781119551034 (hardback) | ISBN 9781119551058 (epdf) | ISBN 9781119551072 (epub) | ISBN 9781119551065 (obook)
Subjects: LCSH: Blockchains (Databases)—Security measures
Classification: LCC QA76.9.B56 M67 2026
LC record available at https://lccn.loc.gov/2025030392

Cover Design: Wiley
Cover Image: © KTSDESIGN/Getty Images
Printed and bound by CPI Group (UK) Ltd, Croydon, CR0 4YY

C9781119551034_100925

To my parents, mentors, and the next generation of engineers. To my parents, thank you for nurturing my curiosity, guiding me with integrity, and inspiring a lifelong commitment to learning. To my mentors, thank you for challenging me to lead with purpose and give back. To future engineers: build with intention, question deeply, and never stop learning. May this book pass on the knowledge and passion I've gained and serve as a foundation for your journey ahead.

—Marco

Contents

Foreword

Blockchain has emerged as one of the most transformative technologies of our time, promising a decentralized future where trust is rooted in code rather than institutions. Yet with this shift comes a sobering reality: decentralizing infrastructure does not decentralize risk. In fact, it often magnifies it, spreading vulnerabilities across complex stacks of smart contracts, wallets, oracles, bridges, and consensus mechanisms.

Having worked in cybersecurity across various sectors, I have seen how blockchain innovation often surpasses its security measures. Many dApps are created with good intentions but lack the necessary architectural principles for robust security against real-world threats.

This book was written to close that gap.

Blockchain Application Security offers a practical, risk-centric approach to defending blockchain ecosystems. It brings together my background in threat modeling, application security, and system architecture to help both developers and defenders think more strategically about the risks they face. From protocol vulnerabilities and DeFi exploits to attack simulations and layered defense, each chapter is grounded in both theory and hands-on relevance.

This book provides practical advice on securing blockchain applications. It includes understanding the technology, applying security-by-design principles, identifying blockchain-specific vulnerabilities, and designing for attack resilience through threat modeling. It also explains how to secure dApp infrastructure based on real attack case studies and offers examples of auditing smart contract vulnerabilities.

Security goes beyond fixing bugs; it safeguards the trust central to decentralized systems. My goal is to help readers not only understand the threats but also anticipate them, build resilience, and contribute to a safer Web3 future.

Whether you're a builder, auditor, architect, or passionate about blockchain security, I hope this book becomes your trusted companion.

— Marco Morana

Field CISO, Lead for Financial Services and Product Security Engineering, Avocado Systems Inc.

Principal Author, *Blockchain Application Security: How to Design Secure and Attack-Resilient Blockchain Applications*

Preface

Blockchain technology has evolved significantly over the past decade, reshaping the way we think about trust, security, and decentralization. I wrote this book to share the lessons I have learned from my early experiments with blockchain security, dating back to 2015. At the time, I was a Senior Application Security Architect at one of the largest financial institutions in the United States, where I was responsible for conducting security reviews of blockchain-based projects. These initiatives were undertaken in collaboration with start-ups developing blockchain technology and banking consortia exploring its potential applications.

With any emerging technology comes both innovation and risk. While blockchain has introduced new capabilities that disrupt traditional business processes, it also presents novel security challenges that organizations must navigate to ensure safe and effective adoption. One of the primary concerns remains the integration of blockchain with legacy systems, a complex but necessary requirement for broader enterprise adoption. Despite the growing enthusiasm around blockchain, real-world implementations must undergo rigorous security assessments to address vulnerabilities and operational risks.

Like all digital innovations, blockchain is not immune to security threats. A well-known example from its early days was the 2016 DAO attack, where a decentralized autonomous organization built on Ethereum was exploited due to multiple security flaws, resulting in a loss of approximately $50 million. This incident highlighted the critical need for proactive security measures in blockchain development, particularly for decentralized applications (dApps) and smart contracts.

In the early stages of private and permissioned blockchains, security assessments were often limited to checklist-based reviews, lacking the depth required to address blockchain-specific risks. The absence of standardized methodologies made it difficult to conduct architecture risk analysis and threat modeling, leaving significant gaps in security assurance. Over time, I observed the growing need for structured security frameworks to guide the secure design, development, and deployment of blockchain applications.

Fast forward to today, blockchain technology has transitioned from being an experimental innovation in financial services to becoming a mainstream platform for Web3.0 applications, including decentralized finance (DeFi), supply chain solutions, and digital identity management. The ecosystem has matured, equipping developers with advanced tools and frameworks to build dApps that integrate with blockchain platforms efficiently. However, this progress has also expanded the attack surface, making security a paramount concern.

This book began as a project in 2018 with the goal of educating readers about the fundamentals of blockchain technology, its use cases, and security considerations for private and permissioned enterprise-grade blockchain platforms. Over time, the scope broadened to address the security challenges of dApps, offering insights into how developers, security professionals, and business leaders can design, build, and secure blockchain-based solutions.

Who Should Read This Book?

This book is designed for stakeholders managing the security risks of dApps and blockchain-based products. Whether you are an architect, engineer, security leader, or business executive, this book provides insights tailored to your role:

- Security architects will gain a comprehensive understanding of designing secure dApps from the ground up. They will learn how to embed security into architecture, design patterns, and APIs, ensuring blockchain applications are resilient against attacks.
- Software security engineers will find practical guidance on securing dApp implementations, identifying vulnerabilities, and testing blockchain applications for security flaws. This book also covers threat modeling, DevSecOps integration, and secure coding practices for smart contracts.
- CISOs and product security business owners will benefit from governance, compliance, and risk management perspectives. This book provides strategies for security oversight, ensuring compliance with regulatory standards and best practices for blockchain security assurance.

How to Navigate This Book?

Should this book be read from beginning to end? That depends on the reader. While the chapters build upon one another, making sequential reading beneficial especially for those new to blockchain security, the content is structured to allow readers to focus on topics most relevant to their roles.

For security architects, Chapter 2 is the core section, covering dApp architectures, security requirements, and security by design principles. This chapter provides a detailed breakdown of securing APIs, protecting secrets, and implementing secure smart contracts. Chapter 3 is also essential, as it provides a comprehensive approach to securing blockchain applications by focusing on vulnerability prevention, threat modeling, and security testing. It introduces structured methodologies to identify, analyze, and mitigate risks in dApps, particularly within DeFi ecosystems. One of the key highlights of this chapter is the dApp DeFi threat modeling use case, providing a practical, step-by-step guide for security architects and engineers. It outlines how to identify attack vectors, assess design flaws, and implement risk mitigation strategies early in development. This structured approach ensures dApps are secure by design, implementation, and testing, rather than addressing security issues postdeployment. Chapter 4 offers practical insights on dApp creation, auditing methodologies, and security implementation strategies.

For software security engineers, the primary focus should be on Chapter 2, which covers API security, key management, and smart contract security best practices. This chapter provides essential insights into securing dApp components, ensuring that authentication, data protection, and cryptographic key handling are implemented securely. Chapter 3 offers an in-depth exploration of smart contract vulnerabilities, detailing common attack vectors such as reentrancy, integer overflows, and logic flaws. It also introduces threat modeling methodologies, helping engineers systematically identify and mitigate risks before deployment. Additionally, this chapter discusses DevSecOps security tooling, highlighting automated vulnerability scanning, continuous security integration, and security validation techniques that can be embedded into the development pipeline.

Chapter 4 is particularly valuable, offering hands-on guidance on smart contract audits, security testing methodologies, and practical implementation strategies. It provides real-world examples of securing dApps, focusing on code auditing, penetration testing, and best practices for secure deployment. This chapter also covers attack simulation exercises, allowing engineers to test security assumptions and refine their defenses against emerging threats in blockchain applications.

For CISOs and heads of product security, Chapter 1 is critical as it establishes the legal, regulatory, and risk considerations associated with blockchain applications. This chapter explores the compliance landscape, including data privacy laws, financial regulations, and jurisdictional challenges, helping organizations align blockchain adoption with corporate risk management strategies.

Chapter 3 is indispensable, providing a deep dive into security incidents, compliance audits, and risk management methodologies for dApps and blockchain platforms. It covers real-world case studies of blockchain security breaches, analyzing attack techniques, lessons learned, and strategic mitigation measures. Additionally, this chapter outlines frameworks for security audits, industry compliance standards, and enterprise-wide blockchain security governance. The risk assessment methodologies presented here help prioritize threats, implement proactive defenses, and establish incident response strategies tailored to blockchain ecosystems.

While Chapter 4 is primarily technical, it is still relevant for CISOs and heads of product security who need a high-level understanding of secure code audits. This chapter provides insights into what security code reviews should focus on, key risk areas in blockchain applications, and the types of vulnerabilities auditors typically uncover. It helps security leaders evaluate and oversee secure coding practices, ensuring that blockchain development teams implement proper security controls.

The appendices, including the Risk Analysis Report, the Risk Mitigation Plan, and the Threat Risk Register, provide examples to support informed, risk-based decisions for managing dApp DeFI technical and business risks. These resources help document security measures, establish risk mitigation strategies, and ensure compliance with regulatory requirements.

By leveraging these chapters, CISOs and heads of product security can build a structured approach to blockchain security governance, compliance oversight, and risk management, while gaining a better understanding of secure code auditing processes.

Final Thoughts

Blockchain technology continues to evolve and disrupt multiple industries, but its success depends on how well we secure and manage its risks. The intent of this book is to provide actionable guidance, real-world examples, and best practices for securing blockchain applications – whether they are enterprise-grade permissioned platforms or dApps in Web3 and DeFi.

As security professionals, engineers, and business leaders, we all share the responsibility of ensuring trust, security, and resilience in blockchain-powered systems. I hope this book serves as a valuable resource in your journey to building secure and attack-resilient blockchain applications.

Let's drive blockchain security forward – creating attack-resilient applications by design!

Acknowledgments

Writing *Blockchain Application Security: How to Design Secure and Attack Resilient Blockchain Applications* has been an incredible journey, one that would not have been possible without the support, mentorship, and collaboration of many individuals along the way.

First and foremost, I want to extend my deepest gratitude to my coauthors, Francesco Piccoli and Harpreet Singh, for their invaluable contributions, insights, and expertise. Your dedication and collaboration throughout this project have made this book a richer and more impactful resource for the blockchain security community. It has been a privilege to work alongside you both.

I would like to express my heartfelt thanks to the book reviewers who generously volunteered their free time and subject matter expertise to review the various chapters of this book. Special thanks to David Kravitz, Steven Wierckx, David Cervigni, Mohd Mehdim, Vijay Dhanasekaran, and Ivor Sasovets for their invaluable contributions that have significantly enriched this work.

A special thanks to Brett Kurzman, our editor at Wiley, whose guidance, patience, and editorial expertise have been instrumental in shaping this book. Your support ensured that our vision for this work was realized in the best possible way.

I would also like to express my appreciation to my colleagues and mentors at Citi, where much of my blockchain security architecture work took shape. I am especially thankful to Gulrez Jamadar, who provided the leadership and guidance that allowed me to explore and design secure blockchain solutions. I also deeply appreciate Adarsh Kochhar, whose technical expertise, collaborative discussions, and shared passion for security architecture were instrumental in shaping my understanding of blockchain security challenges and solutions.

Teams working on information security and compliance risk assessments, as well as blockchain architecture risk analysis, highlighted the importance of applying security by design principles to blockchain technology in decentralized applications. The concepts presented in this book are derived from extensive professional experience in safeguarding the development and implementation of blockchain-based financial networks, as well as evaluating the architecture of digital asset platforms.

I am also grateful to the broader cybersecurity and blockchain security communities. Presenting on blockchain security at CISOs' Forums and participating in roundtables with blockchain security subject matter experts has been incredibly enriching. Their innovative work and continuous efforts to secure decentralized systems have been a constant source of inspiration.

Introduction

Are you excited about the future of trust, security, and transparency in the digital world? The blockchain technology is revolutionizing industries by enabling direct, peer-to-peer transactions without the need for intermediaries, fostering greater efficiency and autonomy.

With its decentralized network and distributed ledger, blockchain ensures that transactions are secure, tamperproof, and transparent, creating an ecosystem of trust and accountability. The possibilities are limitless beyond finance; blockchain is already transforming healthcare, insurance, energy, supply chain management, and even government services.

For technologists, blockchain presents an incredible opportunity to build next-generation applications that redefine how data, identity, and assets are managed. While challenges such as privacy, scalability, and security risks exist, innovative decision models help ensure blockchain is optimally applied to the right use cases. Overall, blockchain technology represents a paradigm shift in trust, security, and transparency in the digital age. Blockchain is transforming peer-to-peer transactions by enabling direct exchanges without the need for intermediaries. By removing reliance on centralized authorities, blockchain fosters trustless interactions, reducing transaction costs and significantly increasing processing speed. This decentralized approach enhances efficiency, security, and transparency, paving the way for a more streamlined and frictionless digital economy.

Blockchain transactions come in various forms, with digital asset exchanges being the most common. These exchanges typically involve cryptocurrencies or tokens, allowing peers to trade assets seamlessly. For instance, a seller can transfer goods to a buyer, who pays using cryptocurrency for the agreed amount. Since transactions occur directly between parties, third-party intermediaries are eliminated, resulting in faster, cost-effective, and more secure exchanges. This decentralized approach enhances efficiency and trust in digital transactions.

Traditional payment transactions rely on centralized payment systems and networks, such as credit card processing networks. In such systems, a cardholder would need to find a seller who accepts credit card payments. This payment agreement requires the involvement of a mutual trusted ecosystem of third parties, including the merchant, the merchant bank that will accept the credit card payment, the payment network that will provide for the clearing and settlement of the transaction, and the issuer bank that issued the credit card to the buyer for making payments from funded accounts. Buyers and sellers can make direct payments without relying on a centralized authority, such as a bank or credit card company. In order to execute transactions on the blockchain, buyers and sellers use digital wallets and a trusted cryptocurrency as a form of payment.

This approach offers several benefits over traditional payment methods, including greater privacy and security, as transactions on the blockchain are immutable and transparent. Additionally, transaction fees on the blockchain tend to be lower than those charged by traditional payment systems. However, blockchain technology is not without its own set of risks and limitations, including the volatility of cryptocurrencies and the potential for fraud and hacking. Using this approach provides numerous advantages compared to conventional payment methods. These benefits include enhanced privacy and security since blockchain transactions are both immutable and transparent. Furthermore, transaction processing costs on the blockchain are usually less expensive than those imposed by traditional payment systems.

By utilizing blockchain technology, the process of clearing and settling transactions is facilitated through the implementation of consensus algorithms that rely on the validation of transactions by other peer users on the blockchain network. This distributed system ensures that the transactions are accurate and reliable, and once they are executed, they are recorded in a tamperproof ledger. This distributed ledger is distributed as a database by default, granting both the seller and the buyer access to read the record of the transaction. Each transaction's record is historically linked to other transactions and stored in blocks of data that are securely linked using cryptographic algorithms that resist attacks.

This level of security and transparency in the blockchain network ensures that the transactions are immutable and cannot be altered by any single entity, providing a high level of protection against fraudulent activities. Furthermore, because the ledger is distributed, there is no central point of control, making it extremely difficult for bad actors to manipulate or compromise the system. The use of blockchain technology in clearing and settling transactions offers a secure and efficient method of recording and verifying transactions, enabling faster and more reliable settlements while reducing the need for intermediaries.

Transactions on the blockchain are verified before the transfer of digital assets between parties. Once the transfer is validated, the time and value of the exchange are recorded on the blockchain. The validation of transactions on the blockchain is decentralized, meaning that a central authority is not required to validate transactions. For example, the origin of the transaction is verified to ensure it comes from an authorized payer who has sufficient funds to complete the transaction, and the payee's identity is validated to ensure they are not blacklisted or sanctioned as a nontrusted beneficiary.

The decentralized validation of transactions on the blockchain is achieved through the use of algorithms that allow other peers on the network to establish the authenticity of the transaction. Once a transaction is validated, it is recorded on the immutable, tamperproof ledger, which is distributed among all the nodes on the network. The ledger is continuously updated in real time, ensuring faster and more efficient settlements without intermediaries.

To ensure the tamperproof nature of the ledger, each new transaction is appended as a block of data to the previous transaction using cryptographic algorithms to create an unbreakable chain. Once a new block of data is added, it cannot be removed or modified, making this operation a write-only operation of a distributed database. The use of blockchain technology enables a secure and efficient way to record and verify transactions while reducing the need for intermediaries.

As a type of data storage technology, the blockchain technology brings about a novel approach in comparison to traditional distributed databases in terms of facilitating trust among peers who are designated to validate transactions before they occur. This is accomplished by recording these transactions in a way that is accessible and verifiable by any nodes that are part of the network. While all nodes can access a copy of the transactional data, certain nodes in the network may be selected based on their computational power to partake in validating transactions. These validations also prevent possible fraudulent activities, such as the same funds being spent twice in more than one transaction by the same peer or node.

Once transactions are recorded in the distributed and decentralized database, it becomes impossible for a fraudster seeking to forge historical records of transactions that occurred in the past to tamper with them. The blockchain is the underlying technology that supports payments with cryptocurrencies such as bitcoins. Payments made by peer nodes using bitcoins are recorded in the blockchain and each record is time-stamped to represent a historical record of transactions that have occurred with bitcoins in the past.

The blockchain can be considered an example of distributed ledger technology (DLT) when it is used to record payments in a distributed ledger. The blockchain is also being referred to as the "chain of blocks" due to the data structure used to record transactions in the distributed ledger. This chain of blocks consists of the historical data records or blocks that are tightly linked together using cryptographic algorithms. Each new block in the blockchain represents a new transaction entry that is appended to the previous one as a new block, and this linkage provides an historical record of blocks/transactions.

Once transactions are recorded in the distributed general ledger/blockchain, a copy of all recorded transactions is shared and distributed to each peer that has access to the blockchain network through their computer/nodes. This ensures that each node in the network has an identical copy of the database, making it a distributed database with read-only control that is decentralized. Every peer node that has an address registered to access the network is also authorized to read the transactions occurring by other peers/nodes without the need for special permissions. In a blockchain network, every transaction is recorded and made visible to all participating nodes/users. This transparency means that anyone on the network can see the details of each transaction, including the amount of assets exchanged, the time and date of the transaction, and the public addresses of the sender and receiver.

This increased visibility and transparency of transactions can improve trust between nodes/users in several ways. Firstly, it makes it more difficult for fraudulent or illegitimate transactions to take place unnoticed, as all transactions are visible to all network participants. Any attempt to forge a transaction or manipulate the blockchain would be quickly detected by other nodes/users who would then flag the transaction as fraudulent. This provides a significant deterrent to bad actors and helps to maintain the integrity of the network. Secondly, the visibility of transactions can help to increase confidence in the network and the transactions being recorded. When all nodes/users can see that a transaction has been validated and recorded by other nodes/users, they can be more confident that the transaction is legitimate and accurate. This can help to build trust between nodes/users and encourage greater participation in the network. The visibility of transactions can help to facilitate greater collaboration between nodes/users. When all nodes/users can see the details of each transaction, it becomes easier to track the flow of assets through the network and to identify patterns or trends in transaction activity. This can be useful for identifying areas of the network that may require additional attention or support and for developing new strategies for optimizing the performance and efficiency of the network as a whole.

The blockchain technology's ability to enable trust among peers to execute and record transactions with digital assets without the need for intermediaries has made it a popular choice not just for peer-to-peer payments but also for any business use cases where a trustworthy record of transactions needs to be validated before they can be executed and recorded. The blockchain use cases can be found in various private and public industry sectors, including healthcare, insurance, energy, government, finance, and accounting and tax preparation sectors.

The use of blockchain technology in the healthcare sector has gained significant attention in recent years due to the potential to enhance the security and privacy of patient data. In traditional healthcare systems, patient records are often scattered across different healthcare providers, making it challenging to access and manage patient data in a secure and efficient manner.

By leveraging blockchain technology, healthcare providers can create a decentralized system to store and manage patient records. Patient data can be stored on the blockchain in a tamperproof and immutable manner, ensuring that the records are secure and cannot be modified or deleted by unauthorized individuals. One of the most significant benefits of using blockchain technology in the healthcare sector is the improved accessibility and sharing of patient records among authorized parties. In traditional healthcare systems, it can be challenging to share patient records among different healthcare providers due to privacy concerns and security risks. By using blockchain technology, healthcare providers can create a permissioned network where authorized parties can access patient records securely and transparently. Patients can also control access to their records, giving them greater control over their healthcare data. In addition, the use of blockchain technology can help to prevent healthcare fraud by providing an auditable record of patient data. Blockchain technology can be used to track the entire life cycle of a patient record, from creation to sharing, ensuring that any modifications or changes to the data are transparent and easily traceable. Overall, the use of blockchain technology in the healthcare sector has the potential to improve the security, privacy, and accessibility of patient data, while also enhancing the efficiency and effectiveness of healthcare delivery.

The use of blockchain technology in the insurance sector offers significant benefits such as the prevention of fraud, streamlining of insurance processes and automation of claims. By using blockchain, insurance companies can record contracts in a tamperproof and immutable way, ensuring transparency and accountability in the process. This helps to prevent fraudulent activities such as falsification of information, double claiming, and other types of scams. In addition, the use of blockchain technology in insurance can help to automate and streamline various processes such as policy management, underwriting, and claims processing.

In addition to the use cases mentioned in the previous examples, the blockchain technology has the potential to revolutionize how governments and enterprises manage identities and personal data of their citizens or customers. With the use of blockchain, citizens' personal data, such as name, address, birthdate, and other sensitive information, can be stored in a secure and decentralized manner on-chain, where it is only accessible by authorized parties. This eliminates the need for centralized storage of personally identifiable information (PII) data, which is often compromised as result of cyberattacks and data breaches. Furthermore, blockchain technology enables the use of Decentralized Identifiers (DIDs) to streamline and secure identity verification processes. Instead of relying on physical documents like passports or driver's licenses, individuals can use cryptographically verifiable DIDs to prove their identity. These identifiers are stored and managed on a blockchain, allowing authorized parties to validate identity credentials without exposing sensitive personal data or requiring access to centralized databases. This approach enhances privacy, reduces fraud, and supports interoperable digital identity frameworks across platforms.

In addition to managing identities, blockchain technology can also be used to improve the efficiency and transparency of government processes. For example, the use of smart contracts on the blockchain can automate and streamline the process of government procurement, ensuring that contracts are executed in a transparent and trustworthy manner. The use of blockchain technology in the government sector has the potential to enhance the security, transparency, and efficiency of government operations, while also empowering citizens by giving them greater control over their personal data.

The financial sector has been a major testing ground for blockchain use cases, including the record of cross-border payments, settlement recording for investment trading, and sharing of financial contract information like letters of credit between banks and their customers. The need for greater efficiency, compared to manual processes and legacy technology solutions, is driving experimentation and adoption of blockchain in finance. The finance sector has a variety of potential use cases for blockchain technology, For example in payment and remittance, blockchain can enable faster, cheaper, and more secure cross-border payments and remittances by removing intermediaries and providing near-instant settlement. In trading and settlements, blockchain can automate and streamline the trading and settlement of financial assets, significantly reducing the time for settling the trades and eliminating errors and delays. By streamlining transaction processing and automating settlement processes, the blockchain can significantly reduce transaction settlement times from days to hours, as seen in trade finance experiments.

In asset management, blockchain can enable the efficient and transparent management of various types of assets, including stocks, bonds, and real estate, providing investors with greater visibility and control over their portfolios. Blockchain technology can be used for identity verification in the finance sector. By using a distributed ledger, individuals can maintain control over their own identity data and share it with financial institutions as needed. This can improve the accuracy of identity verification and reduce the risk of identity theft. Finally, blockchain technology can help financial institutions comply with regulations by providing a secure, immutable ledger of transactions. This can improve transparency and make it easier to audit financial transactions for compliance purposes. The blockchain can also streamline the process of approvals for trades, reducing the potential for manual errors and fraud. In addition, the blockchain can be used as an auditing tool for regulatory reporting. These use cases demonstrate the potential of blockchain technology to enhance efficiency, reduce costs, and increase transparency and trust in financial transactions.

One potential use case for blockchain in tax fraud prevention could be in creating a decentralized and tamperproof system for tracking value-added tax (VAT) payments in a supply chain. VAT is a tax on the value added to goods and services at each stage of production and distribution, and it can be vulnerable to fraud and evasion when businesses fail to report or remit the appropriate amounts.

By using blockchain, all transactions in a supply chain could be recorded and validated in a transparent and secure manner. Each transaction would be time-stamped and linked to the previous one, creating an immutable and auditable record of all VAT payments made by each participant in the chain. This would help to prevent fraud by making it more difficult for businesses to underreport or evade VAT payments, and it would also provide a more efficient way to track and reconcile payments.

Furthermore, the use of smart contracts could automate certain aspects of the VAT collection process, such as verifying the authenticity of VAT invoices and calculating the correct VAT amounts due. This would reduce the risk of errors or discrepancies in reporting, and it would also reduce the administrative burden on businesses and tax authorities alike. Overall, the use of blockchain for VAT tracking and collection could help to improve trust and transparency in the tax system, while also reducing the incidence of fraud and evasion.

The blockchain is a decentralized and distributed technology that is best utilized in cases where there are multiple parties involved in contributing and sharing data. Once data is written on the blockchain, it cannot be deleted, and access to the data does not require permissions. The trust in the blockchain is enabled by technology rather than centralized control, and all writes to the data store can be audited through a tamperproof log. However, the decision to use the blockchain should be made based on specific decision models, as it may not be the best fit for every use case. When centralized control is sufficient and there is no need to distribute copies of data, a centralized system may be more appropriate. Distributed and decentralized architecture may introduce challenges such as scalability, throughput, and privacy that centralized systems may not have. For example, adding additional nodes and peers to the blockchain to increase scalability results in increased costs for additional computer resources necessary to perform validations and enable trust among peers. The computational effort to validate transactions also increases with the size of the blockchain.

Another factor of the decision-making to whether to adopt the decentralized architecture versus centralized is also maintaining confidentiality of the data where access of such data can be controlled by a central entity that will grant others permissions to access the data based upon their role/function. This is in opposition to a use case of the blockchain where each node can access to the data without permissions whose confidentiality of the transaction is maintained by making the buyers and sellers in transactions anonymous. In cases where maintaining privacy of the transactions is a requirement, that is, such data is being considered confidential among the peers transacting, the use of permissionless and anonymous open accessible blockchain might not be the best use case.

Decentralized systems, such as blockchain, are best suited for scenarios where there is a need for multiple parties to contribute and share data, and where the trust between parties needs to be established without a central authority. In such cases, decentralized systems can provide benefits such as increased transparency, reduced risk of fraud, and improved security through cryptography.

However, there are also scenarios where centralized systems can be more suitable. For example, if there is a need for a single point of control or if there is no need to share data among multiple parties, then a centralized system might be a better fit. Centralized systems can provide benefits such as easier management and maintenance, faster processing, and lower costs.

It is important to note that while blockchain has gained popularity as a solution for a variety of use cases, it is not a solution for all types of transactions. Decision models should be used to determine whether blockchain is the best fit for a given scenario. Factors such as the need for decentralization, the need for trust between parties, and the complexity of the transaction should be taken into consideration when making such a decision. Additionally, blockchain technology is still evolving and there are ongoing challenges with scalability, interoperability, and regulation that need to be addressed.

The blockchain is known for providing a tamperproof record of transactions that cannot be altered for fraudulent purposes using cryptographic algorithms, making it more secure than centralized traditional databases. However, this feature alone may not be sufficient to consider the blockchain secure, as it may still be vulnerable to exploitation of design and implementation vulnerabilities.

The blockchain can still be vulnerable to attacks if it is poorly designed and implemented. For example, digital assets such as tokens or cryptocurrencies can be compromised if a hacker gains access to the private keys used to sign transactions. Digital assets, such as tokens or cryptocurrencies, are spent from one peer to another by using signatures that rely on private keys to sign transactions. Private keys enable a buyer to spend a certain amount of cryptocurrency to a seller as the beneficiary of the transaction. These private keys are in essence the assets that need to be protected and kept private by peer/nodes since these enable the possibility to spend these or trade in exchange of money/fiat currency. These private keys must be protected and kept private by peer/nodes since they allow the possibility of spending or trading these assets for money/fiat currency. Often the security of the digital wallets that are used for storage of these private keys represents the weakest link of the security by design of the blockchain: a fraudster who successfully compromises the private keys results in the possible stealing of cryptocurrencies by the fraudster and then spending them.

Storing private keys in online wallets used by investors to trade at cryptocurrency exchanges carries an even higher risk compared to personal digital wallets. This is because online digital wallets are often targeted by fraudsters seeking to exploit vulnerabilities and steal cryptocurrency investors' private keys. These attacks can result in the theft of large amounts of digital assets, causing significant financial losses. To mitigate this risk, it is important for each investor to validate the security of the cryptocurrency exchange before using it to store their private keys. This includes ensuring that the exchange has been subjected to a security audit that includes testing for identifying and remediating digital wallet vulnerabilities. Ideally, it is always preferable to store your keys in secure hardware off-chain most of the time that is not being used for signing transactions on the blockchain. The most secure approach is to rely upon secure hardware off-chain whenever possible. This means keeping the keys offline and disconnected from the Internet, except when signing transactions on the blockchain. This significantly reduces the likelihood of a successful attack against the keys, as it makes them inaccessible to attackers seeking to exploit digital wallet vulnerabilities.

Open permissionless blockchains, such as Bitcoin and Ethereum, are designed to be transparent and publicly accessible to anyone who wants to participate. While this design offers several benefits such as decentralization, immutability, and trustlessness, it also introduces significant data privacy risks. All transactions on an open permissionless blockchain are recorded in a public ledger that can be accessed by anyone. This means that once a transaction is made, it becomes permanently recorded on the blockchain and is visible to anyone with an Internet connection. While the identities of the parties involved in the transaction are not revealed, the transaction amount, time, and location can be easily traced. This presents a risk for individuals and businesses who want to keep their financial transactions private. In an open permissionless blockchain network, it is difficult to control who can access and view the data. This introduces the risk of data breaches and leaks, where sensitive information stored on the blockchain could be accessed and exploited by malicious actors. This is particularly concerning for businesses and organizations that handle sensitive data such as personal information, financial transactions, or intellectual property.

Open permissionless blockchains provide for anonymity of transactions as being executed by addresses that are recorded instead of individual users. This is only in theory as the use of the blockchain is not exactly anonymous but pseudonymous because each user has a public address and this address could be traced back to an IP address or exchange account through proper network analysis and other techniques that can help identify individuals based on their transaction patterns and behaviors. This can be used to deanonymize users and reveal their identities, which can have serious privacy implications. To mitigate these risks, several privacy-enhancing techniques have been proposed, such as zero-knowledge proofs, ring signatures, and homomorphic encryption. These techniques aim to provide stronger privacy guarantees for users of open permissionless blockchains, while still maintaining the benefits of decentralization and trustlessness.

However, these techniques are still in their early stages of development and adoption, and it remains to be seen how effective they will be in protecting data privacy on open permissionless blockchains.

Another challenge to the use of permissionless and open blockchains is posed by scalability, that is, the blockchain growth due to new nodes added to the network has ultimately an impact on performance that can be throughput of transactions per second to execute. The scalability issue arises with the increasing number of nodes and transactions in the blockchain: this is because every newly added transaction as a block to the blockchain increases the computational task to validate this transaction.

In order to address scalability, security, and data privacy limitations of open permissionless blockchain by enterprises in different regulated industry sectors, a new class of "enterprise-based blockchains platforms" has been developed over the years. What differentiates these enterprise-based blockchains from non-enterprise-based blockchains is the adoption of a closed model versus open model and permission-based model versus permissionless model. In a closed model, transactions are not open to anybody to read and to access but only to those who have permission to access them; such permissions are only granted by partners participating in consortiums of trusted business partners.

Permission-based blockchains were created to address certain limitations of public blockchains, particularly with respect to data privacy and access control. While this transparency is a key feature of public blockchains, it can be problematic for certain use cases that require confidentiality of the transactional data. While this transparency is a key feature of public blockchains, permission-based blockchains are more suitable for use cases that require confidentiality and restricted access.

Enterprise-based blockchains are designed to integrate with existing off-chain legacy systems and infrastructures, rather than functioning as stand-alone systems. This is achieved through the use of Blockchain as a Service (BaaS) APIs, which allow client nodes to consume blockchain services such as creating digital assets/tokens, signing and transferring asset values to other peer nodes, and querying transactions. By integrating with legacy systems, enterprise blockchains are able to leverage existing infrastructure while providing enhanced functionality through blockchain technology.

Permission-based blockchains aim to address data privacy limitations by encrypting transactional data, making it accessible only to parties involved in the transaction, such as the buyer and seller. Private blockchains ensure the confidentiality of peer node identities and transaction records. Participation in a permission-based blockchain network requires identification and authorization, ensuring that only specific buyers and sellers can join. This has significant implications for transaction read and write entitlements: in public open blockchains, everyone can read every executed transaction by other nodes in the blockchain, while in permission-based blockchains, access is restricted to authorized parties only.

Scalability is a major challenge faced by the use of permissionless and open blockchains, as the addition of new nodes to the network can impact performance, specifically the throughput of transactions per second. As more nodes and transactions are added to the blockchain, the issue of scalability arises due to the increased computational task required to validate each new transaction as a block in the blockchain. Permission-based private blockchain address the scalability limitations of permissionless blockchain being scalable by design such as on-chain scaling solutions that can increase the block size limit, hence reducing the block verification time and implementation of scalable consensus mechanisms and reduction of the load of the main blockchain by moving transactions off chain, and use solutions such as the Lightning Network.

Despite their differences, both permission-based and permissionless blockchains utilize common security features. One important feature is the immutability property of the blockchain, which is achieved through the implementation of timestamps and cryptographic hashes, ensuring that transactions stored in the blockchain are tamperproof. Another common security feature is transaction validation, which involves reaching consensus among the participating nodes about the validity of transactions before they are permanently recorded. Various consensus algorithms can be employed in both permissionless and permission-based blockchains to determine which transactions are valid and which are not.

Despite both permission-based and permissionless blockchains implementing security by design, they do not ensure complete security or cyber-attack tolerance. The blockchain application must undergo thorough security testing to identify and mitigate any vulnerabilities caused by software vulnerabilities. If these vulnerabilities are exploited by external threat actors, businesses may suffer significant losses, such as fraudulent transfers of digital assets. Therefore, it is essential to identify and remediate blockchain software vulnerabilities before deploying it to conduct high-risk transactions, such as payments or trades and settlements in production environments.

Examining blockchain applications for vulnerabilities is important because these vulnerabilities are often related to the specific blockchain implementation. For instance, the implementation of consensus algorithms needs to be tested for specific attacks that attempt to bypass validations and commit fraudulent transactions without authorization. Distributed and decentralized applications (dApps) have a larger attack surface than centralized applications, so it's important to test their resilience to attacks from any node in the network to achieve "attack resilience." This involves validating blockchains with attack-driven test cases to assess smart contracts for economic exploits like flash loan attacks. It includes simulating consensus manipulation and red team testing to exploit specific vulnerabilities, such as API weaknesses in DeFi frontends, node infrastructure compromises, RPC endpoints, or CI/CD pipelines.

One of the weakest points of the blockchain is the vulnerability of each node's APIs, which use private keys for transaction signing and implementing consensus algorithms. These core blockchain APIs provide critical functionalities that are often targeted for compromising sensitive data, such as private keys. Therefore, it is crucial to safeguard access to the blockchain nodes and protect the APIs and private keys from unauthorized access.

In order for future blockchain applications to be deployed in production environments, they will need to be designed to withstand cyberattacks against data and functionality, providing the desired level of assurance for financial institutions. A comprehensive strategy involves adopting an attack-centric threat modeling approach, utilizing methodologies such as PASTA (Process for Attack Simulation and Threat Analysis) or aligning with MITRE ATT&CK tactics. These frameworks facilitate a structured analysis of potential adversarial exploitations of vulnerabilities within blockchain architecture. By simulating attacker behavior, these approaches assist in identifying design flaws and control gaps, This modeling informs proactive risk mitigation through security-by-design principles based on real-world attack scenarios.

By using a threat model, specific attack scenarios can be derived, such as attempting to steal private keys from digital wallets, launching denial of service attacks on cryptocurrency exchanges, and misappropriating cryptocurrencies.

A key area of focus for blockchain security is the security of programmable smart contracts. These contracts are coded as executable code, which allows for the creation of self-executing business rules within the blockchain. As part of the enterprise and distributed architecture virtual machine, these rules are executed automatically by the computer nodes of the blockchain.

Smart contracts are designed to automate the execution of business rules in the blockchain. These rules are encoded in the smart contract and are executed by the nodes in the network. However, since smart contracts are essentially code, vulnerabilities in the code can have significant consequences. For example, if a smart contract has a vulnerability that allows an attacker to transfer tokens or cryptocurrencies to fraudulent accounts, it can result in financial losses for the affected parties.

Other examples of smart contract vulnerabilities include reentrancy attacks, where an attacker repeatedly calls a function within a contract before the previous call has completed, resulting in unexpected behavior or unauthorized access to funds. Another vulnerability is the lack of input validation, where an attacker can manipulate the input data to the smart contract to achieve unintended results, such as bypassing authentication or authorization checks. Smart contracts can also be vulnerable to integer overflow and underflow, where an attacker manipulates integer values in a contract to overflow or underflow, leading to unexpected behavior or unauthorized access to funds.

Payment blockchain applications rely on smart contracts to automate the settlement of transactions between buyers and sellers by executing specific rules coded into the smart contract. As these smart contracts are essentially lines of code, any vulnerabilities in the code can have a significant impact on the business, such as allowing fraudulent transfers of digital assets such as cryptocurrencies to unauthorized accounts. An example of a smart contract vulnerability that can be exploited for fraud is the reentrancy attack. This attack allows an attacker to repeatedly call a function in a smart contract before the previous function call has completed, allowing them to drain the contract's funds. For example, let's say there is a smart contract that holds funds and allows users to withdraw them. The contract includes a function called "withdraw" that transfers the requested amount of funds to the user's account. However, the contract's code includes a flaw that allows an attacker to call the "withdraw" function multiple times before the previous call has completed. The attacker can then use this vulnerability to repeatedly drain the contract's funds by calling the "withdraw" function and immediately calling it again before the first call has completed. Each call withdraws the requested amount of funds, allowing the attacker to drain the contract's funds until they are empty. This vulnerability was famously exploited in the 2016 DAO hack, where an attacker used a reentrancy attack to drain over $50 million worth of cryptocurrency from a smart contract on the Ethereum blockchain.

To ensure the security of smart contracts, it is crucial to conduct a thorough security review of their source code and implementation of transaction automation. This review should involve a combination of manual and automated testing, such as using static code review tools to identify potential vulnerabilities. Without a comprehensive security review, smart contracts can be vulnerable to various types of attacks, such as the reentrancy attack which exploits a vulnerability in the smart contract code to allow an attacker to repeatedly withdraw funds from the contract. Additionally, it is essential to keep smart contracts updated and regularly audited to ensure that any new vulnerabilities or weaknesses are identified and addressed promptly.

Despite significant investment and progress in the development of enterprise-based blockchain technology, there remain several technology challenges that must be addressed before blockchain applications can be widely adopted across various industries. One such challenge is interoperability, as different blockchain networks often use varying protocols and standards, making it difficult for them to communicate with one another. This lack of interoperability can result in fragmentation and limitations on the ability of different blockchain networks to work together, hindering their ability to provide the desired benefits.

Another challenge is scalability, as the current blockchain technology is still struggling to keep up with the increasing demand for transaction processing as more users join the network. While solutions such as increasing block size limits and implementing off-chain scaling solutions like the Lightning Network have been proposed, they are not yet widely adopted and may require significant development effort.

Moreover, the issue of privacy in the blockchain remains a concern, particularly in permissionless blockchains where anyone can join and participate in the network. While techniques like zero-knowledge proofs and encryption are being used to enhance privacy, there is still a need for further research to develop more robust and effective privacy-preserving techniques.

Smart contracts are an essential part of blockchain technology and have the potential to revolutionize various industries by providing automated, trustless, and efficient transactions. However, as mentioned earlier, smart contracts are essentially pieces of code, and any vulnerability in the code can lead to disastrous consequences, such as loss of funds or sensitive information.

Therefore, it is crucial to invest in the development of tools that can help secure smart contracts. These tools can include automated tools such as static code analyzers, which can scan smart contract code for vulnerabilities and provide developers with a list of potential security issues that need to be addressed. Additionally, other tools can help with dynamic analysis of smart contracts, which involve testing smart contracts in a simulated blockchain environment to identify vulnerabilities that may not be apparent during static analysis. Furthermore, ongoing research and development can help to identify new threats and vulnerabilities to smart contracts, and subsequently, new tools can be developed to address these emerging threats. A continued investment in the development of tools for securing smart contracts is essential for ensuring the security and reliability of blockchain-based applications that rely on them.

Although the industry has invested heavily in the engineering of secure, usable, and scalable enterprise-based blockchain technology, there are still several technological challenges that must be overcome before blockchain applications can be widely adopted across different industry sectors. At the current stage of the evolution of the blockchain technology to be used by businesses, it is necessary to follow a security strategy of "building security in" from the beginning, that is, during the experimentation with BaaS platforms to build enterprise-based applications. To ensure the safe adoption and use of blockchain technology by businesses, it is imperative to implement a security strategy of "security by design" from the outset. This means incorporating security measures into the experimentation phase of using BaaS platforms for developing enterprise-based applications. Building security in from the beginning will help prevent vulnerabilities and potential security breaches as the technology evolves and becomes more widely adopted.

Data privacy is a crucial aspect that new blockchain platforms must prioritize when designing security features. As the decentralized nature of blockchain applications may result in private data being distributed across multiple jurisdictions, ensuring data privacy becomes a challenging task. Therefore, one of the primary objectives of a security by design approach for blockchain platforms is to guarantee privacy for both consumers and business clients.

Protecting personal identifiable information (PII) is of paramount importance when it comes to blockchain applications. As this information includes sensitive data related to an individual's identity, it needs to be safeguarded against unauthorized access and encrypted whenever stored or transmitted. Compliance with data privacy regulations is crucial to ensure that such information is protected at all times. For instance, the GDPR (General Data Protection Regulation) in Europe mandates stringent privacy legislations, and PII data sets need to be handled with utmost care to comply with these regulations.

One example of data privacy affecting blockchain stored data in the United States is the California Consumer Privacy Act (CCPA). The CCPA requires companies that store, sell, or share personal data of California residents to comply with certain data privacy standards. This includes companies that use blockchain technology to store personal data of California residents. Under the CCPA, California residents have the right to request access to their personal data, as well as the right to have their personal data deleted. Companies that use blockchain technology to store personal data must ensure that they are able to provide California residents with access to their data and the ability to have it deleted, while also maintaining the integrity and immutability of the data stored on the blockchain. This can be a difficult balance to achieve, and companies must carefully consider the potential privacy implications of using blockchain technology to store personal data.

Blockchain can still be considered an emerging technology in some regards, as it continues to evolve and expand its applications beyond its original use case for cryptocurrency transactions. While blockchain has been around for over a decade, its use cases are still being explored and developed in various industries, such as finance, healthcare, supply chain management, and more. Blockchain platforms are likely to continue to grow and mature as a technology in the coming years.

The market offers various blockchain platform providers, including both open-source and proprietary options. Some platforms are tailored for enterprise use, while others specialize in dApps and cryptocurrencies. The maturity of the blockchain technology will be determined by how well the existing platforms can tackle ongoing challenges in security, privacy, scalability, and interoperability. It is also crucial to address regulatory frameworks that may hinder the widespread adoption and integration of blockchain into existing systems.

Collaboration among various stakeholders such as businesses, blockchain platform providers, and experimental trials of blockchain for specific use cases has led to significant progress in engineering more secure blockchain applications. Financial institutions have also actively engaged with blockchain platform technology providers to create innovative financial applications for cross-border payments, securities trading, and settlements.

Ensuring the security of blockchain applications and smart contracts is crucial, and therefore, auditing and certification of these applications is imperative. However, it remains a challenging task, as there are no defined security standards for blockchain yet. The security verification process of blockchain platforms, including APIs, applications, and software, is often limited to testing for common vulnerabilities, overlooking threat modeling and architecture risk reviews that could help identify design flaws. Therefore, it is crucial to prioritize auditing and certifying the security of blockchain developed applications and smart contracts to prevent potential vulnerabilities and risks. To ensure attack resiliency, it is important for the secure by design effort of blockchain technology to focus on not only securing new technology features and functionality but also preventing the abuse of these features. When it comes to expected functionalities, it is crucial to implement identity verification for customers, which includes beneficiary accounts with proper controls for Know Your Customer (KYC) compliance to enforce Anti–Money Laundering (AML) checks before facilitating cross-border payments.

To establish trust in the use of blockchain for corporate payments or institutional investments, businesses may require a comprehensive security audit of blockchain applications against industry standards. This audit should include verification of the security of the blockchain's design and architecture, as well as rigorous testing to identify potential vulnerabilities in the software. Furthermore, the audit may also need to address specific industry regulations related to data privacy, cross-border transactions, and financial compliance, depending on the intended use case.

In order to satisfy stringent security and compliance requirements, organizations must ensure their teams are well-versed and properly equipped. This publication is intended for application security professionals, including software developers and architects, who aim to develop secure and resilient decentralized blockchain applications that can withstand regulatory scrutiny and real-world threats. This book is for software developers and application architects interested in designing secure and resilient decentralized blockchain applications. The purpose of this book is to offer a detailed understanding of how the blockchain platform can serve as a secure and resilient technology enabler for decentralized applications (dApps) and various use cases. Instead of designing and developing dApps first and then trying to secure them afterward, this book encourages you to integrate security considerations from the outset – before the design, development, testing, and production deployment stages.

For practitioners in blockchain technology, whether you're a chief information security officer (CISO) or a security engineer/architect working on a new dApp, it's essential to understand the potential risks that come with adopting blockchain technology. This includes understanding the security and privacy risks associated with the storage and transfer of sensitive data on a decentralized and immutable ledger. They should also be aware of the risks associated with the use of cryptocurrencies, including market volatility, fraud, and theft. Additionally, it is important to understand the regulatory and legal risks associated with the use of blockchain technology, as different jurisdictions have different laws and regulations related to blockchain and cryptocurrencies. Finally, if you're an engineering practitioner in blockchain, such as an engineering manager developing a dApp, it's crucial to understand the risks tied to insecure dApp design and development. This includes identifying and remediating smart contract vulnerabilities and following standardized, secure by design practices. Failing to do so could open the door for malicious actors to exploit these weaknesses, potentially resulting in the exfiltration of crypto assets or fraudulent transactions.

For technical practitioners such as software engineers and software architects working on future dApp projects, this book provides the skills needed to identify "security by design" weaknesses in blockchain applications as they are being architected. By integrating "security by architecture" principles and risk-centric design processes like threat modeling, you can detect threats, vulnerabilities, and attacks that may affect various components: from the dApp frontend and digital wallets to the middleware in backend blockchain platforms and the smart contracts powering transactions. Beyond merely identifying these weaknesses, the book also guides you in designing blockchain-based applications that are inherently secure and resistant to attacks. You'll develop a deep understanding of both blockchain technology and its associated risks, as well as learn best practices for architecting and implementing secure blockchain solutions.

For business and technology practitioners seeking to secure the use cases for which their blockchain is designed, this book provides a comprehensive understanding of blockchain security controls and guidance on designing and experimenting with blockchain applications. By reading it, you'll gain invaluable insights into integrating your dApp with blockchain platforms to achieve the desired functionality and use case while ensuring the entire architecture is secure by design and free from common flaws. You'll also learn how to deploy secure smart contracts by identifying and addressing vulnerabilities, a process often referred to as smart code auditing. Overall, this book is essential for anyone looking to develop secure, robust, and decentralized blockchain-based applications.

1

The Blockchain Technology Primer

> *We are at the very beginning of time for the human race. It is not unreasonable that we grapple with problems. But there are tens of thousands of years in the future. Our responsibility is to do what we can, learn what we can, improve the solutions, and pass them on.*
>
> —Richard Feynman

1.1 Introduction

This chapter provides a comprehensive introduction to distributed ledger technology (DLT) and blockchain. It begins by defining DLT, clarifying its relationship to blockchain, and outlining the key characteristics and benefits of DLT. Next, it explores different blockchain network types, comparing public versus private and permissioned versus permissionless networks – followed by a discussion of scalability and interoperability solutions. An overview of blockchain's data structure and various consensus algorithms is included, along with the evolution of cryptocurrencies. The chapter also emphasizes the importance of privacy and identity controls in blockchain technology and concludes by addressing legal and regulatory considerations.

Before delving into how to securely design and implement blockchain-based applications (such as dApps), it's essential to understand the fundamental concepts and constructs of blockchain technology. A strong foundation in how data is stored, validated, and protected on the blockchain is crucial for recognizing potential risks, adhering to best practices, and ultimately building robust and secure solutions. This chapter, therefore, serves as a valuable resource for anyone aiming to deepen their understanding of blockchain technology before moving on to advanced security design and implementation.

Blockchain Application Security: How to Design Secure and Attack Resilient Blockchain Applications, First Edition. Marco Morana, Harpreet Singh and Francesco Piccoli.
© 2026 John Wiley & Sons, Inc. Published 2026 by John Wiley & Sons, Inc.

1.2 Brief History of the Blockchain and Its Evolution

The genesis of blockchain technology as a DLT can be traced back to the publication of the Bitcoin white paper published in 2008 [1]. The paper, titled "Bitcoin: A Peer-to-Peer Electronic Cash System," described a decentralized system for recording and verifying transactions without the need for a central authority.

Since the publication of the Bitcoin white paper in 2008, the technology has evolved and been adapted for a wide range of applications, and it is now seen as a powerful tool for creating secure, decentralized systems. While Bitcoin was the first application of the blockchain, it was initially developed as a decentralized payment system and was not designed to handle the complex needs of enterprise organizations. Over the years, the technology has evolved to meet the needs of businesses and governments, leading to the development of enterprise-based blockchain platforms. These platforms are designed to handle a much wider range of use cases, from supply chain management and logistics to identity verification and voting systems.

One of the key features of enterprise-based blockchain platforms is their ability to be customized to meet the unique needs of different organizations. This customization can include everything from the type of consensus algorithm used to verify transactions, to the specific permissions granted to different users within the network.

Another major advantage of enterprise-based blockchain platforms is their ability to integrate with other technologies, such as cloud computing and the Internet of Things (IoT). By integrating with these technologies, blockchain platforms can create a more robust and seamless ecosystem for businesses and organizations.

Overall, the progress we have made from the initial Bitcoin blockchain to the current enterprise-based blockchain platforms has been significant. These platforms are now being used by a wide range of organizations around the world and are seen as a powerful tool for creating secure, decentralized systems that can be customized to meet the unique needs of different industries and use cases.

1.3 DLT and the Blockchain

It is crucial to have a firm grasp of the basic terminology used to describe blockchain components. When we refer to a technology as "decentralized," it means that it is designed to run on a network of computer hosts, as opposed to "centralized" technology that runs on a central system or computer. In the case of blockchain, each computer host can be considered a node in a network of computers, where every node interacts with one another as peers, possessing the same level of permissions. These interactions among peer nodes are the reason why the blockchain network is described as a peer-to-peer (P2P) network, in which computers share data directly with each other without the need for a central server.

The blockchain technology is often referred to as a form of DLT due to its distributed ledger characteristic. The ledger that stores transaction records is distributed among the nodes in the blockchain network, making it decentralized. This means that there is no central authority that maintains or controls what can be recorded in the database. As a result, multiple copies of the ledger exist and are distributed across the P2P network.

DLT [2] is a type of technology that enables the operation and use of distributed ledgers. Distributed ledgers can be thought of as a type of database architecture that allows multiple parties to share a single database, while each party retains control over their own data. Blockchain is a specific type of DLT that uses cryptographic techniques to create a tamperproof and immutable record of transactions. In a blockchain, each transaction is recorded instead as a block that is linked to the previous block, forming a chain of blocks. This creates a permanent and unalterable ledger of transactions that can be trusted by all parties involved.

A crucial characteristic of blockchain applications is that the control over P2P transactions is "distributed" among peer nodes, rather than being centralized in a single server node. This distributed feature also applies to the storage of application transaction data, which is shared and dispersed among the network's computer nodes, rather than being stored in a central system. The blockchain data store is a unique form of distributed database that records transactions between peer nodes. These transactions serve as historical records of data exchanges, such as the transfer of tokens or digital assets between computer nodes.

In cases where transactions between peer nodes involve financial exchanges, such as payments from one account to another, the recorded data takes on the characteristics of a ledger. This digital representation of a book records the details of the transaction, including the sender's account, the recipient's account, the payment amount, and the date and time of the transaction. The blockchain's digital ledger operates similarly to a traditional ledger, with the added feature of being stored and distributed across the network's nodes. This creates a distributed digital ledger, in which each node has a copy of the database.

Applications that leverage the blockchain as a DLT are enabled to use a digital system of transaction recording, sharing, and verifying data across a network of peer computers that can be mutually trusted, that is without the need for a central authority server. Instead of relying on a single central authority or intermediary, the network is distributed among multiple computers, with each computer having a copy of the ledger. This means that no one entity has control over the database, and all participants have equal access to it.

Applications that integrate the blockchain as a distributed and decentralized database will enable multiple users/parties to have simultaneous access to a common, secure, and tamper-evident database of transactional type of information, without the risk of a single point of failure or the need for first establishing trust between the parties. At its core, DLT is a technology that incorporates users with the creation of a shared digital ledger, where transactions are recorded in real time. This shared database of transactions can be used to record any type of data, from financial transactions to digital identity information and asset ownership.

Another important feature of the DLT is its implicit security. The data stored in the ledger is protected by cryptographic techniques, such as digital signatures and hash functions: These techniques ensure that the data is tamper-evident, meaning that any attempt to modify the data can be detected by other participants in the network. Applications can leverage the DLT also for its transparency and immutability: once a transaction is recorded in the ledger, it cannot be deleted or modified, and all participants in the network can see the transaction. These features make DLTs particularly useful for applications where trust and transparency are critical, such as financial transactions, supply chain management, and digital identity.

The blockchain is a popular and widely used DLT that has the potential to transform many industries by enabling secure and efficient data sharing and collaboration between multiple parties. As the technology continues to evolve, it is likely to become an increasingly important part of the digital infrastructure of the future.

As a specialized form of DLT, the blockchain integrates cryptographic security, decentralized consensus, and immutability to maintain a shared, tamper-resistant ledger across multiple nodes, all without relying on a centralized authority. The blockchain can be considered as a type of distributed and decentralized database that allows storing of transaction records with each record linked to the previous one in a chain of blocks.

By employing cryptographic algorithms, each newly added block of transactions becomes part of an immutable record, distributed across the network and shared by all participants. Because the ledger is essentially read-only, it fosters trust among parties who can rely on its tamper-resistant nature. This makes blockchain especially suited for applications requiring secure and dependable transactions, such as financial transfers or supply chain management.

Another key feature of the blockchain is its ability to enable transaction automation with the use of smart contracts. Smart contracts are self-executing contracts that are programmed to execute automatically when certain conditions are met. These contracts are stored on the blockchain and can be used to automate various processes, such as payments or the transfer of ownership.

Overall, the blockchain is a powerful tool for creating secure, decentralized, and transparent systems. Its unique features make it well-suited for a wide range of applications, from financial services to supply chain management to healthcare. As a type of DLT, it is one of many technologies that are driving the development of new, decentralized systems and applications.

One of the essential characteristics of a blockchain database is its distributed and decentralized ledger, which only allows records to be written in append-only mode. This attribute is similar to that of log files, where modifications are protected by the append-only mode. Once this attribute is set, logs cannot be deleted or modified, and writes are only allowed to append to the end of the file, ensuring a tamperproof record of events or transactions. In the case of the blockchain, the "append-only" feature is one of the unique features that makes it "immutable" when recording a historical record of transactions. It only allows new transactions to be added but not changing or deleting previously recorded transaction records.

To achieve this append-only feature, cryptography is used to ensure that the record of transactions previously occurred is not forgeable and can be implicitly trusted. This trust in the transactions recorded in the blockchain's distributed and decentralized database is achieved differently from traditional databases. In traditional databases, trust is delegated to a central authority that is granted special entitlements such as administrator entitlements to read, write, delete, and modify the data recorded in the database. The inherent trust offered by the use of distributed and decentralized ledger technology makes the blockchain a "trustless" technology that can operate without the need for centralized trust. This feature is achieved without requiring the system to trust either a central authority or the peer/node systems to trust each other.

Blockchain technologies, as DLTs, enable the creation of a new class of decentralized applications, known as dApps. These dApps run on multiple computers, have decentralized controls, and record transactions in a distributed ledger that is immutable and trusted. Transactions that are recorded in the blockchain include the transfer of digital assets, such as cryptocurrencies and tokens, between peer nodes. These cryptocurrencies or tokens can represent digital assets for value exchanges, such as payments from a buyer node/user to seller node.

In the blockchain, transactions are collected and appended to blocks, which are then broadcasted to other nodes in the P2P network. A transaction usually references the previous transaction outputs as new transaction inputs and dedicates all input cryptocurrency values to the new outputs. Once these transactions are validated by selected nodes in the network, they are recorded in the distributed ledger and become irreversible. A copy of this ledger is distributed among the nodes that are part of the payment network.

Distributed applications can use the blockchain to record any type of digital asset for value exchange transactions, such as P2P payments with bitcoins or other cryptocurrencies. By using blockchain technology, dApps can achieve decentralized control, immutability, and trust without relying on a centralized authority.

In summary the blockchain is a type of DLT that uses cryptography to maintain a secure, transparent, and decentralized database of transactions among multiple parties. By functioning as a distributed digital ledger, blockchain supports a range of use cases and enables several unique key features, such as:

1. **Decentralization:** Unlike traditional databases that are managed by a central authority, blockchain is a P2P network of computers, where each computer (node) has a copy of the database. This ensures that there is no single point of failure or control, and that the data is distributed and accessible to everyone on the network.
2. **Transparency:** Since the ledger is distributed and every transaction is recorded and can be viewed by anyone in the network, there is a high level of transparency in the system. This can help increase trust and accountability among users.
3. **Security:** The blockchain uses advanced cryptography to ensure the integrity and security of the data stored in the database. Each transaction is cryptographically verified and linked to the previous one, forming a chain of blocks (hence the name "blockchain"). Once a block is added to the chain, it cannot be altered or deleted without the consensus of the network, making it virtually tamperproof.
4. **Efficiency:** By eliminating the need for intermediaries, such as banks or other financial institutions, transactions can be processed faster and with lower fees. This can be especially beneficial for cross-border transactions, which can often be slow and expensive.
5. **Traceability:** Every transaction is recorded and time-stamped, creating a permanent and easily auditable record of every asset transfer. This can help reduce fraud, increase accountability, and provide a clearer picture of how assets are moving through the system.
6. **Transaction Automation:** Blockchain technology enables the use of smart contracts, which are self-executing contracts with the terms of the agreement between buyer and seller being directly written into lines of code. These contracts can be programmed to automatically trigger certain actions or events based on predefined conditions, eliminating the need for intermediaries or third parties.
7. **Asset Digitization:** Among the most prominent applications of blockchain is its use in storing and transferring digital assets, such as cryptocurrencies. By leveraging blockchain technology to create decentralized digital currencies, cryptocurrencies like Bitcoin and Ethereum have gained widespread popularity, largely due to their secure and decentralized nature.

These unique attributes make the blockchain a powerful tool for building a more secure and efficient digital world. Overall, blockchain technology holds the potential to transform numerous sectors from finance and banking to healthcare and supply chain management thanks to its defining qualities of decentralization, security, transparency, smart contracts, and cryptocurrency.

1.4 Blockchain Networks

A blockchain network is a decentralized digital ledger system that records transactions across numerous computers using a series of interconnected blocks. This structure ensures all network participants maintain identical copies of the ledger, significantly hindering any retroactive record alterations without collective agreement. Noted for their transparency, security, and tamper resistance, blockchain networks support a wide range of applications beyond cryptocurrencies, including supply chain oversight, secure voting mechanisms, and digital contracting. The foundational technology of blockchain securely stores data in a verifiable and permanent way, accessible only to authorized users.

We can delve deeper into the various components that constitute different types of blockchain networks. A crucial distinction in blockchain technology lies between public and private networks. Each type has distinct features and applications. Public blockchain networks are accessible to anyone; any individual can join as a node and view the blockchain's contents, fostering an environment of openness and inclusivity.

The most well-known example of a public blockchain is the Bitcoin network. In a public blockchain, transactions are recorded in a transparent, decentralized way, without the need for a central authority or intermediary. This makes public blockchains particularly well-suited for use cases where transparency and decentralization are important, such as in the case of cryptocurrencies. However, public blockchains also have some drawbacks. Because they are open to everyone, they are often less secure than private blockchains, as they are more susceptible to attacks and hacking attempts. Additionally, the decentralized nature of public blockchains can make them slower and less efficient than private blockchains, particularly when it comes to processing large volumes of transactions.

In contrast, a private blockchain network is restricted to a specific group of users or organizations. Private blockchains are often used for business-to-business (B2B) transactions, where the parties involved need to ensure the privacy and security of the information being exchanged. In a private blockchain, access to the network is restricted, and transactions are validated by a select group of nodes that are trusted by the participants.

Private blockchains offer several advantages over public blockchains. Because they are restricted to a specific group of users, they are more secure and less susceptible to attacks. Additionally, because the number of nodes in the network is limited, private blockchains can be faster and more efficient than public blockchains, particularly when it comes to processing large volumes of transactions. However, one potential drawback of private blockchains is that they can be less transparent and decentralized than public blockchains, which can make them less well-suited for certain use cases where transparency and decentralization are important.

The key differences between public and private blockchain networks come down to access, security, transparency, and efficiency. Public blockchains are open to everyone, more transparent, and less secure, while private blockchains are restricted to a specific group of users, less transparent, and more secure. Both types of networks have their own unique characteristics and use cases, and the choice of which to use will depend on the specific needs of the organization or use case in question.

Some examples of public blockchains are:

1. **Bitcoin** [3]: The first and most well-known public blockchain network, Bitcoin is a decentralized digital currency that allows for P2P transactions without the need for intermediaries.
2. **Ethereum** [4]: Another popular public blockchain network, Ethereum is a decentralized platform that allows developers to build and deploy dApps and smart contracts.
3. **Litecoin** [5]: A decentralized cryptocurrency that was designed to be faster and more efficient than Bitcoin.
4. **Ripple** [6]: A public blockchain network that was developed for financial institutions. Ripple's digital asset, XRP, is used to provide liquidity for cross-border transactions.

Some examples of private blockchains are:

1. **Hyperledger Fabric** [7]: An open-source blockchain platform that was developed by the Linux Foundation, Hyperledger Fabric is a permissioned blockchain network that is designed for enterprise use cases.
2. **Corda** [8]: Another open-source blockchain platform that was developed for enterprise use cases, Corda is a permissioned blockchain network that is used to facilitate B2B transactions.
3. **Quorum** [9]: A permissioned blockchain platform that was developed by JPMorgan Chase, Quorum is used by financial institutions to handle sensitive data and transactions.
4. **R3 Corda Enterprise** [10]: An enterprise version of Corda that is designed for businesses that require more advanced features and support.

Overall, public blockchains are used for a variety of purposes, from digital currencies to dApps and smart contracts, while private blockchains are used by businesses and organizations for internal processes and transactions that require higher levels of privacy and security.

Another categorization for blockchains is to be a type of hybrid blockchain that is a combination of both public and private blockchain networks. It allows organizations to take advantage of the benefits of both types of networks, while mitigating their respective drawbacks.

There are two main types of hybrid blockchain networks: the public–private hybrid and the interchain hybrid.

1. **Public–Private Hybrid:** In this type of hybrid blockchain, there are two separate networks: a private blockchain network and a public blockchain network. The private blockchain network is used to handle sensitive data and transactions, while the public blockchain network is used to handle less sensitive data and transactions that need to be publicly visible. The two networks are connected by a bridge, which allows data to be transferred between them. This type of hybrid blockchain is often used by organizations that need to maintain the privacy and security of their data and also need to interact with the wider public blockchain ecosystem.
2. **Interchain Hybrid:** In this type of hybrid blockchain, different blockchain networks are interconnected to allow for the exchange of data and assets. This allows organizations to take advantage of the strengths of different blockchain networks and protocols, while still maintaining the security and privacy of their data. The interchain hybrid can be used to enable interoperability between different blockchain networks or to provide a bridge between a private blockchain network and a public blockchain network. This type of hybrid blockchain is often used in industries such as finance, where different blockchain networks may be used to handle different types of assets.

Some examples of hybrid blockchain networks include the following:

- **Dragonchain** [11]: A public–private hybrid blockchain platform that was developed by Disney. Dragonchain allows businesses to build and deploy their own blockchain applications on a private network, while also allowing them to interact with the public blockchain network.
- **Polkadot** [12]: An interchain hybrid blockchain platform that allows for the interoperability of different blockchain networks. Polkadot uses a unique consensus algorithm that allows for high scalability and security.
- **Ripple** [13]: A public–private hybrid blockchain platform that was developed for financial institutions. Ripple's private blockchain network is used to handle transactions between banks, while its public blockchain network is used to provide liquidity for its digital asset, XRP.

Overall, hybrid blockchain networks offer a flexible and secure way for organizations to take advantage of the benefits of both public and private blockchain networks. They are increasingly being used by businesses and governments around the world to build and deploy blockchain applications that are tailored to their specific needs.

Another important differentiation for blockchain between permissioned and permissionless is in relation to the type or the level of access and control that users have within the network. In a permissioned blockchain, access to the network is restricted to a group of approved participants. These participants are typically known entities, such as companies or organizations, that are trusted to participate in the network. Participants must have permission from the network administrator or owner to join and interact with the network. This means that permissioned blockchains are often used in enterprise settings or for private transactions that require higher levels of security and privacy.

Some specific characteristics of permissioned blockchains include:

- **Controlled Access:** Only approved participants are allowed to join and interact with the network.
- **Centralized Governance:** The network is often controlled by a central entity or group of entities that oversee the network's operations and make decisions about its future direction.
- **Higher Scalability:** Because the number of participants is limited, permissioned blockchains can often handle more transactions per second (TPS) than permissionless blockchains.

Examples of permissioned blockchains include Hyperledger Fabric, Corda, Quorum, and R3 Corda Enterprise. In a permissionless blockchain, anyone can join and participate in the network without needing approval from a central authority. This means that the network is completely decentralized, with no single entity or group controlling it. Permissionless blockchains are often used for public transactions, such as digital currencies or dApps that do not require high levels of privacy or security.

As opposed to "permission-based blockchain," you have permissionless. A permissionless blockchain's unique characteristics are as follows:

- **Open Access:** Anyone can join and participate in the network without needing approval.
- **Decentralized Governance:** The network is run by a community of users who make decisions through a consensus mechanism.
- **Lower Scalability:** Because anyone can join the network, permissionless blockchains can be slower and less efficient than permissioned blockchains.

Examples of permissionless blockchains include Bitcoin, Ethereum, and Litecoin. Overall, the choice between permissioned and permissionless blockchains depends on the specific use case and requirements of the application. Permissioned blockchains offer higher levels of privacy and security, while permissionless blockchains offer complete decentralization and transparency.

Permissioned blockchains are generally considered more secure for private and sensitive transactions, as they have controlled access and centralized governance. This means that only trusted entities are allowed to participate in the network, reducing the risk of fraud or malicious behavior. Additionally, permissioned blockchains often use more advanced consensus mechanisms, such as Practical Byzantine Fault Tolerance (PBFT), that can ensure greater security and faster transaction processing. Ultimately, the security of a blockchain network depends on a variety of factors, including the consensus mechanism, the level of decentralization, the type of data being stored, and the specific use case.

Both permissioned and permissionless blockchains have their own security benefits and drawbacks, and the choice between them depends on the specific needs and requirements of the application. In a permissioned blockchain, while the network is controlled by a central entity or group, the transactions are still recorded in a decentralized manner across the network. This means that it is difficult for any single entity to manipulate or tamper with the data on the blockchain, providing a high level of immutability. Similarly, in a permissionless blockchain, the decentralized nature of the network ensures that the data on the blockchain is transparent and immutable. Transactions are validated and recorded through a consensus mechanism that involves multiple nodes across the network, making it very difficult for any one node to manipulate the data. In terms of security, both permissioned and permissionless blockchains have their own strengths and weaknesses, and the choice between them depends on the specific use case and requirements of the application. Both types of blockchains offer specific advantages depending on the requirements and use cases, such as security levels, privacy needs, and the degree of decentralization desired. The choice between permissioned and permissionless depends on the application's specific needs and the balance between control and openness that is required. We provide in Table 1.1 a summary of the features and implementations among permissioned and permissionless blockchains.

1.4.1 Nodes

In a blockchain network, nodes are essential participants that store a complete copy of the blockchain and communicate with other nodes to ensure the integrity of the network. These nodes can be run by individuals, organizations, or machines, depending on the specific blockchain network. Nodes can also have

Table 1.1 Comparison of permissioned and permissionless blockchains.

Feature/ requirement	Permissioned blockchains	Permissionless blockchains
Access control	Only approved participants can join	Anyone can join without needing approval
Governance	Managed by a central entity or group	Run by a community of users via consensus
Scalability	Handles more TPS due to limited participants	Slower and less efficient due to unlimited participants
Security and privacy	Higher levels of security and privacy, suitable for enterprise use	Less controlled but maintained by mass consensus
Use cases	Often used for private transactions requiring high security	Used for public transactions and dApps
Decentralization	Centralized control over operations and decisions	No central control; decisions made by consensus
Consensus mechanism	Advanced mechanisms like PBFT	Mechanisms involve all participants, like PoW or PoS

different roles and responsibilities within the network, depending on their capabilities and permissions.

Nodes are responsible for maintaining the infrastructure and ensuring the security and reliability of a blockchain network. They perform a set of tasks, such as validating transactions, storing data, broadcasting information, and even mining new blocks. Nodes validate incoming transactions before adding them to the blockchain, check whether the transaction meets the required conditions, store a copy of the entire blockchain in a decentralized manner, and communicate with other nodes to share information about new transactions, blocks, and other network updates.

On the other hand, users of a blockchain network are individuals who interact with the network to perform specific actions or transactions. For example, a user might send or receive cryptocurrency, vote in a decentralized election, or create a smart contract. While all nodes are also users of the blockchain network, not all users are nodes. Users interact with the blockchain network through a variety of interfaces, such as wallets, dApps, or command-line tools. They do not necessarily need to store a copy of the entire blockchain or communicate with other nodes, as these tasks are handled by the nodes themselves.

Nodes can be differentiated between two main types: full nodes and light nodes. A full node, also known as a "full client," is a node that stores a complete copy of the blockchain and participates in validating transactions and blocks on the network. Full nodes require significant computational power and storage space to operate, and they can be resource-intensive. A light node, also known as a "light client," is a node that stores only a subset of the blockchain data and relies on full nodes to validate transactions and blocks. Light nodes are typically less resource-intensive than full nodes and can be run on devices with limited computational power or storage space.

The main difference between full nodes and light nodes is the amount of data they store and the amount of processing power they require. Full nodes store a complete copy of the blockchain and are responsible for validating transactions and blocks, which requires significant computational power and storage space. Light nodes, on the other hand, store only a subset of the blockchain data and rely on full nodes for validation, which requires less computational power and storage space.

In terms of security, both full nodes and light nodes can be considered secure. However, full nodes provide a higher level of security and decentralization than light nodes. Full nodes are responsible for enforcing the rules and protocols of the network, ensuring that no single entity has control over the blockchain data. In contrast, light nodes rely on full nodes for validation, which introduces an element of centralization.

In summary, nodes are critical participants in a blockchain network that store a copy of the blockchain and communicate with other nodes to ensure the integrity of the network. Users of a blockchain network interact with the network to perform specific actions or transactions. Full nodes store a complete copy of the blockchain and are responsible for validating transactions and blocks, while light nodes store only a subset of the blockchain data and rely on full nodes for validation. Full nodes provide a higher level of security and decentralization than light nodes, but they also require more resources to operate. The choice between full nodes and light nodes depends on the specific use case and requirements of the application.

1.4.2 Scalability Components

Scalability is a critical factor for blockchain networks, especially as the technology continues to gain widespread adoption and usage. A blockchain network's scalability refers to its ability to handle a growing number of transactions and users without compromising its performance or security. The importance of scalability in blockchain networks is evident from the fact that many early blockchain projects faced significant challenges in this regard. For example, the Bitcoin blockchain can only handle a limited number of TPS, leading to delays and high fees during periods of high network usage. Similarly, the Ethereum network faced scalability issues during the 2017 Initial Coin Offering (ICO) boom [14], with congestion becoming a common problem.

To address these issues, many new blockchain projects have focused on scalability as a key design principle. These projects use various approaches, such as sharding, sidechains, and off-chain processing, to increase their transaction throughput and reduce the burden on the main blockchain. For example, the Ethereum network is currently undergoing a major upgrade to transition to a more scalable architecture known as Ethereum 2.0 [15].

The importance of scalability in blockchain networks can be understood in the following ways:

1. **User Adoption:** Scalability is essential for blockchain networks to achieve widespread user adoption. If the network cannot handle a growing number of users and transactions, it may become too slow or too expensive for users to use effectively. This could lead to users abandoning the network in favor of more scalable alternatives.
2. **Decentralization:** Decentralization is a fundamental aspect of blockchain technology, which means that no single entity or group should have control over the network. However, decentralization can be challenging to achieve in a highly scalable network. A blockchain network that relies on a small number of nodes to process transactions may become vulnerable to attacks or manipulation.
3. **Security:** Blockchain networks rely on cryptography and consensus algorithms to ensure the security and integrity of their data. However, these mechanisms can become less secure if the network is too scalable. For example, a highly scalable network may be more vulnerable to 51% attacks, in which an attacker gains control of a majority of the network's computing power.
4. **Use Cases:** Scalability is essential for blockchain networks to support a wide range of use cases effectively. For example, a blockchain network that is designed for supply chain management must be able to handle a large number of transactions in real time, while a blockchain network designed for storing digital identities must be able to handle a large volume of data securely.

Scalability is a critical factor for blockchain networks to achieve widespread adoption, maintain decentralization and security, and support a wide range of use cases. As blockchain technology continues to evolve, scalability will remain a critical design consideration for future blockchain projects. Both permissioned and permissionless blockchains have implemented various scalability solutions to improve network performance and handle increasing transaction volumes. Permissioned blockchains often use more traditional scaling approaches, such as increasing block size limits, while permissionless blockchains have explored newer solutions, such as Layer 2 scaling solutions like the Lightning Network [16] or sharding. Layer 2 solutions are built on top of Layer 1 blockchains to improve scalability and reduce transaction fees by handling transactions off-chain or in a secondary framework. Layer 1 refers to the base blockchain network, such as Bitcoin, Ethereum, or Solana. It handles all transactions and consensus mechanisms natively.

Ultimately, the scalability of a blockchain network will depend on the specific implementation and the trade-offs made between security, decentralization, and scalability.

The Lightning Network offers a promising solution to the scalability issues faced by blockchain networks, and it has the potential to significantly increase transaction throughput and reduce transaction fees. However, it is important to note that the Lightning Network is still in its early stages of development and deployment, and there are still technical and operational challenges that need to be addressed before it can be widely adopted.

The Lightning Network is a Layer 2 scaling solution that can be implemented on top of a blockchain network. It is not tied to a specific blockchain platform and can be used by any blockchain network that supports its implementation. Currently, the Lightning Network has been implemented on top of Bitcoin's blockchain and is available as a payment option on some Bitcoin wallets and exchanges. In addition, other blockchain platforms such as Litecoin, Bitcoin Cash, and Ethereum have also implemented or are working on implementing the Lightning Network. It is worth noting that the Lightning Network is still a developing technology and its implementation may vary depending on the specific blockchain platform.

Private enterprise-based blockchains have different scalability needs and requirements than public blockchains. They often prioritize scalability solutions that can handle a high volume of transactions in a private, permissioned environment. As the blockchain grows larger, it requires more storage and processing power to maintain, which can limit the number of nodes that can participate in the network to validate transactions to reach consensus. Many blockchain networks use consensus mechanisms such as Proof of Work (PoW) or Proof of Stake (PoS) to verify transactions and ensure the integrity of the blockchain. However, these mechanisms can become less effective as the number of nodes in the network grows, leading to longer transaction times and higher energy consumption.

To address these trade-offs, blockchain developers are exploring new consensus mechanisms, such as Proof of Authority (PoA) or Proof of Elapsed Time (PoET), that can provide high levels of security with lower resource requirements. They are also exploring new approaches to scaling, such as sharding, sidechains, and state channels, that can increase the number of transactions that can be processed while maintaining security. Some of the scalability solutions available for private enterprise-based blockchains include:

1. **Sharding** [17]: Sharding involves dividing the blockchain network into smaller groups or partitions called shards, each of which can process transactions independently. This approach can significantly improve scalability by allowing multiple transactions to be processed simultaneously. Private enterprise-based blockchains, such as Hyperledger Fabric, have implemented sharding to improve their scalability.
2. **Sidechains** [18]: Sidechains are separate blockchains that are linked to the main blockchain network. They can be used to process transactions independently of the main chain, which can help reduce congestion on the main chain and improve scalability. Private enterprise-based blockchains, such as Quorum, have implemented sidechains to improve their scalability.
3. **Off-Chain Transactions** [19]: Off-chain transactions involve moving some transactions off the blockchain network to reduce congestion and improve scalability. This approach can be used to handle a large volume of transactions without overburdening the main blockchain network. Private enterprise-based blockchains, such as Corda, have implemented off-chain transactions to improve their scalability.
4. **Federated Consensus** [20]: Federated consensus involves a group of trusted nodes that work together to validate transactions and reach consensus. This approach can help improve scalability by reducing the number of nodes required for consensus and increasing the speed of transaction processing. Private enterprise-based blockchains, such as Ripple, have implemented federated consensus to improve their scalability.

Overall, private enterprise-based blockchains have more flexibility in terms of scalability solutions because they operate in a private, permissioned environment. This allows them to experiment with different approaches to improve their scalability based on their specific needs and requirements.

Scaling blockchain networks while maintaining data structure integrity and security is a complex and challenging task. Security and scalability are often considered trade-offs in blockchain networks because the mechanisms that provide security can also hinder scalability, and vice versa. Security is a critical aspect of blockchain networks. The blockchain's DLT, cryptographic hash functions, and digital signatures make it virtually impossible for unauthorized users to tamper with the data. On the other hand, scalability is essential to the success of blockchain networks, as it enables more users to participate and more transactions to be processed. However, scaling blockchain networks can be challenging because each node on the network must verify every transaction to make sure these can be trusted and not double spent, which can slow down processing times and increase the size of the blockchain.

1.4.3 Interoperability Components

Blockchain interoperability refers to the ability of different blockchain networks to communicate and interact with each other seamlessly. It is becoming increasingly important for several reasons. Blockchain interoperability is essential for the growth and adoption of blockchain technology. It enables the exchange of value and information between different blockchain networks, facilitates the development of more sophisticated dApps, and helps to address the issue of scalability. As such, it is an area of active research and development in the blockchain community, and we can expect to see significant progress in this area in the coming years.

Interoperability allows for the exchange of information and value between different blockchain networks, which expands the use cases and applications of blockchain technology. For instance, a user on one blockchain network could easily and securely exchange value with a user on a completely different blockchain network, without the need for intermediaries. Interoperability also enables the development of dApps that can leverage the features and functionalities of multiple blockchain networks. This could lead to the creation of more complex and sophisticated dApps that can interact with multiple blockchains and offer unique features and capabilities. Finally, interoperability helps to address the issue of scalability by enabling the use of multiple blockchain networks to process transactions simultaneously. This can help to reduce congestion on individual blockchain networks and improve overall throughput and performance.

There are several approaches to achieving blockchain interoperability, including sidechains, cross-chain bridges, atomic swaps, and interoperability protocols. These approaches allow for the transfer of data and value between different blockchain networks, while ensuring security, privacy, and trust.

In terms of private enterprise-based blockchains, interoperability solutions are becoming increasingly important as more organizations adopt blockchain technology for their operations. Private blockchains are often designed for specific use cases and may not be compatible with other blockchain networks, which can limit their usefulness. Interoperability solutions can enable private blockchains to interact with other blockchain networks, which can help to expand their use cases and increase their value proposition.

Interoperability is achieved through the use of standardized protocols and interfaces that allow different blockchain networks to communicate and share data with each other. These protocols can be implemented on both public and private blockchains, as long as the networks are designed to be compatible with each other.

In the case of public blockchains, interoperability can be achieved through the use of cross-chain protocols such as Cosmos [21], Polkadot, and Chainlink [22]. These protocols allow different public blockchain networks to communicate and share data with each other, creating a more interconnected and accessible blockchain ecosystem.

In the case of private blockchains, interoperability can be achieved through the use of enterprise-focused solutions such as Hyperledger Fabric and Corda. These platforms provide tools and protocols that enable different private blockchain networks to communicate and share data with each other, creating a more interconnected and efficient enterprise blockchain ecosystem.

Interoperability solutions are important for blockchain networks and dApps because they enable the exchange of assets and data across different blockchain networks. This can help to improve scalability, reduce transaction fees, and enable the creation of more complex dApps that span multiple blockchain networks. Examples of different types of interoperable solutions are:

1. **Cross-Chain Communication:** Cross-chain communication is the process of enabling different blockchain networks to communicate with each other. This can be achieved through various methods, including relay chains, sidechains, and blockchain bridges. The goal of cross-chain communication is to enable the transfer of assets and data across different blockchain networks. This can help to improve scalability and reduce transaction fees, as well as enable the creation of more complex dApps that span multiple blockchain networks. Cross-chain communication can be implemented using various techniques, such as sidechains, relay chains, or pegged chains, and can be used by both public and private blockchains to enable the transfer of assets and data between different blockchain networks.
2. **Atomic Swaps:** Atomic swaps are a type of P2P exchange that enables the exchange of cryptocurrencies between different blockchain networks without the need for a centralized exchange. Atomic swaps use smart contracts to lock up assets on one blockchain network and release assets on another blockchain network in exchange. Atomic swaps can help to improve liquidity and enable decentralized trading between different cryptocurrencies. Atomic swaps, which allow for the exchange of assets between different blockchain networks without the need for a trusted intermediary, have been implemented by various public blockchains, including Bitcoin and Litecoin, as well as by some private blockchains.
3. **Bridge Protocols:** Bridge protocols are a type of software that enables communication between different blockchain networks. These protocols act as intermediaries that enable the transfer of assets and data between different blockchains. Bridge protocols can be designed to support specific blockchain networks or to enable cross-chain communication between any blockchain networks. Some examples of bridge protocols include Polkadot, Cosmos, and Aion. Bridge protocols, which act as a bridge between different blockchain networks and enable interoperability, can also be used by both public and private blockchains.

Overall, the choice of interoperable solution depends on the specific use case and requirements of the application, as well as the capabilities and limitations of the underlying blockchain networks.

Interoperability standards are another type of solution aimed at facilitating interaction between different blockchains. Interoperable standards are important because they provide a common framework for different blockchains to interact with each other. This reduces the complexity and cost of building custom integrations and enables more seamless communication and interoperability between different blockchain networks. These standards consist of technical specifications or protocols that define how blockchains can communicate and operate with one another. By providing a common framework for interaction, interoperability standards enable different blockchain platforms to exchange data and assets in a consistent and standardized manner, eliminating the need for custom-built integrations for each individual connection. One example of an interoperable standard is the Interledger Protocol (ILP) [23]. ILP is an open-source protocol that provides a standardized way to transfer value across different payment networks, including blockchains. ILP enables interoperability by abstracting the underlying payment networks, allowing them to communicate with each other in a standardized way. Another example of an interoperable standard is the Token Taxonomy Framework (TTF) [24]. TTF is an open-source framework that provides a common language and taxonomy for defining tokens and smart contracts. TTF enables interoperability by providing a standard way to define and manage tokens across different blockchain platforms.

1.4.4 Platforms

Blockchain and other DLT platforms share several key similarities and differences. Both are decentralized, meaning they operate without a central authority to validate transactions or maintain the system. They are also designed to be immutable, ensuring that once a transaction is recorded, it cannot be altered or deleted. Additionally, both types of platforms offer transparency, allowing visibility into the transaction history, and they rely on consensus mechanisms to validate transactions and uphold the system's integrity.

However, there are notable differences in how they implement security and privacy controls. Blockchain typically secures transactions through a decentralized verification process involving multiple nodes, making it difficult for any single entity to manipulate data. This contrasts with some DLTs that may use different consensus algorithms or have elements of centralization, potentially affecting their security level. Furthermore, while blockchain uses public–private-key cryptography to secure transactions and authenticate users, other DLTs may employ different cryptographic techniques or key management strategies, impacting their security and user accessibility. Some DLT platforms might also offer enhanced privacy features that are not typically found in blockchain, such as the ability to conduct transactions privately or use advanced anonymity technologies.

Most blockchain wallets have built-in security features to protect the private key, such as multifactor authentication (MFA) and hardware wallets. In DLT platforms, there may be different ways to secure access to the network or to protect keys in wallets, depending on the platform's design. Blockchain technology is often associated with the idea of pseudonymity, where users are identified by public keys instead of real-world identities. This can provide some level of privacy, but transactions are still visible on the blockchain for anyone to see. Some blockchain platforms, such as Monero and Zcash, use advanced cryptographic techniques to enhance privacy and provide more anonymity. Other DLT platforms may have different privacy controls, depending on the platform's design and intended use case, regulatory considerations, and consensus mechanisms. For example, permissioned DLTs used in enterprise environments often implement strict identity management, access controls, and data confidentiality measures to comply with legal and operational requirements. These platforms may leverage technologies such as Zero-Knowledge Proofs (ZKPs), private channels, or off-chain data storage to enhance privacy while maintaining compliance and performance.

Differences among blockchain platforms can be identified in terms of implementation of smart contracts, the deployment environment that is supported such as deployments in the cloud and the service model such as by use of Application Programming Interfaces (APIs) such as Blockchain as a Service (BaaS). In terms of smart contracts, different blockchain platforms use different programming languages and frameworks to write and execute smart contracts. For example, Ethereum uses Solidity, while Hyperledger Fabric supports various programming languages such as Go, JavaScript, and Java.

In terms of deployment in the cloud, blockchain platforms can be deployed in different cloud environments such as Amazon Web Services (AWS), Azure, and Google Cloud. Some blockchain platforms also offer their own cloud-based solutions, such as Microsoft's Azure Blockchain Service and IBM's Blockchain Platform.

Several blockchain platforms support deployment in the cloud and offer different deployment models, including Software as a Service (SaaS) and Platform as a Service (PaaS). Here are some examples:

- **Ethereum:** Ethereum can be deployed in the cloud using services such as Microsoft Azure, AWS, and Google Cloud Platform. These cloud providers offer both SaaS and PaaS models, allowing users to choose the level of control and customization they need.

- **Hyperledger Fabric:** Hyperledger Fabric can also be deployed in the cloud using services such as IBM Cloud, AWS, and Microsoft Azure. These cloud providers offer both SaaS and PaaS models, allowing users to deploy Fabric networks in a way that best fits their needs.
- **Corda:** Corda is designed to run in a variety of deployment models, including cloud-based environments. Corda can be deployed using services such as Microsoft Azure and AWS, which offer both SaaS and PaaS models.
- **Ripple:** Ripple can be deployed in the cloud using services such as Microsoft Azure and AWS. These cloud providers offer both SaaS and PaaS models, allowing users to choose the level of control and customization they need.

In terms of the differences between SaaS and PaaS, SaaS is a model in which the software is hosted and managed by a third-party provider and made available to customers over the Internet. PaaS is a model in which the cloud provider offers a platform for customers to develop, run, and manage their own applications. In the case of blockchain platforms, SaaS models typically involve preconfigured networks that users can join and interact with, while PaaS models offer more flexibility and control over the network configuration and deployment.

Regarding the use of APIs, most blockchain platforms provide APIs that developers can use to interact with the blockchain network and integrate into their applications. These APIs may differ in terms of functionality and ease of use depending on the platform.

BaaS platforms provide various types of APIs that allow developers to interact with blockchain networks without the need to manage their own infrastructure.

Here are some of the different types of APIs that are commonly exposed by BaaS platforms:

1. **Blockchain Network APIs:** These APIs provide access to the core functionalities of the underlying blockchain network, such as creating and managing smart contracts, creating new transactions, querying blockchain data, and monitoring network health.
2. **Identity Management APIs:** These APIs enable developers to create and manage digital identities on the blockchain network, which can be used to authenticate users and ensure the security of transactions.
3. **Payment and Settlement APIs:** These APIs provide payment processing and settlement functionality on the blockchain network, enabling developers to create and manage payment systems and applications.
4. **Tokenization APIs:** These APIs enable the creation and management of tokens on the blockchain network, which can represent anything from virtual currencies to physical assets.

5. **Analytics APIs:** These APIs provide access to real-time and historical data on blockchain network activity, including transaction volume, network congestion, and smart contract activity.
6. **Governance APIs:** These APIs enable the creation and management of decentralized autonomous organizations (DAOs) and other governance mechanisms on the blockchain network.
7. **Integration APIs:** These APIs enable developers to integrate blockchain functionality into existing applications and systems, such as enterprise resource planning (ERP) software, customer relationship management (CRM) systems, and supply chain management tools.

Overall, the APIs provided by BaaS platforms offer a range of powerful tools and functionalities for developers looking to build blockchain-based applications and services. The types of APIs provided by BaaS platforms can vary depending on the underlying blockchain network that the platform supports. However, many of the APIs listed are common among both public and private blockchains. For example, blockchain network APIs, identity management APIs, payment and settlement APIs, tokenization APIs, and analytics APIs are generally available on both public and private blockchains. These APIs provide core functionality that is required for any blockchain network, regardless of whether it is public or private.

Governance APIs and integration APIs may be more common on private blockchains, as they are often used by businesses and organizations to manage internal processes and systems. However, some public blockchain networks also offer these types of APIs. It's also worth noting that some blockchain networks may provide additional APIs or functionalities that are specific to their network. For example, Ethereum, one of the most popular blockchain networks, offers APIs for managing dApps and interacting with its unique programming language, Solidity. In summary, while the specific APIs offered by BaaS platforms may vary depending on the underlying blockchain network, many of the core APIs are common across both public and private blockchains.

1.4.5 dApps

Businesses today can leverage different blockchain platforms to design and build dApps, each with their own unique features and capabilities. From a software developer perspective, the general steps involved in building applications that use blockchain platforms are:

1. **Choose a Blockchain Platform:** There are many blockchain platforms available, each with their own unique features and capabilities. Developers must choose the blockchain platform that best suits their application's needs.

2. **Design the dApp Architecture:** Developers must design the architecture of their application, including how it will interact with the blockchain platform. This includes designing smart contracts, which are self-executing contracts with the terms of the agreement between buyer and seller being directly written into lines of code.
3. **Develop the Application:** Developers must write the code for their application, including the front-end interface and the back-end logic. This code must be designed to interact with the blockchain platform via APIs.
4. **Test the Application:** Developers must thoroughly test their application to ensure that it works as expected and that it interacts correctly with the blockchain platform.
5. **Deploy the Application:** Once the application has been tested and is ready for deployment, it must be deployed to the appropriate environment, such as a public blockchain network or a private blockchain network.
6. **Maintain and Update the Application:** Once the application is live, developers must maintain and update it to ensure that it remains secure, functional, and up-to-date with the latest blockchain platform features.

Overall, building applications that use blockchain platforms requires a combination of blockchain-specific skills, such as smart contract development, as well as general software development skills.

1.4.6 Practical Examples

Blockchain application development has moved beyond the experimental phase and is now in the mainstream adoption phase by businesses. Many businesses are recognizing the potential benefits of blockchain technology, such as increased security, transparency, and efficiency, and are actively seeking out ways to incorporate it into their operations. As a result, there are now several types of mature blockchain platforms available that software developers can leverage to develop their dApps.

These platforms offer a variety of features and capabilities, including support for smart contracts, privacy features, scalability, and more. They also have strong communities of developers and users, making it easier for developers to get started and find support when needed. Additionally, many of these platforms have a strong track record of security and reliability, which is crucial for businesses that need to trust the platform with their sensitive data and transactions.

Overall, with the increasing adoption of blockchain technology by businesses, software developers now have several mature blockchain platforms to choose from, which can help them build dApps more easily and effectively.

There are several factors that can contribute to the popularity of a blockchain platform:

1. **Features and Capabilities:** The features and capabilities of a blockchain platform are a major factor in its popularity. Platforms that offer unique or advanced features, such as support for smart contracts, privacy features, or scalability, are more likely to attract developers and users.
2. **User and Developer Community:** A strong and active user and developer community can greatly contribute to the popularity of a blockchain platform. A large community can provide support, resources, and ideas for developers, as well as drive adoption of the platform among users.
3. **Security and Reliability:** Blockchain platforms that are secure and reliable are more likely to gain trust and adoption from users and businesses. A platform with a strong track record of security and reliability is more likely to be adopted by businesses and other organizations that need to trust the platform with their sensitive data and transactions.
4. **Ease of Use and Accessibility:** Blockchain platforms that are easy to use and accessible to a wide range of users are more likely to be popular. Platforms that provide easy-to-use tools, documentation, and resources for developers and users can help lower the barrier to entry and increase adoption.
5. **Cost and Licensing:** The cost and licensing of a blockchain platform can also impact its popularity. Platforms that are open source or have low licensing fees are more likely to attract developers and businesses that are looking to keep their costs low.

Overall, a successful blockchain platform needs to offer unique and valuable features, have a strong community, be secure and reliable, be easy to use and accessible, and have reasonable costs and licensing terms. Here are some examples of popular DLT platforms and networks that are widely used by businesses and software developers:

1. **Ethereum:** Ethereum is a decentralized blockchain platform that enables the creation of dApps and smart contracts. It is one of the most widely used blockchain platforms and has a large and active developer community.
2. **Hyperledger Fabric:** Hyperledger Fabric is an open-source DLT platform that is designed for use in enterprise settings. It is highly customizable and supports modular architecture, allowing businesses to build their own blockchain networks with tailored features and functions.
3. **Corda:** Corda is an open-source DLT platform that is designed specifically for use in business environments. It uses a "notary" system to ensure the validity of transactions and is highly customizable to meet the needs of individual businesses.

4. **Ripple:** Ripple is a blockchain-based payment protocol that is designed to enable fast and secure cross-border payments. It is widely used by financial institutions and other businesses that need to transfer money across borders.
5. **Quorum:** Quorum is an enterprise-focused blockchain platform that is based on the Ethereum codebase. It is designed to provide privacy and security for businesses and supports the creation of smart contracts and dApps.
6. **Stellar:** Stellar [25] is a blockchain-based payment protocol that is designed to enable fast and affordable cross-border payments. It is widely used by businesses and nonprofits that need to transfer money across borders and has a large and active developer community.

These are just a few examples of the many DLT platforms and networks that are popular among businesses and software developers. Each platform has its own unique features and strengths, and the choice of platform depends on the specific needs of the business or application being developed. Here are some examples of dApps built on top of blockchain technology, which use DLT ecosystems:

1. **Decentralized Finance (DeFi) Applications:** DeFi applications are financial applications built on blockchain networks, allowing users to access financial services without intermediaries, such as banks. Examples of DeFi dApps include lending and borrowing platforms, Decentralized Exchanges (DEXs), and prediction markets.
2. **Supply Chain Management:** Blockchain technology can be used to track and verify the movement of goods throughout a supply chain. This helps to ensure the authenticity and integrity of products, reduce counterfeiting, and increase transparency. For example, the VeChain blockchain platform has developed several supply chain dApps, including "My Story" and "Producers Market."
3. **Gaming and Collectibles:** Blockchain technology can be used to create unique digital assets, such as nonfungible tokens (NFTs), which can be traded and collected. Gaming and collectible dApps, such as CryptoKitties, have gained significant popularity in recent years.
4. **Identity Verification and Management:** Blockchain technology can be used to securely store and manage personal identity information, which can be used for authentication and authorization purposes. uPort is an example of a dApp that allows users to create and manage their digital identity on the Ethereum blockchain.
5. **Decentralized Social Networking:** Decentralized social networking dApps use blockchain technology to ensure user privacy and security and to give users control over their own data. Examples of decentralized social networking dApps include Mastodon, Steemit, and Minds.

These are just a few examples of the many ways that blockchain technology can be used to create innovative and dApps within DLT ecosystems.

In some industries such as the financial sector, many applications have been developed leveraging enterprise-focused blockchain platforms that have been developed by start-ups. Here are some examples:

1. **Axoni** [26]: Axoni is a New York-based start-up that has developed a blockchain platform for enterprise use. Its platform is designed to be highly scalable and to support complex financial instruments, such as derivatives and securities.
2. **ConsenSys** [27]: ConsenSys is a blockchain start-up that has developed a suite of blockchain tools and applications for use in a variety of industries. Its platform includes tools for developing and deploying smart contracts, as well as a number of prebuilt applications for industries such as finance and supply chain management.
3. **Digital Asset** [28]: Digital Asset is a blockchain start-up that has developed a platform for building distributed applications. Its platform is designed to enable the creation of complex, multiparty workflows and supports a wide range of use cases, including financial services, supply chain management, and healthcare.
4. **Chain** [29]: Chain is a blockchain start-up that has developed a platform for building blockchain networks for enterprise use. Its platform is designed to enable the creation of secure and scalable networks that can be used for a wide range of applications, from financial services to healthcare.

These enterprise-focused blockchain platforms are designed to meet the specific needs of businesses and other organizations. They typically offer features such as scalability, security, and the ability to support complex financial instruments and multiparty workflows.

1.5 The Blockchain Data Structure

Before we dwell into the details of blockchain essential components. It is necessary to have a good grasp of how blockchain's data structure works. This knowledge is also essential for identifying and developing potential new use cases for blockchain technology.

The data structure of a blockchain is a critical aspect of its design, as it enables the system to maintain the integrity and security of the transactional data being recorded. At its core, a blockchain is a decentralized and distributed ledger of transactions that is maintained by a network of computers, or nodes, rather than a central authority. This means that each node on the network has a copy of the same ledger, and that any changes to the ledger must be validated and agreed upon by a consensus of the nodes in the network.

The structure of each block in the blockchain is designed to ensure that the data stored within it is tamperproof and resistant to modification. Each block contains a pointer to the hash of the previous block, which allows the blocks to be chained together in a secure and immutable way. In addition to the previous block hash, each block contains a timestamp and a nonce, which are used to validate and verify the block before it can be added to the chain. The block also contains a hash structure or the root of a Merkle tree that contains all the transactions stored in that block.

The use of cryptography, including one-way hashes and digital signatures, further enhances the security and integrity of the blockchain. Each transaction in the blockchain is signed with a digital signature, which is hashed and encrypted using a private key and can be verified by any node on the network using the corresponding public key. This ensures that transactions are authentic and cannot be tampered with or modified without detection. Overall, the data structure of the blockchain is a critical component of its design, as it enables the system to maintain the integrity and security of the transactional data being recorded. By using a decentralized, distributed ledger that is tamperproof and resistant to modification, the blockchain has the potential to revolutionize industries and transform the way we think about data security and integrity.

In the context of distributed ledgers, the importance of the blockchain data structure cannot be overstated. The blockchain is a decentralized, distributed database that stores a continuously growing list of records, or blocks, which are linked and secured using cryptography. It is designed to provide a tamperproof, transparent, and auditable ledger of transactions that can be trusted without the need for a central authority or intermediary.

The data structure of the blockchain is critical to achieving these goals. Each block in the chain contains a hash of the previous block, along with a timestamp and a nonce, as well as the transactions that have been verified and added to the block. This creates a chain of blocks that cannot be tampered with or altered, as any attempt to modify one block will change the hash of that block, which will in turn change the hash of all subsequent blocks.

This tamperproof quality is critical in ensuring the integrity of the distributed ledger. Without it, the distributed ledger would be vulnerable to attacks and fraud, and it would be impossible to trust the data stored in it. The blockchain data structure enables the creation of a decentralized, transparent, and secure ledger that can be used to store and verify all kinds of data and transactions, from financial transactions to medical records to supply chain data.

In addition to its tamperproof quality, the blockchain data structure also enables consensus and validation of transactions without the need for a central authority. The use of a consensus algorithm, such as PoW or PoS, enables nodes in the network to agree on the validity of transactions and blocks, ensuring that the distributed ledger remains accurate and trustworthy.

1.5.1 Hash Functions

Hash functions play a critical role in the functioning of blockchain technology. They are used in several aspects of the blockchain, including in the creation of blocks, the validation of transactions, and the maintenance of the integrity of the entire blockchain.

When a block is created in the blockchain, a hash function is used to create a unique identifier for the block. This identifier, known as the block hash, is created by taking the contents of the block and applying a cryptographic hash function to it. The resulting hash is a fixed-length string of characters that serves as a unique identifier for that block.

In addition, each block in the blockchain contains a reference to the hash of the previous block. This creates a chain of blocks, where each block is linked to the previous one, and allows for the creation of a secure and tamperproof ledger.

Hash functions are also used to validate transactions in the blockchain. When a transaction is created, it is first hashed using a cryptographic hash function. This creates a unique identifier for the transaction that can be used to verify its authenticity. The transaction is then broadcast to the network, and nodes in the network validate it by verifying its hash against its contents.

The use of hash functions in the blockchain is essential to ensure the security and integrity of the entire system. The one-way nature of hash functions makes it impossible to reconstruct the original data from the hash, which means that once data is hashed, it cannot be modified without detection. This feature, combined with the use of a chain of blocks, creates a system that is resistant to tampering and provides a secure and decentralized ledger for transactions.

A hash function is a mathematical function that takes in an input (also called a message) of arbitrary size and produces a fixed-size output, called a hash, that is typically much smaller in size. Although the same input will always produce the same output, it is computationally infeasible to find two inputs that will produce the same output. This property is known as collision resistance, and it is a critical characteristic of hash functions used in blockchain.

In the context of blockchain, hash functions are used extensively to ensure the integrity and security of data stored on the network. The most commonly used hash function in blockchain is the Secure Hash Algorithm 256 (SHA-256), which is a widely used cryptographic hash function that produces a fixed-size output of 256 bits. Other commonly used hash functions include RIPEMD-160, used in Bitcoin's address format, and Keccak-256, used in Ethereum.

Hash functions are used in several ways in blockchain, including:

1. **Generating Block Hashes:** Each block in a blockchain contains a hash of the previous block, which creates a chain of blocks. This hash is generated by applying a hash function (such as SHA-256) to the header of the previous block.
2. **Creating Unique Transaction Identifiers:** Each transaction in a block is hashed to create a unique identifier that can be used to reference the transaction.
3. **Enable Transaction Verification:** Transactions in a block are typically organized into a Merkle tree structure, where each leaf node represents a transaction and each nonleaf node is the hash of its children. The root of the Merkle tree is included in the block header, which allows nodes to quickly verify that a transaction is included in a block without having to download the entire block.

Hash functions are a fundamental building block of blockchain technology, providing the necessary cryptographic security and integrity to ensure that the data stored on the network cannot be tampered with or modified.

Hash functions have several unique properties that make them useful in the context of blockchain, tokens, and dApps:

1. **Deterministic:** Hash functions always produce the same output for a given input. This property is important for ensuring consistency in the blockchain, where the same input data should always produce the same hash value.
2. **One Way:** Hash functions are designed to be one way, meaning that it is easy to compute the hash output of an input but it is computationally infeasible. The one-way property is used explicitly to ensure the difficulty of Bitcoin PoW.
3. **Fixed-Size Output:** Hash functions always produce a fixed-size output, regardless of the size of the input data. This property is important for optimizing the storage and transmission of data on the blockchain.
4. **Collision Resistant:** Hash functions used in blockchain are designed to be collision resistant, meaning it is computationally infeasible to find any two different inputs that produce the same hash value. Additionally, second preimage resistance ensures that given a specific input (like a transaction), it's infeasible to craft a different input that produces the same hash; this is critical for protecting against tampering or forgery. This ensures uniqueness and integrity of data on the blockchain, such as ensuring each block has a unique hash.

In terms of algorithms, some of the most commonly used hash functions in the blockchain include SHA-256 and SHA-3. These algorithms are known for their cryptographic security and have been extensively tested and validated in the context of blockchain applications. Other hash functions, such as Keccak, BLAKE2, and Scrypt, are also used in certain blockchain applications.

Hash functions are an integral part of the blockchain technology as they provide several benefits such as data integrity, immutability, and security. The use of hash functions in blockchain ensures that data stored in blocks cannot be modified without detection. This property makes it useful for creating a distributed ledger, as it ensures that the records stored on the blockchain cannot be tampered with.

Additionally, hash functions are used in creating digital signatures and cryptographic keys that are used for secure transactions and identity verification. Here are some examples of how hash functions are used in blockchain technology:

1. **Block Hashing:** Hash functions are used in the creation of blocks in a blockchain. Each block contains a hash of the previous block's header, which helps maintain the integrity and security of the blockchain. This linking of blocks using hash functions creates an unbreakable chain of data that cannot be tampered with.
2. **Mining:** In blockchain-based systems that use PoW consensus algorithms, hash functions are used in the mining process. Miners use specialized computers to solve complex mathematical puzzles that involve hashing. The first miner to solve the puzzle and find a hash preimage that meets the difficulty target is rewarded with new coins and the right to create the next block in the blockchain.
3. **Digital Signatures:** Hash functions are used in digital signatures to ensure the integrity and authenticity of data. In a blockchain-based system, a digital signature is created by applying a private key to the hash of a message: this creates a unique digital fingerprint that can be verified by anyone with the corresponding public key. Digital signatures are used to authenticate transactions and ensure that they have not been tampered with.
4. **Smart Contracts:** Hash functions are used in the creation and execution of smart contracts. Smart contracts are self-executing programs that are stored on a blockchain. Hash functions are used to create a unique identifier for each smart contract, which is used to refer to the contract and execute its code.
5. **Tokenization:** Hash functions are used in the creation and management of tokens on a blockchain. Each token is represented by a unique hash value, which is used to verify ownership and transfer of the token. This ensures that each token is unique and cannot be duplicated or counterfeited.

The use of hash functions in blockchain networks can be traced back to the creation of Bitcoin, the first blockchain-based cryptocurrency. The creator of Bitcoin, who used the pseudonym Satoshi Nakamoto, implemented the SHA-256 hash function as a fundamental component of the Bitcoin protocol.

SHA-256 was developed by the National Security Agency (NSA) in the United States and is a widely used cryptographic hash function. It takes an input of any size and produces a fixed-size output of 256 bits, making it a one-way function that is computationally infeasible to reverse. This makes SHA-256 ideal for use in a blockchain network where data integrity and immutability are critical.

Since the creation of Bitcoin, other blockchain networks have emerged that use different hash functions. For example, Ethereum, the second-largest blockchain network, uses Keccak-256 (also known as SHA3-256), a variant of the SHA-3 family of hash functions. Other popular blockchain networks such as Litecoin and Bitcoin Cash also use variants of the SHA-256 hash function.

Hash functions are used in popular blockchain networks:

1. **SHA-256:** The SHA-256 hash function is used in the Bitcoin network to create digital signatures for transactions and to secure the blockchain.
2. **SHA-512:** The SHA-512 hash function is used in the Ripple network to create digital signatures and to secure the blockchain.
3. **Keccak-256 (SHA-3):** The Keccak-256 hash function, also known as SHA-3, is used in the Ethereum network to create addresses and digital signatures and to secure the blockchain.
4. **Scrypt:** The Scrypt hash function is used in the Litecoin network to secure the blockchain and to mine new blocks.
5. **CryptoNight Hash:** The CryptoNight hash function is used in the Monero network to provide privacy and anonymity to transactions.

These are just a few examples, as there are many other blockchain networks that use different types of hash functions for various purposes.

1.5.2 Digital Signatures

Overall, digital signatures play a crucial role in the security and trustworthiness of blockchain networks, ensuring that transactions are authenticated, verified, and tamperproof. In the context of blockchain applications, digital signatures play a crucial role in ensuring the integrity and authenticity of transactions. Digital signatures are essentially hashed data that are processed with the private key of the sender of a message and can be verified by the receiver of the message using the corresponding public key pair.

To verify a digital signature, the recipient of the message decrypts the encrypted hash using the public key and compares it with the hash of the provided message.

Since the public key is available for verification, any computer node can request the public key to verify the signature. In Bitcoin transactions, the public key is derived from the user's key pair and used to verify transaction signatures, but it is not attached to a certificate or validated by a certificate authority (CA). Instead, trust is ensured through cryptographic verification and blockchain consensus, not external attestation.

The Bitcoin protocol uses hashes and digital signatures for the initial authorization of payments between peer nodes. In a Bitcoin transaction, there are typically both input and output values. The transaction references previous transaction outputs as new transaction inputs and dedicates all input Bitcoin values to new outputs.

In the context of Bitcoin transactions, the inputs represent the bitcoins owned by the payer (buyer), while the outputs represent the bitcoins being transferred to the recipient (seller). To complete a transaction, the buyer must prove ownership of the input bitcoins by providing a digital signature, which is verified using their public key. This cryptographic validation enables the buyer to spend their bitcoins and transfer them to the seller. In Bitcoin, the public key is derived from the user's key pair and used to verify transaction signatures and trust is established through cryptographic verification and blockchain consensus, without relying on external attestation.

It's important to note that in Bitcoin transactions, the actual transaction data is not encrypted, which means that it's possible for other nodes on the network to view every transaction that's included in a block.

Validating the signature of a Bitcoin transaction is critical for ensuring transaction integrity, as it detects any tampering with the transaction data and enforces nonrepudiation. Signature verification serves as proof that the transaction originated from the payer (buyer), since only the buyer possesses the private key used to sign the transaction. This private key is never shared, meaning no other node can forge or replicate the same signature to authorize the transaction.

Digital signatures play a crucial role in ensuring the integrity and authenticity of transactions in the Bitcoin protocol. To achieve transaction integrity and enforce nonrepudiation, the validation of the digital signature is a critical control. It not only detects tampering of data in the block but also requires signature validation as proof of origination of the transaction by the source node seller, whose private key cannot be used by another node to sign the same transaction.

Before a buyer can pay bitcoins to a seller, they must prove ownership of the inputs by providing a valid digital signature, which is verified by the network using the buyer's public key. Unlike traditional systems, the Bitcoin protocol does not rely on CAs or digital certificates; instead, trust is established through decentralized cryptographic verification performed by the network's consensus mechanism.

It's worth noting that transactions in the Bitcoin protocol are not encrypted, so it's possible for other nodes to view every transaction included in a block. Digital signatures are a crucial component of blockchain security. They enable users to prove their identity, ensure transaction validity, and create a tamperproof record of all transactions on the blockchain.

Digital signatures are used in blockchain to ensure that transactions are valid, tamperproof, and originated from a known source. When a user initiates a transaction on the blockchain, they use their private key to sign the transaction data. This creates a digital signature that proves that the transaction came from that user and hasn't been tampered with. The digital signature is then broadcast to the network along with the transaction data, and every node on the network can verify the authenticity of the signature using the user's public key.

Once a node receives a transaction with a digital signature, it verifies the signature to ensure that it's valid and hasn't been tampered with. If the signature is valid, the transaction is added to the blockchain, and it becomes part of the permanent record. Because digital signatures use public-key cryptography, it's practically impossible to forge or fake a signature without access to the user's private key. This means that transactions on the blockchain are highly secure and resistant to tampering or fraud.

Understanding how digital signatures work in the blockchain is important to ensure the security and integrity of transactions between nodes.

In simple terms, a digital signature is created by hashing the transaction data and then signing that hash using the sender's private key. Anyone can verify the signature using the sender's public key by checking that the decrypted hash matches the transaction hash. In the Bitcoin protocol, ownership of bitcoin is proven by demonstrating control over the private key associated with the public key linked to a specific Bitcoin address. This cryptographic proof allows the network to verify and authorize the transfer of value without relying on third-party attestation.

Digital signatures are used to ensure the authenticity, integrity, and nonrepudiation of transactions on the blockchain. Here are a few examples of how digital signatures are used in other types of blockchain platforms:

1. **Bitcoin Digital Signatures:** In the case of Bitcoin, digital signatures are used for the initial authorization of payments between peer nodes. Before a seller can sell their bitcoins to a buyer, they must prove ownership by validating the digital signature that allows them to spend their bitcoins. Transactions are not encrypted in Bitcoin, so every node can view all transactions included in a block. The validation of the signature is crucial to ensuring the integrity of the transaction and to prevent repudiation.

2. **Ethereum Digital Signatures:** Ethereum uses digital signatures to validate the transactions made on its platform. Like Bitcoin, each transaction is signed with the private key of the sender and can be verified using their public key.
3. **Ripple Digital Signatures:** Ripple uses a unique digital signature scheme called the "Hash Tree," which allows for faster and more efficient verification of transactions. In this scheme, each transaction is hashed and combined with other transactions to create a Merkle tree. The final hash of the tree is then signed with the sender's private key and can be verified by anyone with access to their public key.
4. **Hyperledger Digital Signatures:** Hyperledger Fabric uses a combination of digital signatures and access control lists to ensure that only authorized parties can access and modify data on the blockchain. Each transaction is signed with the sender's private key and must be approved by all parties involved in the transaction before it can be added to the blockchain.
5. **Corda Digital Signatures:** Corda uses digital signatures to ensure the privacy and security of its transactions. Each transaction is signed by the sender and recipient of the data, as well as any intermediaries involved in the transaction. The signatures are then stored off-chain and can be verified by anyone with access to the transaction data.

Digital signatures are also essential to consensus algorithms as they help to ensure that transactions and blocks being added to the blockchain are authentic and have not been tampered with. In most consensus algorithms, nodes in the network have to agree on the validity of transactions before adding them to the blockchain.

In addition to securing transactions, digital signatures can also be used in consensus algorithms that require multiple validators to verify the authenticity of a transaction or block. For example, in a PoS blockchain, validators are required to stake a certain amount of cryptocurrency to participate in the consensus process. Validators take turns proposing new blocks and validating transactions within those blocks. When a validator proposes a new block, they must include a digital signature to prove that they're the rightful owner of the stake and have the authority to propose new blocks. Other validators then verify the digital signature to ensure that the block was proposed by a legitimate validator. If the signature is invalid or the validator is attempting to cheat, other validators will reject the block and refuse to add it to the blockchain.

Similarly, in a Byzantine Fault Tolerance (BFT) consensus algorithm, validators use digital signatures to prove that they've received and verified a transaction or block. When a validator receives a new transaction or block, they sign it with their private key to indicate that they've verified it. Other validators then verify the signature to ensure that the transaction or block is valid and hasn't been tampered with. If enough validators agree that the transaction or block is valid, it's added to the blockchain. However, if a validator attempts to cheat or provides an invalid signature, the other validators will reject the transaction or block and prevent it from being added to the blockchain.

Public-Key Cryptography: Public-key cryptography is an essential tool for securing transactions on the blockchain and enabling secure communication between parties. It enables users to prove their identity and ensures that transactions are authentic and tamperproof, enhancing the security and trustworthiness of the blockchain network. The blockchain systems rely on cryptographic verification of key ownership rather than trusted third-party attestation via Public Key Infrastructure (PKI), which involves digital certificates and CAs. The simple use case of public-key cryptography is to use a pair of two keys, consisting of a public key and a private key, to encrypt and decrypt messages. The public key is available to anyone, while the private key is kept secret and known only to the owner. Messages encrypted with the public key can only be decrypted with the corresponding private key, and vice versa.

Another use case for public-key cryptography is to use it for digital signatures for transaction verification by ensuring that transactions are authentic and have not been tampered with. When a user creates a digital signature, he will use their private key to encrypt a message that includes the transaction data. The encrypted message can only be decrypted using the user's public key, which serves as a proof of identity. Other users on the blockchain can verify the digital signature by using the sender's public key to decrypt the message and confirm that it matches the original transaction data. If the signature is valid, it indicates that the transaction came from the expected sender and hasn't been tampered with.

Public-key cryptography also allows for secure communication between parties who have never met or established a prior relationship. Because the public key can be freely shared without compromising the security of the private key, it's possible for two parties to exchange encrypted messages without ever having to exchange a secret key.

While public-key cryptography is currently one of the most secure methods for protecting sensitive information and ensuring the integrity of transactions, it's widely believed that quantum computing poses a significant threat to the security of public-key cryptography as currently implemented. Quantum computing is a new computing paradigm that relies on the principles of quantum mechanics to perform computations much faster than classical computers. While quantum computing is still in its early stages of development, it's widely believed that once it reaches a certain level of maturity, it will be able to break many of the cryptographic algorithms currently in use, including those used in public-key cryptography. One of the key challenges that quantum computing poses for public-key cryptography is that it can easily factor large numbers, which is the basis of many public-key cryptography algorithms, including Rivest–Shamir–Adleman (RSA) and Diffie–Hellman. For instance, the security of Diffie–Hellman key exchange is based on the mathematical difficulty of the discrete logarithm problem, which is computationally infeasible to solve with classical computers. However, a sufficiently advanced quantum computer could efficiently solve the discrete logarithm problem using Shor's algorithm, thereby breaking the security guarantees of Diffie–Hellman and similar schemes. This would allow an attacker to derive private keys from public information, compromising the confidentiality of encrypted communications. As quantum computing matures, its ability to factor large numbers and solve discrete logarithms threatens to undermine the foundations of current public-key cryptography, including RSA, Diffie–Hellman, and Elliptic Curve Cryptograph (ECC), necessitating a transition to quantum-resistant algorithms to maintain data security and trust in digital transactions.

To address this challenge, researchers are exploring new cryptographic algorithms that are resistant to quantum computing, such as lattice-based cryptography, code-based cryptography, and hash-based cryptography. These new algorithms rely on mathematical problems that are believed to be hard even for quantum computers to solve, offering a potential solution to the challenge of quantum computing.

There are several digital signature algorithms (DSAs) that are commonly used in popular blockchain networks. Here are a few examples:

1. **ECDSA:** The Bitcoin network and Ethereum use the Elliptic Curve Digital Signature Algorithm (ECDSA) to create and verify digital signatures. ECDSA is a variant of the DSA that uses elliptic curve cryptography to provide a higher level of security with smaller key sizes.
2. **EdDSA:** Edwards-Curve Digital Signature Algorithm (EdDSA) is a variant of the Schnorr signature algorithm that uses elliptic curve cryptography to provide faster signature generation and smaller key sizes. Specific instances of EdDSA are implemented in blockchain platforms such as Ripple and Stellar to create and verify digital signatures.

1.5.3 Block Structure

To fully understand the blockchain, it is important to have a basic understanding of several crypto constructs, including hashes and digital signatures, consensus algorithms, cryptocurrencies, digital tokens, smart contracts, digital wallets, and identities and privacy-enhancing controls. These concepts are essential to understanding how the blockchain works and how it can be used in a variety of applications.

It is also critical to understand the block structure of the blockchain because it is the fundamental building block of the entire blockchain system. Each block contains important data that ensures the integrity and security of the transactional data being recorded. The block structure is designed in a way that makes it tamperproof and resistant to modification, which is one of the key features that makes blockchain technology so valuable.

Specifically, by understanding the block structure, you can understand how the blockchain is able to create a secure and decentralized ledger of transactions. You can also understand the role of the different components of a block, such as the previous block hash, timestamp, nonce, and Merkle tree root. This understanding is important for anyone who wants to work with blockchain technology, whether as a developer, investor, or user. By understanding the block structure, you can understand how the blockchain is able to create a secure and decentralized ledger of transactions.

The core data structure of a blockchain is a chain of blocks that contains a sequence of transaction data, along with some metadata, such as a timestamp and a hash value. Each block in the chain is connected to the previous block through a cryptographic hash value, creating an unbreakable link between them. This ensures that any attempt to tamper with the data in one block will be immediately detected and invalidated, as it will change the hash value of that block, and all subsequent blocks.

The block itself typically contains a list of transactions that have been validated by a network of participants, also known as nodes or miners. These transactions can involve the transfer of digital assets, such as cryptocurrencies, or any other type of data that needs to be securely recorded and verified.

The block data structure is a crucial component of the blockchain, providing the foundation for the tamperproof and secure recording of transactional data. Each block in the chain contains a block header, which includes the SHA-256 hash of the previous block, a timestamp, a nonce for proof of validation, and the hash structure or root of a Merkle tree that stores all the transactions in that block.

The tamperproof quality of the blockchain is due to the one-way encrypted hashes that link the transactional data in the new block together. The timestamp and nonce are critical components of the block header that enable validation of new blocks before they can be added to the chain. The timestamp provides a source of variation and is checked for validity, while the nonce is the answer to the PoW puzzle that nodes must solve to add a valid block to the chain.

Additionally, each block contains the root of a Merkle tree, which is updated every time a new transaction is accepted and stored in the block. The Merkle tree is a data structure that enables efficient verification of the transactions in a block, further enhancing the security of the blockchain.

The security of the blockchain is also based on the law of incentives, which makes it less likely for an attacker to attempt to break the chain due to the substantial investment in computer resources required to alter the data in previous or current blocks.

Blockchain Block Data Structure: The block structure in a blockchain is important because it enables the tamperproof and secure storage of transactional data. The use of cryptographic hashing and linking of blocks in a chain creates an immutable record of transactions that cannot be altered without detection. This is a key feature of blockchains and is what enables their use in a wide range of applications such as cryptocurrencies, supply chain management, and digital identity systems.

A block in a blockchain is a unit of data that contains a group of transactions. Each block has a specific structure that includes a header and data. The header includes important information such as the encrypted SHA-256 hash of the previous block, a timestamp, a nonce for proof of validation of the block, and the hash structure or root of a Merkle tree that contains all the transactions stored in that block. The data contained in the block is the group of transactions that have been verified and added to the blockchain.

The basic data structure of the blockchain is critical to maintaining the integrity of the transactional data being recorded.

What Is Stored in Each Block: To fully understand the blockchain, it is important to have a basic understanding of what is stored in each block. Each block contains several key pieces of information:

- **Encrypted SHA-256 Hash of the Previous Block:** One of the most important features of the blockchain is its ability to be tamperproof or resistant to modification of the data being recorded. This is accomplished through the use of one-way hashes that are linked together, making it impossible to alter any data without detection. The hash function is a one-way encryption that ensures that once data has been hashed, it cannot be decrypted back to its original value. In a blockchain, each new block contains a pointer to the hash of the previous block, allowing them to be chained together as a chain of blocks or blockchain that cannot be tampered with. In addition to the previous block hash, there are other critical pieces of information included in each block that enable the validation of new blocks before they can be added to the chain.
- **Timestamp:** The timestamp and the nonce are two critical pieces of information that are used in the validation of new blocks. The timestamp is the current time in seconds since 1 January 1970, and it is checked for validity to ensure that new blocks are added in the correct order. The timestamp is also used as a source of variation in the block hash, making it more difficult for

an adversary to manipulate the blockchain. The nonce is a number used once that is included in the block as the answer to the PoW puzzle. In the case of Bitcoin, node miners must solve this puzzle in order to add a valid block to the blockchain.

- **Merkle Tree:** Each block also contains the root of a Merkle tree, which is a data tree structure that is also constructed using hashing. The Merkle tree is updated every time a new transaction is accepted and stored in the block.
- **Unspent Transaction Outputs:** In the context of the blockchain block, it is important to explain UTXOs that stands for unspent transaction outputs. UTXOs are essentially the "coins" in the system, which represent a certain amount of cryptocurrency that can be spent. When a transaction occurs on the blockchain, the inputs to that transaction are UTXOs that have been previously created and remain unspent. For example, let's say that Alice has 5 bitcoins in her UTXO set, and she wants to send 2 bitcoins to Bob. She would create a transaction that references her UTXO of 5 bitcoins as the input to the transaction, and specifies that 2 bitcoins should be sent to Bob's address as the output. The remaining 3 bitcoins would then be sent back to a new UTXO at Alice's address, which she can then spend in a future transaction. However, if Alice chooses to allocate strictly less than 3 bitcoins back to herself, say, 2.9 bitcoins, the remaining 0.1 bitcoin is implicitly offered as a transaction fee, which incentivizes miners to include her transaction in the next block. Each block on the blockchain contains a set of UTXOs that represent the state of the system after all transactions in that block have been processed. When a new block is added to the blockchain, the UTXOs in that block become the input for the next set of transactions, and the UTXOs from the previous blocks are considered "spent" and no longer available for use. UTXOs are a key part of the security of the blockchain, as they enable the system to track the movement of coins and prevent double-spending. By ensuring that each UTXO can only be spent once, the blockchain can maintain a trusted ledger of transactions without the need for a central authority to manage the process.

Anatomy of Block: Understanding the details of a block in the blockchain is important for security of the blockchain since that is dependent on the details of each block. Each block of the blockchain contains a cryptographic hash of the previous block, which creates a chain of blocks that cannot be altered without being detected. By understanding the details of each block, developers can build more complex and sophisticated applications that can interact with the blockchain. It is important that the anatomy of a block in the blockchain is designed to ensure the integrity and security of the data being recorded. By including metadata, transactional data, and cryptographic hashes, the blockchain enables transparent and secure transactions without the need for intermediaries.

The Block Header: The block header contains important metadata about the block, including:

1. **Version Number:** The version number specifies the software version used to create the block.
2. **Timestamp:** The timestamp indicates the exact time the block was created.
3. **Nonce:** The nonce is a random number generated by the miner that is used in the mining process to generate a valid hash.
4. **Merkle Root:** The Merkle root is the hash of all the transactions included in the block.
5. **Block Size:** The block size is the total size of the block in bytes.
6. **Transaction Counter:** The transaction counter is the number of transactions included in the block.
7. **Transactions:** Transactions are the actual data included in the block. Each transaction contains:
   a. **Inputs:** Inputs are the previous transaction outputs (UTXOs) that are being spent in the current transaction.
   b. **Outputs:** Outputs are the new UTXOs that are created as a result of the transaction.
   c. **Scripts:** Scripts are instructions that specify how the transaction can be spent.
   d. **Block Hash:** The block hash is a unique identifier for the block, calculated by hashing the block header using a cryptographic hash function (such as SHA-256).
   e. **Previous Block Hash:** The previous block hash is the hash of the previous block in the blockchain. This links the blocks together in a chain, making the blockchain tamper-evident.

By understanding how the data is stored and validated in each block, developers can ensure the security and integrity of the blockchain. Each block in the blockchain contains transactional data, and understanding the structure of the block can help developers create new applications and functions on top of the blockchain.

1.5.4 Merkle Trees

Merkle trees are a data structure used in cryptography that are often employed in the construction of blockchains. In a blockchain, a Merkle tree is used to efficiently summarize all the transactions that have been included in a block. A Merkle tree is a binary tree in which every nonleaf node is labeled with the cryptographic hash of the labels of its child nodes. Each leaf node represents a transaction, and its label is simply the hash of the transaction data. To construct the tree, the transactions are first hashed individually, and then pairs of hashes are combined and hashed again until only a single hash remains at the root of the tree.

In the context of blockchains, a Merkle tree is used to enable efficient verification of the transactions contained in a block. Each block contains a header that includes a hash of the root of the Merkle tree representing all the transactions included in the block. To verify a particular transaction, a node in the blockchain can simply request the relevant branch of the Merkle tree from the block that contains the transaction, and then verify that the transaction's hash is included in the branch and that the root hash of the branch matches the root hash included in the block header.

By using a Merkle tree, blockchains can provide a succinct and efficient way to prove that a particular transaction is included in a block without requiring every node in the network to store a complete copy of the blockchain. This helps to reduce the computational and storage requirements of participating in a blockchain network. In a blockchain, cryptographic hash functions and digital signatures play important roles in ensuring the integrity and authenticity of the data stored in each block. A hash function is a mathematical function that takes an input (such as a message or a block of data) and produces a fixed-size output, which is usually a string of numbers and letters. The output of a cryptographic hash function, often called a hash or digest, is deterministically derived from the input data, meaning the same input always produces the same output. While it is computationally infeasible to find two different inputs that produce the same hash (a collision), hash outputs are not truly unique due to the fixed-length output and the infinite number of possible inputs.

In a blockchain, hash functions are used to create a digital fingerprint of each block. The block header contains a hash of the entire block, including the previous block's hash, timestamp, and a Merkle tree of all the transactions in the block. Any change to the contents of the block, such as the addition or removal of a transaction, would result in a different block hash. This makes it very difficult for a malicious actor to tamper with the contents of a block without being detected.

A digital signature is a mathematical scheme for verifying the authenticity and integrity of digital messages or documents. It uses a combination of public and private keys to create a unique signature that can only be created by the owner of the private key. The signature can be verified by anyone using the corresponding public key. In a blockchain, digital signatures are used to ensure that transactions are valid and that they have been authorized by the owner of the sender's address. When a transaction is created, the sender's private key is used to create a digital signature that is included in the transaction data. When the transaction is verified by the network, the digital signature is checked using the sender's public key to ensure that it is valid. A critical characteristic in this process, particularly in Bitcoin, is that the public key used to verify the transaction signature must correspond to the Bitcoin address of the source UTXO; this correspondence is a required condition for the transaction to be considered valid by the network.

By using cryptographic hash functions and digital signatures, blockchains can provide a secure and tamperproof way to store and transfer digital assets. The hash functions create a unique fingerprint of each block, which ensures that any changes to the block contents are easily detected. The digital signatures ensure that transactions are valid and authorized by the sender, which helps to prevent fraudulent activity on the network.

1.5.5 Fundamental Blockchain Elements

By definition, the blockchain is essentially a distributed database that uses cryptography to secure and verify transactions. It consists of a series of blocks linked together in a chain, with each block containing a set of transactions, a timestamp, and a hash of the previous block.

Let's examine the fundamental functional elements of the blockchain and explore some instances of their application. These components include:

1. **Data Blocks:** The basic building block of the blockchain is the block, which contains a set of transactions, a timestamp, a cryptographic hash of the previous block, and a nonce value. Each block in the blockchain contains a set of transactions. For example, in the case of Bitcoin, each block can contain up to 1 MB of transaction data, and can include thousands of individual transactions. Other blockchain networks may have different block sizes or limits.
2. **Cryptography and Crypto-Algorithms:** Cryptography is used in the blockchain to secure and protect data. This includes digital signatures, public-key cryptography, and hash functions. Digital signatures are critical to the blockchain as they provide a way to verify the authenticity of transactions and ensure that they have not been tampered with. In the blockchain, each transaction is signed using a digital signature that is generated by the sender's private key. This signature is then verified by the network using the sender's public key to confirm that the transaction has not been altered and was indeed signed by the sender. Public-key cryptography is commonly used in blockchains for authentication and digital signatures, but its application for encryption varies across different blockchain platforms. For example, in Bitcoin, when a user wants to send a transaction to another user, they create a message that includes the recipient's public key and the amount of Bitcoin they want to send. The sender then uses their own private key to create a digital signature for the transaction, which is included with the message. The transaction is then broadcast to the network, and all other participants

use the sender's public key to verify the digital signature and ensure that the transaction was indeed signed by the sender. Once the transaction is verified, it is added to the blockchain, and the recipient can access the Bitcoin sent to them. Public-key cryptography plays a critical role in ensuring the security and integrity of the blockchain by enabling secure and reliable communication between network participants. Each block in the blockchain is identified by a unique cryptographic hash, which is generated by hashing the contents of the block. This ensures that any change to the contents of the block will result in a different hash value. The hash of each block is included in the next block in the chain, creating a secure, tamperproof ledger. For example, in the Bitcoin blockchain, the SHA-256 hashing algorithm is used to generate the hash values. The Blockchain hash function is used to convert input data into a fixed-size string of characters. The hash function used in the blockchain is typically SHA-256. Each block in the blockchain is identified by a unique cryptographic hash, which is generated by hashing the contents of the block. This ensures that any change to the contents of the block will result in a different hash value. The hash of each block is included in the next block in the chain, creating a secure, tamperproof ledger. For example, in the Bitcoin blockchain, the SHA-256 hashing algorithm is used to generate the hash values.

3. **Consensus Algorithms:** The consensus mechanism is the algorithm used to ensure that all nodes in the network agree on the state of the blockchain. This can be achieved through PoW, PoS, or other consensus algorithms. Many blockchain networks, such as Bitcoin and Ethereum, use the consensus algorithm PoW to validate new blocks and maintain the integrity of the blockchain. This involves solving complex mathematical puzzles in order to create a new block and earn a reward. The difficulty of the puzzles is adjusted over time to ensure that new blocks are created at a predictable rate.
4. **Computer Nodes:** Nodes are the individual computers or devices that make up the blockchain network. Nodes are responsible for validating transactions, creating new blocks, and maintaining the integrity of the blockchain. There are many different types of nodes, including full nodes, light nodes, and mining nodes.
5. **Smart Contracts:** Smart contracts are self-executing contracts that automatically enforce the terms of an agreement between parties. They are coded onto the blockchain and can be used to automate a wide range of transactions and agreements. Smart contracts are self-executing contracts with the terms of the agreement between buyer and seller being directly written into code, which automatically executes the terms of the agreement when certain conditions are met. They can be used in a wide range of applications, to enable secure and transparent automation of complex processes.

6. **P2P Networks:** The P2P network is a fundamental component of the blockchain, enabling the decentralized and secure operation of the technology. The P2P network is used by nodes in the blockchain network to communicate with each other and propagate information such as transactions and newly mined blocks. In a blockchain network, every node maintains a copy of the ledger and communicates with its peers to keep its copy of the ledger up-to-date. This is in contrast to traditional client-server architectures, where a centralized server maintains the database and clients request data from the server. The P2P network is important because it allows the blockchain to operate in a decentralized and trustless manner, with no central authority controlling the network. Each node in the network is responsible for validating transactions and blocks, and no single node has the power to make changes to the ledger without consensus from the network.

Understanding how blocks are linked together in the blockchain is important because it is the foundation of the blockchain's tamperproof and immutable nature. The linking of blocks using cryptographic hashes creates a secure and unbreakable chain of blocks where each block is connected to the previous block and all subsequent blocks. If an attacker attempts to modify data in a block, the hash of the modified block will be different from the original block's hash, breaking the chain of hashes linking the blocks. Any nodes in the network will detect the discrepancy between the modified block's hash and the hash stored in the next block, and the network will reject the altered block, preserving the integrity of the blockchain. This makes it virtually impossible for anyone to tamper with any data recorded in the blockchain without being detected. Additionally, understanding how blocks are linked together helps in understanding the consensus mechanism used in the blockchain. Consensus mechanisms are used to validate and confirm transactions and blocks before they are added to the blockchain. Different blockchain platforms use different consensus mechanisms, and understanding the linking of blocks is important to understand how each consensus mechanism works.

How Blocks Are Linked Together: Knowing how blocks are linked together in the blockchain is crucial for developers who want to build blockchain-based applications. It helps in designing and implementing smart contracts and other dApps that run on top of the blockchain. In a blockchain, each block contains a cryptographic hash of the previous block, which creates a chain of blocks, hence the name "blockchain." The linking of blocks using hashes provides an essential feature of the blockchain's tamperproof nature. Let's take a closer look at how blocks are linked together using hashes:

1. **Genesis Created Block:** The first block in the chain, also known as the genesis block, is usually hardcoded into the blockchain's source code. It does not have a previous block hash as it is the first block in the chain.
2. **Added Blocks:** Each subsequent block contains a hash of the previous block's header. The header includes information such as the previous block's hash, the current block's hash, the timestamp, and other relevant data.
3. **Hashes:** To create the hash, the header data is put through a cryptographic hashing algorithm, such as SHA-256, which produces a fixed-length output. This hash output is unique to the input data and cannot be reversed, meaning that it is practically impossible to re-create the original input data from the hash. The hash of the previous block is then included in the current block's header, creating a link between the two blocks. This process is repeated for each new block that is added to the chain, creating a chain of blocks with each block linked to the previous one.
4. **Tip of the Chain:** The latest block in the chain, known as the "tip" of the chain, contains a hash of the previous block in the chain, and this process continues as new blocks are added.

The linking of blocks using hashes creates a tamperproof nature of the blockchain. For example, in the Bitcoin blockchain, each block header contains the SHA-256 hash of the previous block's header. The Ethereum blockchain also uses a similar method to link blocks using hashes, but it uses a different hashing algorithm known as Keccak-256.

The method of linking blocks together using hashes is not unique to Bitcoin's blockchain. It is a fundamental aspect of most blockchain platforms, including Ethereum, Ripple, Litecoin, and many others, in Ethereum; each block contains a header with similar information to Bitcoin, including the previous block's hash, a timestamp, and a nonce value. However, Ethereum's block structure is more complex, with additional fields for storing information about smart contracts, gas limits, and other data. In Ripple, the linking of blocks is achieved through a unique consensus mechanism called the Ripple Protocol Consensus Algorithm (RPCA). In this system, rather than miners solving cryptographic puzzles to add new blocks, a set of trusted validators reaches consensus on the next block in the chain. Litecoin, which is similar to Bitcoin, also uses the same method of linking blocks through a hash pointer in each block header. However, Litecoin has a shorter block time than Bitcoin, meaning that new blocks are added to the chain more frequently.

Regardless of the specific blockchain platform, the method of linking blocks using hashes is a critical component of ensuring the integrity and security of the blockchain. By creating a chain of blocks that is linked together through cryptographic hashes, each block in the chain becomes dependent on the previous block, making it extremely difficult to tamper with the data stored in the blockchain.

One example of a consensus algorithm that uses digital signatures is the PoS algorithm. In PoS, nodes that hold a certain amount of cryptocurrency can participate in the consensus process by staking their coins as collateral. When a node proposes a new block, it must include a digital signature of the block's hash using its private key. Other nodes can then verify the signature using the proposer's public key to ensure that the block has not been tampered with.

Another example is the Delegated Proof of Stake (DPoS) algorithm used by EOSIO blockchain protocol and software developed by Block.one. In DPoS, token holders elect a fixed number of block producers who are responsible for adding new blocks to the blockchain. Each block producer has its own private key that is used to sign the blocks it produces. Other nodes on the network can then verify the block producer's signature to ensure the block is authentic.

1.5.6 Inherent Security Risks of Blockchain Technology

Security considerations in blockchain data structure involve identifying potential risks to the confidentiality, integrity, and availability of the data stored on the blockchain. This includes identifying common threats and attacks that could compromise the blockchain's security and countermeasures that can be implemented to mitigate these risks and ensure the security of the blockchain. From the perspective of a software developer developing dApps and smart contract software as well as for a security architect designing decentralized blockchain platform–based applications, it is important to understand how countermeasures against threats to the blockchain are embedded in the design of the blockchain data structure. The blockchain is designed to be a decentralized and transparent ledger that stores information in a secure and tamperproof manner but applications-based blockchain can be attacked and, if compromised, can lead to a loss of trust of the whole application and undermine the trust by clients and customers using these applications.

By understanding countermeasures against threats in the design of the blockchain data structure, it is possible to ensure that the blockchain applications are designed to be as secure as possible from the outset. This means that potential security risks are identified and addressed before they can be exploited by attackers. Embedding countermeasures in the design of the blockchain data structure also ensures that security is integrated into the entire blockchain system, from the network architecture to the consensus mechanism.

Hash functions and digital signatures are inherent security features of the blockchain and are part of the security-by-design approach used in blockchain technology. In a blockchain, each block contains a cryptographic hash of the previous block, which ensures that the blocks are linked in a tamper-evident chain. Any modification to a block would change its hash value, which would be immediately detected by the network. Additionally, the use of digital signatures provides a secure mechanism for validating transactions and ensuring that they cannot be tampered with. Together, these features create a secure, decentralized system that is resistant to attacks. By designing blockchain technology with security in mind, these inherent security features help to mitigate the risks associated with malicious actors, data tampering, and other security threats.

Hash functions are mathematical functions that take an input (such as a message or file) and produce a fixed-size output (the hash or digest) that represents the original data. A secure hash function is one that is computationally infeasible to find two different inputs that result in the same output. Hash functions can be used to verify the integrity of data, as any modification to the original data will result in a different hash value. This can help prevent attacks such as tampering with data in transit or at rest, as any modifications to the data will be detected.

The Bitcoin protocol uses hashes and digital signatures for initial authorization of payments among peer/computer nodes. When a peer node initiates a Bitcoin transaction, they verify the signature of the message that represents the transaction, which was signed by the computer node. The use of digital signatures ensures that the transaction is authentic and cannot be tampered with, providing a secure mechanism for initiating payments in a decentralized network.

Each Bitcoin transaction contains both an input and an output. The input of a transaction refers to the Bitcoin value of a previous transaction output, which is being spent in the current transaction. The output of a transaction specifies where the Bitcoin value from the input is being sent, typically to a new owner's address. This process is repeated in subsequent transactions, as each new output can become an input for another transaction, creating a chain of transactions that are verified and secured by cryptographic hashes and digital signatures.

Digital signatures are cryptographic mechanisms that use public-key cryptography to provide authenticity and nonrepudiation of data. A digital signature involves a signer (the sender of the data) applying a mathematical algorithm to the data using their private key, which generates a unique signature. The signature can then be verified by anyone with access to the signer's public key, which confirms that the data was indeed sent by the signer and has not been tampered with. This can help prevent attacks such as identity spoofing, where an attacker impersonates a legitimate sender, and replay attacks, where an attacker intercepts and retransmits data. By using hash functions and digital signatures in combination, data can be protected against a range of attacks, including tampering, identity spoofing, and replay attacks. However, it is important to use secure algorithms and practices when implementing these mechanisms to ensure their effectiveness.

In the context of a blockchain application, it is important to understand the basic factors of risks and what is the difference between threats, vulnerabilities, and attacks. Security considerations in blockchain data structure involve identifying common threats and attacks that could compromise the integrity, confidentiality, and availability of blockchain data. A threat is a potential danger to the security of the blockchain. It is a situation or event that could exploit a vulnerability and cause harm to the blockchain. Examples of threats in the context of blockchain include the possibility of a network outage event, threats directed toward compromising transactions and the private keys by attacks against digital wallets. An attack, on the other hand, is an intentional and malicious attempt to exploit a vulnerability in the blockchain system. Attacks can be carried out by threat actors such as individuals or groups with the intent of gaining unauthorized access to confidential information such as transactional data and Personally Identifiable Information (PII) data as well as attacks directed toward disrupting the functioning of the blockchain network. Examples of attacks on blockchain security include Distributed Denial of Service (DDoS) attacks, double-spending attacks, or Sybil attacks. A vulnerability is a possible condition for an exploit by an attack such as technical weakness or design flaw in the blockchain system that can be exploited by attackers to compromise the security of the blockchain applications. Vulnerabilities can have different root causes and can arise due to coding errors, misconfigurations, or other mistakes. Examples of vulnerabilities in the context of blockchain include weaknesses in the consensus algorithm, flawed smart contracts, or improper key management practices.

In summary, a threat is a potential danger, a vulnerability is a condition for an exploit such as weakness in the system that can be exploited, and an attack is the intentional exploitation of a vulnerability. By understanding the difference between these concepts, blockchain developers and administrators can identify potential threats, quantify the risks of vulnerabilities being identified with security tests, and fix the vulnerabilities and implement appropriate security measures to protect the blockchain from attacks. When developing a distributed and decentralized blockchain-based application, it is important to have a general understanding of threats, vulnerabilities, and attacks to best leverage the countermeasures to these threats and attacks and to make sure that vulnerabilities in the source code, including smart contract code, are identified by security tests and vulnerability scanning tools and are remediated before the blockchain application is released into the production environment.

When dealing with threats, vulnerabilities, and attacks on blockchain, it's crucial to develop and articulate a risk mitigation strategy to minimize the risk exposure of applications. This approach involves identifying potential security threats, assessing vulnerabilities within the system, and implementing measures to prevent attacks, thereby safeguarding the blockchain application from potential breaches and ensuring its integrity and security.

Most Common Attacks to the Blockchain: Before embarking on the design and development of distributed applications, architects and software developers should familiarize themselves with the most common attacks against blockchain platforms. This foundational knowledge is crucial for building robust and secure systems:

1. **Monopoly Attacks:** A monopoly attack that is also referred to as 51% attack, occurs when a single entity or group controls over 50% of the mining power in a blockchain network that relies on the PoW consensus mechanism such as Bitcoin. Since the PoW consensus involves miners competing to solve complex mathematical puzzles in order to validate transactions and add new blocks to the blockchain, it is often referred to as mining, and the miners who participate in it are rewarded with new cryptocurrencies such as Bitcoins or new Ethers. Despite the fact that PoW consensus mechanism is designed to make it difficult for attackers to manipulate the blockchain data, this attack is possible because some cryptocurrencies have a relatively small number of miners, making them more susceptible to a 51% attack. In addition, there are specialized mining hardware and mining pools that can make it easier for an attacker to control a large portion of the network's hash rate. To mitigate the risk of a 51% attack, some PoW blockchains implement additional security measures, such as checkpointing, which involves periodically recording the blockchain's state to prevent a complete rewrite. Additionally, some PoW blockchains are exploring the use of alternative consensus mechanisms, such as PoS, which can be less susceptible to 51% attacks. Other blockchain platforms, such as Ripple and Stellar, use different consensus mechanisms that do not rely on data mining. For example, Ripple uses a consensus protocol called the RPCA, which involves a group of trusted validators reaching a consensus about the validity of transactions. Stellar, on the other hand, uses a consensus protocol called the Stellar Consensus Protocol (SCP), which uses a federated voting process to reach consensus.
2. **Double-Spending Attacks:** This involves spending the same cryptocurrency twice by creating two conflicting transactions. Double-spending attacks can occur in any decentralized digital currency system, including Bitcoin and other cryptocurrencies, and it can be a significant issue for the system's integrity. Countermeasures include implementing a consensus mechanism that ensures only one transaction is valid. In a typical transaction, a sender transfers a certain amount of cryptocurrency to a receiver. The transaction is then verified and added to a block in the blockchain by miners. This ensures that the transaction is legitimate and that the sender cannot spend the same cryptocurrency again. However, a double-spending attack can occur when

a malicious user tries to spend the same cryptocurrency twice by manipulating the system. This can be achieved through several methods, such as using a compromised or fake wallet, exploiting a vulnerability in the network, or launching a 51% attack. To prevent double-spending attacks, cryptocurrency networks use various techniques, including:

a. **Confirmation Time:** Most cryptocurrency networks require a certain number of confirmations from miners before a transaction is considered valid. The more confirmations a transaction has, the harder it is to reverse, making it less vulnerable to double-spending attacks.
b. **Consensus Algorithms:** Blockchain networks use consensus algorithms to ensure that all transactions are verified and agreed upon by the network's nodes. The most popular consensus algorithms used in cryptocurrencies are PoW and PoS.
c. **Address Reuse Prevention:** Users can prevent double-spending by avoiding reusing the same cryptocurrency address multiple times. Most wallets on UTXO-based blockchains (like Bitcoin) automatically generate a new address for each transaction to enhance privacy and prevent address reuse. In contrast, account-based blockchains such as Ethereum often use a fixed wallet address, where all transactions are tied to a single account, and privacy must be managed through other means.
d. **Centralized Transaction Processing:** Some cryptocurrencies rely on centralized transaction processing, where a trusted third party verifies and processes transactions. While this approach is less decentralized, it can be more efficient in preventing double-spending attacks.
e. **Smart Contracts:** Smart contracts are self-executing contracts with the terms of the agreement between buyer and seller being directly written into lines of code. They can help prevent double-spending attacks by automatically verifying and executing transactions based on predefined rules.

Preventing double-spending attacks is essential for ensuring the integrity of decentralized digital currency systems. Cryptocurrency networks use various techniques, including confirmation times, consensus algorithms, address reuse prevention, centralized transaction processing, and smart contracts, to prevent double-spending and ensure that transactions are valid and secure.

3. **Eclipse Attacks:** This occurs when an attacker isolates a node from the blockchain network and feeds it with fake data to manipulate its view of the blockchain. This can prevent the victim node from receiving valid information and can allow the attacker to perform a range of attacks, including double-spending attacks, Sybil attacks, and denial-of-service attacks. Eclipse attacks can take place in a variety of decentralized networks, including blockchain networks

and P2P networks. The attacker typically starts by creating multiple fake identities or nodes in the network, which are used to control the flow of information to the victim node. The attacker can then use these fake nodes to isolate the victim node and manipulate its behavior. Countermeasures include implementing a P2P network architecture and maintaining a diverse set of nodes. Here are some techniques that can be used to prevent Eclipse attacks:

   a. **Randomized Node Selection:** Randomized node selection can be used to prevent attackers from isolating a victim node by ensuring that connections are established randomly rather than in a predictable manner.
   b. **Peer Discovery Mechanisms:** Peer discovery mechanisms can be used to ensure that new connections are established with legitimate nodes in the network, rather than with fake nodes created by the attacker.
   c. **Network Partitioning:** Network partitioning can be used to prevent attackers from isolating a victim node by dividing the network into smaller, more manageable segments that are easier to defend.
   d. **Network Monitoring:** Continuous monitoring of network traffic can help identify and block malicious nodes before they can cause significant damage.
   e. **Encrypted Connections:** Encrypted connections can be used to prevent attackers from eavesdropping on network traffic and manipulating the behavior of legitimate nodes.
   f. **Consensus Algorithms:** Consensus algorithms can be used to prevent Eclipse attacks by ensuring that decisions are made based on the agreement of the majority of participants, rather than by a small group of malicious actors.

4. **Sybil Attacks:** This involves creating multiple identities or nodes to gain control of the blockchain network. A Sybil attack is a type of attack on a computer network in which an attacker creates multiple fake identities, or "Sybils," to control a disproportionately large portion of the network. The goal of a Sybil attack is to disrupt or manipulate the network's operation by taking advantage of its distributed nature. In a Sybil attack, the attacker creates a large number of fake identities, each of which appears to be a legitimate user or node on the network. By controlling a large number of nodes, the attacker can influence the network's behavior, including controlling the consensus process, monopolizing resources, or influencing the flow of information. To prevent Sybil attacks, various techniques are used, including:
   a. **Identity Verification:** To prevent Sybil attacks, some networks require users to verify their identities using various methods, such as email verification, phone verification, or even biometric authentication. Identity verification can help prevent Sybil attacks in several ways. First, it can

help distinguish legitimate users from fake ones, making it easier to detect and block Sybil attacks. Second, it can make it more difficult for attackers to create multiple fake identities, as they would need to create multiple email addresses, phone numbers, or biometric profiles. Third, identity verification can help establish a reputation system for users, which can further help identify and block Sybil attacks. One way to prevent fake identities from being created is to require strong identity verification, such as government-issued IDs or biometric authentication. This can make it more difficult for attackers to create multiple fake identities. However, it's important to note that strong identity verification is not a foolproof solution to prevent Sybil attacks. Attackers can still find ways to bypass identity verification systems, such as by using stolen or fake identity documents or by exploiting vulnerabilities in the verification process. MFA, such as requiring a password and a security token or a biometric factor, can help prevent attackers from creating fake identities. Techniques such as public-key cryptography and digital signatures can help verify the identity of participants in the network, making it more difficult for attackers to impersonate legitimate users.

b. **Reputation Systems:** Reputation systems can help identify and block Sybil attacks by tracking the history of each user or node on the network. This can help distinguish legitimate users from fake ones and prevent the attacker from taking control of the network. Reputation systems can mitigate the risk of Sybil attacks by tracking the history of each user or node on the network and assigning a reputation score based on their behavior. This can help distinguish legitimate users from fake ones, and prevent attackers from taking control of the network. By using a reputation system, a user or node is assigned an initial reputation score based on their identity verification or other factors. The user or node's reputation score is updated based on their behavior on the network. Positive behavior, such as contributing to the network, following the rules, and engaging in good faith transactions, can increase the reputation score. Negative behavior, such as spamming, cheating, or engaging in malicious activities, can decrease the reputation score. The reputation score can be used to determine access to network resources, such as transaction validation, voting power, or storage capacity. Users or nodes with higher reputation scores can have more access to resources than those with lower reputation scores.

c. **Decentralized Control:** This can reduce the risk of Sybil attacks by distributing control over the network among multiple nodes or participants, rather than relying on a centralized authority. This makes it more difficult for attackers to take control of the network by creating multiple fake identities. Consensus algorithms can help prevent Sybil attacks by making it more difficult for attackers to create multiple fake identities and control the network. By requiring participants to contribute a certain amount of computational power, stake, or trust, consensus algorithms can ensure that decisions are made based on the agreement of the majority of participants, rather than by a small group of malicious actors.

5. **DoS Attacks:** This involves overwhelming a blockchain network with a flood of invalid transactions, rendering the network unable to validate new transactions. Countermeasures include implementing transaction fees and limiting the number of transactions per block. DDoS attacks against the blockchain are a type of attack in which an attacker floods the network with a large number of requests, overwhelming the system and causing it to become unresponsive or unavailable. These attacks can prevent users from accessing the blockchain or performing transactions and can also lead to the loss of funds or other critical data. DDoS attacks against the blockchain can take several forms, including:
   a. **Layer 3 and 4 Attacks:** These attacks target the network infrastructure, including routers, switches, and firewalls. The goal is to overwhelm the network with a large volume of traffic, making it difficult for legitimate traffic to reach its destination.
   b. **Layer 7 Attacks:** These attacks target the application layer of the blockchain, including the consensus algorithm and the transaction processing system. The goal is to exhaust the processing power of the system, making it impossible to validate new transactions or blocks.

Some examples of DDoS attacks against the blockchain include the attack on Ethereum Classic [30] in 2019, which resulted in a 51% attack, and the attack on Bitcoin in 2018, which caused transaction confirmation times to increase significantly.

Attack Risk Mitigations: There are several techniques that can be used to mitigate DDoS attacks against the blockchain:

1. **Distributed Network Infrastructure:** A distributed network infrastructure can help prevent DDoS attacks by spreading the load across multiple nodes in the network. This can make it more difficult for attackers to overwhelm the system.
2. **Cloud-Based Protection:** Cloud-based protection services can be used to detect and mitigate DDoS attacks in real time. These services can help filter out malicious traffic and prevent it from reaching the blockchain.
3. **Traffic Throttling:** Traffic throttling can be used to limit the amount of traffic that can reach the blockchain. This can help prevent the system from becoming overwhelmed and allow legitimate traffic to continue flowing.

4. **Consensus Algorithm Design:** The design of the consensus algorithm can also play a role in mitigating DDoS attacks. Algorithms that require less processing power and are less susceptible to attack can help prevent DDoS attacks from causing significant damage.
5. **Encrypted Connections:** Encrypted connections can help prevent attackers from intercepting and manipulating network traffic. This can make it more difficult for attackers to launch DDoS attacks against the blockchain.

In summary, DDoS attacks against the blockchain are a serious threat that can cause significant damage to decentralized networks. They can be mitigated through the use of distributed network infrastructure, cloud-based protection services, traffic throttling, consensus algorithm design, and encrypted connections. It is important to have a comprehensive strategy in place to prevent and mitigate DDoS attacks and ensure the security and integrity of decentralized networks.

In Table 1.2, we summarize common blockchain attacks and their countermeasures.

Table 1.2 Common blockchain attacks and their countermeasures.

Attack type	Description	Countermeasures
Monopoly attacks (51% attack)	Occurs when a single entity or group gains control over more than 50% of the hashing power in a PoW blockchain network, allowing them to manipulate the recording of new blocks and perform double spending	1. Checkpointing to periodically secure blockchain state 2. Switching to alternative consensus mechanisms like PoS 3. Implementing more decentralized mining protocols to prevent mining pool dominance
Double-spending attack	This attack involves spending the same digital currency twice; it's a risk mainly in blockchain networks where transaction confirmations are delayed	1. Increased confirmations before a transaction is accepted 2. Using more robust consensus mechanisms 3. Enhanced network monitoring
Eclipse attack	In this attack, a node is isolated from the rest of the network, feeding it false information and potentially leading to erroneous transaction confirmations	1. Robust peer discovery mechanisms 2. Node randomization 3. Continuous monitoring and automatic reconnection strategies

Sybil attack	The attacker creates multiple fake identities to gain a majority influence in the network, which can be used to disrupt network operations or manipulate consensus	1. Identity verification procedures 2. Reputation systems and resource testing 3. Using a decentralized and diverse node network
DoS and DDoS attacks	These involve overwhelming the network with a large number of useless requests to prevent legitimate network activities and services	1. Rate limiting and transaction fee implementations 2. Distributed network architecture 3. Advanced network monitoring and response systems

1.6 Consensus Algorithms

Digital signatures, hashing of block data, and the PoW consensus algorithm are the pillars that ensure the integrity and immutability of the blockchain data. Any attempt to tamper with the data would require the attacker to create a new chain of blocks that is longer than the existing one, which is computationally prohibitive. This means that the attacker would need to re-create all the blocks from the point of modification up to the latest block, rebuild the Merkle tree, and redo all the PoW. The blockchain's immutability is a result of this tamperproof design, making it a reliable and permanent storage of data that cannot be retroactively modified or deleted.

Consensus algorithms are a crucial aspect of the blockchain platform as they play a vital role in ensuring the security, reliability, and immutability of transactions. In a decentralized and distributed network such as the blockchain, there is no central authority to validate transactions. Instead, multiple nodes in the network work together to validate and confirm each transaction. This process requires a consensus algorithm, which is a set of rules and procedures used to ensure that all nodes in the network agree on the validity of each transaction.

Consensus algorithms are important because they prevent fraudulent activities, such as double-spending and tampering with transaction records. In addition, they help to maintain the integrity of the blockchain network by ensuring that all nodes have the same copy of the ledger, and that any changes to the ledger are agreed upon by the majority of the nodes in the network.

1.6.1 Different Types of Consensus Algorithms

In the world of cryptocurrencies, transactions require an additional layer of validation before they can be recorded in the digital ledger, known as the blockchain.

Unlike traditional financial systems, this validation is not performed by a centralized authority but rather by a decentralized network of peer nodes.

PoW: In the case of Bitcoin, the network's peer nodes, also known as miners, use consensus algorithms such as PoW to validate pending transactions. This process involves computational power and cost, and as a reward for their efforts, miners can create new bitcoins. In this way, the blockchain system operates in a trustless and decentralized manner, creating a new paradigm for transaction validation and recordkeeping.

Once a seller signs a transaction to transfer cryptocurrency to a buyer's address, an additional validation is required before the transaction can be added to the digital ledger. The blockchain's unique feature is that this validation is performed by some of the decentralized and distributed network's peer nodes, eliminating the need for a trusted authority. In the Bitcoin protocol, certain peer nodes, known as miners, can create new bitcoins as a reward for validating transactions that require computational power and cost to perform. These miners use consensus algorithms such as PoW to validate pending transactions.

The PoW is responsible for ensuring that the pending block of transactions waiting to be recorded in the blockchain is legitimate. This verification process is conducted by miner nodes and serves to prevent recording errors and fraudulent activities, such as double-spending of Bitcoins. The PoW involves solving complex cryptographic puzzles, and once a solution, or "nonce," is found, it can be easily validated by other nonminer nodes in the network. The nonce value is used to achieve consensus among the participating nodes and determine whether the new block of transactions can be added to the blockchain.

Ultimately, the PoW guarantees the reliability and validity of the new block and its associated transactions. Before a transaction can be recorded in the distributed ledger and the value transferred, it must undergo further validation by the nodes in the network. In the Bitcoin protocol, this validation is achieved through consensus via the PoW nonce. This validation is necessary to prevent fraudulent activities, such as double-spending, where a peer attempts to spend the same cryptocurrency to two different peers simultaneously. The PoW validation of new blocks is a significant innovation of the Bitcoin protocol, introduced by its inventor, Mr. Satoshi Nakamoto. It enables decentralized nodes to validate transactions without the need for a central authority.

To complete a PoW, a number, known as the nonce, must be found that, when hashed twice with SHA-256, produces a result smaller than the predetermined difficulty target. This can be explained simply as finding the solution to a puzzle. For instance, if the challenge is to find a hash digest that begins with five zeros (randomly chosen), the nonce must be discovered to solve the puzzle. As an example, let's consider the equation SHA-256("ABC" + nonce) = hash digest that begins with "00000." Here, the nonce is combined with the hash of the previous block, which is "ABC." The resulting concatenated string is then hashed, and a puzzle is presented where the goal is to find a hash that begins with a predetermined number of leading zeros, say five zeros. The first miner who discovers a hash with the required number of zeros is deemed to have solved the PoW. In the given example, the nonce represents the number of attempts required to solve the puzzle, which can be a daunting task, often involving millions of attempts. Once the nonce is discovered, it serves as proof that the PoW has been completed. This proof is then propagated to all the other nodes in the P2P network and can be readily verified.

When miners validate new blocks using PoW, they are rewarded with transaction fees and newly released bitcoins. The PoW consensus model is probabilistic, meaning that the miner assigned to perform the PoW is determined based on the likelihood of winning, which is directly proportional to the computational power of the node. During the race to solve the cryptographic puzzle, there can be a race condition where two miners solve the PoW simultaneously. In such cases, the Bitcoin protocol assigns the win to the miner that has solved the longest block, resolving the conflict. Although PoW is difficult to compute, requiring significant computing power, it is easy to validate by other nodes once it has been performed. The nonce serves as a PoW, and once the puzzle has been solved by a miner, it can be easily checked by other nodes. The new block is then propagated as the legitimate new block in an updated ledger. The PoW algorithm ensures that a new block added to the blockchain is legitimate and can be easily validated by other nodes on the network, while being difficult to compute. Once the nonce is found, other nodes can quickly validate the proof by computing the hash with the nonce. Additionally, the PoW algorithm is also equipped to handle disputes, such as when two miners find a solution at the same time. In this scenario, the miner who validates the longest chain will be selected as the winner.

To participate in the PoW as a validator node, a node needs to download mining software and the complete history of blockchain transactions, and have sufficient computational power (such as a computer rig of Application-Specific Integrated Circuits [ASICs] or Graphics Processing Units [GPUs]) to carry out the mining process. The PoW algorithm ensures that any added block is a legitimate transaction and can be validated by all network nodes, although the computation process itself is difficult.

The security of the PoW consensus protocol is based on economic incentives: it is more profitable for a miner to follow the protocol and earn fees in Bitcoins than to attempt to attack it by compromising the integrity of the blockchain (e.g. through forking the chain) in order to commit fraud (e.g. double-spending Bitcoins). The computational cost of attacking the blockchain is much greater than that incurred in following the protocol and serves as a deterrent against such attacks. The PoW achieves security by making it more advantageous to conduct mining operations and earn fees than to attempt to break the chain through forking for double-spending. The PoW consensus algorithm's security is established through a "computational power deterrent," as the cost of attacking the blockchain protocol and circumventing the consensus validation, such as attempting to fork the chain to double-spend bitcoins, would incur a substantially higher computational cost than performing a mining operation for a bitcoin mining fee.

The PoW algorithm offers an elegant mathematical solution to the challenge of achieving consensus among peers in a distributed and decentralized network, without the need for a central authority to act as a clearinghouse. However, the use of PoW has limitations due to the associated computational and power consumption costs, as well as scalability issues that depend on block size. Transaction validation with PoW can take anywhere from 10 to 60 minutes, which impacts its throughput. Additionally, there is the potential for a 51% attack, where a group of miners collude to control more than 50% of the network's mining computing power. In such an attack, the colluding miners could prevent new transactions from being confirmed, putting payments between some or all users on hold. They could also reverse completed transactions, resulting in double-spending of bitcoins.

Alternative approaches and algorithms to the PoW have been developed to overcome its limitations. While the PoW is computationally intensive and requires significant power consumption, it ensures the validation and attack-proof correctness of transactions in a decentralized and distributed network. However, these limitations have led to the development of alternative approaches that aim to achieve the same goals while reducing power consumption and vulnerability to 51% miner attacks. Despite these limitations, the PoW remains a popular algorithm and is used in Bitcoin blockchain and many other types of permissionless blockchains.

Various consensus algorithms have been developed as alternatives to the PoW, which are being adopted by both permissionless and permissioned blockchains. Notable examples include PoS, Proof of Identity (PoI), Proof of Activity, PoET, Paxos, PBFT, Round Robin (RR), and RAFT. One of the main limitations of the PoW is that the required computational power increases with the blockchain's growth, making it less attractive for attackers to forge the blockchain and forcing miners to continually deploy more computer resources to compete to solve the PoW challenges.

PoS: With the alternative PoS consensus algorithm, "the probability of creating a block and receiving the associated reward is proportional to a user's ownership stake in the system." The rationale behind PoS is that users with the highest stakes in the system have the greatest interest in maintaining a secure network since they will suffer the most if the cryptocurrency's reputation and price decline due to attacks. In a PoS algorithm, the stake often refers to the amount of cryptocurrency invested in the system, which is determined by tracking transactions to a specific address. Once staked, the cryptocurrency cannot be spent, and this is the determining factor in determining the likelihood of a node being able to publish a new block.

The way in which nodes are selected to publish their block varies depending on the different variations of PoS consensus algorithm, such as chain-based, using BFT, coin age, or delegation/voting system. In chain-based PoS, the blockchain network considers all users with a stake and chooses among them based on their ratio of stake to the overall amount of cryptocurrency staked. For example, if a user holds 35% stake in the entire blockchain, they will be selected 35% of the time. The PoS consensus model has been implemented in cryptocurrency blockchains such as Casper and Krypton, as well as in permissionless blockchain Ethereum.

Consensus algorithms such as PoS face the challenge of resolving disputes among participating computing nodes in the network when one node produces different results from the others. This challenge is known as the Byzantine generals communication problem, which involves communicating a command from one node to a distributed set of other nodes and resolving disputes when a command is not received by one of the nodes due to a different output.

DPoS Algorithm: One additional type of consensus algorithm that is worth mentioning is the DPoS algorithm. DPoS is a variation of the PoS algorithm in which a small group of nodes is selected to validate transactions and add new blocks to the blockchain. These nodes are known as block producers, and they are chosen through a voting process in which token holders in the network elect the nodes they believe will act in the best interest of the network. DPoS is often used in high-throughput, permissionless blockchains such as EOS and Tron. One of the main advantages of DPoS is that it allows for faster transaction processing times compared to other consensus algorithms. However, some critics argue that DPoS may be more susceptible to centralization, as the token holders with the most voting power can potentially control the selection of block producers.

PBFT: The PBFT is an algorithm used to resolve disputes among network participants in a set of computing nodes when one node generates different output than others. The PBFT algorithm is also employed in Hyperledger blockchain. In a PBFT-based PoS algorithm, all staked users will cast their votes to decide whether to add a proposed block or not, and the PBFT will resolve any disputes that may arise. With PBFT PoS, all staked nodes have a say in deciding who will add a new block to the blockchain. The vote is a cryptographically signed message that can either grant or revoke rights to delegated nodes to validate and add blocks. PBFT PoS is used in permissionless cryptocurrencies such as Bitshares, Steem, Cardano, and EOS.

RR: The RR consensus algorithm operates by allowing each node to take turns creating blocks within a predetermined time frame. This approach helps to prevent block publication halts, even if one of the selected nodes is not ready to publish a block. The algorithm has undergone extensive testing and has been used for distributing Internet traffic by load balancers among a group of web servers. RR ensures that no single node creates a majority of the blocks, but it does require trust to be established among the selected nodes. As a result, it is primarily used to establish consensus in permission-based blockchains.

PoA: PoA consensus algorithms, also known as PoI, rely on the trust of nodes that perform block validations. This trust is established through proof of the identity of these nodes. Nodes that can provide PoI build a reputation among the blockchain network for being trustworthy in validating blocks. This reputation, in addition to the PoI, which is established by checking identification documents, also depends on the behavior of the node that is scrutinized for its trustworthiness. Nodes with the highest reputation scores are chosen as validation nodes, while nodes with low scores are less likely to be selected. PoI is commonly used in permission-based and hybrid blockchains such as Ethereum, Kovan, and Chain.

Hybrid consensus algorithms combine both PoW and PoS algorithms and are often referred to as Proof of Activity algorithms. One such algorithm is PoET, which allows nodes with block publishing capability to request wait times from a secure hardware time source. The time source generates a random wait time and returns it to the requesting node. During this wait time, the node remains idle. Once the wait time has elapsed, the node can create the block. If any nodes are still idle, they will stop waiting. The algorithm ensures that each node waited the allotted time and did not start early. PoET runs on secure hardware within Trusted Execution Environment (TEE) chips, which are tamperproof and trusted in execution by the chip manufacturer. IBM Hyperledger Sawtooth is an example of a blockchain platform that uses PoET.

RAFT: The RAFT consensus algorithm is a distributed consensus algorithm designed for managing replicated state machines. It was introduced in 2014 by Diego Ongaro and John Ousterhout. The basic idea behind the RAFT algorithm is to elect a leader node among the participating nodes in the network. The leader node is responsible for maintaining a consistent state across all nodes in the network. The leader node receives client requests, appends them to its log, and sends the resulting log entries to other nodes in the network for replication. The other nodes then append the entries to their own logs and apply them to their state machines. One of the key features of RAFT is its use of a strong leader. Unlike other consensus algorithms, such as the PBFT, RAFT uses a leader to maintain consistency. If a node fails or becomes unresponsive, the leader can step down and a new leader can be elected. RAFT also uses a concept called "term" to ensure that there is only one active leader at any given time. Each term begins with a new leader election, during which the nodes exchange votes to determine the new leader. Once a leader is elected, it sends out heartbeat messages to other nodes to maintain its leadership status.

Paxos: The Paxos consensus algorithm is a well-known algorithm used for achieving consensus among a group of distributed nodes. It was introduced in 1990 by Leslie Lamport and is widely used in various distributed systems to ensure consistency and fault tolerance. In the Paxos algorithm, nodes communicate with each other through messages to agree on a value. The algorithm operates in rounds, where each round consists of several phases. The phases are as follows:

1. **Prepare Phase:** A node sends a proposal to all the other nodes, which includes a proposal number. The proposal number is a unique identifier for the proposal.
2. **Promise Phase:** When a node receives a proposal, it sends a message back to the proposer, promising to accept any proposal with a proposal number greater than the one it has already seen.
3. **Accept Phase:** If the proposer receives promises from a majority of nodes, it can send an accept message to all the nodes, including the value it wants to be chosen. If a node receives an accept message from a proposer, it will accept the value if it has not already accepted another value.
4. **Learn Phase:** When a node receives a message that a value has been accepted, it can add that value to its local state.

The Paxos algorithm ensures that only one value is chosen by the nodes and that all nodes agree on the chosen value. If multiple proposals are made, the algorithm guarantees that only one proposal will be chosen, even if there is a failure of some nodes in the network. While the Paxos algorithm has been widely used, it can be complex to implement and difficult to understand. In recent years, alternative consensus algorithms such as RAFT have gained popularity due to their relative simplicity and ease of implementation.

In Table 1.3, we provide a comprehensive overview of various blockchain consensus algorithms, highlighting their operational mechanisms and contrasting their benefits and limitations to help in evaluating the most suitable consensus method for different blockchain applications.

1.6.2 Deterministic Versus Nondeterministic Consensus Algorithms

In a consensus algorithm, the nodes in a network must agree on a single valid state of the ledger. There are two types of consensus algorithms: deterministic and nondeterministic.

Table 1.3 Common types of consensus algorithms.

Type of consensus algorithm	Description	Implemented in	Advantages	Disadvantages
PoW	Utilizes computational power to solve cryptographic puzzles. Miners compete to complete transactions on the network and receive rewards in return. PoW is secure but energy-intensive, exemplified by Bitcoin.	Bitcoin, Ethereum (currently transitioning to PoS), Litecoin	Highly secure against attacks, decentralized validation process	High energy consumption, slow transaction processing times
PoS	Validators are chosen to create a new block based on the number of coins they hold and are willing to "stake" as collateral. More energy-efficient than PoW and used by networks like Ethereum 2.0.	Ethereum 2.0, Cardano, Tezos	Energy-efficient compared to PoW, lower risk of centralization	"Nothing at stake" problem where validators might support multiple blockchain forks
DPoS	A variation of PoS where stakeholders vote on a small number of delegates who then validate transactions and create blocks. It enhances transaction speed and efficiency, used by blockchains like EOS and Tron.	EOS, Tron, Lisk	High transaction throughput, more democratic as stakeholders have a say in governance	Risk of centralization due to limited number of validators
PoET	Utilizes a fair lottery system where the participating node with the shortest wait time wins the right to create the next block. It requires less energy consumption and is used in permissioned blockchain networks like Hyperledger Sawtooth.	Hyperledger Sawtooth	Low power consumption, fair and random process	Requires TEEs, which could be a security risk
PBFT	Achieves consensus even if some nodes (less than one-third of the network) act maliciously or are faulty. This algorithm is used in systems requiring high throughput and low latency, such as Hyperledger Fabric.	Hyperledger Fabric, Stellar, Ripple	Works well in environments with less trust among nodes, high throughput and low latency	Scalability issues as it requires a lot of communication between nodes

RR	A method where each participant takes turns in a fixed sequence, which can prevent monopolization of the block creation process. This method is straightforward and often used in private blockchains.	Tendermint, other private/consortium blockchains	Simple and fair, low cost in terms of computational power	Not suitable for public blockchains due to trust requirements
PoA	Nodes are validated and selected as trustworthy entities to create blocks based on their identity and reputation. Often used in permissioned networks where high throughput and scalability are needed, such as VeChain.	VeChain, POA Network, GoChain	Fast and scalable, less energy-intensive than PoW	Highly centralized, depends on the integrity and security of the few validators
Proof of activity	A hybrid approach combining PoW and PoS. Blocks are initially mined using PoW, then validators are chosen via PoS to sign the new blocks, ensuring both energy efficiency and security.	Decred	Combines security of PoW with efficiency of PoS, suitable for networks transitioning from PoW to PoS	Can be complex to implement, still partly relies on energy-intensive PoW
PoI	Validators are chosen based on their verified identities, reducing the risk of fraudulent activities. This approach is useful in consortium blockchains where participants are known and trusted entities.	Various industry-specific blockchain applications	Reduces the risk of fraudulent blocks, ideal for regulatory compliance	Centralization concerns, privacy issues due to the disclosure of validators' identities
RAFT	A distributed consensus algorithm designed for easier management of a replicated log in distributed systems. RAFT is known for its simplicity and efficiency in handling a distributed network where nodes can fail.	Used in distributed databases and some blockchain implementations	Easy to understand and implement, provides a clear method for leader election	Leader-based system may be a bottleneck, vulnerable to leader failures
Paxos	Ensures that network participants agree on a single value in the presence of failures. It is known for its robustness and fault tolerance, commonly used in distributed systems to achieve consensus despite node failures.	Used in various distributed systems	High fault tolerance, well-studied with proven effectiveness in distributed systems	Complex to understand and implement, inefficient in terms of message overhead, leading to slower decisions

Deterministic consensus algorithms rely on a predefined set of rules to determine which block in a chain is valid. The rules are known in advance, and each node can independently verify the validity of a block without relying on the trustworthiness of other nodes. This means that once the rules are agreed upon, the consensus process becomes more efficient and predictable. Examples of deterministic consensus algorithms include Ripple's XRP Ledger and Stellar.

On the other hand, nondeterministic consensus algorithms rely on a dynamic set of rules that can change over time, and the validity of a block is determined by a voting process among the nodes. This means that the consensus process can be more flexible and also more complex and less efficient than deterministic consensus algorithms. Examples of nondeterministic consensus algorithms include PoW, PoS, and PBFT. Nondeterministic consensus algorithms are often used in permissionless blockchains, where the participants are unknown and untrusted, and there is a need for a more flexible and resilient consensus process.

Deterministic consensus algorithms are often used in permissioned blockchains, where the participants are known and trusted, and there is a need for a high degree of predictability and efficiency in the consensus process. For instance, Ripple, which is based on hash trees, has implemented ad hoc consensus models that designate specific server nodes as validators for new transactions. These "deterministic" algorithms specify a set of block signers beforehand who validate the block generated by a designated block generator. Each block signer signs the block after validation, and the majority of validators need to agree for a transaction to be valid. An agreement can be reached by an 80% threshold among the node validators to accept the new block, and other nodes in the network can check the validation before propagating the new blocks.

1.7 Cryptocurrencies

Cryptocurrencies are a type of digital asset that can be used for value exchange in P2P transactions using the blockchain as a platform. Essentially, cryptocurrencies are a form of digital currency that use encryption techniques to secure and verify transactions and control the creation of new units. Cryptocurrencies use the blockchain as the underlying platform for secure, transparent, and immutable recordkeeping of transactions between parties, without the need for a central intermediary.

One of the main advantages of cryptocurrencies' reliance on the blockchain as the underlying platform is the P2P method of conducting financial transactions. This allows users to transact directly with each other without the need for a central intermediary, which can make transactions faster, cheaper, and more efficient. Additionally, cryptocurrencies can be used for a variety of purposes beyond simple financial transactions, such as smart contracts and dApps.

Cryptocurrencies have gained popularity in recent years due to their decentralized nature and the potential for anonymity and security in transactions. The fact that cryptocurrencies can be bought and sold on exchanges, and their value is determined by supply and demand in the market. Overall, cryptocurrencies have the potential to transform traditional financial systems and enable greater financial inclusion and accessibility. However, the technology is still relatively new and evolving, and there are ongoing debates and challenges around regulation, scalability, and energy consumption.

Cryptocurrencies have come a long way since their inception in 2009, with the creation of Bitcoin by an anonymous individual who went by the pseudonym Satoshi Nakamoto. Cryptocurrencies refer to virtual or digital currencies that represent value assets and are secured through encryption techniques. Cryptography is used to ensure the secure transfer of assets among peers, with digital signatures being a prime example. In the case of Bitcoin, transactions are cryptographically signed records that reassign ownership of bitcoins to a node address. These addresses are used to send and receive asset-value transactions, with private keys being used to sign transactions and transfer the value of assets to new addresses.

Ownership of cryptocurrencies is still based solely on the possession of private keys. Private keys are the cryptographic keys that allow the holder to access and transfer the cryptocurrency associated with a particular address on the blockchain. As long as a person holds the private key associated with a particular address, they have full control over the cryptocurrency associated with that address. This means that ownership is not tied to any personal information, and anyone who has the private key can transfer the cryptocurrency to another address without any intermediaries.

Because the security of cryptocurrency is linked to the secure storage of the private keys, it is crucial to keep the private keys secure, as anyone who gains access to them can transfer the associated cryptocurrency to another address without permission. This is why many cryptocurrency owners choose to store their private keys in secure digital wallets or even offline hardware wallets to minimize the risk of theft or loss.

In the early days, Bitcoin was mainly used by tech enthusiasts and libertarians who believed in the idea of a decentralized, P2P currency that was not controlled by any central authority following an anonymous open use case with anyone with a computer and public IP being able to join and participate in the consensus process. In the case of Bitcoin, some nodes have mining capabilities, allowing them to validate transactions and create new bitcoins as a reward.

Early Bitcoin transactions were slow and costly for several reasons. Firstly, the network was still in its early stages and did not have as many users or nodes as it does today. This meant that transactions took longer to propagate across the network and get confirmed by other nodes. Secondly, the mining difficulty of the

Bitcoin network was much lower in the early days, which meant that mining new blocks was relatively easy and could be done with lower-powered hardware. As the network grew and more miners joined, the mining difficulty increased, making it more challenging and time-consuming to mine new blocks. Thirdly, the block size limit of the Bitcoin blockchain was initially set to just 1 MB. This limited the number of transactions that could be included in each block and led to a backlog of unconfirmed transactions, causing delays and higher transaction fees.

Bitcoin transactions were slow and costly, and there were concerns about the security and stability of the network, the potential for 51% attacks, and the risk of hacking or theft of Bitcoin wallets increased with the increase of usage and the use of online wallets by digital exchanges for trading cryptocurrencies. This has led to high-profile hacks of exchanges and wallets, resulting in the loss of millions of dollars' worth of Bitcoin. The scalability of the Bitcoin network has also been a concern. As the number of transactions on the network grows, the 1 MB block size limit and the time it takes to confirm transactions can create delays and increase transaction fees. While the development of solutions such as the Lightning Network and Segregated Witness (SegWit) have helped to address these issues, there is still ongoing debate and development around scaling solutions for the network.

Over time, Bitcoin and other cryptocurrencies have evolved, with improvements to the underlying technology and the emergence of new use cases. The development of blockchain technology, which is the backbone of most cryptocurrencies, has allowed for faster and more secure transactions, as well as greater transparency and immutability. New cryptocurrencies have also emerged, offering different features and use cases. For example, Ethereum introduced the concept of smart contracts, which allow for the execution of programmable agreements without the need for intermediaries. This paved the way for dApps and entire ecosystems such as DeFi and NFTs.

Other cryptocurrencies, such as Ripple (XRP), have focused on improving the speed and efficiency of cross-border payments. Ripple (XRP) was designed for high-speed, low-cost cross-border payments, catering primarily to financial institutions. Litecoin introduced faster block times and a different hashing algorithm to serve as a more efficient P2P payment system.

Polkadot and Cosmos have focused on interoperability, aiming to connect multiple blockchains and allow them to communicate seamlessly. Chainlink provided decentralized oracles, bridging off-chain data with on-chain smart contracts. Privacy-focused cryptocurrencies like Monero and Zcash emerged to offer enhanced anonymity and confidential transactions, addressing concerns around traceability in public blockchains.

Another significant innovation came with Solana, which was designed to address the scalability limitations faced by earlier blockchains like Bitcoin and Ethereum. Solana introduced a unique consensus mechanism known as Proof of

History (PoH), which provides a verifiable timestamping method to order transactions efficiently before they are validated by PoS. This hybrid model allows Solana to achieve high throughput handling thousands of TPS with low latency and minimal transaction costs. Solana's architecture makes it especially suitable for high-performance dApps, including real-time DeFi platforms, NFT marketplaces, and on-chain gaming. Despite some concerns about decentralization and network stability, Solana has been influential in pushing the boundaries of what's possible in terms of blockchain speed and user experience.

These innovations reflect a broader trend: cryptocurrencies are no longer just digital money, they're programmable financial and data platforms, each tailored to different technical challenges and real-world applications.

Regulation of cryptocurrencies has also evolved significantly over the past decade, with many jurisdictions now recognizing digital assets as a legitimate asset class. Governments and regulatory bodies have begun to establish clearer legal frameworks to govern their use, taxation, and exchange, aiming to reduce fraud, improve consumer protection, and enable broader adoption within traditional financial systems. For example, in the United States, the Internal Revenue Service (IRS) classifies cryptocurrencies as property for tax purposes, while the Security and Exchange Commission (SEC) has taken an active role in determining whether certain tokens are securities subject to registration requirements. The European Union (EU), through the upcoming Markets in Crypto-Assets (MiCA) regulation, is working to harmonize crypto regulations across member states, focusing on stablecoins, issuers, and service providers.

Cryptocurrency exchanges have also become subject to stricter regulatory oversight. Leading exchanges such as Coinbase in the United States and Binance in various jurisdictions are required to comply with Know Your Customer (KYC) and Anti–Money Laundering (AML) obligations, similar to traditional financial institutions. In some countries like Japan, crypto exchanges must be registered with the Financial Services Agency (FSA) and adhere to capital and cybersecurity requirements.

At the same time, the market has seen a marked increase in participation from institutional investors, who are entering the space through regulated investment vehicles such as Bitcoin ETFs, crypto custody solutions, and blockchain-related equity funds. Large asset managers like BlackRock, Fidelity, and Goldman Sachs have begun offering or exploring cryptocurrency exposure for their clients, signaling a shift toward mainstream financial integration of digital assets.

Overall, while blockchain technology has made significant strides in recent years, there are still several security challenges that need to be addressed to ensure the long-term security and stability of the platform. To address privacy concerns, there are several projects underway to develop more privacy-focused blockchains and privacy-enhancing technologies such as ZKPs. Several projects are underway to improve the scalability of blockchains, including sharding, sidechains, and off-chain scaling solutions like the Lightning Network.

To address interoperability limitations between different blockchain networks, there are today several projects underway to develop interoperability solutions, such as cross-chain bridges and atomic swaps. Finally, to address cybersecurity concerns, there are ongoing efforts by blockchain security practitioners to improve the security of blockchain-based applications through better code auditing, security testing, and the use of advanced security technologies like Multiparty Computation (MPC). For example, in a DEX, MPC can be used to securely compute the exchange rate between two cryptocurrencies without revealing the users' private keys or the amounts being exchanged. This allows for secure and private transactions without relying on a centralized intermediary.

Finally, one of the limitations to widespread consumer adoption of blockchain-based applications is that there is often no clear governance structure in place. This can make it difficult to make decisions about the future direction of the blockchain, such as how to handle protocol upgrades or how to handle disputes. To address this, there are several projects underway to develop governance models for blockchains, including on-chain governance and off-chain governance models. Without a defined governance model, it can be difficult to reach consensus on critical decisions such as protocol upgrades, dispute resolution, or changes to economic parameters. This can lead to uncertainty, fragmentation, or even hard forks outcomes that can erode trust among users and investors.

To address this challenge, several blockchain projects have introduced or are experimenting with formal governance mechanisms, including both on-chain governance (e.g. Polkadot, Tezos, Decred), where voting and decision-making happen directly on the blockchain, and off-chain governance (e.g. Bitcoin, Ethereum), which relies on community discussions, developer consensus, and social signaling. Hybrid models are also emerging, aiming to balance decentralized participation with structured decision-making, which is essential for long-term sustainability and user confidence.

1.7.1 Cryptocurrencies Use Cases

The first and most well-known cryptocurrency is Bitcoin, which was created in 2009 by an unknown individual or group using the pseudonym Satoshi Nakamoto. Bitcoin is a decentralized digital currency that uses encryption techniques to regulate the generation of units of currency and verify the transfer of funds. It operates on a public ledger called the blockchain, which records all transactions made using the currency.

As Bitcoin gained popularity, other cryptocurrencies began to emerge. These cryptocurrencies are often referred to as "Altcoins." Altcoins, or alternative cryptocurrencies, are any cryptocurrency that is not Bitcoin. There are many altcoins, such as Ethereum, Ripple, Litecoin, and Bitcoin Cash. Altcoins [31] can have different features and use cases than Bitcoin and may offer different advantages or disadvantages. For example, some Altcoins use different mining algorithms that make them more accessible to individual miners, while others may have faster transaction times or more advanced smart contract capabilities.

In addition to Altcoins, Stablecoins [32] evolved as a type of cryptocurrencies that are designed to maintain a stable value relative to a specific asset or group of assets that is pegged to a stable asset, such as a fiat currency like the US dollar or a commodity such as gold. This is done to reduce the volatility of the cryptocurrency and make it more useful as a medium of exchange. Stablecoins can be useful for people who want to avoid the volatility of other cryptocurrencies, but they may have their own risks, such as the risk that the stable asset they are pegged to will lose value. Tether and US Coins are examples of stablecoins and are pegged to the US dollar, with each USDT. These stablecoins aim to provide stability in their value, which can make them useful for various applications such as trading and remittances.

Finally, there are utility tokens, which are cryptocurrencies that are designed to be used within a specific ecosystem or platform. For example, the Basic Attention Token (BAT) is a utility token used within the Brave web browser ecosystem to reward users for viewing advertisements. Another popular token example is Ether, the utility token for the Ethereum blockchain, and is used to pay for transaction fees and smart contract execution. Utility tokens can have value beyond their use on a specific platform, but their value is usually tied to the success of that platform. Utility tokens can be useful for people who want to access a specific service or platform.

Each type of cryptocurrency has its own use cases, pros, and cons. Altcoins can not only offer different features and capabilities than Bitcoin but may also be more volatile or less established. Stablecoins can reduce volatility but may have their own risks. Utility tokens can be useful for accessing specific services or platforms but may not be useful as a general medium of exchange. Ultimately, the choice of cryptocurrency will depend on the individual's needs and preferences.

1.7.2 Use of Cryptocurrencies and Security Risks

Cryptocurrencies can be traded on various online exchanges, which act as intermediaries between buyers and sellers. These exchanges allow users to buy and sell cryptocurrencies for other digital assets, fiat currencies, or other cryptocurrencies.

Some of the popular cryptocurrency exchanges include Binance, Coinbase, Kraken, and Bitfinex, among others.

Cryptocurrency exchanges act as a platform where buyers and sellers can trade digital coins with each other, while the exchange takes a commission for each transaction. Because many cryptocurrency exchanges rely on use of online wallets to store the private keys needed to access and transact with cryptocurrency, these quickly become the target of cyberattacks seeking to exploit vulnerabilities for theft of cryptocurrencies, including phishing and scamming attempts aimed to spread malware or install keyloggers to steal login credentials for online wallets. A common attack pattern has been to exploit vulnerabilities in blockchain software to compromise digital wallets and private keys, allowing fraudsters to transfer cryptocurrency to fraudulent account node addresses under their control.

As of 2025 [33], cryptocurrency exchange hacks continue to pose significant threats, with billions of dollars in digital currencies stolen over the years with peaks of $3.7 billion in losses in 2022 and $3.3 billion in 2021. High-profile incidents include breaches at digital exchanges such as Bitstamp, Mt. Gox, Bitfinex, NiceHash, Coincheck, and Zeal. In 2024 losses due to crypto hacks resulted in $2.2 billion. These figures underscore the ongoing challenges and vulnerabilities faced by digital exchanges in securing assets against sophisticated cyber threats.

Besides hacking of digital wallets for fraud and theft, use of cryptocurrency by hackers also allows them to monetize their activities such as in the case of ransomware attacks, where fraudsters hold the victim's data hostage until a ransom is paid in cryptocurrencies. Since the transfer of money using cryptocurrencies is almost completely anonymous and can be done through cover channels like Tor, it can be challenging for law enforcement to trace the transfer to the destination fraudulent computer node. Another type of attack that can be monetized by use of cryptocurrencies is cryptojacking, where attackers divert computer CPU resources to mine cryptocurrencies. This technique typically leverages other attacks such as phishing or social engineering and malware-based attacks to gain unauthorized access to computer resources and install mining software.

Blockchain technology also allows for the creation of tokens that represent a particular asset or utility and can be traded like other cryptocurrencies. These tokens can be issued on top of another blockchain and can range from commodities to loyalty points to even other cryptocurrencies. Tokens can also be created through an ICO, which is similar to an Initial Public Offering (IPO) for start-up companies. During an ICO, investors can purchase crypto coins using fiat currency or preexisting digital tokens like Ether, and in return, they are given a specific cryptocurrency token created through the use of smart contracts. Smart contracts are programmable computer code that runs on a decentralized computer network, automating the execution of transactions based on a set of rules.

A successful example of an ICO project is the Ethereum platform, which issued Ethers as its coin tokens.

The use of smart contracts to create new tokens for ICOs has also made them a prime target for fraudsters and scammers. These individuals may exploit vulnerabilities in the smart contract code to implement Ponzi schemes or other fraudulent high-yield investment programs where the contract owner not only collects fees for others to invest but can also steal cryptocurrency from the contract. Additionally, ICOs are often used to create "pump and dump" schemes, where scammers talk up the value of the ICO to generate interest and drive up the price, only to quickly dump the coins for profit. As ICOs are not regulated securities, they lack the oversight required for regulated security investments by bodies such as the SEC. This lack of regulation increases the risk that investors will lose their entire investment, making it essential to carefully research any ICO before investing.

1.8 Digital Wallets

1.8.1 Introduction

Digital wallets are an essential component of the blockchain technology ecosystem; they are used to securely store and manage the user's digital assets such as private keys and to retrieve them to sign transactions on the blockchain network, allowing the user to send and receive digital assets. At a high level, a digital wallet is a software application that allows users to securely store, manage, and use their digital assets, such as cryptocurrencies, tokens, and digital certificates. In addition to storing and managing digital assets, digital wallets may also provide other features such as the ability to exchange or trade digital assets, view transaction history, and receive or send payments. Some digital wallets may also support the integration of various dApps, enabling users to access a wide range of services directly from their wallet.

Digital wallets support secure operations with private keys such as signing that in the case of use of cryptocurrencies occurs when there is transfer of cryptocurrency from seller to buyer. This transfer is facilitated by the use of private keys that are securely stored in the digital wallet and enable the signing of the amount of cryptocurrency value to be transferred. In the case of Bitcoin protocol for example, the signed transfer results in the ownership of a sum of bitcoins being transferred to the buyer's address, which is the hash of the ECDSA public key. When a digital wallet is created, a root master key pair is generated, which can then be used to create public and private key pairs. Initially, a public–public–private key pair stored in the wallet does not represent any value and only holds cryptocurrency when a certain amount is sent to the wallet address.

Digital wallets typically utilize encryption techniques to protect the user's private keys, which are used to access their digital assets and authorize transactions. The wallet provides a means for the private keys to be protected by either software or hardware encryption.

Software Wallets: Software encryption is used in software wallets to protect the private keys stored on the device or computer where the wallet application is installed. Software digital wallets are in essence software applications that are installed on a computer or mobile device. These wallets are typically free to download and use and offer a convenient way to manage digital assets. Software wallets can be further classified into two types: hot wallets and cold wallets. Hot wallets are always connected to the Internet as online or web wallets, and "cold" wallets are disconnected from the P2P network and kept offline. When it comes to hot web wallets, the private keys are managed by a third party, typically through a secure tamperproof hardware device with strong access controls, such as a cloud storage provider. In this case, the security of the private keys is the responsibility of the third-party digital wallet provider, who acts as the custodian of the keys on behalf of the owner. These providers may rely on fourth-party cloud hosting services, and cryptocurrency exchanges often offer digital wallets as a service and act as custodians of investors' private keys. Cold wallets, on the other hand, are offline and provide an extra layer of security as they are not susceptible to hacking attempts. However, they are less convenient to use since they require physical access to the wallet to perform transactions.

"Cold" digital wallets enable users to store their keys offline, with the use of a separate physical medium like a USB drive or a secure hardware device like an hardware wallet. Cold digital wallets offer a user an added layer of security against potential online attacks, as the storage devices can be disconnected from the Internet, creating a "moat" or air gap. Not all the cold wallets are necessarily software or hardware based, for example when private keys are printed in paper under the custody of the user that can still be considered another form of "cold" wallet, in which the private keys and wallet address are printed on paper. However, paper wallets do not provide the same level of security as encrypted USB drives or secure hardware devices, as the keys are printed in clear and can be stolen by anyone who can read the paper. Additionally, the security of paper wallets is only as strong as the printer used to print them, which can be vulnerable to Internet connections, WiFi, and memory attacks. By using USB drives to store the keys, users can physically unplug and disconnect them from the device running the node to sign transactions and transfer cryptocurrencies to another node or device. When the USB drive is encrypted, it is also protected from potential compromise in the event of loss or theft.

Once a user installs a software wallet, they become the custodian of the private keys and retain full control over them. The software wallet can be installed on a desktop computer or a mobile device and allows the user to interact with the P2P network/blockchain as a computer node. The private keys are stored on the user's personal device and can be accessed through the software wallet. By signing transactions, a node can transfer UTXOs to another node by sending them to the beneficiary public addresses. Software wallets provide users with the ability to sign transactions, allowing them to perform various functions within the blockchain network. Software wallets provide users with the ability to sign transactions, allowing them to perform various functions within the blockchain network.

Note: The UTXO model is used in blockchains like Bitcoin, but not all blockchains follow this model; for example, Ethereum uses an account-based model instead.

In a decentralized P2P computer network, users have the option to install either a full node digital wallet or a light node digital wallet when interacting with other users. A full node digital wallet requires the user to download a complete copy of the distributed ledger or blockchain that contains all the executed transactions. For example, a full node digital wallet for Bitcoin would contain a copy of all transactions that have taken place since the creation of the first Bitcoin in 2009. This type of digital wallet can also participate in the validation of transactions and, with enough computing power, can also function as a miner node. However, due to the complete copy of transaction history that it contains, a full node wallet requires a significant amount of memory storage (approximately 145 GB) to download the complete copy of the Bitcoin distributed ledger.

In contrast, software digital wallets running as light nodes do not require the user to download the complete copy of the distributed ledger and can perform limited checks to validate transactions using Simplified Payment Verification (SPV). Light node wallets typically do not validate each transaction stored in every block of the blockchain but only validate the transactions initiated by the wallet and provide the user confirmation that the PoW has been done by checking the PoW values before adding the transactions to the blockchain. However, light node wallets cannot check other transactions in the block because they do not have access to those blocks, potentially allowing invalid blocks to be added. This limitation of light nodes does not have much impact on the security of the digital wallet itself, except for the validation of the "best header chain," which is the chain of valid block headers that has the most cumulative PoWs.

In a decentralized P2P computer network, users have the option to install either a full node digital wallet or a light node digital wallet when interacting with other users. A full node digital wallet requires the user to download a complete copy of the distributed ledger or blockchain that contains all the executed transactions. For example, a full node digital wallet for Bitcoin would contain a copy of all transactions that have taken place since the creation of the first Bitcoin in 2009. This type of digital wallet can also participate in the validation of transactions and, with enough computing power, can also function as a miner node. However, due to the complete copy of transaction history that it contains, a full node wallet requires a significant amount of memory storage (approximately 145 GB) to download the complete copy of the Bitcoin distributed ledger.

Software wallets can be vulnerable to attacks due to operating system vulnerabilities that allow attackers to gain unauthorized access to both the device and the digital wallet software. Although MFA or two-factor authentication (2FA) is typically required to access digital wallets, they can still be compromised by various forms of attacks, such as social engineering. For example, attackers may send an email attachment that installs key-logging malware, enabling them to steal the user's credentials and gain access to their digital wallet. These vulnerabilities highlight the importance of keeping software wallets and operating systems up-to-date with the latest security patches and using caution when opening emails or downloading attachments from unknown sources.

Hardware Wallets: A significant improvement in the security of digital wallets comes from the use of hardware wallets, which are designed specifically for secure storage and management of private keys. By comparison with software wallets, hardware wallets use hardware encryption to protect the private keys stored on the device. Hardware digital wallets are in essence physical devices that store private keys offline. These wallets provide a high level of security as they are not connected to the Internet and, therefore, are less vulnerable to cyberattacks. Hardware encryption is implemented using specialized hardware components, such as secure enclaves or tamper-resistant chips, which provide a higher level of security compared to software-based encryption. Hardware wallets require a user to physically connect the device to a computer or mobile device to sign transactions, which provides an added layer of security against unauthorized access.

Hardware wallets feature built-in antitampering technology and store private keys in a protected area of a microcontroller, making it impossible to transfer them out of the device in plaintext. Unlike software wallets, secure hardware devices are immune to computer viruses that can attack the software on the device. They can come in the form of a plug-in computer card or an external device that attaches directly to a computer or network server. Even when attached to a computer, hardware wallets can still be considered as "cold" storage, as the keys stored on the device never leave it, even when data is transmitted by the hardware to a computer through a cable. Despite hardware wallets being considered more secure than software wallets because of use of hardware encryption and use of cold storage, there are still risks involved. Attackers may exploit design flaws in the hardware, such as the implementation of nonsecure random algorithms for the creation of the seed to generate private keys, which can subject them to cryptographic attacks. Additionally, vulnerabilities in the hardware and firmware can also pose security risks. The security of the hardware itself also depends on supply chain security, which is only as trustworthy as the vendor of the secure hardware devices and the hardware production process that is followed to prevent the introduction of backdoors by malicious actors.

Although many digital exchanges now store private keys in cold wallets, there is still a risk of compromise due to the fact that these wallets must operate as hot wallets in order to facilitate automated disbursements to customers who wish to withdraw their funds. These hot wallets, which have live private keys and are controlled through APIs to sign transactions and pay customers, are vulnerable to attacks. Even with the use of cold wallets, there is still a risk during the signing of transaction data using private keys, as malware can monitor the transactions and swap the recipient address with one under the control of an attacker. The risk lies in the fact that malware can alter the recipient address before the transaction is signed, making the signature valid for a malicious transaction. Therefore, the integrity of the transaction must be verified just before signing, not after.

In addition, hardware wallets can also be vulnerable to social engineering attacks where a user is tricked into using a web interface under the control of an attacker and sending money to a fraudulent address. It is important to remain vigilant and take precautions to protect against such attacks, such as only using trusted web interfaces and verifying the accuracy of recipient addresses before sending transactions.

Other types of digital wallets that are important to understand from the perspective of implementation of blockchain applications are custodial and noncustodial wallets.

Custodial Wallets: Custodial wallets are often considered more convenient and user-friendly than noncustodial wallets, as the service provider takes care of many of the technical aspects of managing cryptocurrency. However, they also come with a higher level of risk, as the user's assets are held by a third party and are subject to the security measures and policies of the service provider. Additionally, custodial wallets may not offer the same level of anonymity and privacy as noncustodial wallets, as the service provider may be required to collect and share user information with regulatory authorities. Custodial wallets are a type of cryptocurrency wallet that is managed and maintained by a third-party service provider, such as a cryptocurrency exchange or a financial institution. In a custodial wallet, the private keys of the user's cryptocurrency holdings are held by the service provider, rather than the user themselves.

This means that the user does not have complete control over their cryptocurrency assets and must rely on the custodial service provider to keep their assets safe and secure. The service provider is responsible for managing the security and storage of the private keys and handling transactions on behalf of the user.

Noncustodial Wallets: Noncustodial wallets are a type of cryptocurrency wallet that allows users to have complete control over their cryptocurrency assets and private keys. In a noncustodial wallet, the user holds their own private keys and is solely responsible for the security and management of their cryptocurrency holdings. Unlike custodial wallets, noncustodial wallets do not rely on third-party service providers to manage and secure the user's cryptocurrency assets. This means that the user is not subject to the policies and security measures of a service provider and has the ability to directly interact with the blockchain network. Noncustodial wallets typically provide a higher level of security and privacy compared to custodial wallets, as the user has full control over their private keys and can store them offline in a hardware wallet or other secure storage method. Additionally, noncustodial wallets offer a greater level of anonymity, as the user is not required to provide personal information to a third-party service provider. However, noncustodial wallets may be more technically challenging to use compared to custodial wallets, as the user is responsible for managing their own private keys and securing their cryptocurrency assets. Mistakes such as losing or misplacing private keys can lead to irreversible loss of funds; so it is important for users to understand the risks and take necessary precautions when using a noncustodial wallet.

1.8.2 Security Features of Digital Wallets

Digital wallets employ multiple layers of security controls to ensure the secure storage and management of digital assets such as private keys used to sign and authorize transactions. Specifically, these are the security controls that digital wallets have implemented to protect from nonauthorized access to digital assets that are stored and secured in the digital wallets.

Digital wallets use various forms of authentication to verify the identity of users such as "at minimum" passwords to access their account and, as stronger authentication than passwords, biometric authentication (such as fingerprint or facial recognition or voice recognition) and MFA such as a code sent to a user's mobile device. Digital wallets may also use security tokens to authenticate users. These tokens are physical devices that generate a unique code that is required for authentication.

Encryption is used to protect the private keys. In digital wallets, symmetric encryption may be used to encrypt sensitive data such as private keys, digital certificates, and other digital assets. Asymmetric encryption may be used to protect communication between the wallet and other entities, such as a merchant's server.

Deterministic Wallets: In the realm of security, a fundamental principle holds that the strength of a chain is only as robust as its weakest link. Within a digital wallet, the weakest link could potentially lie within the key generation process. Many digital wallets in use today are classified as "deterministic," meaning that private and public keys are produced from a known value known as the master seed. Digital wallets that rely on deterministic key generation have private keys generated from a single master seed, while public keys are derived from the corresponding private keys. This approach allows for straightforward backup and restoration of keys since only the knowledge of the master seed is required. For Bitcoin wallets specifically, the wallet's public keys are used to generate unique addresses. In other words, addresses are digital representations derived from the ECDSA public key, typically through a multistep hashing process that includes SHA-256 and RIPEMD-160, followed by encoding (e.g. Base58Check in Bitcoin).

In digital wallets, public addresses enable the receipt of funds, while private keys facilitate the transfer of Bitcoins to a designated beneficiary address. After all generated addresses have been utilized in transactions, new keys must be created, and a new backup of those keys must be established. Unfortunately, randomly generated keys cannot be re-created, and a new set of private and public keys must be produced, rendering the old backup with the keys obsolete. To address these limitations, deterministic wallets were created. These wallets employ a master key seed, which can generate an unlimited number of private keys in this type of wallet; the master seed is the only secret required for key regeneration and backup.

Hierarchical Deterministic (HD) Wallets: In modern wallets, the master seed keys are derived from human-readable passphrases, such as a 12-word mnemonic phrase. These passphrases are the sole secret that wallet owners must remember to create a backup of their digital wallet. The 12-word mnemonic passphrase is utilized to generate a master key, which sequentially and hierarchically creates private keys and Bitcoin addresses. This hierarchical key generation in digital wallets is why they are also referred to as HD wallets. While HD wallets provide convenience, their security may be compromised since only knowledge of the master secret seed is necessary to access the private keys required for Bitcoin spending. To address this issue, some HD wallets separate the generation of private and public keys. Although HD wallets have a master public key that can be used to create Bitcoin addresses, it cannot be utilized to derive the private keys. Certain HD wallet implementations keep the private keys offline, using cold hardware wallets, and never expose them online. In contrast, the master public key is the sole component stored online. The utilization of such an HD hardware wallet implementation provides an increased level of security compared to online wallets, where both private and public keys are stored online, thereby exposing them to attacks.

Multi-Signature (Multi-Sig) Wallets: Digital wallets can be made more secure by implementing multi-sig, which necessitate one or more signatures and private keys to approve a transaction and authorize the transfer of an asset to a beneficiary address. The Bitcoin protocol provides support for various numbers, or a "quorum," of eligible signers, including "m" number of signers out of the total eligible "n" signers. For instance, a 1 of 2 digital wallet needs only one of the two eligible user signers to execute a transaction. On the other hand, a 2 of 2 digital wallet requires both eligible user signers to approve the transaction. This feature enhances the security of digital wallets by making it more difficult for hackers to access the private keys required to transfer assets from the wallet.

Multi-sig wallets can also function as single-user, multiple-key signing wallets where private keys can be stored in different devices, such as a mobile wallet and a server. To sign a transaction, the user would need to access and use both the mobile-stored key and the server-stored key. This type of wallet is inherently more secure than a single-key/device user since an attacker would need to compromise multiple devices where the keys are stored to compromise the transactions. Multi-sig wallets can support different use cases that increase the security of transactions. For instance, a 2 of 3 multi-sig scheme can be used for improved security of a hot digital wallet. In this scenario, one key is stored in a cold wallet as a backup, one key is stored in the hot wallet for signing, and a third key is used by an intermediary wallet that can only sign and authorize. Even if one of the keys is compromised, a multi-sig wallet can still recover one of the keys required for signing from a backup, and it would still require another key, such as the one stored by the intermediary wallet, to sign a transaction.

Moreover, an intermediary digital wallet whose signing is required to authorize a transaction can also enforce specific rules for transactions. For example, it can validate the beneficiary address against a blacklist of addresses to prevent money laundering or validate that the amount of the transaction does not exceed a specific limit. Smart contracts can be used to define the rules for authorizing signatures in a "m" of "n" multi-sig wallet. The group of asset owners can set the quorum, which is the minimum number of eligible users out of a total of "n" required to sign a transaction. They can also specify the rules that need to be validated for committing the transaction through a smart contract.

A potential risk associated with multi-sig wallets that depend on smart contracts is the possibility of software bugs being introduced. Exploitation of such bugs could enable a fraudster to bypass the additional required signers to authorize a transaction or manipulate certain functions that govern the digital wallets, such as setting themselves as owners and carrying out transactions. One significant challenge of smart contracts is that once stored on the blockchain, vulnerabilities in the code persist indefinitely, meaning that a flawed multi-sig software in a digital wallet cannot be patched or fixed. Consequently, assets stored in such a wallet may be stolen or become inaccessible until the flawed smart contract is no longer used (e.g. by invoking a kill function in the smart contract), and a new version of the digital wallet that does not have the bug is released for storing new cryptocurrency assets. If one or two of the partial private keys needed for a quorum of “m” out of “n” signers are compromised, the attacker cannot sign the transaction without compromising the distributed processing of partial keys/shares stored by multiple users and devices involved in the signing, verification, and authorization of the transaction. It is important to note that the attacker will never have access to the full private keys.

1.9 Digital Transactions

Transactions serve as the fundamental building blocks of the blockchain. It is crucial to comprehend how transactions facilitate the exchange of value, whether it be through tokens or cryptocurrencies, which are utilized to transact value between parties. As a software developer, understanding the various types of cryptocurrency transactions is essential for building blockchain-based applications that are efficient, secure, and take full advantage of the unique features of the technology.

Transactions in a blockchain are composed of inputs and outputs, and each input references an unspent output from a previous transaction, forming a chain of transactions that can be traced back to the original issuance of the cryptocurrency. To be valid, a transaction must meet several criteria, such as having the correct digital signatures and referencing valid unspent outputs.

As a blockchain architect, understanding blockchain transactions is crucial for designing and implementing efficient and secure blockchain systems. A blockchain transaction is a record of a transfer of assets (usually cryptocurrencies) from one entity to another on a blockchain network. It contains inputs, outputs, and a digital signature that verifies the identity of the sender. To design a blockchain system, the architect needs to consider various factors such as the type of transaction, the consensus mechanism, the block size, and the network scalability. For example, choosing the right transaction type (e.g. pay-to-public-key-hash [P2PKH], pay-to-script-hash [P2SH], SegWit) can affect the transaction fees, transaction speed, and overall security of the network.

The architect also needs to choose a consensus mechanism that can ensure the validity of transactions and prevent double-spending. Depending on the consensus mechanism, the transaction confirmation time and the transaction throughput may vary. Furthermore, the blockchain architect needs to consider the block size and the network scalability to optimize the transaction processing speed and reduce the network congestion. This may involve implementing advanced techniques such as sharding, or Layer 2 scaling solutions.

As a software developer, understanding how transactions work in a blockchain is crucial for building secure and reliable dApps. A transaction is a fundamental building block of a blockchain and represents the transfer of value from one account to another. As a developer, you may need to interact with the blockchain by creating and broadcasting transactions, querying transaction data, and integrating transactions into your applications. Understanding the intricacies of transactions in a blockchain is essential for building secure and reliable dApps that can interact with the network in a trustworthy manner.

To enable Bitcoin transactions, certain critical elements are required, including cryptography, data structures, and scripting. In the context of Bitcoin transactions, a transaction represents the transfer of Bitcoin value from one address to another. Transactions are stored on the blockchain in a chronological order, forming a ledger of all the Bitcoin transactions that have ever occurred.

If we were to dissect a typical Bitcoin transaction, we would find three primary components: the header, inputs, and outputs. Understanding the fields within these sections is crucial for further discussion. It is important to note that these fields are found within a "raw transaction," which is broadcasted between peers when a transaction is initiated.

The header of a Bitcoin transaction contains several key fields that are important to understand:

- **Hash:** This is a hash value generated over the entire transaction, which serves as both a pointer and a means of verifying the integrity of the data. We will delve deeper into this concept in the following section.
- **ver:** This field indicates the version number that should be used to validate the transaction. The latest version was introduced in a soft fork that was activated in December 2015.
- **vin_sz:** This field indicates the number of inputs to the transaction, while vout_sz counts the number of outputs.
- **lock_time:** This field specifies the earliest time at which a block can be added to the blockchain. It is either the block height (see the following note) or a Unix timestamp.

Note: In the context of Bitcoin and similar blockchains, block height refers to the number of blocks preceding a particular block in the blockchain, starting from the genesis block (which has a block height of 0).The block height defines a position in the blockchain, and if lock_time uses it, it sets a minimum block number that must be reached before the transaction becomes valid.

The input section of a Bitcoin transaction contains the following key fields:

- **Previous Output Hash:** This field contains a hash pointer that references a previously UTXO, which represents the funds being spent in the current transaction. This field is an index into the list of outputs of the previous transaction and represents the specific UTXO being spent.

- **ScriptSig:** This field contains a spending script that provides proof that the creator of the transaction is authorized to spend the funds referenced in the previous two fields.

The output section of a Bitcoin transaction consists of two key fields:

- **Value:** This field represents the amount of Satoshi being spent in the transaction, where 1 BTC is equal to 100,000,000 Satoshi.
- **ScriptPubKey:** This field contains the second of two scripts included in a Bitcoin transaction. It is a script that points to the recipient's hashed public key. We will explore this field further in the final section of this article.

Table 1.4 helps visualize a structured BTC transaction data structure and the values.

The Transaction Flow: In a Bitcoin transaction, inputs represent the funds being spent. Unlike traditional bank accounts, where balances are adjusted directly, Bitcoin operates using the UTXO model. This means that transactions consume existing UTXOs and generate new ones rather than modifying an account balance. For example, Alice has a UTXO worth 50 BTC. She initiates a transaction to:

Table 1.4 Bitcoin transaction data structure.

Section	Field	Example value
Header	Hash	0000000000000000000a16...
	Version (ver)	2
	Input count (vin_sz)	1
	Output count (vout_sz)	2
	Lock time (lock_time)	700000
Inputs	Previous output hash	6b9f8c2a5b1d...
	Index	0
	ScriptSig	483045022100fa...
Outputs	Value (Satoshis)	300000000 (3 BTC)
	ScriptPubKey	76a91488ac...
	Value (Satoshis)	1800000000 (18 BTC)
	ScriptPubKey	76a914c7b5...

- Send 30 BTC to Bob,
- Receive 18 BTC as change, and
- Pay 2 BTC as a transaction fee to miners.

When this transaction is processed, Alice's original 50 BTC UTXO is fully spent. This results in the creation of two new UTXOs:

- Bob receives a new UTXO of 30 BTC (the output to Bob's address).
- Alice receives a new UTXO of 18 BTC (her change).

Since Alice's original 50 BTC UTXO has been consumed, her new wallet balance is now 18 BTC *(read the following note for clarity)*, as this is the only remaining UTXO she owns after the transaction. Once Alice creates the new transaction, she will broadcast it to the Bitcoin network. The transaction is then verified by other nodes on the network and eventually validated by the consensus of the minders and then recorded to the blockchain. The transaction is depicted in Figure 1.1.

Note: After the transaction, the original 50 BTC UTXO is fully spent and replaced with two new UTXOs: one worth 30 BTC (sent to Bob) and one worth 18 BTC (change sent back to Alice's address). While this 18 BTC becomes a new UTXO Alice owns, her total wallet balance depends on all other unspent UTXOs she controls, not just this one. Therefore, her wallet balance may be higher than 18 BTC if she holds other UTXOs that were not involved in this transaction.

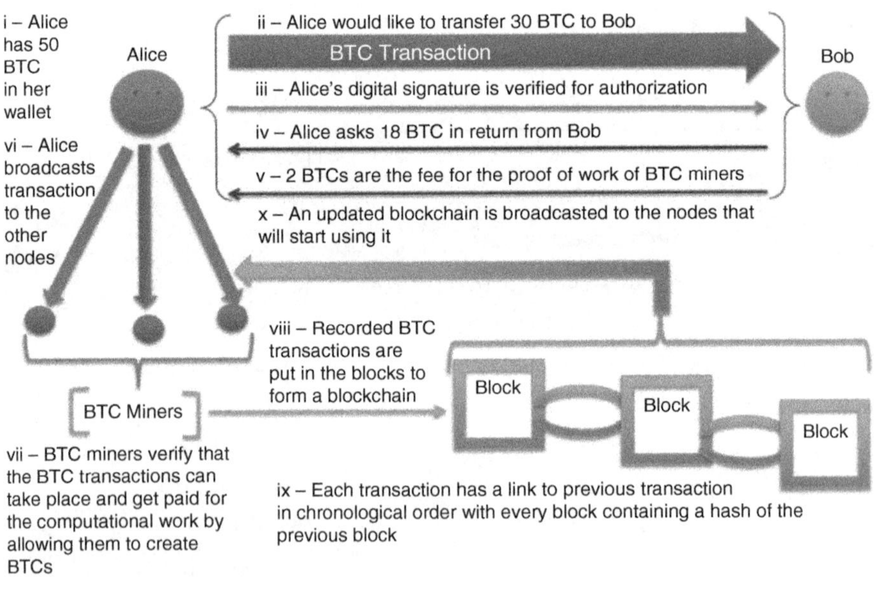

Figure 1.1 Bitcoin transaction flow example.

In a Bitcoin transaction, the input represents the source of the funds, which in this case is Alice's Bitcoin address, while the output represents the recipient, which is Bob's Bitcoin address. When a user initiates a transaction, they assert ownership over a specific output and provide cryptographic proof that they are authorized to spend the funds.

For example, if Alice transfers 30 BTC to Bob, Bob's wallet will receive a new UTXO of 30 BTC. In return, Alice requests 18 BTC back as change and pays 2 BTC as a mining fee for transaction validation via PoW. As a result, after the transaction is processed, the net output from Alice to Bob is 10 BTC, Alice retains 18 BTC in a new UTXO, and 2 BTC goes to miners as a transaction fee.

If Alice wishes to make a payment, the inputs serve as proof that she was allocated a particular sum of money, while the outputs determine how the funds are distributed. In this case, Bob receives 30 BTC, and Alice's new wallet balance is defined by her 18 BTC UTXO, which she can spend in future transactions.

Transaction Verification Checks: A critical task performed by a Bitcoin network is to ensure the validity of this incoming transaction. This involves verifying that the data has not been tampered with, that no new money is being created, and that only authorized recipients are spending UTXOs, among other factors. The verification of transactions involves several factors, including but not limited to:

- **Checking UTXO Availability:** Each transaction must reference existing UTXOs from the UTXO pool. The system verifies that all outputs being claimed as inputs are present in this pool. Once a UTXO is spent, it is removed from the pool and cannot be claimed again.
- **Signature Verification:** Each input in a Bitcoin transaction must have a valid cryptographic signature proving ownership. This requires executing the locking and unlocking scripts (ScriptPubKey and ScriptSig) associated with the input and confirming that they return true, ensuring that the spender has the right to use the funds.
- **Preventing Double Spending:** To maintain transaction integrity, the network ensures that no UTXO is being spent more than once in the same transaction. While similar to UTXO availability checks, this step specifically prevents any input from appearing multiple times in the same transaction.
- **Ensuring Nonnegative Output Values:** All output values in a transaction must be greater than or equal to zero. This prevents the creation of invalid transactions that could introduce negative values, which would be nonsensical in a financial system.
- **Checking Input–Output Consistency:** The total input value of a transaction must be greater than or equal to the total output value. If the sum of inputs exceeds the sum of outputs, the difference is considered a transaction fee, which is collected by the miner who includes the transaction in a block.

There are several types of transactions that are possible in Bitcoin. Here are some examples:

1. **P2PKH Transactions:** these transactions are the most common type of Bitcoin transactions. In a P2PKH transaction, the sender uses the recipient's Bitcoin address, which is derived by hashing their public key. The recipient's actual public key is not revealed at the time of receipt. Instead, it is only disclosed later when the recipient spends the funds by providing the unhashed public key and a valid digital signature. This process ensures that the output can only be spent by the owner of the corresponding private key.
2. **P2SH:** This transaction type allows the recipient to specify a script instead of a public key hash (PKH) to receive the payment. The sender must provide a script that satisfies the conditions specified in the recipient's script.
3. **Multi-Sig:** This transaction type requires multiple private keys to authorize a transaction, providing increased security.
4. **Pay-to-Witness-Public-Key-Hash (P2WPKH):** This transaction type is part of the SegWit upgrade and uses a different address format to reduce the transaction size and improve scalability.
5. **Pay-to-Witness-Script-Hash (P2WSH):** Similar to P2WPKH, this transaction type uses a witness script instead of a PKH to receive payments.
6. **Timelock:** This transaction type allows the sender to specify a time delay before the recipient can spend the funds, providing an extra layer of security.
7. **Atomic Swap:** This transaction type enables two parties to exchange different cryptocurrencies without the need for a trusted third party.

P2PKH is one of the most commonly used transaction types in Bitcoin. It is the standard transaction type for sending Bitcoin from one address to another, and it is estimated that over 90% of all Bitcoin transactions use this format. P2PKH transactions are also supported by most Bitcoin wallets and exchanges, making them a widely accepted and trusted transaction type.

As a developer, it is important to understand how P2PKH transactions work because it is one of the most commonly used transaction types in Bitcoin. This means that if you are building any software or applications that interact with the Bitcoin network, you will likely need to handle P2PKH transactions in some way.

The basic P2PKH transaction is a type of transaction in the Bitcoin network that involves transferring funds from one party to another using the recipient's PKH. In this type of transaction, the sender creates a new output that is "locked" with the recipient's PKH. To spend this output, the recipient must provide a valid signature using their private key, which corresponds to the PKH.

To create a P2PKH transaction, the sender typically constructs a transaction with at least one input and one output. The input refers to a previously unspent output (UTXO) that the sender wants to spend, while the output is typically created with a script that includes the recipient's PKH. To spend the output, the recipient must provide a signature and a public key that corresponds to the PKH. To ensure the validity of a transaction, a scriptSig/scriptPubKey pair must collectively evaluate as true. The spender of a transaction supplies a scriptSig that is executed alongside the scriptPubKey of the claimed transaction output. The stack is shared by both scripts. By verifying that the scriptSig and scriptPubKey pair evaluates to true, the P2PKH transaction ensures that only the redeemer with the matching public/private key pair can claim and spend the bitcoin. If all the other transaction criteria are met, the transaction is considered valid and can be added to the blockchain.

1.9.1 Transaction Automation with Smart Contracts

Smart contract transactions involve the execution of computer programs that are stored and run on a decentralized network of computers. As described by Szabo [34] in 1996, smart contracts are transaction protocols that automatically enforce the terms of a contract. The main objectives of smart contracts are to fulfill contractual obligations, such as payment terms and confidentiality agreements, while minimizing the need for intermediaries and reducing transaction costs. Smart contracts can be used to create and trade synthetic assets, including derivatives and bonds, with complex payment structures that can be analyzed and executed with low transaction costs. This enables the creation of new financial instruments that are more efficient, secure, and accessible to a wider range of investors. As a blockchain architect, it is important to understand how smart contracts work and their potential applications in various industries.

Automated transactions through smart contracts allow for the self-enforcement of business rules and agreements among multiple parties. These rules are coded into a smart contract executable, typically written in a high-level language like Solidity and executed on a distributed computing platform like the Ethereum Virtual Machine (EVM). Smart contract rules can automate the execution of transactions, including the transfer of cryptocurrency like Ether. Users can interact with a smart contract by sending transactions to its unique address, which triggers the miners to execute the contract code and reach consensus on its output.

The blockchain platform hosting the smart contract can implement a Turing-complete state machine, allowing for the computation of any computable task given sufficient resources. A Turing-complete state machine is a computational model that is capable of performing any computation that can be performed by a Turing machine, which is a theoretical computing machine that can simulate any algorithmic process. In simpler terms, it is a model of computation that can perform any computational task that is possible to perform in a general-purpose computer.

In the context of blockchain and smart contracts, a Turing-complete state machine is implemented in the blockchain platform (e.g. EVM) and allows for the execution of complex computations and programs in a decentralized and trustless environment. This means that smart contracts written in high-level programming languages like Solidity can be executed on the blockchain by the nodes in the network, without the need for a centralized authority to verify and validate the execution. The state machine is responsible for maintaining the current state of the system and executing the necessary computations to transition to the next state based on the inputs received from the transactions on the blockchain. As the contract executes, the state of the state machine governing its execution changes. Overall, smart contracts enable automated and secure transactions with predetermined business rules and agreements, without the need for trusted intermediaries.

Smart contracts make use of the blockchain's immutability properties, meaning that once the smart contract rules are compiled and stored in the blockchain, they cannot be further modified. This makes the smart contract executable, which is stored in the blockchain and can be executed on a distributed and decentralized platform, immutable. The immutability of the smart contract executable guarantees its trustworthiness, as the rules coded in the smart contract cannot be tampered with once they are stored in the blockchain.

As a blockchain architect, it is important to understand how smart contracts can be used to automate business transactions. The rules encoded in a smart contract can automate payment transactions between a buyer and seller, as well as enforce trading rules and automate payments in financial instruments. For example, a smart contract could be programmed to automatically pay a seller when specific conditions are met or to calculate and transfer premiums to a swap dealer on a quarterly basis.

Before a smart contract is coded and entered into the blockchain, the parties involved must agree on the programmable rules governing the contract. Once the rules are compiled and stored in the blockchain, they become immutable and ready for execution by the distributed virtual machine computer. The smart contract can be triggered by specific inputs, such as an external event like a natural disaster in the case of an insurance policy.

Smart contracts are characterized by their ability to self-enforce rules and execute transactions automatically, as well as their transparency due to the distributed nature of the blockchain. As a blockchain architect, it is important to understand how to apply smart contract logic to automate business transactions and ensure trustworthiness and transparency in the process.

Security Concerns with Smart Contracts: Smart contracts provide an efficient, secure, and transparent way to automate transactions and execute business rules, which can lead to increased efficiency, reduced costs, and improved trust among parties. Smart contracts are stored on a decentralized blockchain, making them highly secure and resistant to tampering or hacking attempts. On the other hand, smart contracts are computer programs, and as such, they are susceptible to bugs and vulnerabilities like any other software. If a smart contract has a vulnerability, it could be exploited to steal funds or cause other damage. At high level, these are the main security concerns for smart contracts:

1. **Dependency from External Data Sources:** Smart contracts are designed to be self-executing and deterministic, executing predefined logic without external intervention. However, many smart contracts depend on real-world data to make decisions, data that resides outside the blockchain. This reliance introduces the need for oracles, which serve as trusted intermediaries to bridge on-chain logic with off-chain information. Oracles deliver external data such as asset prices, weather events, or identity verification, enabling smart contracts to respond to real-world conditions. However, if these oracles or their data sources are compromised, the smart contract's execution can be manipulated, potentially leading to unintended or malicious outcomes.
2. **Vulnerabilities in the Smart Contract Code:** Smart contracts can be targeted by attackers who try to exploit vulnerabilities in the smart contract code or find ways to manipulate the contract execution. Malicious actors can attempt to steal funds, disrupt the contract execution, or exploit weaknesses in the underlying blockchain.
3. **Use of Vulnerable Libraries:** If smart contracts use libraries that have not been properly audited or tested for bugs and vulnerabilities, they may be susceptible to exploitation by attackers trying to compromise the smart contract code.
4. **Business Logic Flaws:** Business logic flaws in smart contracts can be exploited by attackers to steal or manipulate funds or data. Using formal verification techniques are necessary to help identify and correct these flaws before the contract is deployed, increasing its security and reducing the risk of attacks.
5. **Immutable Code for Vulnerabilities:** While immutability is a key feature of blockchain technology, it also means that once a smart contract is deployed, it cannot be changed. If a vulnerability is discovered after deployment, it can be difficult or impossible to fix without deploying a new contract.

Smart contracts have critical aspects that need to be considered to ensure their security. One such aspect is the presence of vulnerabilities in the software that attackers can exploit for fraudulent activities. To mitigate these risks, developers need to follow best practices for smart contract development, including conducting thorough code reviews and security audits, testing their contracts extensively, and keeping up-to-date with the latest security research and best practices.

Bugs in smart contract code can be exploited by hackers to gain ownership of crypto assets worth millions of dollars. For example, a coding error in a smart contract code can be invoked by attackers to withdraw all collected fees, making them the owner of the contract.

The DAO hack [35] is an example of a reentrancy vulnerability that was exploited to steal US$60 million worth of Ether. To prevent fraudsters and scammers from exploiting known vulnerabilities, smart contracts need to be audited for security regularly. This requires developing threat models and use and abuse cases, identifying vulnerabilities through static analysis tools and manual code review, and maintaining a central repository source documenting known vulnerabilities. Ad hoc tools like symbolic execution tools can also be used for scanning smart contract software to identify common bugs. Bug bounties can also incentivize security experts to find and report smart contract code bugs.

1.9.2 Token Transactions

Token transactions in the context of blockchain platforms refer to the transfer of digital tokens between parties on the blockchain. These tokens represent units of value that can be used as a medium of exchange, store of value, or other purposes depending on the specific blockchain platform and the design of the token. Token transactions are recorded on the blockchain ledger and can be verified and traced by anyone on the network. Token transactions are often facilitated by smart contracts that enforce rules and logic around the transfer and ownership of the tokens.

Some examples of token transactions include:

1. **ICOs:** A company creates and sells a new cryptocurrency token to raise funds for a new project or business venture.
2. **Utility Tokens:** Tokens that represent access to a particular service or product, such as a blockchain-based platform or application.
3. **Security Tokens:** Tokens that represent ownership of an asset, such as shares in a company or real estate.
4. **Stablecoins:** Tokens that are pegged to the value of a fiat currency or commodity, such as gold, in order to reduce price volatility.
5. **NFTs:** Tokens that are unique and represent a specific digital asset, such as artwork, music, or collectibles.

Token transactions can be implemented on both public and private blockchain platforms, and examples of public implementations include the Ethereum platform of ERC-20 tokens [36]. These tokens are fungible tokens that represent assets such as digital currencies, utility tokens, and assets like real estate, stocks, and commodities. Some examples of ERC-20 tokens include Tether (USDT), Chainlink (LINK), and Uniswap (UNI). Similarly to ERC-20 tokens, Binance Smart Chain (BSC) are used on the BSC and used for a variety of purposes such as governance, payments, and staking. Some examples of BEP-20 tokens include PancakeSwap (CAKE) and Binance Coin (BNB).

Private implementations may be used for specific purposes such as within a supply chain or loyalty program. Smart contracts can be used to automate the transfer of tokens and enforce specific rules and conditions around their use. In Hyperledger Fabric, chaincode (the platform's version of smart contracts) defines business logic and can be used to implement token issuance, transfer, and management. This enables organizations to represent digital assets or tokens within a private, permissioned blockchain network. Tokens can be used for various purposes such as rewards, incentives, and loyalty points. Tokens can be issued on the Corda blockchain to represent assets such as digital currencies, commodities, and securities. Corda also allows for the creation of fungible tokens and NFKs, depending on the use case.

In the context of DeFi [37], token transactions refer to the exchange, lending, borrowing, or other financial activities involving tokens (digital assets) on decentralized platforms using smart contracts. These transactions can involve a variety of tokens such as stablecoins, governance tokens, utility tokens, and other types of tokens used in DeFi protocols.

Token transactions in DeFi are typically executed on DEXs or other DeFi platforms using Automated Market Makers (AMMs) or order books. These transactions are recorded on a blockchain, ensuring transparency and immutability. One of the main benefits of using tokens in DeFi is that they enable interoperability between different protocols and platforms. For example, a user can borrow one type of token from one DeFi protocol and then use it to provide liquidity in another protocol.

Security Risks with Token Transactions: In general, there are several security issues and risks that architects and software developers should consider when designing and implementing transactions that use tokens. Some of these include:

1. **Token Theft:** Tokens can be stolen if the smart contract is not properly secured, or if the token itself has a vulnerability that allows an attacker to steal it.
2. **Token Replay Attacks:** Token transactions can be replayed, which means that a malicious actor could potentially execute the same transaction multiple times in order to steal tokens.

3. **Token Freezing:** If the smart contract governing the token is not properly secured, an attacker could freeze the token, making it unavailable to legitimate users.
4. **Smart Contract Vulnerabilities:** Smart contracts that govern token transactions can have vulnerabilities that allow attackers to exploit the contract and steal tokens.
5. **Malicious Token Contracts:** Malicious actors may create fake token contracts that look legitimate but are designed to steal tokens or other sensitive information.
6. **Insider Attacks:** Individuals with insider access to the system, such as employees or contractors, may attempt to steal tokens or exploit vulnerabilities in the system.
7. **Regulatory Compliance:** Depending on the jurisdiction and use case, transactions involving tokens may be subject to regulatory compliance requirements, which can add complexity and increase risk.

In addition to these security risks associated with digital tokens, there are also several security risks associated with token transactions in specific types of distributed applications like DeFi. These security risks might be associated with smart contract vulnerabilities and business logic flaws, flash loan attacks, oracle manipulation, and market manipulation. Therefore, architects and software developers must ensure that their DeFi applications are designed with robust security measures and that they are thoroughly audited and tested before deployment.

To mitigate token transaction security risks, architects and software developers should carefully design and implement token transactions, including proper security measures such as encryption, access controls, and secure coding practices. It is also important to perform thorough testing and auditing of the system to identify and remediate any vulnerabilities before they can be exploited by attackers. Additionally, compliance with relevant regulations should be carefully considered and implemented as necessary.

1.10 Privacy Controls

In designing a blockchain system, architects and software developers must consider the level of transaction privacy required for their specific use case. Depending on the needs of the application, a private blockchain or a public blockchain may be more suitable.

For applications that require maximum transaction privacy, a private blockchain can provide more control and security. In a private blockchain, only authorized participants can access and view transaction data, which can protect sensitive information from being publicly available. This type of blockchain is often used by organizations that require secure and private data management, such as financial institutions and government agencies.

Privacy controls can be applied to ensure transaction privacy over public blockchain. For example, in the DEX Uniswap [38], users can trade tokens in a trustless and transparent way on a public blockchain. However, to protect the privacy of their trading activity, they can use tools such as private wallets, VPNs, and other anonymization techniques to obscure their IP address and prevent their identity from being linked to their trades.

On the other hand, for applications that require transparency and trust in the transaction history, a public blockchain may be more appropriate. In a public blockchain, transactions are visible to all participants, which can enhance transparency and accountability. This type of blockchain is often used for applications such as cryptocurrency transactions, where it is important for all parties to be able to see the transaction history.

Privacy Enhancing Controls: Implementing privacy enhancing controls can be a viable approach to ensure transaction privacy in permissionless blockchains used for P2P payments where individual buyers and sellers use peer nodes without any involvement of third-party clearinghouses such as interbank payments or bank card issuers and payment processors as in the case of credit card payments. However, in the case of permission-based blockchains where nodes are registered to known users who are authorized to join, identity verification is required, and user identities must be revealed. For example, this is the case of a financial use case, where transactional data privacy needs to comply with regulations that require the protection of customer data and financial transactions. For example, in a trade finance application, privacy controls may be implemented to protect confidential information related to a trade transaction, such as the buyer and seller information, trade documents, and financial details. One approach is to use a private or permissioned blockchain to limit access to the network to only trusted parties and to use cryptography to ensure that only authorized parties can view specific transactions or data.

Privacy controls may also be implemented at the transaction level, such as through the use of ZKPs, which allow parties to prove that they have access to certain information without revealing the information itself. In certain types of financial transactions, revealing user identities may be necessary to comply with regulatory requirements and to prevent financial crimes such as money laundering. One reason to reveal user identities, as is done in traditional payments, is to identify the beneficiary of a transaction and trace the payment's origin back to the seller's identity. This identification of buyers, sellers, payees, and beneficiaries allows for compliance with regulations, including AML and Office of Foreign

Assets Control (OFAC) sanctions checks on blacklisted beneficiaries, to prevent various types of crimes (e.g. money laundering, cyber extortion, profit from ransomware). One approach to enabling these controls is to use off-chain methods. Blockchain applications can utilize the KYC functionality of the financial services platform they integrate with to verify the identity of transaction beneficiaries before the transactions are signed and validated by trusted peer nodes and asset values are transferred to the beneficiary.

For DeFi and dApp applications, privacy controls such as ZKPs and off-chain computation can be used to ensure that sensitive information is not visible to all participants in the network. It is essential for architects and software developers to carefully consider the level of transaction privacy needed for each specific use case to design the appropriate privacy controls into the blockchain.

Implementing KYC controls on-chain can be achieved through a BaaS approach, which allows sharing user identities among banks without the need for each bank to collect the same information from customers. The ideal scenario would be to give the owner of the identity information control over who can access their identity, such as a bank that requests it. Government registries and tax authorities can also validate and confirm user identities. Consortiums of banks, start-ups, and financial regulators are currently experimenting with blockchain-managed identities for various use cases. With this approach, duplication of confidential PII and identity documentation (e.g. passports) is eliminated since different banks, legal entities, and tax authorities no longer need to collect the same information for KYC.

However, storing user identities in a private and permission-based blockchain presents challenges due to the personal identification data (e.g. SSN) and identification documentation (e.g. driver's license, passports) falling under the scope of data privacy regulations in each country where the information is stored, including the nodes where this information is permitted for access. Protecting personal identification data is not only a policy and standards requirement but also an implementation necessity.

Encryption of the data at rest and in transit, along with access controls such as strong authentication and authorizations, must be enforced to ensure its security. However, storing personal identity information in all replicated copies of the blockchain increases the potential for attackers to target this data.

Compliance with privacy regulations such as General Data Protection Regulation (GDPR) may require the right to be forgotten, but when encrypted identity information is stored in the immutable blockchain, it is uncertain whether deleting the encryption keys would meet this requirement.

A potential solution to the challenges posed by privacy regulations is to avoid storing personal identity data in the blockchain altogether and instead use alternative methods to validate identities. This ensures the confidentiality and privacy of user PII. A self-sovereign identity (SSI) platform could address these challenges by allowing personal identification documents and PII, such as TaxIDs, to be shared only when the owner decides to do so, thereby enforcing the principle of minimum privilege and reducing the risk of wide distribution of confidential information.

An example of an SSI platform could leverage standard PKI. Initially, a user would generate a unique identifier, such as a digital identity ID, based on their public and private key. This digital identity would be securely stored in their digital wallet using hardware encryption and never shared. If a user needs attestation of their identity from an identity authority, they would present their public key and provide identity information documentation, such as a copy of their passport, to the identity authority, such as a government authority. The identity authority would attest the user's identity by signing the request as proof of attestation. This attestation would consist of a digitally signed piece of information that can be shared with third parties, such as banks, to validate the attestation with the public key and accept transactions signed by the user with their private key.

The decision to use a public or private blockchain and the selection of privacy controls is heavily influenced by the particular use case and regulatory standards. It is crucial for software developers and architects to meticulously evaluate the required level of privacy and to construct the blockchain system accordingly.

1.10.1 Anonymity Versus Pseudonymity of Blockchain Transactions

To maintain privacy on the blockchain, users' identities are kept anonymous behind the addresses used for transactions. This anonymity is nearly guaranteed as user identities are not revealed and transactions are identified only by their addresses. Access to the decentralized P2P network is permissionless, allowing any system with a public IP address to join. Once a user has a registered node address, they can obtain a copy of the distributed ledger and access all transactions, including the buyers' and sellers' addresses and transaction amounts. User identities are not required to register a node, and personal details such as PII are not recorded. Node addresses, which are SHA-256 hashes of ECDSA public keys, do not contain any reference to user identities.

This model of privacy is more accurately described as pseudonymity, rather than full anonymity. Pseudonymity in blockchain means that users are represented by cryptographic public addresses rather than real names, but these addresses can still be linked to a real-world identity through external data sources, network monitoring, or blockchain analysis techniques. While no personal information is embedded in the addresses themselves, patterns of behavior, reused addresses, or interactions with regulated services (such as crypto exchanges) can create identifiable fingerprints.

Despite the fact that permissionless blockchains are designed to provide privacy through anonymity, there is no guarantee that user privacy can be ensured. In fact, it is possible to trace public addresses back to an IP address or a user account with a cryptocurrency exchange account, which may ultimately lead to an actual identity. Law enforcement agencies, for example, can determine user accounts by gathering forensic information about transactions and the IP addresses of computers, websites, and emails used during registration of the cryptocurrency address. Various network analysis techniques, such as transaction graphs and taint analysis, can be used to gain information about bitcoin users and trace association between a user address and earlier transaction addresses, including the type of bitcoin exchange used. It is worth noting that these analysis techniques are also used by malicious actors, and unless anonymity-enabled channels like Tor or 12p are used to hide IP addresses, a persistent and motivated attacker may be able to associate the buyer and seller of IP addresses with their bitcoin transactions.

In light of these considerations, permissionless blockchains should not be considered truly anonymous, but rather "quasi" anonymous due to the possibility of tracing users through network analysis. However, efforts have been made to enhance user privacy in some cryptocurrency-based protocols such as Zcash. These protocols use ZKP algorithms to enable privacy by inference, allowing a statement to be proven true or false without revealing any information. Ring signatures are another privacy-preserving control that can be implemented to protect the privacy of payees by making it impossible for a payment verifier to identify them. In a ring signature, a group of people endorses a digital signature, making it impossible to determine who actually made the signature. Monero is an example of a permissionless blockchain that has implemented ring signatures to enhance user privacy.

1.10.2 Techniques for Enhancing Transaction Privacy

As a security architect or engineer working with blockchain technology, it is important to understand the different techniques available for enhancing transaction privacy, including mixing, ring signatures, and ZKPs. It is also important to understand the strengths, limitations, and potential risks associated with each technique, in order to design secure and effective blockchain systems. Techniques available for enhancing transaction privacy include mixing, ring signatures, and ZKPs.

Mixing, also known as coin shuffling, is a technique used to break the link between the input and output addresses of a transaction. This is achieved by mixing the transaction inputs with those of other transactions, thereby making it difficult to trace the flow of funds. There are various ways of achieving mixing, including centralized mixing services and decentralized mixing protocols. As a security architect or engineer, it is important to understand the strengths and limitations of each mixing method, as well as the potential risks and attack vectors associated with mixing services.

One example of a mixing implementation is in the Bitcoin blockchain, which uses a process called CoinJoin. CoinJoin is a protocol that allows users to combine their Bitcoin transactions with those of other users in a way that makes it more difficult to trace the transactions to individual users. In CoinJoin, a group of users pool their transactions together into a single transaction. This transaction includes multiple inputs and outputs, with each user's input and output being mixed in with those of the other users. This creates a "mixing pool" that obscures the source and destination of each individual transaction, making it more difficult for anyone to trace the flow of funds.

Ring Signatures: Other blockchain platforms may use similar mixing techniques, although they may be implemented in slightly different ways. For example, the Monero blockchain uses a technique called ring signatures, which allows a user to sign a message on behalf of a group, making it difficult to determine which member of the group actually signed the message. This can be used to obscure the source of a transaction in a similar way to CoinJoin. Ring signatures are a type of digital signature that allow a user to sign a message anonymously on behalf of a group. This means that a verifier cannot determine which member of the group actually signed the message, thereby preserving the anonymity of the signatory. Ring signatures can be used in cryptocurrency transactions to hide the identity of the sender, by creating a ring of possible senders that includes not only the actual sender but also other unrelated addresses. As a security architect or engineer, it is important to understand the cryptographic principles behind ring signatures, as well as their strengths and limitations. A few examples of blockchain platforms that have implemented ring signatures as part of their privacy protocols are Monero, Bytecoin, PIVX, Dash, and Komodo:

- **Monero** [38]: Monero is a privacy-focused cryptocurrency that uses ring signatures to keep transactions confidential. Ring signatures allow a sender to mix their transaction with other transactions on the network, making it difficult to trace a transaction back to a specific sender.
- **Bytecoin** [39]: Bytecoin is another privacy-focused cryptocurrency that uses ring signatures. Like Monero, Bytecoin uses ring signatures to mix transactions and keep them confidential.
- **PIVX** [40]: PIVX is a privacy-focused cryptocurrency that uses ring signatures as part of its privacy protocol. PIVX enhances user privacy by implementing zk-SNARKs (zero-knowledge succinct noninteractive arguments of

knowledge), a form of ZKP that allows users to prove the validity of a transaction without revealing any underlying information, such as the sender, receiver, or transaction amount.
- **Dash** [41]: Dash is a cryptocurrency that uses a variation of ring signatures called PrivateSend. PrivateSend allows users to mix their transactions with other users' transactions on the network, making it difficult to trace a transaction back to a specific user.
- **Komodo** [42]: Komodo is a cryptocurrency and a multichain platform that emphasizes security, scalability, and interoperability. While Komodo used Delayed Proof of Work (dPoW) as its consensus mechanism, leveraging the Bitcoin blockchain to notarize blocks and enhance security, it also incorporates privacy features such as ring signatures (read the following note) to obscure transaction origins. Unlike some privacy-focused cryptocurrencies that are built entirely around privacy by default, Komodo provides optional privacy through its privacy coins. These features use ring signature techniques to mix transactions and enhance anonymity, making it difficult to trace the origin of funds.

Note: It's important to note that ring signatures are cryptographic methods that allow a signer to prove that a transaction was approved by someone in a group, without revealing exactly who. This helps maintain plausible deniability and on-chain privacy for users opting in. Thus, while Komodo's core consensus relies on a hybrid dPoW model for security, it supplements privacy through the use of ring signatures for select transactions or assets within its ecosystem.

ZKPs: ZKPs are a cryptographic technique that allow one party to prove to another party that they know a piece of information, without revealing the information itself. In the context of cryptocurrency transactions, ZKPs can be used to prove ownership of a private key, without revealing the key itself. This can be used to prove that a transaction is valid, without revealing any identifying information about the sender or recipient. As a security architect or engineer, it is important to understand the different types of ZKPs, as well as their applicability to different use cases.

For software developers and architects working on blockchain projects, it is important to be aware of ZKP implementations found in popular blockchain platforms for design and implementation of privacy-enhancing features for a blockchain project and understand the limitations that may impact the security or performance of their own implementation.

Some examples of ZKP implementations found in popular blockchain platforms include Zcash, which is a privacy-focused cryptocurrency that uses ZKPs to enable shielded transactions. Zcash's ZKP system is called zk-SNARK, which allows users to prove ownership of a private key without revealing the key itself. This allows transactions to be verified without revealing any sensitive information.

Ethereum has implemented a ZKP system called zk-rollups, which allow for scaling and privacy enhancements. With zk-rollups, transactions are bundled together and verified off-chain, with a ZKP being submitted to the blockchain to prove the validity of the transaction bundle. This reduces the amount of data that needs to be stored on the blockchain and improves transaction throughput.

Stellar has implemented a ZKP system called Bulletproofs, which enables confidential transactions on the Stellar network. Bulletproofs use a range proof to prove that a transaction is valid without revealing the actual transaction amount. This allows for confidential transactions while maintaining the transparency of the blockchain.

Finally, while Bitcoin does not have a native implementation of ZKPs, there are several projects exploring this technology for Bitcoin. One example is the zk-SNARK implementation called ZK-Proofs, which is being developed by the Zcash team for use on the Bitcoin blockchain.

1.11 Identity Controls

In the context of identity management, blockchain needs identities to ensure that only authorized users can participate in the network and perform specific actions such as accessing or modifying data. Blockchain provides a distributed and decentralized platform, and identities are necessary to establish trust and security within the network. By verifying the identity of users and ensuring that they have the necessary permissions to perform certain actions, blockchain can prevent unauthorized access and manipulation of data.

In a public blockchain, identity controls are essential to ensure that users are who they claim to be and to prevent malicious actors from gaining unauthorized access to the network. Public blockchains are open to anyone, which means that anyone can participate in the network. This makes identity controls even more critical to ensure the security and integrity of the network. Public blockchains typically use cryptographic methods to verify the identity of users, such as public–private key pairs or digital signatures.

In contrast, in a private blockchain, identity controls may be less critical, as the network is only open to a select group of authorized participants. Private blockchains are often used in enterprise settings, where participants are known and trusted. In this context, identity controls may be more focused on ensuring that only authorized participants have access to the network and that their actions are tracked and auditable. Private blockchains may use traditional identity verification methods, such as usernames and passwords or biometric authentication, in addition to cryptographic methods.

In general, identity controls help to establish trust and security within the blockchain ecosystem by verifying the identity of users and ensuring that they have the necessary permissions to perform certain actions. For example, identity controls help to prevent unauthorized access to the blockchain and ensure that data is secure. By verifying the identity of users, the blockchain can ensure that only authorized parties have access to sensitive information. Identity controls are also essential for ensuring compliance with regulatory requirements. Many industries are subject to strict regulatory requirements, and identity controls can help ensure that the blockchain meets these requirements. Identity controls can also help to establish trust and reputation within the blockchain ecosystem. By verifying the identity of users and ensuring that they have the necessary permissions, the blockchain can establish trust among participants. From a software developer and architect perspective, it is important to know about identity verification methods in blockchain. Understanding identity verification methods in blockchain is essential for building secure, compliant, and user-friendly blockchain applications. Identity verification methods are important in general because they help ensure the security and privacy of online transactions and interactions.

As more and more activities are conducted online, the need for reliable and secure identity verification methods becomes increasingly important. For example, in e-commerce, identity verification methods can help prevent fraud and protect both buyers and sellers. In online banking, identity verification methods are essential for ensuring that only authorized users can access and transfer funds. Furthermore, identity verification methods can also help protect sensitive data and ensure compliance with regulatory requirements. For example, in healthcare, identity verification methods are critical for protecting patient data and ensuring compliance with Health Insurance Portability and Accountability Act (HIPAA) regulations. Overall, identity verification methods play a crucial role in ensuring the security, privacy, and compliance of online interactions. As technology continues to evolve and more activities are conducted online, the need for reliable and secure identity verification methods will only continue to grow.

1.11.1 Identity Verification Methods

Identity verification methods are important for ensuring the security, authenticity, and integrity of transactions on the blockchain. By using these methods, blockchain platforms can help to build trust and confidence among users, while also complying with regulatory requirements and protecting against fraud. For example, in order to ensure authenticity, identity verification methods like digital signatures can help to ensure that transactions on the blockchain are authentic. Digital signatures provide a way for users to sign transactions using their private key, which can then be verified using their public key. This helps to ensure that only authorized users are able to initiate transactions on the blockchain. Bitcoin uses digital signatures to verify the identity of users and authorize transactions. Each user has a unique digital signature that is linked to their public key, which is used to verify the authenticity of transactions.

Identity verification methods can also be used to maintain privacy by giving control to the users regarding who can access their identity information. For example, SSI systems allow users to control their own identity information and share it with others as needed, without relying on a centralized authority. This can help to protect the privacy of users while still ensuring that they are properly identified and verified. Verifying the identity of users in blockchain platforms can help to prevent fraudulent transactions from occurring by enabling trust among parties in a blockchain transaction where this is an asset for value exchange. For example, in the case of P2P payments, a validation of the identity of a beneficiary can occur by implementation of KYC controls. By requiring users to undergo a KYC process, blockchain platforms can help to prevent fraud by ensuring that only legitimate users are able to access the platform and participate in transactions. KYC can help to prevent several types of fraud on the blockchain, including identity theft, money laundering, and terrorist financing. By verifying the identity of users, blockchain platforms can make it more difficult for fraudsters to use the platform for illegal activities. KYC is also often required by regulatory bodies in order to comply with AML and counterterrorist financing (CTF) regulations. Blockchain platforms such as Ethereum have implemented a KYC process for users who want to participate in token sales or other activities on the platform. Ripple has also implemented a KYC process for users who want to participate in certain activities on the platform.

Identity verification controls may not make sense with anonymous open blockchains, as the purpose of these blockchains is often to provide a high degree of privacy and anonymity to users. In anonymous open blockchains, users are not required to reveal their real-world identity in order to participate in transactions. Instead, they are identified by a pseudonymous address or public key that is not linked to their real-world identity. In anonymous blockchain networks, the focus is on maintaining the privacy and anonymity of users. This means that there is typically no requirement for users to disclose their personal identity or other identifiable information. Instead, the focus is on maintaining the identity of nodes or IP addresses within the network. Nodes or IP addresses are typically used to identify and track transactions on the blockchain, and they play a critical role in maintaining the security and integrity of the network. However, there may be situations where identity verification controls could still be useful in anonymous open blockchains. For example, some blockchain platforms may require users to verify their identity in order to access certain features or services, even if they are using an anonymous open blockchain. In these cases, identity verification could help to prevent fraud and ensure the integrity of transactions. Additionally, some blockchain platforms may use identity verification controls to verify the identity of nodes on the network, even if the users themselves are anonymous. This can help to prevent malicious actors from participating in the network and compromising its security.

Permission-based blockchains differ from anonymous-based blockchains in terms of their approach to identity verification. In permission-based blockchains, identity verification controls are typically required in order to establish trust among participants and prevent unauthorized access to the network. Participants must obtain permission from a central authority or network administrator in order to join the network, and their identity must be verified in order to prevent fraud and other security threats. In permission-based blockchains, identity verification is often required to ensure that only authorized parties can participate in the network and users are typically required to undergo a formal identity verification process before they are granted access to the network.

Permission-based blockchains such as Hyperledger Fabric, for example, require users to be authenticated and authorized before they can participate in transactions. Hyperledger Fabric supports a range of identity verification methods, including digital signatures, MFA, and identity federations. Identity federations are networks of trusted organizations that work together to authenticate and authorize users. In a permission-based blockchain, identity federations can be used to verify the identity of participants and ensure that they are authorized to participate in transactions. Besides digital signatures and MFA some permission-based blockchain platforms such as Corda also use reputation-based systems. Reputation-based systems help to establish trust among participants by assigning reputation scores based on their behavior and performance on the network. In a permission-based blockchain, reputation-based systems can be used to verify the identity of participants and ensure that only trustworthy individuals are allowed to participate in transactions.

This may involve providing personal identification information, such as a government-issued ID, as well as other forms of authentication, such as MFA or biometric verification. The use of identity verification in permissioned blockchain networks helps to ensure that only authorized parties can access and modify data on the blockchain. This can help to prevent fraud, ensure compliance with regulatory requirements, and maintain the overall security and integrity of the network.

Identity verification can be implemented using smart contracts on a blockchain platform. Smart contracts are self-executing programs that run on a blockchain, and they can be programmed to enforce rules and conditions related to identity verification. For example, a smart contract could be used to verify a user's identity by requiring them to provide certain information, such as their name, address, and government-issued ID. The smart contract could then use this information to verify the user's identity against a trusted identity database, such as a government registry.

Once the user's identity is verified, the smart contract could then grant them access to certain resources or execute a transaction on their behalf. The smart contract could also enforce rules related to data privacy and security, such as limiting access to certain data or requiring MFA for sensitive transactions. One example of an enterprise blockchain platform that uses smart contracts for identity verification is the Hyperledger Fabric framework. Hyperledger Fabric is an open-source permissioned blockchain platform that was developed by the Linux Foundation's Hyperledger project. Hyperledger Fabric uses smart contracts called "chaincode" to implement business logic and enforce rules for transactions on the blockchain. Chaincode can be used to implement identity verification by requiring users to provide certain information and validating it against a trusted identity source. For example, an organization using Hyperledger Fabric might require users to provide their government-issued ID and other personal information before being granted access to certain resources on the blockchain. The organization could create a chaincode program that uses this information to validate the user's identity and grant them access to the resources if the identity is verified.

1.11.2 Privacy-Preserving Identities

In the context of blockchain and distributed systems, it is often desirable to preserve the privacy of users while still allowing for identity verification. This can be a challenging problem, as traditional identity verification mechanisms often require the disclosure of sensitive personal information. One approach to addressing this challenge is to use techniques such as ZKPs and selective disclosure. These techniques allow users to prove their identity or ownership of a particular asset without revealing any additional information beyond what is strictly necessary. In order to protect the blockchain user's privacy and reduce the risk of identity theft, fraud, and other security threat controls such as ZKPs, selective disclosure, and other privacy-preserving techniques, users can authenticate their identities and ownership of assets without revealing sensitive personal information.

ZKPs: ZKPs are a class of cryptographic protocols that allow one party to prove to another that they know a particular secret without revealing any information about that secret. ZKPs can be used to provide proof of ownership or identity without revealing any additional personal information. For example, a ZKP can be used to prove that a particular user has the authority to perform a certain action on a blockchain without revealing the user's actual identity.

There are several blockchain platforms today that have implemented ZKPs [43], such as Ethereum, Zcash, Hyperledger Fabric, Corda, and MS azure blockchain. In the case of Ethereum, developers can use Solidity to create ZKP-based applications that enable users to prove ownership of certain information without revealing the information itself. Zcash is a privacy-focused cryptocurrency that uses ZKPs to shield transaction details from public view while still maintaining a decentralized ledger. The Zcash protocol uses a specific type of ZKP called a zk-SNARK to prove that a transaction is valid without revealing any of the details of the transaction. Hyperledger Fabric also supports the use of ZKPs for privacy-preserving identity verification and authentication. The platform has a modular architecture that allows developers to integrate ZKPs into their applications as needed. Corda supports the use of ZKPs for a variety of applications, including anonymous identity verification, selective disclosure of transaction details, and private asset transfers. Microsoft's BaaS platform supports the integration of ZKPs into blockchain applications. Azure Blockchain allows developers to create private, permissioned networks and deploy ZKP-based smart contracts for secure, private transactions.

Decentralized Identity (DID) Platform and Selective Disclosure: These are techniques that allow users to disclose only a subset of their personal information to third parties hence preserving privacy while allowing for identity verification. For example, a user might disclose their age and gender to a healthcare provider while withholding more sensitive information such as their medical history. Selective disclosure can be implemented using a variety of technologies, including attribute-based credentials and digital signatures. Selective disclosure is a technique used to provide only certain pieces of information to a requester, while keeping other information private. This can be useful for preserving privacy while still allowing for identity verification.

Here are some examples of blockchain platforms that use selective disclosure:

1. **Sovrin** [44]: Sovrin is a DID platform that uses selective disclosure to allow users to share only the information they choose with third parties. It uses ZKPs to ensure that the information is accurate without revealing any unnecessary details.
2. **uPort** [45]: uPort is a DID platform that allows users to create and manage their own identities. It uses selective disclosure to allow users to share only the information they choose with third parties. This can help to prevent unnecessary data exposure and preserve user privacy.
3. **Hyperledger Indy Identities and Claims** [46]: Hyperledger Indy is a blockchain platform that focuses on DID and verifiable claims. It uses selective disclosure to allow users to share only the information they choose with third parties, helping to preserve privacy and prevent unnecessary data exposure.

4. **Civic** [47]: Civic is a blockchain-based identity verification platform that uses selective disclosure to allow users to share only the information they choose with third parties. It uses MFA and biometric verification to ensure the identity of the user while preserving their privacy.

Selective disclosure can be implemented through various techniques, including ZKPs and digital signatures. Architects designing blockchain platforms need to know about selective disclosure as another approach to preserving privacy while allowing for identity verification because it enables users to disclose only the minimum necessary information to prove their identity or authorize a transaction, while keeping the rest of their information private. This can be important in situations where users may not want to disclose their full identity or information, such as in financial transactions or healthcare data sharing.

1.11.3 Identity and Access Management

Identity and Access Management (IAM) is an important aspect of securing any system, including blockchain platforms and distributed applications that leverage BaaS. IAM provides a framework for controlling user access to resources within a system, ensuring that only authorized users can access sensitive data or perform critical actions. In the context of blockchain platforms, IAM can be used to control access to blockchain nodes, smart contracts, and other resources. This can help prevent unauthorized access, tampering, or other malicious activities that could compromise the integrity and security of the blockchain network. Similarly, IAM is critical for distributed applications that leverage BaaS, as these applications may rely on external services for authentication and authorization. Without proper IAM controls, these applications may be vulnerable to attacks such as identity theft, session hijacking, and other security breaches.

Integrating with existing IAM solutions is important while using BaaS platforms. Today most organizations already have established IAM systems in place to manage their users' identities and access to various resources. By integrating the BaaS platform with the existing IAM solution, the organization can leverage their existing investment in IAM and avoid duplicating efforts in managing identities across different platforms. Integrating with existing identity providers (IdPs) can simplify the user experience by allowing users to access the BaaS platform with their existing credentials, eliminating the need to create and manage separate login details.

Many blockchain platforms today have implemented support for standard IdPs and authentication/authorization methods to enable seamless integration with existing identity management systems and user workflows. This allows organizations to leverage their existing investment in identity management infrastructure and policies while also benefiting from the security and transparency advantages offered by blockchain technology.

AWS [48] provides access control to its blockchain platform, Amazon Managed Blockchain, through integration with AWS IAM. Additionally, it uses API keys to authenticate users and authorize access to blockchain network resources via APIs. These API keys are unique tokens generated by the platform to identify users and track their interactions with blockchain services.

Ethereum, one of the most popular public blockchain platforms, has also implemented support for various identity standards, such as OpenID Connect and OAuth 2.0. OAuth 2.0 is a widely used authentication and authorization protocol that allows users to grant third-party applications access to their resources without sharing their credentials. Ethereum also allows for integration with external identity management systems through the use of smart contracts and DID frameworks such as uPort or Sovrin. These frameworks allow for the creation of self-sovereign identities that can be used across different blockchain networks and platforms.

The IBM Blockchain Managed Fabric [49] provides integration with LDAP, Active Directory, and other identity management systems. It also supports the use of identity federation to allow multiple organizations to collaborate on the same blockchain network while maintaining their own identity management systems. Hyperledger Fabric also provides support for external identity management systems, such as LDAP or Active Directory, which can be used to authenticate users before they are allowed to access the blockchain network. This allows organizations to leverage their existing identity management infrastructure and integrate it with the blockchain platform. Integrating with external identity management systems or federating with other organizations' IdPs can help to improve the security and efficiency of P2P transactions on blockchain networks, and there are several platforms that support these features. Another important characteristic to look for in a blockchain platform is the support of Security Assertion Markup Language (SAML)-based single sign-on (SSO). IBM Blockchain Platform supports SAML-based SSO to allow users to log in to the platform using their organization's existing identity management system. SAML is an XML-based protocol for exchanging authentication and authorization data between parties and is commonly used for web-based SSO.

R3 Corda allows for integration with external identity management systems, including Active Directory, LDAP, and Kerberos. It also supports identity federation through the use of "identity notaries" that validate the identity of users from different organizations on the same blockchain network.

In many cases where APIs are used to access BaaS platforms, the types of APIs and authentication methods vary depending on the specific platform, its integration capabilities with IdPs, and the supported access control mechanisms.

In summary, understanding identity controls is crucial for architects because it enables them to design more secure, scalable, and interoperable distributed applications. By leveraging the appropriate identity controls, architects can help to ensure that their applications are more resilient to security threats, more easily integrated with other systems, and more able to protect sensitive data. By implementing strong identity verification, authentication, and authorization controls, architects can ensure that only authorized users or systems can access the application and its associated resources. This helps to mitigate the risk of unauthorized access, data breaches, and other security incidents that could compromise the integrity and confidentiality of sensitive data.

Additionally, understanding the available identity controls enables architects to design more robust and scalable systems. For example, by using standardized identity protocols such as OAuth or OpenID Connect, architects can implement IAM solutions that are interoperable with a wide range of other systems and applications, making it easier to integrate and exchange data securely. Integrating with existing IdPs and authentication/authorization methods is an important aspect of enterprise blockchain adoption, as it allows organizations to leverage their existing identity management infrastructure and policies while also ensuring secure and reliable access to blockchain-based systems and applications.

1.11.4 Decentralized Identities (DIDs)

DID is a concept that allows individuals to have control over their own identity information, instead of relying on centralized authorities or third-party providers. This means that users can store their identity information, such as their name, age, and other personal details, in a decentralized system that is secured by blockchain technology.

From the perspective of a blockchain architect developing dApps, this means that users are allowed to authenticate themselves and access dApps without relying on centralized IdPs. Instead, users can present their identity information to the dApp as needed, using selective disclosure techniques such as ZKPs.

This approach provides users with greater control over their own identity information, which is especially important in today's world where identity theft and data breaches are becoming increasingly common. By decentralizing identity, individuals can protect their personal information from being accessed and exploited by unauthorized parties, while still being able to authenticate themselves and access the services they need. Since privacy regulations and compliance requirements vary by jurisdiction, implementing a DID system in compliance with these regulations can be a complex task. It requires careful consideration of the specific legal and regulatory requirements of the jurisdictions where the system will be used, as well as the technical and operational requirements for ensuring security, privacy, and user control.

DID can help comply with GDPR [50] and US Consumer California Consumer Privacy Act (CCPA) [51] by giving users control over their own personal data. With DID, users are able to store their identity information in a secure, decentralized manner and selectively disclose it to others as needed. This means that users can give consent for their data to be used by organizations and can revoke that consent at any time. In the context of GDPR, that is a comprehensive data protection regulation that applies to all businesses operating within the EU or processing data related to EU citizens, where it requires that data controllers ensure that personal data is processed in a transparent, lawful, and fair manner and that individuals have the right to access, correct, and delete their personal data. DID can be used to enable "data portability," which is one of the key rights of data subjects under GDPR. Data portability allows users to obtain and reuse their personal data across different services, which can be made possible by giving users control over their own identity information.

In the context of CCPA that is a data privacy regulation that applies to businesses operating in California or processing data related to California residents, where it gives California residents the right to know what personal information is being collected about them, the right to request that their personal information be deleted, and the right to opt-out of the sale of their personal information. DID can help organizations comply with the law's requirement to provide consumers with the right to know what personal information is being collected about them and the right to request that their personal information be deleted. DID can enable individuals to manage their own data and selectively disclose it to organizations they trust, making it easier for organizations to identify and delete an individual's personal information upon request.

Compliance with data privacy regulations such as GDPR and CCPA is an important aspect to consider when choosing a blockchain platform offering BaaS that has implemented DID controls that are compliant with GDPR and CCPA. For example, IBM Blockchain Platform and Oracle Blockchain Platforms offer DID solutions that are both compliant with GDPR and CCPA. These platforms provide tools for developers to create applications that are compliant with data privacy regulations while leveraging the benefits of DID controls.

1.12 Legal and Regulatory Considerations

The extent to which blockchain applications are regulated as financial applications depends on several factors, including the specific use case, the jurisdiction in which the application operates, and the nature of the financial activities being conducted. Blockchain applications that involve activities such as cryptocurrency exchanges, ICOs, digital asset trading, or providing financial services are more likely to be subject to financial regulations. These regulations can include licensing requirements, registration obligations, reporting and compliance measures, and consumer protection provisions. Examples of financial regulations that may apply include securities laws, AML and KYC requirements, payment services regulations, and derivatives regulations. Blockchain-based tokens and offerings are subject to securities regulations in many jurisdictions. If a token is considered a security, it must comply with securities laws, including registration requirements, investor protection measures, and disclosure obligations.

It's important to note that the regulatory landscape for blockchain technology and cryptocurrencies is still evolving and can vary significantly between jurisdictions. Regulatory authorities are actively working to adapt existing regulations or introduce new ones to address the unique aspects and challenges of blockchain-based financial applications. Compliance with these regulations is crucial to ensure the legality and legitimacy of financial activities conducted through blockchain applications. Among the challenges that blockchain applications face are compliance with data privacy and data protection laws, including regulations on cross-border data transfers and adherence to jurisdiction-specific legal requirements.

For example, financial transactional data stored on a blockchain is distributed across multiple nodes, making it difficult to control and protect sensitive information. Ensuring data privacy and protection becomes more complex, as it requires implementing robust encryption mechanisms, access controls, and compliance with data privacy regulations. Transferring financial transactional data across borders may be subject to legal restrictions, particularly when it involves personal or sensitive financial information. Data protection laws, such as the GDPR in the EU, impose restrictions on the transfer of personal data to countries without adequate data protection measures. Blockchain applications need to address these data transfer challenges to ensure compliance with relevant regulations.

As blockchain networks operate across borders, complying with various jurisdictional regulations becomes challenging. Financial transactions may be subject to different regulatory frameworks and reporting requirements in different countries. It can be difficult to harmonize these regulations and ensure compliance across all jurisdictions involved. Addressing these challenges requires a combination of technical, legal, and regulatory measures. Implementing strong data privacy and protection mechanisms, engaging in transparent and responsible data handling practices, and collaborating with regulators to establish appropriate frameworks are essential steps for blockchain applications to overcome these challenges and ensure compliance with financial regulations.

Besides financial regulations that apply when dealing with financial transactions that involve digital assets for value exchanges, blockchain applications might fall in scope for legal regimes, which can vary depending on the specific use case, jurisdiction, and the nature of the activities being conducted. Blockchain applications often involve the use of smart contracts, which are self-executing contracts with terms and conditions written in code. Contract law principles, such as offer, acceptance, consideration, and enforceability, apply to these smart contracts. Blockchain applications may incorporate innovative technologies, algorithms, or protocols that could be subject to intellectual property protection. Copyright, patent, and trademark laws are relevant to the protection and licensing of blockchain-related intellectual property. Blockchain applications must adhere to consumer protection and fair trading laws to ensure transparency, accuracy, and fairness in their operations. These laws may cover areas such as advertising, marketing practices, consumer rights, and dispute resolution. Blockchain applications involved in international trade or transactions may need to comply with trade laws, import/export regulations, and sanctions regimes imposed by different countries.

Given the evolving legal landscape and varying regulations across jurisdictions, it is crucial to incorporate appropriate controls in the design and implementation of blockchain applications. By considering legal and regulatory requirements from the early stages of development, blockchain applications can be designed to align with applicable laws and regulations, mitigating potential compliance risks.

Blockchain technology itself can provide benefits in terms of traceability and transparency, which can aid regulators in enforcing regulatory reporting requirements. The immutable and decentralized nature of blockchain can enable more efficient and reliable recordkeeping, making it easier to trace and verify transactions. This can contribute to improved regulatory oversight and enhance transparency in areas such as financial transactions, supply chain management, or healthcare data. Additionally, blockchain applications can be designed to include features that support regulatory compliance. For example, smart contracts can be programmed to automatically enforce compliance rules and regulations, ensuring that transactions meet specific requirements. By integrating compliance measures directly into the blockchain architecture, regulators can have greater visibility and assurance of adherence to applicable regulations. However, it is important to note that while blockchain technology can facilitate traceability and reporting, it does not eliminate the need for regulatory oversight. Compliance with specific regulatory requirements still needs to be ensured, and regulatory bodies play a crucial role in monitoring and enforcing compliance within the blockchain ecosystem.

Collaboration between blockchain developers, businesses, and regulators is vital to strike a balance between innovation and compliance. Ongoing dialogue and cooperation can help shape regulatory frameworks that address the unique characteristics of blockchain technology while safeguarding consumer protection, data privacy, and other regulatory objectives. The use of cryptocurrencies and blockchain technology has brought about a transformative shift in the financial landscape, introducing new possibilities and challenges. As these technologies continue to gain momentum, regulatory frameworks are being developed to address the unique characteristics and potential risks associated with cryptocurrencies and blockchain-based transactions.

Regulations impacting the blockchain use of cryptocurrencies aim to strike a delicate balance between fostering innovation and ensuring consumer protection, market integrity, and compliance with existing financial laws. These regulations vary across jurisdictions, reflecting the diverse approaches taken by governments and regulatory bodies worldwide.

Consumer protection regulations play a crucial role in safeguarding individuals who engage in transactions involving cryptocurrencies. These regulations are designed to mitigate the risks associated with scams, fraud, and theft that users may encounter in the cryptocurrency space. They aim to ensure that consumers are adequately protected by promoting transparency, fairness, and the disclosure of relevant information. By implementing these regulatory measures, authorities work toward combating fraudulent activities and creating a safer environment for cryptocurrency users.

One example of consumer protection regulations in the context of cryptocurrency transactions is the requirement for adequate disclosures. In many jurisdictions, regulations mandate that cryptocurrency businesses and service providers provide clear and transparent information to users regarding the risks, terms, and conditions associated with using cryptocurrencies.

For instance, regulators may require cryptocurrency exchanges to disclose information about their fees, withdrawal policies, and security measures. They may also mandate clear communication of the risks involved in cryptocurrency investments, such as volatility, potential loss of funds, and the absence of regulatory protections. This enables users to make informed decisions and understand the potential risks associated with engaging in cryptocurrency transactions. Furthermore, consumer protection regulations may also establish guidelines to prevent deceptive practices and combat fraudulent activities in the cryptocurrency space. Regulators may take action against fraudulent ICOs, Ponzi schemes, or misleading marketing practices by imposing penalties and sanctions on offenders. These measures aim to safeguard consumers by promoting fair and transparent practices within the cryptocurrency industry.

Several prominent cryptocurrency exchanges, including Coinbase, Binance, and Kraken, prioritize consumer protection by providing detailed disclosures to their users. These platforms disclose information on fees, transaction limits, security measures, and the risks involved in trading cryptocurrencies. They also emphasize transparency by sharing trading rules, security practices, and risk factors specific to their platforms. Additionally, these exchanges have implemented measures to combat fraudulent activities, ensure regulatory compliance, and prevent money laundering. Overall, these efforts aim to promote consumer awareness and informed decision-making within the cryptocurrency industry.

Among the blockchain platforms, Ethereum and Cardano are blockchain platforms that prioritize transparency, disclosure, and consumer protection. Ethereum provides extensive documentation, whitepapers, and resources to educate users about its technology, features, and associated risks. Users can review smart contracts and make informed decisions about participating in dApps built on Ethereum. Cardano, overseen by the Cardano Foundation, offers comprehensive information on its features, security protocols, and governance mechanisms. They also engage in educational initiatives to help users understand the risks and benefits of participating in the Cardano ecosystem. These platforms demonstrate efforts to provide users with the necessary disclosures for informed decision-making in cryptocurrency transactions. Users should review the provided information and exercise due diligence before engaging in cryptocurrency-related activities.

Besides consumer protection regulations, securities regulations have a critical impact on trading cryptocurrencies when they are classified as securities. When a cryptocurrency is deemed a security, it falls under the purview of securities regulations, which govern the issuance, trading, and transfer of securities. Securities regulations impose specific compliance requirements on entities involved in the issuance and trading of securities. This includes obligations related to registration, disclosure, reporting, and compliance with antifraud provisions. Market participants, such as cryptocurrency exchanges, brokers, and token issuers, need to adhere to these regulations to ensure compliance and protect investors. Securities regulations also prioritize investor protection by establishing standards for transparency, disclosure, and fair practices. Issuers of securities are required to provide accurate and timely information to investors, enabling them to make informed investment decisions. Regulations also aim to prevent fraudulent activities, market manipulation, and insider trading, safeguarding the interests of investors.

When a cryptocurrency is considered a security, the issuance and trading of that cryptocurrency must also comply with securities regulations. This includes following proper procedures for conducting ICOs [52], Security Token Offerings (STOs) [53], or other forms of token sales. Compliance with securities regulations helps ensure that offerings are conducted in a fair and regulated manner, reducing the risk of fraudulent or illegal activities. Securities regulations also contribute to the overall integrity and stability of the cryptocurrency market. By enforcing rules and regulations, they discourage illicit practices, market manipulation, and fraud that could undermine market confidence. Compliance with securities regulations helps foster trust in the market and enhances its credibility among institutional investors and traditional financial institutions. Securities regulations provide a framework that outlines the rights and responsibilities of issuers, investors, and intermediaries in the trading of securities. This clarity helps mitigate risks, reduces legal uncertainties, and facilitates the integration of cryptocurrencies into the broader financial ecosystem.

Architects designing blockchain-based applications need to have a comprehensive understanding of the regulatory framework that applies to cryptocurrencies. This knowledge is crucial for ensuring that the applications they design comply with investor and consumer protection regulations. By understanding the applicable laws and regulations, architects can incorporate the necessary features and mechanisms into their designs to address investor protection, transparency, compliance protocols, privacy and data protection, and smart contract security. Having this understanding allows architects to navigate the regulatory landscape effectively and develop blockchain-based applications that align with the regulatory requirements, promoting trust and compliance within the industry.

Understanding the impact of consumer data privacy protection laws on blockchain technology is crucial for architects designing distributed blockchain applications. Consumer data privacy protection laws, such as GDPR, CCPA, Personal Information Protection and Electronic Documents Act (PIPEDA) [54], and Personal Data Protection Act (PDPA) [55], impose legal obligations on organizations handling personal data. Architects must understand these laws to ensure that the blockchain applications they design comply with the requirements. Noncompliance can result in severe financial penalties, legal liabilities, reputational damage, and loss of user trust.

Consumer data privacy laws aim to protect individuals' personal information and provide them with certain rights and controls over their data. Architects need to consider privacy-enhancing mechanisms, such as data minimization, consent management, pseudonymization, and encryption, to ensure that personal data stored on the blockchain is adequately protected. By understanding the laws, architects can design privacy-centric features and controls into their applications.

Consumer data privacy laws grant individuals rights over their data, including the right to access, rectify, delete, and restrict the processing of their personal information. Architects need to ensure that blockchain applications provide mechanisms for users to exercise these rights and that the necessary transparency and accountability measures are in place.

Global Privacy Regulations: Blockchain architects and software developers implementing blockchain-based distributed applications in different countries/

jurisdictions should familiarize themselves with the following consumer data privacy protection laws that may apply to the use of blockchain technology:

1. **GDPR:** GDPR is a comprehensive data protection regulation in the EU that sets out the rights and obligations for handling personal data. It applies to organizations processing personal data of individuals residing in the EU, regardless of the organization's location.
2. **CCPA:** CCPA is a data privacy law in California, United States, that grants consumers certain rights over their personal information. It applies to businesses that collect or sell personal data of California residents and meet specific criteria.
3. **PIPEDA:** PIPEDA is a federal privacy law in Canada that regulates the collection, use, and disclosure of personal information by private sector organizations in the course of commercial activities.
4. **PDPA:** PDPA is a data protection law in Singapore that governs the collection, use, and disclosure of personal data by organizations. It applies to both private sector organizations and public agencies.
5. **Australian Privacy Act:** The Australian Privacy Act is a comprehensive privacy law in Australia that regulates the handling of personal information by Australian government agencies and businesses with an annual turnover above a certain threshold.
6. **Brazil General Data Protection Law (LGPD):** LGPD is a data protection law in Brazil that establishes rules for the collection, use, and processing of personal data. It applies to organizations operating in Brazil and processing personal data of Brazilian residents.

These laws impose obligations on organizations regarding data collection, consent, data subject rights, data security, breach notifications, and cross-border data transfers. Architects and developers should understand the key principles and requirements of these laws to ensure compliance when designing and implementing blockchain-based applications that involve the processing of personal data. It is recommended to consult legal experts with expertise in data privacy and blockchain to navigate the complexities of these regulations effectively.

KYC and AML verifications [56] are regulatory requirements requiring businesses to verify the identities of their customers and ensure that they are not involved in money laundering or other illegal activities. KYC/AML verification might be required in blockchain platforms that deal with cryptocurrencies or other financial assets.

In the context of Bitcoin, one of the reasons why AML and KYC measures are important is to address the potential risks of illicit activities associated with cryptocurrencies. Bitcoin and other cryptocurrencies have gained attention for their pseudonymous nature, which can make them attractive for money laundering, terrorist financing, and other financial crimes.

AML regulations aim to prevent and detect these illicit activities by imposing requirements on cryptocurrency exchanges and other service providers to implement robust measures for customer identification, transaction monitoring, and reporting suspicious activities. These regulations help ensure that individuals or entities engaging in cryptocurrency transactions are properly identified, and their activities are monitored to detect any suspicious patterns or potential money laundering activities. KYC procedures are an integral part of AML efforts and involve the verification of the identity of customers before they can engage in cryptocurrency transactions. This typically involves collecting and verifying personal information and documentation to establish the identity and legitimacy of the individuals or entities involved. KYC controls can be implemented as effective controls in the case of blockchain. While blockchain technology is known for its decentralized and pseudonymous nature, KYC controls provide a means to verify the identity of users and establish a level of trust and compliance with regulatory requirements.

Many cryptocurrency exchanges today require users to go through a KYC process before they can create an account and start trading. Users are typically required to provide identification documents such as passports or driver's licenses, proof of address, and sometimes even a selfie for facial verification. The exchange verifies this information to ensure the identity of the user and comply with regulatory requirements. In some cases, ICOs and STOs that are fundraising mechanisms conducted on blockchain platform offerings require participants to complete a KYC process before they can participate. This helps to establish the identity of the contributors and comply with regulatory obligations. Some DeFi platforms, which operate on blockchain, are starting to implement KYC controls. They may require users to complete a KYC process if they want to access certain features or services, such as borrowing and lending platforms. The KYC information is typically stored off-chain in a secure manner, with the user's identity being verified by a trusted third-party service.

The main innovation today that can help in the implementation of KYC controls entails blockchain-based identity solutions that aim to provide individuals with SSI and control over their personal information. These solutions often include a KYC component where users can securely store and share their verified identity information when required. Users have the ability to selectively disclose their information while maintaining control over their data. Blockchain-based identity solutions offer innovative approaches to implementing KYC controls while maintaining privacy and giving individuals control over their personal information. These solutions leverage the decentralized and tamper-resistant nature of blockchain technology to provide self-sovereign identities.

With blockchain-based identity solutions, individuals can securely store their verified identity information on the blockchain. This information is cryptographically protected and can only be accessed and shared with the user's explicit consent. Users have the ability to selectively disclose specific identity attributes or credentials when interacting with service providers or complying with KYC requirements. These solutions often employ the use of decentralized identifiers (DIDs) and verifiable credentials. DIDs are unique identifiers associated with individuals, and verifiable credentials are digitally signed attestations about specific identity attributes or qualifications. By utilizing these technologies, individuals can maintain control over their personal data while still meeting KYC obligations.

Blockchain-based identity solutions also address some of the challenges of traditional KYC processes, such as the need for repeated identity verification for different services. Once an individual's identity is verified and recorded on the blockchain, it can be reused across various blockchain-based applications and services, eliminating the need for redundant verification processes.

KYC processes play a crucial role in addressing cross-border regulations by ensuring compliance with local and international laws related to AML and countering the financing of terrorism. KYC processes provide a framework for verifying customer identities, conducting due diligence, and mitigating risks associated with cross-border transactions. By complying with KYC requirements, businesses can contribute to a safer and more secure global financial system while complying with cross-border regulations.

KYC can help in the context of cross-border regulations. For example, KYC procedures involve verifying the identity of customers and assessing their risk profiles. By gathering information about customers' identities, business relationships, and transactions, businesses can evaluate the level of risk associated with cross-border transactions. This helps identify and mitigate potential risks related to money laundering or terrorist financing. Cross-border transactions are subject to regulatory frameworks and reporting obligations in different jurisdictions. KYC procedures enable businesses to comply with these regulations by ensuring that customers' identities are verified, necessary documentation is provided, and due diligence is conducted. This helps in preventing illegal activities and adhering to regulatory requirements across borders. KYC processes also facilitate international collaboration and information sharing among financial institutions and regulatory authorities.

Through standardized KYC procedures and data sharing mechanisms, businesses can ensure the authenticity and integrity of customer information, even in cross-border transactions. This helps in detecting and preventing fraudulent activities across multiple jurisdictions. Adhering to robust KYC practices enhances the reputation and trustworthiness of businesses operating in cross-border transactions. In certain cases, cross-border transactions may require enhanced due diligence measures based on the jurisdiction, type of transaction, or the risk profile of the customer. KYC procedures enable businesses to apply additional scrutiny and collect more information when necessary. This helps in identifying and managing higher-risk transactions and customers involved in cross-border activities.

KYC processes also facilitate international collaboration and information sharing among financial institutions and regulatory authorities. Through standardized KYC procedures and data sharing mechanisms, businesses can ensure the authenticity and integrity of customer information, even in cross-border transactions. This helps in detecting and preventing fraudulent activities across multiple jurisdictions.

By implementing effective KYC controls, businesses demonstrate their commitment to combating financial crimes and maintaining a high level of integrity. This, in turn, builds confidence among customers, regulatory authorities, and international partners.

Moreover, KYC plays a vital role in AML efforts. AML refers to a set of laws, regulations, and procedures designed to prevent the illegal generation of income through activities such as money laundering and terrorist financing. KYC processes are an essential component of AML frameworks; for example, KYC frameworks require businesses to verify the identity of their customers. This involves collecting relevant information and documentation to establish the identity of individuals or entities involved in financial transactions. By ensuring the accuracy and authenticity of customer identities, KYC helps prevent criminals from using false identities to carry out illicit activities. KYC processes involve evaluating the risk profile of customers based on factors such as their identity, transaction patterns, and business relationships. This risk assessment enables businesses to identify and categorize customers based on their potential exposure to money laundering or terrorist financing risks. By assigning risk ratings, businesses can focus their resources and implement appropriate AML measures for high-risk customers.

Overall, KYC plays a crucial role in AML by ensuring the identification and verification of customers, assessing their risk profiles, conducting due diligence, monitoring transactions, and complying with regulatory obligations. AML regulations require businesses to establish and maintain robust KYC processes as part of their overall compliance efforts. By implementing and adhering to KYC procedures, businesses demonstrate their commitment to complying with AML regulations and contribute to the global fight against financial crimes. Failure to comply with KYC requirements can result in legal and reputational consequences for businesses. Compliance helps businesses detect and prevent money laundering and terrorist financing activities, therefore contributing to a safer and more transparent financial system.

1.13 Conclusions

In this chapter we provided a comprehensive introduction to DLT and blockchain technology. Understanding the intricacies of DLT, blockchain data structure, security, consensus algorithms, cryptocurrencies, privacy controls, identity management, and regulatory compliance is essential for comprehending and effectively utilizing blockchain technology.

In this chapter we discussed how DLTs form the foundation of blockchain technology encompassing a broader range of decentralized ledger systems.

Understanding DLT concepts, such as consensus mechanisms and network architectures, is crucial for grasping the fundamentals of blockchain. The blockchain is a specific type of DLT that uses a chain of blocks to store transactional data in a decentralized and immutable manner. There are different types of blockchain networks, including public and private, permissioned and permissionless, and hybrid networks. Applications that are based upon blockchain technology offer several benefits, including increased transparency, enhanced security, improved efficiency, and reduced costs for transactions.

A deep understanding of the blockchain data structure is necessary to appreciate how transactions are organized and secured. Knowledge of blocks, hashes, Merkle trees, and the linkage of blocks enables effective analysis and development of blockchain applications. The blockchain data structure is crucial for ensuring the integrity and security of transactions recorded on the blockchain. The block structure in blockchain consists of blocks that store transactional data with Merkle trees facilitating efficient data verification. The chain structure in blockchain refers to the linking of blocks using cryptographic hashes, creating an immutable and tamper-resistant record. In the blockchain data structure, cryptographic hash functions play a vital role in blockchain technology by providing data integrity and ensuring the uniqueness of each block. Digital signatures are used in blockchain to verify the authenticity and integrity of transactions, enhancing security.

One of the critical aspects of the blockchain that is dealt with in this chapter is the security of the blockchain, including an understanding of the inherent security features, security, threats, attacks, and mitigation strategies. Understanding of the security features of the blockchain requires familiarity with cryptographic principles, hash functions, and digital signatures. Understanding how these work helps in implementing robust security measures to safeguard against tampering and unauthorized access. One inherent security feature of the blockchain is to provide controls for ensuring transactions are validated before these are recorded in the immutable blockchain data ledger. The blockchain architecture consists of a distributed and decentralized network of computer nodes where validations for transactions rely on consensus algorithms to achieve agreement among network participants in validating and adding new blocks to the blockchain. In this chapter we provide a description of how different consensus mechanisms like PoW, PoS, and PBFT can be used to achieve consensus for validation of transactions.

In this chapter we also introduced the concept of cryptocurrencies as digital assets built on blockchain technology that can be used to enable P2P transactions and value for digital asset exchanges. Since cryptocurrencies are an integral part of blockchain ecosystems, it is important to have an understanding of the underlying concepts such as transaction validation, mining, and digital wallet management. From the perspective of securing digital assets such as cryptocurrencies while in storage, we specifically discussed how digital wallets provide a secure means of storing and managing cryptocurrencies, offering various features and levels of security. We also discussed how blockchain transactions with cryptocurrencies can occur in various contexts, including Bitcoin transactions, smart contract interactions, and token transfers.

Privacy is a critical consideration in blockchain applications. Privacy controls in blockchain involve trade-offs between public and private networks, as well as techniques like mixing, ring signatures, and ZKPs. Knowledge of such techniques enables architects to design systems that balance transparency with confidentiality, depending on the use case requirements.

Another critical control of the blockchain are identity controls: these aim to establish reliable and secure identity verification methods, preserving privacy and decentralizing identity management. Blockchain-based identity management systems offer self-sovereign and DID solutions. Understanding identity verification methods, privacy-preserving identities, and DID frameworks helps in designing secure and user-centric identity management systems.

Finally, we discussed regulatory compliance for the blockchain. Legal and regulatory considerations are significant in the blockchain space, impacting areas such as cryptocurrencies, consumer data privacy, and cross-border compliance. Compliance with regulations, including AML and KYC requirements, is crucial for blockchain applications. Since blockchain technology operates within legal and regulatory frameworks, comprehending the impact of regulations related to cryptocurrencies, consumer data privacy, and cross-border transactions allows architects to design blockchain solutions that meet compliance requirements and ensure the protection of user rights.

1.14 Future Directions and Trends in Blockchain Technology

In the past 10 years, blockchain technology has made significant strides toward mainstream adoption but is not yet universally adopted across industries. While blockchain has gained attention and interest from various sectors, its widespread adoption is still in progress. The Hype Cycle for Blockchain and Web3 provides a general framework for understanding the adoption and maturity of different aspects of blockchain technology. According to the information provided by Gartner [57], it suggests that blockchain platforms are estimated to take between two and five years to reach the plateau of productivity. This indicates that these platforms are currently in the early or middle stages of development and adoption, with more time needed for widespread implementation and mainstream use. On the other hand, digital wallets and consensus algorithms are projected to reach the plateau of productivity within less than two years. This suggests that these areas are experiencing faster progress and are closer to achieving widespread adoption and maturity.

While cryptocurrency trading has been one of the early and prominent use cases for blockchain technology, there is still a vast untapped potential for transformative applications in various industries. The true "killer use cases" of blockchain are yet to emerge, and they have the potential to significantly improve our lives. One example is the streamlined insurance claims process enabled by the combination of Web3, artificial intelligence (AI), and IoT. Blockchain technology's immutability, transparency, and smart contract capabilities can enable automated and efficient claims processing, reducing bureaucracy and delays. With AI and IoT, data can be collected and analyzed in real time, enabling accurate assessment of damages and automated payout processes. This has the potential to revolutionize the insurance industry and provide faster, more convenient settlements for individuals.

In addition to insurance, blockchain technology holds promise in areas such as supply chain management, healthcare data interoperability and privacy, DID management, DeFi, digital voting systems, intellectual property rights management, and many more. The combination of blockchain with other emerging technologies can unlock new possibilities and create innovative solutions that leapfrog traditional systems.

However, it's important to note that widespread adoption of these use cases will require not only technological advancements but also regulatory frameworks, industry collaboration, and user trust. Over time, as the technology matures, scalability improves, and regulatory clarity emerges, we can expect to see more real-world applications that leverage the full potential of blockchain and its integration with other transformative technologies.

So, while we may not have witnessed all the "killer use cases" of blockchain technology just yet, the convergence of Web3, AI, IoT, and other technologies holds great promise for revolutionizing various aspects of our lives, making processes more efficient, transparent, and secure. It's indeed just a matter of time before we see these advancements come to fruition.

Several industries have embraced blockchain technology and implemented it in various applications. For example, the financial services sector has seen significant adoption of blockchain for cross-border payments, smart contracts, and trade finance. Supply chain and logistics companies have explored blockchain for enhanced traceability and transparency. Governments have also shown interest in utilizing blockchain for digital identity management and recordkeeping.

However, challenges such as scalability, interoperability, regulatory frameworks, and user education still exist, which can hinder the widespread adoption of blockchain technology. Additionally, there is a need for further research and development to address these challenges and improve the usability and efficiency of blockchain solutions.

The future of blockchain technology is expected to witness several trends and directions that will shape its development and adoption. Efforts to improve interoperability between different blockchain platforms and enhance scalability are crucial for broader adoption. Solutions like sharding, sidechains, state channels, and Layer 2 protocols aim to address scalability challenges and enable more efficient processing of transactions. In the future, blockchain is likely to integrate with other emerging technologies, such as the IoT, AI, and DeFi. This convergence has the potential to unlock new use cases and create synergies between different technologies. In the finance industry, DeFi has gained significant attention and is expected to continue growing. DeFi platforms enable traditional financial services like lending, borrowing, and trading through smart contracts, eliminating intermediaries, and offering greater accessibility and transparency. Several central banks worldwide are exploring the concept of issuing their own digital currencies. Central Bank Digital Currencies (CBDCs) seek to combine the advantages of blockchain technology, such as transparency and efficiency, with the stability and regulatory oversight of traditional fiat currencies. By doing so, they aim to modernize payment systems, enhance financial inclusion, and enable real-time, secure transactions under the control of central banks.

As blockchain technology matures, there will be a growing need for regulatory frameworks and standardization to address legal, compliance, and security concerns. Clear guidelines and standards can provide clarity, promote trust, and facilitate broader adoption.

Future trends in blockchain security are essential to address the evolving challenges and ensure the integrity and trustworthiness of blockchain networks. Privacy is a crucial aspect of blockchain technology. Future developments will focus on enhancing privacy controls, such as ZKPs, homomorphic encryption, and secure MPC. These techniques enable transactions and data to remain confidential while still being verifiable and auditable. With the proliferation of different blockchain networks, ensuring secure communication and interoperability between them becomes vital. Future trends will include the development of cross-chain communication protocols, atomic swaps, and bridge technologies to enable secure and efficient transactions across multiple blockchains.

Future advancements in consensus algorithms will play a crucial role in the security of blockchain networks. Future trends will focus on refining existing consensus algorithms and developing new ones that are more secure against attacks and can accommodate diverse use cases and network requirements. Another area of potential advancement comprises the convergence of AI and machine learning (ML) techniques to strengthen blockchain security. AI-powered security solutions can detect and prevent attacks, identify patterns of fraudulent behavior, and enhance threat intelligence, enabling proactive measures to protect blockchain networks.

One promising path for improving blockchain security is through advancements in smart contract security and auditing: future trends will involve the implementation of improved auditing and testing mechanisms, formal verification techniques, and secure coding practices to minimize vulnerabilities and ensure the reliability and safety of smart contracts. As blockchain technology continues to evolve, regulatory frameworks and governance models will be established to address legal and compliance aspects. Future trends will include the integration of regulatory compliance features directly into blockchain protocols, enabling secure and compliant transactions while maintaining transparency and auditability.

2

Designing Secure Decentralized Applications

Good security is good engineering – built in from the start, not bolted on at the end.

—Mary Ann Davidson

2.1 Introduction

Decentralized blockchain applications have advanced significantly in security, thanks to contributions from highly regulated sectors such as finance, healthcare, and government, as well as from innovative blockchain start-ups. This chapter explores the critical aspects of designing secure blockchain applications by covering security architecture principles and best practices in security engineering. In industries with stringent regulatory requirements, blockchain solutions must be designed, implemented, and rigorously tested to ensure robust security controls that comply with technology standards and control policies.

By definition, decentralized applications, commonly known as dApps, are software applications that operate on decentralized networks, such as blockchain platforms. dApps leverage the decentralized blockchain to distribute application logic and data across a network of nodes that participate in the blockchain network. Like in the general case of blockchain-based applications, a dApp is interacting with the blockchain platform that integrates with each node of the blockchain platform to maintain a local copy of the distributed ledger ensuring each node has the transparency, immutability, and security of the recorded peer-to-peer (P2P) asset for value exchange transactions. Like in a traditional blockchain application, the consensus mechanism of the underlying blockchain network ensures that all nodes agree on the state of the application and validate transactions before these are recorded, enabling trust among the peers/nodes.

Blockchain Application Security: How to Design Secure and Attack Resilient Blockchain Applications, First Edition. Marco Morana, Harpreet Singh and Francesco Piccoli.
© 2026 John Wiley & Sons, Inc. Published 2026 by John Wiley & Sons, Inc.

Just like with any cloud, web, or mobile app, designing a distributed application starts with understanding the business need it's meant to solve. This entails considering both functional and nonfunctional requirements. A thorough analysis of the business case is essential to determine if a dApp is suitable, particularly when data sharing and transaction processing need to occur without a central control repository, as in the case of distributed computing or distributed ledger for transactions. Once the business case is solidified, it becomes possible to assess the associated risks and derive security requirements, including security control objectives.

After documenting the security requirements, the subsequent step is the architecture design. This chapter delves into the process of designing blockchain applications from a security standpoint, exploring various design patterns and data flow diagrams commonly employed in enterprise-based decentralized blockchain architectures. The discussion commences with the fundamentals of designing decentralized blockchain applications, encompassing both off-chain and on-chain application components. Off-chain components include crypto wallets, web, and mobile applications that access on-chain functionalities via Blockchain as a Service (BaaS) and Application Programming Interfaces (APIs).

dApps can integrate with blockchain platforms as clients of a BaaS. Most blockchain platforms offer prebuilt blockchain infrastructure and services that developers can leverage to build, deploy, and manage their dApps without needing to set up and maintain their own blockchain network. APIs allow dApps to interact with the underlying blockchain network by leveraging a standardized set of functions and protocols that the APIs provide, enabling developers to access and interact with the blockchain's features and data. Given that blockchain applications integrate with blockchain platform APIs, the security of these applications inherently relies on the security of the APIs themselves. It is imperative to ensure that the integration of APIs is secure, preventing malicious users or attackers from exploiting vulnerabilities and attack flaws.

dApps leverage blockchain APIs to access the blockchain platform type of functionalities such as for retrieving information from the blockchain, such as transaction details, account balances, smart contract data, or block information, using blockchain APIs. This allows dApps to display real-time data and provide users with up-to-date information about the blockchain state. dApps must ensure that data retrieved from the blockchain through APIs is done securely. This involves implementing appropriate authentication and access control mechanisms to prevent unauthorized access or tampering with sensitive information. Implementing robust data validation processes helps detect any tampering or manipulation attempts that could compromise the accuracy and trustworthiness of the displayed information.

The security of dApps inherently depends on how securely the dApps integrate with the blockchain platform and how well potential vulnerabilities being identified in blockchain components like smart contracts have been remediated. When the dApp interacts with the blockchain platform by means of middleware, the deployment of secure communication protocols like Transport Layer Security (TLS), version 1.3, helps protect against traffic spoofing attacks such man-in-the-middle attacks. The initial use case of the dApp is typically to send cryptographically signed transactions to the blockchain network for verification. These transactions could involve transferring digital assets (such as cryptocurrencies or tokens), interacting with smart contracts, or updating data on the blockchain. dApps must ensure that transactions sent through blockchain are processed securely. dApps should validate and sanitize user inputs before submitting transactions. Thorough input validation helps prevent common vulnerabilities such as injection attacks or unintended actions due to malformed or malicious inputs. By implementing proper validation mechanisms, dApps can enhance the integrity and reliability of transactions.

dApps can leverage cryptographic algorithms, such as digital signatures, to secure transaction signatures before submitting them to the blockchain platform. This ensures the authenticity and integrity of the transactions and protects against unauthorized modifications during transit. The blockchain platform has mechanisms to confirm and verify signed transactions and the recording of them after verification. This helps ensure that transactions are properly processed and recorded, reducing the risk of transactional discrepancies or failures. By following security by design principles, dApps might implement security controls as early as during the requirements phase such as robust authentication, input validation, transaction signatures to drive by design the overall security, and trustworthiness of the dApp's transactional functionalities.

One critical component of any dApp is the use of smart contracts for automating business functionalities, but not all dApps utilize smart contracts as it may just utilize other aspects of blockchain technology, such as decentralized storage, consensus mechanisms, or blockchain-based identity management, to achieve their objectives. The majority of dApps that leverage smart contracts as a core automation component often involve activities such as token transfers, use cases for decentralized finance (DeFi) protocols, decentralized exchanges (DEXs), and gaming where smart contracts govern the rules, ownership, and interactions between participants. In these dApps smart contracts play a crucial role in automating key business functionalities, such as consensus algorithms for transaction validation and identity verification of nodes for transaction signing and secure storage of private keys within digital wallets. Smart contracts used in dApps might also represent a primary target for malicious exploits due exploitation of smart contract vulnerabilities. In such a case, conducting thorough security audits and code reviews for smart contracts is essential to identify vulnerabilities, including potential risks such as reentrancy attacks, integer overflow/underflow, or logic flaws. Implementing secure coding practices, following standardized frameworks, and adopting formal verification techniques help enhance the security of smart contracts. This chapter addresses the prevention of design flaws in smart contracts and explores the design of countermeasures for detecting and preventing attacks that seek to exploit smart contract functionalities.

dApps facilitating token transfers or DeFi protocols require integration with a critical component: the digital/crypto wallets. For maintaining the security of the dApp, it is crucial to ensure secure wallet integrations, such as implementing hardware wallets, multi-signature (multi-sig) wallets, or secure key management systems. Digital wallets are critical for the security of the dApp since they are meant to protect users' private keys and prevent unauthorized access or theft of funds. Besides digital wallets, dApps may integrate with external systems or oracles to access off-chain data or interact with traditional financial systems. Implementing secure integration mechanisms, validating data inputs, and ensuring the authenticity and integrity of data retrieved from external sources mitigate risks associated with untrusted data or potential manipulation. Given the financial nature of these dApps, compliance with regulatory requirements and industry best practices is vital. Incorporating proper auditing mechanisms, transaction monitoring, and compliance controls helps detect suspicious activities, ensure regulatory compliance, and prevent money laundering or fraud. Finally establishing a robust security incident response plan helps in timely detection, containment, and mitigation of security breaches or vulnerabilities. Having processes in place to handle security incidents, communicate with affected users, and apply necessary fixes or patches enhances the resilience of the dApp.

In this chapter, we also delve into digital assets, such as tokens and nonfungible tokens (NFTs) used within dApps. Securing these digital assets is of paramount importance to prevent unauthorized access, theft, or fraudulent activities. We explore various security measures to safeguard tokens, such as implementing secure token standards (e.g. Ethereum Request for Comments [ERC]-20, ERC-721) and employing audited and battle-tested token contracts. Additionally, we discuss the importance of secure token wallets, multi-sig schemes, and secure token-based transactions. NFTs are unique digital assets that represent ownership or proof of authenticity for items such as digital art, collectibles, or virtual real estate. We examine the security considerations specific to NFTs, including secure storage and transfer mechanisms, ownership validation, and preventing counterfeit or unauthorized duplication of NFTs. Implementing secure marketplaces and using robust smart contracts to enforce ownership rights are also discussed.

Many dApps use native tokens, such as cryptocurrencies or utility tokens, to incentivize network participants, facilitate transactions, and enable specific functionalities within the application. As tokenization refers to the process of representing real-world assets or rights as digital tokens on the blockchain, in this chapter we will explore the security aspects of tokenization, including the initial token issuance, subsequent token transfers, and the overall life cycle management of tokens. Secure mechanisms for token creation, verification, and redemption are covered to ensure the integrity and security of tokenized assets. In this chapter, we also provide an overview of cryptocurrency transactions and their significant role in dApps. Cryptocurrency transactions are a fundamental aspect of many dApps, enabling the transfer of digital assets, execution of smart contracts, and participation in DeFi protocols. Securing these transactions is crucial to prevent unauthorized access, mitigate risks, and maintain the integrity of the dApp ecosystem.

In this chapter, we will also delve into the topic of securing digital identities and access controls within dApps. Digital identities play a crucial role in verifying and authenticating users, ensuring that only authorized individuals can access and interact with the dApp's functionalities and resources. We discuss the risks associated with compromised identities, such as identity theft or impersonation, and the potential impact on the overall security and reputation of the dApp.

We explore various authentication mechanisms that can be employed in dApps to verify the identity of users. Role-based Access Control (RBAC) is a widely adopted access control model that assigns roles to users and grants permissions based on those roles. We delve into the concept of RBAC in the context of dApps, discussing how it can be used to enforce granular access controls, manage user privileges, and ensure that users only have access to the resources and functionalities they require. We also explore best practices for implementing RBAC in dApps to prevent unauthorized actions and protect sensitive data.

In the context of dApps, Decentralized IDentity (DID) solutions offer enhanced privacy and control over personal data. We discuss the principles behind DID and explore frameworks like self-sovereign identity (SSI) or DID protocols that enable users to manage and control their digital identities without relying on centralized authorities. We explore the security benefits and considerations when integrating DID solutions into dApps. Understanding the importance of digital identity security, exploring authentication mechanisms, leveraging RBAC, considering DID solutions, and implementing access control auditing and monitoring practices contribute to the overall security and trustworthiness of dApps.

Finally, in this chapter, we will dive deep into the topic of secure coding and defensive programming of smart contracts within dApps. Smart contracts are self-executing agreements with the terms of the agreement directly written into code. As smart contracts govern the rules, ownership, and interactions between participants in various dApps, ensuring their security is of utmost importance. Since any vulnerabilities or bugs in the code can have severe consequences. We discuss the potential risks associated with insecure smart contracts, such as fund loss or unauthorized access, and the need for proactive security measures during the development process. In this chapter, we will explore the unique security considerations specific to smart contracts. This includes understanding the underlying blockchain platform's limitations and features, avoiding common vulnerabilities (e.g. reentrancy attacks, integer overflows, or access control issues), and following best practices for secure coding. We discuss techniques for input validation, secure state management, error handling, and secure contract upgradability.

Defensive programming involves anticipating and mitigating potential issues through proactive coding practices. We delve into defensive programming techniques applicable to smart contracts, such as input sanitization, boundary checks, fail-safe mechanisms, and proper exception handling. By adopting defensive programming principles, developers can reduce the attack surface and increase the resilience of smart contracts against potential threats.

Formal verification and auditing play a vital role in ensuring the correctness and security of smart contracts. We explore techniques such as formal verification tools, static code analysis, and manual code reviews to identify vulnerabilities and potential security flaws. We discuss the importance of engaging external auditors or security experts to conduct thorough audits of smart contracts before deployment. We delve into the security considerations during the deployment and upgradability of smart contracts. This includes ensuring the secure transmission of contract bytecode, verifying contract authenticity, and implementing secure contract migration or upgradability mechanisms. We discuss the challenges associated with contract upgradability and explore approaches like proxy contracts or upgradeable smart contract patterns. We discuss various testing techniques, such as unit testing, integration testing, and stress testing, to ensure the reliability and security of smart contracts. Additionally, we explore the use of testnets (test networks) for deploying and testing smart contracts in a controlled environment before their deployment in the production environment.

Reading this chapter on designing secure dApps would be beneficial for software developers and architects for the following reasons:

1. **Understanding the Security Requirements:** The chapter emphasizes the importance of analyzing the business case and deriving security requirements for dApps. It helps developers and architects gain insights into identifying potential risks and designing robust security controls.
2. **Designing Secure Architectures:** The chapter explores various design patterns and data flow diagrams used in decentralized blockchain architectures. It covers both off-chain and on-chain components, including the integration of dApps with blockchain platforms. This knowledge enables developers and architects to design secure and reliable architectures.
3. **Securing Blockchain–dApp Integration:** As dApps rely on blockchain APIs to retrieve and interact with blockchain data, the chapter emphasizes the need for secure integration. It provides guidance on implementing secure

coding practices, encryption, and secure communication protocols to prevent attacks and vulnerabilities associated with API usage.

4. **Ensuring Transaction Security:** dApps involve sending transactions through blockchain interfaces (e.g. middleware) and the chapter discusses the importance of secure transaction processing. It covers topics such as input validation, transaction signatures, and confirmation mechanisms, helping developers and architects enhance transaction integrity and reliability.
5. **Smart Contract Security:** For dApps utilizing smart contracts, the chapter highlights the vulnerabilities and risks associated with smart contract functionalities. It discusses security audits, secure coding practices, and formal verification techniques to identify and prevent attacks on smart contracts.
6. **Secure Digital Asset Management:** The chapter addresses the security measures required to protect tokens and NFTs within dApps. It covers topics such as secure token standards, wallet integrations, and secure storage mechanisms, providing guidance on safeguarding digital assets.
7. **Digital Identity and Access Control:** Understanding the importance of secure digital identities and access controls within dApps is crucial. The chapter explores authentication mechanisms, RBAC, and DID solutions, helping developers and architects implement robust security measures for user verification and authorization.
8. **Secure Coding and Defensive Programming:** As smart contracts are integral to many dApps, the chapter focuses on secure coding and defensive programming practices. It discusses common vulnerabilities, secure contract upgradability, formal verification, auditing, and testing techniques, enabling developers and architects to write secure smart contracts.

By reading this chapter, software developers and architects can gain a comprehensive understanding of designing secure dApps, mitigating risks, implementing best practices, and ensuring the integrity, confidentiality, and reliability of dApps.

2.2 Decentralized Applications

dApps are software applications that operate on decentralized networks, such as blockchain platforms. Unlike traditional applications that rely on a central authority or server, dApps leverage the decentralized nature of blockchain technology to eliminate the need for a single point of control or failure.

Unlike traditional applications, dApps distribute their logic and data across a network of nodes within the blockchain. This decentralized architecture ensures transparency, immutability, and security. The consensus mechanism employed by the underlying blockchain network guarantees agreement among nodes and validates transactions.

dApps revolutionize the way software applications are provisioned to users. Unlike centralized apps, which are owned and controlled by a single entity and rely on servers for operation, dApps operate on blockchain or P2P networks. In a centralized app, users download the app and interact with it through a central server owned by the entity. Data exchange occurs between the user's device and the company's server. On the other hand, dApps eliminate the need for a central authority. They leverage blockchain or P2P networks, enabling direct transactions between users. Instead of relying on a single entity, users engage with each other in a P2P manner.

To access and utilize a dApp, users often pay the developer with cryptocurrency. This payment grants users the ability to download and utilize the program's source code, which is commonly implemented as a smart contract. Smart contracts enable secure and private transactions without the need to disclose personal information. By decentralizing control and fostering direct interactions, dApps empower users and promote trust, transparency, and security in software applications.

The development of dApps gives software developers and architects the opportunity of harnessing the power of blockchain technology to create innovative and disruptive solutions. Unlike traditional applications, dApps operate on a decentralized network, distributing their functionality across multiple nodes instead of relying on a single computer or server. By utilizing blockchain, businesses can ensure that their dApps are open source, transparent, and resistant to censorship. This means that users can directly interact with the application without the need for intermediaries, fostering a P2P environment.

One of the key advantages of dApps is their potential to disrupt traditional industries. By enabling direct P2P interactions and transactions, dApps eliminate the need for a central authority. This has far-reaching implications, as it empowers users to engage in secure and trustless interactions, bypassing traditional intermediaries and their associated costs. Moreover, dApps operate on a foundation of cryptocurrency, which serves as a means of exchange within the application. This introduces new possibilities for financial inclusion and novel business models, as transactions can occur directly between users, facilitated by the blockchain. dApps have a wide range of use cases, including finance, supply chain management, gaming, DEXs, and social networks. By leveraging the benefits of decentralization, dApps aim to provide increased security, transparency, and user control compared to traditional centralized applications.

dApps are well-suited for various use cases that can benefit from the unique characteristics and capabilities of blockchain and decentralized technologies. Here are some common use cases where dApps have demonstrated value:

1. **Financial Services and Payments:** dApps can facilitate P2P transactions, remittances, cross-border payments, decentralized lending, DEXs, and decentralized asset management. They provide transparent, efficient, and secure financial services without the need for intermediaries.
2. **Supply Chain and Logistics:** dApps can improve transparency, traceability, and trust in supply chain management by recording and verifying transactions, product provenance, and inventory management across multiple stakeholders. This enables real-time tracking, reduces fraud, enhances product authenticity, and streamlines logistics operations.
3. **Identity Management:** dApps can offer SSI solutions, enabling individuals to control and manage their digital identities securely. This empowers users to share selective personal information while protecting privacy and reducing reliance on centralized identity providers.
4. **Voting and Governance:** dApps can provide transparent and tamperproof voting systems for elections, decision-making processes, and governance of organizations or communities. They offer enhanced security, immutability, and verifiability of votes, ensuring trust in the democratic processes.
5. **Intellectual Property (IP) Rights:** dApps can facilitate the registration, management, and protection of IP rights, such as copyrights, patents, and trademarks. They provide an immutable record of ownership, enabling efficient licensing, verification, and enforcement of IP rights.
6. **Gaming and Virtual Assets:** dApps can enable ownership, trading, and interoperability of virtual assets in gaming and virtual worlds. They introduce provable scarcity, verifiable ownership, and decentralized marketplaces, empowering gamers and creators with true ownership and value exchange.
7. **Decentralized Content Sharing and Publishing:** dApps can facilitate decentralized content sharing platforms, allowing creators to publish, monetize, and protect their digital content. They enable direct interaction between creators and consumers, eliminating intermediaries and fostering fairer revenue distribution models.
8. **Decentralized Computing and Storage:** dApps can harness the power of decentralized computing resources, such as blockchain-based smart contracts or P2P networks, to provide secure and reliable computation, storage, or dApps hosting.

It's important to note that while these use cases demonstrate the potential of dApps, the suitability of a dApp for a particular use case depends on various factors, including the specific requirements, ecosystem dynamics, scalability needs, regulatory considerations, and user adoption. Each use case should be evaluated individually to determine the feasibility and benefits of a decentralized solution.

As a software developer, starting your dApp journey with popular blockchain platforms can be a great way to enter the world of dApp development. Here are some steps to get you started:

1. **Learn About Blockchain Technology:** Build a solid understanding of blockchain technology fundamentals, including its core concepts, decentralized architecture, consensus mechanisms, and smart contracts. This knowledge is essential for developing dApps. For a comprehensive introduction to blockchain data structures and components, it's recommended to first read Chapter 1.
2. **Choose a Blockchain Platform:** Select a popular blockchain platform that aligns with your project requirements and development skills. Ethereum is a widely adopted platform with extensive developer resources, while other platforms offer unique features and ecosystems.
3. **Study the Platform's Documentation:** Dive into the platform's documentation and developer resources to understand its core concepts, programming languages, and tools. This will help you familiarize yourself with the platform's specific development environment and programming frameworks.
4. **Learn the Smart Contract Programming Languages:** Depending on the chosen platform, learn the programming languages used for developing smart contracts and dApps. Solidity is the most commonly used language for Ethereum, while other platforms may have their own languages or offer compatibility with existing languages.
5. **Develop a Simple dApp:** Start by developing a simple dApp to grasp the development process and familiarize yourself with the platform's tools and frameworks. Focus on building the core functionalities of your dApp using smart contracts and test it on a local development environment.
6. **Utilize Developer Tools and Libraries:** Leverage the available developer tools, libraries, and frameworks provided by the chosen platform. These tools can simplify development tasks, enhance security, and improve efficiency.
7. **Engage with the Web3 and dApp Developer Community:** Join developer communities, forums, and social media groups related to the chosen platform. Engage in discussions, ask questions, and share your progress and challenges. The blockchain community is known for its collaborative nature, and you can learn a lot from experienced developers in the field.
8. **Test and Deploy Your dApp:** Thoroughly test your dApp for functionality, security, and performance. Once you're satisfied with the results, deploy your dApp to the blockchain network of your chosen platform. This will make it accessible to users and allow them to interact with your dApp.

Remember that developing dApps requires not only programming skills but also a deep understanding of the decentralized ecosystem and its unique challenges including security challenges. Learning about security by design of dApps and reading the chapter is important for software developers and architects because it equips them with the knowledge and skills to address the unique security challenges of dApps, protect user privacy and assets, mitigate financial risks, comply with regulations, and gain a competitive advantage in the market.

Specifically, there are many reasons to read this chapter and learn about designing secure dApps:

1. **Establish Confidence and Trust:** The adoption of dApps is growing rapidly, and users are increasingly concerned about the security of their data and assets. By understanding security by design principles, you can build dApps that inspire trust and confidence among users, leading to greater adoption and success. Users are more likely to choose dApps that prioritize security and privacy, and they are more likely to recommend such applications to others. By reading the chapter, you can gain valuable insights and knowledge to enhance the security of your dApps, strengthening your reputation and attracting more users.
2. **Address Security Challenges:** dApps operate on a blockchain or P2P network, which introduces unique security challenges. These include smart contract vulnerabilities, consensus mechanism attacks, private key management, and the potential for malicious actors. By reading the chapter, you can gain insights into these challenges and learn best practices to mitigate risks.
3. **Protect Users Privacy and Their Digital Assets:** Users expect dApps to provide a high level of privacy and security for their personal information and digital assets. By incorporating security by design principles, you can ensure that user data is protected, transactions are secure, and privacy is maintained. This builds trust among users and enhances the reputation of your dApp.
4. **Manage Financial Risks:** Many dApps involve the use of cryptocurrencies or digital assets. Security vulnerabilities can lead to financial losses for users and damage the reputation of the dApp. By understanding security by design principles, you can identify and address potential vulnerabilities early in the development process, reducing the likelihood of financial risks.
5. **Comply with Regulations:** Depending on the nature of your dApp and the jurisdictions it operates in, there may be legal and regulatory requirements to comply with. Security by design principles can help you incorporate necessary security measures and ensure compliance with applicable regulations, safeguarding your dApp from legal issues

This chapter aims to guide you in designing attack-resilient dApps by providing practical insights, security principles, and best practices. It offers tools and techniques that developers can implement to enhance security, mitigate risks, and protect user privacy and assets. By following the concepts outlined here, developers will gain the knowledge and skills needed to design and build robust, secure, and resilient dApps.

2.2.1 dApp Architectures

By definition, a dApp is an application that operates on a decentralized network, such as blockchain to enable secure, tamperproof transactions and interactions among participants without the need for a trusted third party. In the traditional architecture of the World Wide Web, a client-server network model is commonly used. In this model, the server is responsible for storing and managing all the required information, while clients (such as web browsers) send requests to the server to access that information.

The centralized nature of the server allows for easy updates and maintenance since all the required data is stored in one place. Administrators or authorized personnel have control over the server and its database, granting them the ability to manage permissions and make updates to the information as needed. This architecture has been widely used because it offers centralized control and efficient management of data. However, it also has limitations, such as being vulnerable to single points of failure, data breaches, and censorship. It relies on trust in the server and administrators, and users have limited control over their data.

dApps, on the other hand, aim to address these limitations by utilizing decentralized networks, such as blockchains. dApps distribute data and operations across a network of nodes, removing the need for a central server. This architecture provides benefits like transparency, immutability, and resistance to censorship, as no single entity has control over the entire system.

We can introduce dApps architecture by comparing the architecture of a traditional Web 2.0 application to a Web 3.0 dApp in simple terms. Imagine a website like an online store or a social media platform. It consists of a frontend, a backend, and a database.

A simplified traditional Web 2.0 architecture consists of three main deployable components:

1. **Frontend:** This is the user interface (UI) of the application that users interact with. It is typically built using HTML, CSS, and JavaScript and runs in a web browser.
2. **Backend:** The backend is responsible for processing user requests, handling business logic, and interacting with the database. It includes application servers and other server-side technologies that run on a centralized server.
3. **Database:** The database stores and manages the application's data. It can be a relational database or other forms of persistent storage.

Now, let's compare this to a Web 3.0 dApp. A Web 3.0 dApp at high level can be described as consisting in two deployable components but with some key differences:

1. **Frontend:** Similar to a traditional Web 2.0 application, the frontend is the UI. However, in a dApp, the frontend can be built using web technologies like HTML, CSS, and JavaScript, but it can also interact directly with a blockchain network.
2. **Backend:** The backend of a dApp is powered by the blockchain platform, which interacts either directly with the dApp frontend or indirectly through a PFC-JSON middleware. The blockchain backend comprises two key components:
   a. **Enterprise Virtual Machine (EVM):** In a Web 3.0 dApp, backend logic is typically implemented using smart contracts running on a blockchain platform. Smart contracts are self-executing agreements with predefined rules and conditions stored on the blockchain. They enable decentralized and trustless execution of business logic, reducing the need for intermediaries.
   b. **Blockchain Network:** Unlike traditional backends that rely on a centralized server and database, a dApp leverages a blockchain network for data storage, transaction validation, and decentralization. The blockchain acts as a distributed ledger, recording and verifying every interaction within the dApp to ensure transparency, security, and immutability.

The fundamental difference of a dApp lies in its decentralization approach, which disrupts the traditional architecture of web applications. In the traditional model, the web server, acting as the first tier, handles the initial processing and serves web pages. The application server, representing the second tier, houses the business logic of the application. Lastly, the database, the third tier, stores and manages the application's data.

In contrast, dApps leverage blockchain technology as a platform for decentralization. This replaces the need for a centralized web server, application server, and database. Instead, the blockchain serves as a distributed database, effectively eliminating the centralized architecture. The blockchain network collectively maintains and validates the application's data and logic, ensuring transparency, integrity, and security.

In the architecture of Web 2.0 applications, the business logic of the application typically resides in the application server, which is the second tier of the architecture. However, in the Web 3.0 dApp architecture, this traditional approach is revolutionized. Instead of relying on an application server, the business logic is now implemented through smart contracts. Smart contracts play a crucial role in Web 3.0 dApps by replacing the application server as the entity responsible for executing and enforcing the application's business rules. These smart contracts are self-executing agreements written in code and deployed on the blockchain. They define the conditions, actions, and outcomes of interactions within the dApp.

By utilizing smart contracts, the dApp architecture achieves a decentralized and trustless execution of the application's business logic. The smart contracts are distributed across the blockchain network and executed by the participating nodes, ensuring transparency and integrity in the application's operations. Additionally, the immutability and security properties of the blockchain provide further assurance that the business logic is executed as intended without the need for a central authority.

In Web 2.0 applications primarily rely on centralized servers and databases, where the authority and control over data and transactions are centralized within the organization or service provider. Consensus algorithms, on the other hand, are a fundamental component of decentralized blockchain networks, which are the underlying technology of Web 3.0 and dApps. Consensus algorithms play a crucial role in achieving agreement and trust among the distributed nodes in the network, ensuring the integrity and security of the blockchain. Consensus algorithms are mechanisms that enable the distributed nodes in the network to reach a consensus or agreement on the validity of transactions and the state of the blockchain. In the context of dApps, consensus algorithms ensure that all participating nodes in the network follow the same set of rules and agree on the execution and outcome of smart contracts. By achieving consensus, the network can verify the validity of transactions, maintain the integrity of the distributed ledger, and ensure that the smart contracts' results are agreed upon by the majority of the network.

Web 3.0 empowers developers to leverage blockchain technology in building dApps that operate on a distributed state machine. This state machine is maintained by anonymous nodes spread across the Internet. By referring to a "state machine," I am referring to a mechanism that upholds a particular program's current state and governs the future states it can assume. Blockchains serve as state machines instantiated with an initial genesis state and governed by strict rules, often referred to as consensus mechanisms, which dictate the valid state transitions. What truly sets Web 3.0 apart is the absence of a single controlling entity for this decentralized state machine. Instead, the responsibility for maintaining and securing the state machine is shared collectively among all participants in the network. This distributed governance model ensures that no single entity has undue control or influence over the system.

The architecture diagram presented offers a comprehensive depiction of the various components and their interactions within a dApp, as seen from the perspective of a dApp architect. By incorporating the information from the architecture diagram, we can gain valuable insights into the design and functionality of the dApp.

Figure 2.1 showcases a typical dApp architecture. At the center of the dApp, we have the "frontend" component, which represents the UI through which the users interact with the application. It serves as the gateway for user input and displays the output generated by the application.

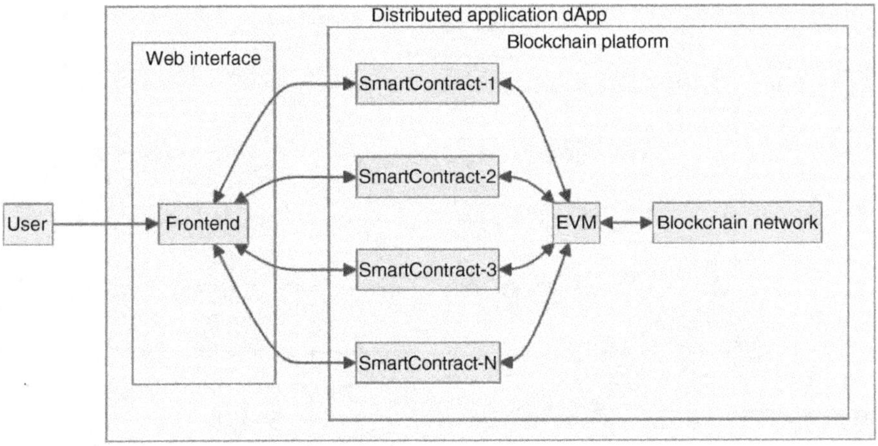

Figure 2.1 dApp high-level architecture.

Surrounding the frontend is the "web interface" subgraph, which encompasses the user-facing components responsible for rendering the dApp in a web browser or other UI environments. This subgraph ensures a seamless and user-friendly experience for interacting with the application. The core of the dApp lies within the "blockchain platform" subgraph. This subgraph comprises various components that enable the decentralized nature of the application. The interaction between the frontend and the smart contracts (represented by SmartContract-1, SmartContract-2, SmartContract-3, and SmartContract-N) is depicted by bidirectional arrows, indicating that the frontend communicates with these smart contracts for specific functionalities with inputs and output data.

The smart contracts are deployed on the EVM, denoted by the bidirectional arrows connecting the smart contracts to the EVM. The EVM provides the execution environment for the smart contracts, allowing them to perform their defined operations and enforce business logic. Finally, the EVM communicates with the underlying "blockchain network" through bidirectional arrows. This interaction ensures the execution of transactions, consensus mechanisms, and the overall integrity of the dApp.

From a software developer perspective, developers can now write smart contracts that define the application logic and deploy them onto the decentralized state machine. This means that every individual who wishes to build a blockchain application can deploy their code onto this shared state machine, ensuring transparency, immutability, and a level playing field for all participants. The shift from centralized control to decentralized governance and the ability to deploy smart contracts onto this shared infrastructure represent the core differentiators of Web 3.0 for architects and software architects in the realm of dApps.

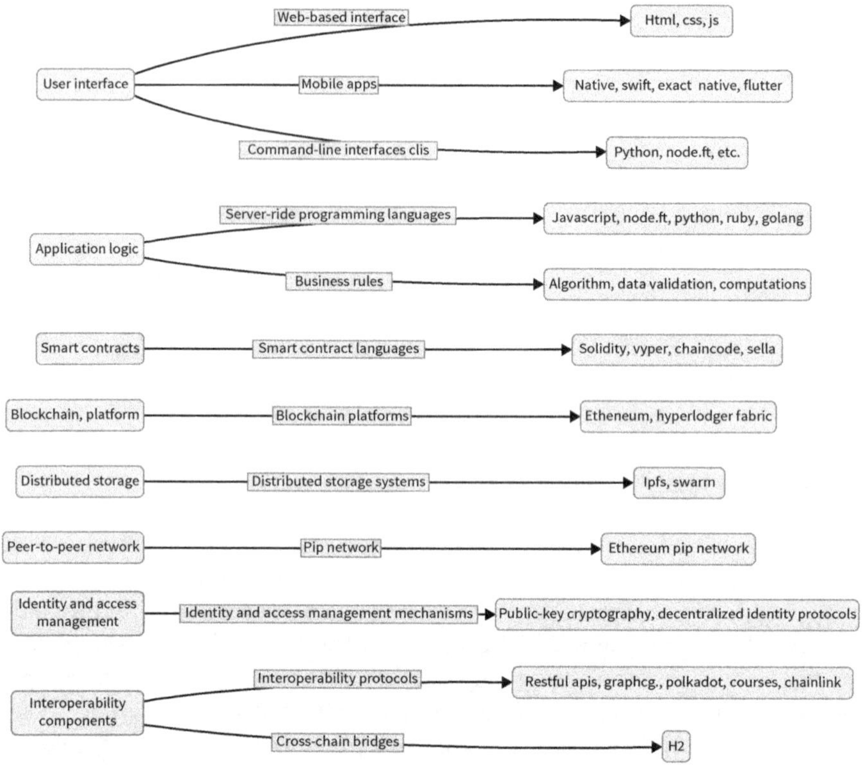

Figure 2.2 Technology stack for developing dApps.

Figure 2.2 provides an overview of the components involved in a dApp architecture. At the center of the architecture, we have the "user interface" component responsible for interacting with users. It encompasses various interfaces, including a web-based interface using HTML, CSS, and JavaScript; mobile apps developed using native technologies or frameworks like Swift, React Native, or Flutter; and command-line interfaces (CLIs) built with programming languages such as Python or Node.js. These interfaces enable users to interact with the dApp through different platforms and devices. The "application logic" component encompasses the core functionality and processing of the dApp. It relies on server-side programming languages like JavaScript, Node.js, Python, Ruby, or Golang to handle server-side operations. Additionally, it incorporates business rules such as algorithms, data validation, and computations to ensure the proper functioning of the application.

The "smart contracts" component represents the autonomous and self-executing contracts deployed on the blockchain. Smart contract languages like Solidity,

Vyper, Chaincode, or Scilla are used to write and develop these contracts, which enforce predefined rules and execute specific functions.

The "blockchain platform" component serves as the foundation for the dApp's decentralized architecture. It includes blockchain platforms like Ethereum or Hyperledger Fabric, providing the infrastructure for storing transactional data, executing smart contracts, and maintaining the consensus mechanism.

The "distributed storage" component refers to systems such as IPFS (InterPlanetary File System) or Swarm, which enable decentralized storage of data, files, or other media associated with the dApp.

The "peer-to-peer network" component represents the decentralized network used for communication and for transaction propagation. In this case, the Ethereum P2P network is specifically mentioned, as it plays a crucial role in facilitating the distributed nature of the dApp.

The "identity and access management" component encompasses mechanisms such as public-key cryptography and DID protocols, ensuring secure and authenticated access to the dApp for users.

The "interoperability components" component focuses on enabling interoperability between different blockchain networks and systems. It includes interoperability protocols like RESTful APIs, GraphQL, Polkadot, Cosmos, or Chainlink, which facilitate data exchange and communication between different blockchain ecosystems. Additionally, cross-chain bridges can be utilized to establish connectivity and facilitate the transfer of assets or data across disparate blockchains.

From a software developer's perspective, dApps can be viewed at a high level as comprising several key components. Each component is built using a variety of software frameworks, programming languages, and third-party libraries and serves a distinct function (see the following note), such as:

1. **UI:** The UI is the layer that interacts with users. It can be a web-based interface, a mobile app, or a CLI. Users interact with the UI to perform actions and view information. Examples of implementations of UI layer in a dApp architecture consist of web-based interface: HTML, CSS, and JavaScript frameworks like React, Angular, or Vue.js. Mobile apps are built using native languages like Swift (iOS) or Java/Kotlin (Android) or cross-platform frameworks like React Native or Flutter. CLIs can be implemented using Python, Node.js, or other scripting languages.
2. **Application Logic:** The application logic layer contains the business rules, algorithms, and processing logic of the dApp. It handles user requests and data validation and performs necessary computations. This layer interfaces with the blockchain layer and smart contracts for executing transactions and retrieving data. Examples of implementations of the application layer in a dApp architecture include server-side programming languages like JavaScript

(Node.js), Python (Django/Flask), Ruby (Ruby on Rails), or Golang. Business rules can be written using custom algorithms, data validation logic, and application-specific computations.

3. **Smart Contracts:** Smart contracts are self-executing contracts or scripts that reside on the blockchain. They define the rules and logic of the dApp, enforce the terms of transactions, and hold the state of the application. Smart contracts are written in specific programming languages such as Solidity for Ethereum. Examples of smart contracts in a dApp architecture include technologies such as Solidity that is the most commonly used language for Ethereum smart contracts. Other blockchain platforms use various languages like Vyper (Ethereum), Chaincode (Hyperledger Fabric), or Scilla (Zilliqa).
4. **Blockchain Platform:** The blockchain layer is the underlying distributed ledger technology (DLT) that the dApp operates on. It consists of a decentralized network of nodes that maintain a copy of the blockchain and participate in the consensus mechanism. The blockchain stores all transactions and data in a tamper-resistant and transparent manner. Examples of implementations of blockchain platforms upon which it is possible to build dApps are the Ethereum blockchain with its public or private network.
5. **Distributed Storage:** dApps may utilize distributed storage systems to store large files, documents, or data that is not suitable for storing directly on the blockchain. Examples of such systems include IPFS or Swarm. Distributed storage ensures data availability, redundancy, and fault tolerance.
6. **P2P Network:** The P2P network allows communication and coordination among the nodes participating in the dApp. Nodes connect directly to each other, enabling the decentralized nature of the application. The network facilitates data propagation, transaction validation, and consensus mechanisms. An example of a P2P network is Ethereum's P2P network, which facilitates communication between nodes running Ethereum software.
7. **Identity and Access Management (I&AM):** To authenticate users and manage their identities, dApp may employ various mechanisms such as public-key cryptography, user-controlled private keys, or DID protocols. These mechanisms ensure secure user access and protect user privacy. I&AM can be implemented by leveraging Public Key Infrastructure (PKI) public-key cryptography and cryptographic protocols like Rivest–Shamir–Adleman (RSA), Elliptic Curve Cryptography (ECC), or Edwards-curve Digital Signature Algorithm (EdDSA) for user identity and secure access. Other examples of DID protocols that can be leveraged by dApps include Sovrin, uPort, or SelfKey, which provide DID management.
8. **Interoperability Components:** dApps can interact with other dApps, blockchain networks, or external systems. Interoperability is facilitated through standardized protocols, cross-chain bridges, or APIs that allow data and value

transfer across different platforms. Examples of APIs and standards that can be leveraged by dApps are RESTful APIs, GraphQL, or specific blockchain interoperability protocols like Polkadot, Cosmos, or Chainlink. Cross-chain bridges are examples of protocols that facilitate communication and value transfer between different blockchain networks.

Note: The choice of specific technologies for implementation is influenced by the project's requirements, developer preferences, and the compatibility with the selected blockchain platform. This means that different dApps may use varying programming languages, frameworks, and libraries depending on factors such as scalability needs, security considerations, and integration capabilities with the underlying blockchain infrastructure.

2.2.2 Comparison of dApps with Traditional Centralized Applications

A dApp is an application that operates on a decentralized network, typically a blockchain, where no single entity or central authority has control over the application. dApps are designed to be open source, transparent, and trustless, meaning that they do not rely on a central server or intermediary for operation and data storage. They leverage blockchain technology to enable secure, tamperproof transactions and interactions among participants without the need for a trusted third party.

In a centralized control app, the application's operations, data storage, and decision-making are controlled by a central authority or server: the central authority has the power to modify, access, or restrict the application's functionalities and user data. Users rely on the central server control that is where the application logic resides to validate transactions, manage identities, and enforce rules. Examples of this architecture are traditional web applications where a central server controls all aspects of the application. Centralized applications rely on client-server architectures where a server application typically relies on a centralized server or a cluster of servers to handle data storage, process user requests, and execute the application's logic. The server manages the data, executes operations, and interacts with clients or users. Users connect to the server to access the application's features and functionalities.

In a dApp, the application's logic, data storage, and operations are distributed across a network of nodes. The blockchain serves as a decentralized database that stores the application's data and executes the application's logic through smart contracts. Users interact directly with the blockchain and participate in the consensus mechanism, removing the need for a central server. When users interact with a dApp, they do so directly with the blockchain. They can submit transactions, read data, and invoke the functionality provided by the smart contracts. This direct interaction removes the need for a central server or intermediary, as users can interact with the application and its data in a P2P manner. Additionally, in a decentralized network, consensus mechanisms are employed to ensure agreement among the participating nodes on the validity of transactions and the state of the blockchain. Consensus mechanisms, such as Proof of Work (PoW) or Proof of Stake (PoS), allow nodes to collectively validate transactions and achieve a shared and agreed-upon state of the blockchain.

By distributing the application's logic, data storage, and operations across a network of nodes and leveraging the blockchain as a decentralized database, dApps offer benefits such as increased transparency, immutability, and resilience. They eliminate the reliance on a central server, enhance security, and enable P2P interactions among users without the need for trust in a single authority. However, they also come with challenges such as scalability limitations and potentially slower transaction speeds compared to centralized applications.

dApps offer several security advantages compared to traditional client-server architectures. One of these advantages is application resilience: where in a client-server architecture, the central server acts as a single point of failure, that is, if the server is compromised or experiences downtime, it can disrupt the entire system, compromising data availability and functionality. In dApps even if some nodes go offline or are compromised, other nodes can continue to maintain the application's functionality and data availability.

A central point of failure in a centralized application can also be a server that is targeted for data compromise of denial of service (DoS). In a client-server model, user data is stored on the server. This centralization poses a higher risk of data breaches or unauthorized access if the server's security measures are inadequate. In dApps, the data is cryptographically secure and distributed across multiple nodes, making it more difficult for attackers to compromise or alter the data; in addition, the blockchain ensures data integrity, as records cannot be tampered with once they are added to the blockchain.

Another advantage that dApps have by comparison with centralized-based applications is enabling trust and transparency: where in a client-server architecture, users must trust the central server and its administrators to manage and protect their data, dApps trust is distributed across the network rather than relying on a central authority. The transparency of the blockchain allows users to independently verify transactions and data integrity without relying on blind trust. dApps provide increased transparency in terms of data handling and execution of smart contracts, as all actions are recorded on the blockchain and can be audited.

However, while dApps offer significant security advantages, it's important to note that they are not immune to all security risks. dApps face their own set of challenges, such as vulnerabilities in smart contracts, potential attacks on individual nodes, or security issues in specific blockchain implementations. The objective of this book is to delve into the realm of secure dApps and outline approaches to ensure their security from the very foundation. By focusing on secure design, meticulous implementation, and comprehensive testing, this book aims to identify and address design flaws through threat modeling and architecture reviews in the early stages. Additionally, it emphasizes the importance of simulating and testing vulnerabilities in smart contracts and potential attacks on individual nodes, enabling the implementation of effective countermeasures to prevent and detect such attacks. Through these measures, the book aims to equip developers with the knowledge and tools necessary to build dApps that prioritize security and resilience.

2.2.3 Analysis of Use Cases for Blockchain and dApps

dApp development is regarded as a groundbreaking field that holds immense potential to reshape the way society engages in trade and interactions. This potential stems from the underlying blockchain technology, which empowers dApps to facilitate secure and trustless exchanges between mutually distrusting entities. By leveraging the principles of decentralization, dApps enable users to transact and interact directly, without the need for intermediaries or trusted third parties. This fundamental characteristic of dApps, made possible by blockchain technology, has garnered significant praise as it opens up new possibilities for transparent, efficient, and inclusive systems of value exchange and collaboration.

As software developers and architects consider the adoption of dApps for their use cases, it is crucial to approach the decision with a clear understanding of blockchain technology and its applicability. While blockchain technology is still relatively new, many organizations are eager to incorporate it into their businesses due to the fear of missing out on its potential benefits. However, a common mistake is approaching the technology with a mindset of "we must use blockchain somewhere, so where can we apply it?" This approach often leads to frustrations and limitations as blockchain cannot be universally applied.

To ensure a more effective approach, it is essential to begin by acquiring a comprehensive understanding of blockchain technology and its distinctive features. This understanding forms the basis of Chapter 1 in our book, which serves as a tutorial on blockchain technology. This chapter aims to provide readers with the fundamental knowledge of how blockchain works and how it can be leveraged in various applications. By delving into the core concepts and mechanisms of blockchain technology, readers will be equipped with the necessary foundation to make informed decisions and explore its potential applications.

By taking the time to understand blockchain's strengths and limitations, developers can then critically evaluate their systems and determine if a decentralized approach through dApps is a suitable fit. This approach allows for a more targeted and informed decision-making process, ensuring that the adoption of blockchain technology is done in a way that aligns with the organization's goals and requirements.

In summary, instead of approaching the integration of blockchain technology as a universal solution, it is essential for software developers and architects to first comprehend the technology and then carefully assess the suitability of systems, both new and existing, to determine where a dApp architecture can be effectively applied. Once software developers and architects have grasped the foundational concepts of blockchain technology, the subsequent crucial step is to identify the precise areas where blockchain is most suitable.

It is vital to acknowledge that not all systems, be they existing or newly developed, will seamlessly align with the principles and requirements of the blockchain paradigm. By recognizing this, developers can avoid the misconception that blockchain is a universal solution and instead focus on identifying use cases that genuinely benefit from the unique characteristics of blockchain technology. This discerning approach ensures that resources are allocated effectively and blockchain is implemented where it can truly add value and enhance the desired functionalities and attributes of the system.

The US Department of Homeland Security (DHS), specifically the Science and Technology Directorate [58], offers valuable assistance in evaluating the need for blockchain technology in development initiatives. National Institute of Standards and Technology (NIST) [59] has developed a comprehensive flowchart that aids in the assessment process (Figure 2.3).

From the perspective of software engineers and architects, it is crucial to consider certain factors when assessing the suitability of blockchain technology solutions for a given project. These factors include:

- **Decentralization:** Assess whether the project truly requires a decentralized approach, considering the nature of participants and the desired level of trust among them. Determine if the benefits of decentralization outweigh the potential trade-offs.
- **Security:** Analyze the type of security that is required for the use case and whether it is better served by the use of blockchain-based application, for example if there is a need to ensure the integrity and confidentiality of data and transactions.
- **Performance:** Consider the performance characteristics of the blockchain platform, such as transaction processing speed and latency. Assess whether the performance meets the project's requirements and user expectations.
- **Integration:** Examine the needs of integrating the blockchain solution with existing legacy systems and infrastructure or whether the dApps can be developed as standalone without the need to be interoperable with other technologies, databases, and APIs.
- **Scalability:** Evaluate whether the blockchain technology can handle the anticipated number of participants and the associated transaction volume

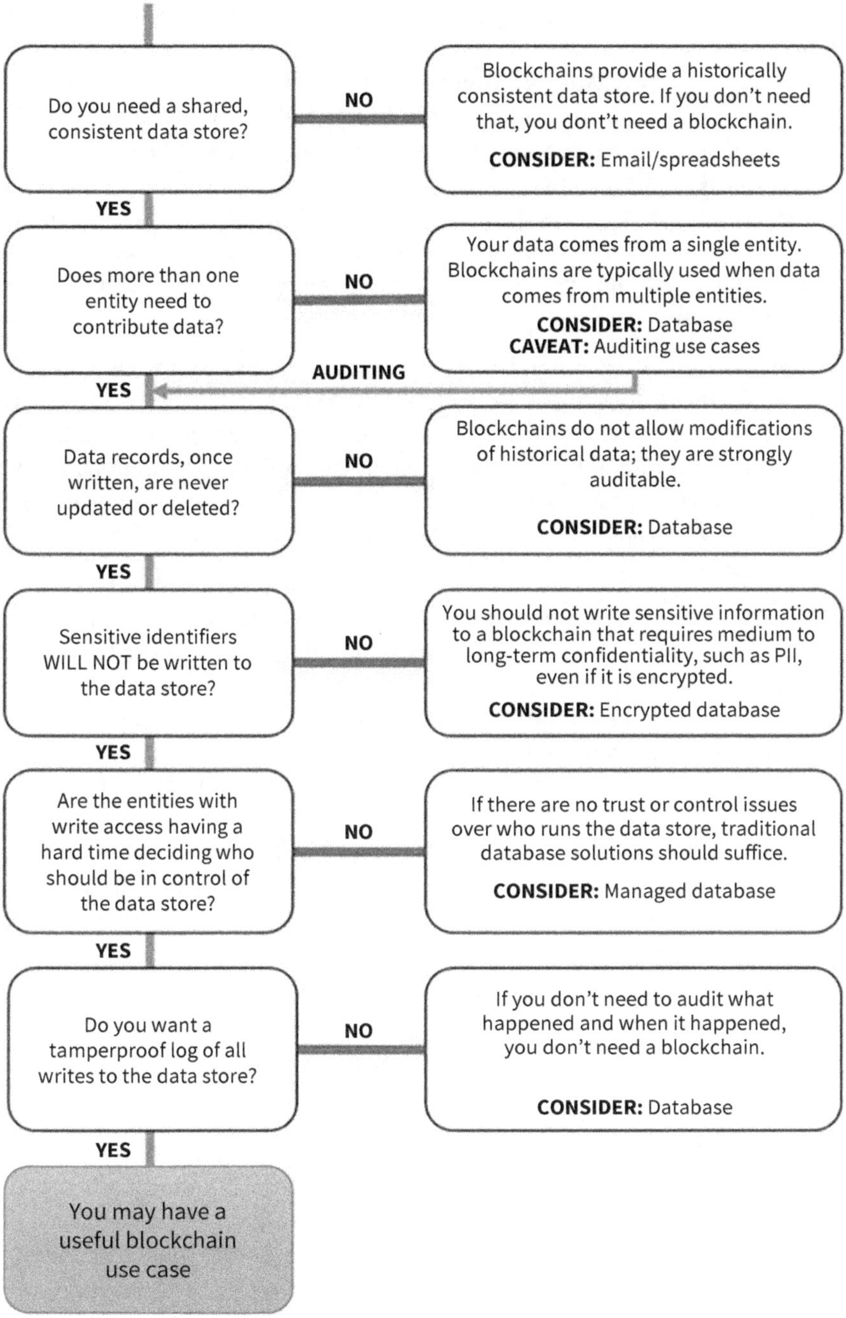

Figure 2.3 DHS decision tree for selection of blockchain use cases.

efficiently. Consider the network's capacity and consensus mechanisms to ensure scalability.

- **Development and Maintenance:** Evaluate the availability of development tools, libraries, and frameworks for building and maintaining blockchain applications. Consider the level of community support, documentation, and developer resources.
- **Cost-effectiveness:** Assess the overall cost implications of adopting a blockchain solution, including development costs, infrastructure requirements, and ongoing maintenance expenses. Consider whether the benefits justify the investment.

By carefully evaluating these factors, software engineers and architects can determine whether leveraging blockchain technology is a suitable choice for their project. Determining whether a dApp or a centralized application is the right fit for the business requirements should be one of the initial considerations in the development process. Evaluating the use case and aligning it with the advantages and limitations of dApps can help in making an informed decision. Here's a suggested approach:

- **Understand the Use Case:** Clearly define the specific problem or business requirements that need to be addressed. Consider factors such as the nature of the application, the need for trust and transparency, data ownership and privacy concerns, potential network effects, and the involvement of multiple parties. You shall use the NIST comprehensive flowchart to determine whether the business use case is better supported by a dApp versus a traditional centralized application.
- **Assess the Benefits of Decentralization:** Evaluate the potential benefits that decentralized technology offers in addressing the identified use case. These may include enhanced security, immutability of data, elimination of central points of failure, trustless interactions, disintermediation, and the ability to operate in a P2P network.
- **Consider Centralized Alternatives:** Assess whether a centralized application can adequately address the identified use case. Centralized applications may offer benefits such as simplicity, easier implementation, lower costs, and centralized control over data and operations.
- **Analyze Scalability and Performance Requirements:** Consider the scalability and performance needs of the application. Evaluate whether a decentralized architecture can handle the expected transaction volumes and processing requirements. Centralized solutions may offer better scalability and performance in certain scenarios.
- **Evaluate Trust Requirements:** Determine the level of trust required in the application. If the use case involves interactions among multiple parties

who do not inherently trust each other, a decentralized approach may provide advantages by leveraging consensus mechanisms and smart contracts to establish trust and facilitate secure transactions.

- **Consider Regulatory and Compliance Factors:** Assess whether regulatory compliance requirements, such as data protection, privacy, or financial regulations, favor a centralized or decentralized approach. Some industries or jurisdictions may have specific regulations that impact the choice of architecture.
- **Assess Network Effects:** Consider the potential network effects that can be achieved through a dApp. If the success of the application relies on a large and active user base, network effects may be better facilitated through a decentralized network, where participants are incentivized to contribute and grow the ecosystem.
- **Evaluate Development and Maintenance Considerations:** Consider the development complexity, skill requirements, and ongoing maintenance efforts associated with dApps. Decentralized technologies, such as blockchain, can introduce additional complexities and development challenges compared to traditional centralized development approaches.

By considering these factors and conducting a thorough analysis, you can determine whether a dApp aligns well with the identified use case and business requirements. It's important to note that the decision may not always be binary and can involve hybrid solutions or gradual adoption of decentralized technologies over time.

2.3 Security Requirements

In the software development life cycle, an essential initial step before designing an application is to define the functional and nonfunctional requirements. These requirements form the foundation of the development process, outlining what the application should do and how it should perform. Among these requirements, security requirements play a critical role in ensuring the application's integrity, confidentiality, and availability. By clearly defining security requirements upfront, development teams can integrate security considerations into the design and implementation phases effectively.

2.3.1 Elicitation of Security Requirements

When it comes to defining security requirements for a software application, the process of requirement elicitation plays a crucial role. Elicitation involves gathering information and insights from stakeholders to identify the specific security needs and objectives of the system. This process allows development teams to understand potential risks, threats, and vulnerabilities, enabling them to establish effective security measures.

Here's an explanation of the steps involved in eliciting security requirements:

1. **Identify Stakeholders:** Identify the relevant stakeholders who have a vested interest in the security of the system. This may include business owners, users, system administrators, IT security personnel, compliance officers, and regulatory bodies.
2. **Conduct Interviews and Workshops:** Engage with the stakeholders through interviews or workshops to understand their perspectives on security. Ask questions about their concerns, expectations, and any specific security requirements they may have.
3. **Analyze Business and Functional Requirements:** Review the business and functional requirements documentation to identify any security implications or dependencies. Consider how security requirements align with the overall goals and objectives of the system.
4. **Conduct a Threat Modeling Whiteboarding Exercise:** Perform an initial threat modeling exercise to brainstorm potential threats and vulnerabilities that the dApp broken down in essential assets/components (i.e. UI, frontend, backend, smart contracts, etc.) may face once it is being deployed. This involves identifying components and data assets, analyzing potential threats and attack vectors, and discussing the impact of those threats.
5. **High-Level Risk Assessment:** Assess the risks associated with the identified potential threats and vulnerabilities. Prioritize these risks based on their criticality by estimating the likelihood and potential impact on the dApp and your business. This analysis helps in determining the level of security scrutiny and the deep dives in the security measures and controls that are needed.
6. **Identify Security Controls:** Based on the identified threats and risks, determine the security controls and countermeasures necessary to mitigate or reduce those risks. This may include access controls, encryption, authentication mechanisms, intrusion detection systems (IDSs), and incident response procedures.
7. **Consider Regulatory and Compliance Requirements:** Take into account any regulatory or compliance requirements that the system must adhere to. These requirements may come from industry-specific standards, data protection laws, privacy regulations, or security frameworks.
8. **Involve Security Subject Matter Experts (SMEs):** Collaborate with security professionals, such as information security officers or cybersecurity consultants, to gain their expertise in identifying and documenting the dApp security requirements. They can provide valuable insights and guidance on industry best practices and emerging threats.

9. **Document Security Requirements:** Document the elicited security requirements in a clear and concise manner. Use a structured format, such as a security requirements specification document, to ensure that the requirements are easily understood, traceable, and testable.
10. **Validate and Review:** Validate the elicited security requirements with the stakeholders and security experts to ensure accuracy and completeness. Conduct reviews and iterations as necessary to refine and improve the security requirements.

The first step in the process is to "identify stakeholders," which involves identifying the individuals or groups who have an interest in the security of the project or system (Figure 2.4).

Next, "conduct interviews and workshops" with the identified stakeholders to gather information about the project's requirements and security considerations.

Once the information is collected, the next step is to "analyze business and functional requirements" to understand the specific needs and objectives of the system from a security standpoint.

Threat modeling is the subsequent step, where potential threats and vulnerabilities to the system are identified and analyzed. Following threat modeling, a "risk assessment" is conducted to evaluate the potential impact and likelihood of the identified threats. This assessment helps prioritize security efforts based on the level of risk. Based on the identified risks, the next step is to "identify security controls" that can mitigate or address the identified threats. These security controls may include encryption, access controls, monitoring systems, or other relevant measures.

Considering "regulatory and compliance requirements" is crucial to ensure that the system adheres to any applicable laws, regulations, or industry standards. To ensure comprehensive security coverage, it is essential to "involve security experts" who possess the necessary expertise and knowledge in security practices and standards. The outcomes of the previous steps are then used to "document security requirements" that capture the specific security measures and controls needed for the system.

Finally, the last step is to "validate and review" the documented security requirements to ensure they align with the stakeholders' expectations, business objectives, and industry best practices.

This step-by-step process illustrates the sequence of steps involved in eliciting security requirements for your dApp. Starting from identifying stakeholders, the process progresses through conducting interviews and workshops, analyzing business and functional requirements, performing an initial threat modeling session, conducting a high-level risk assessment, identifying critical security controls, and considering regulatory and compliance requirements by involving

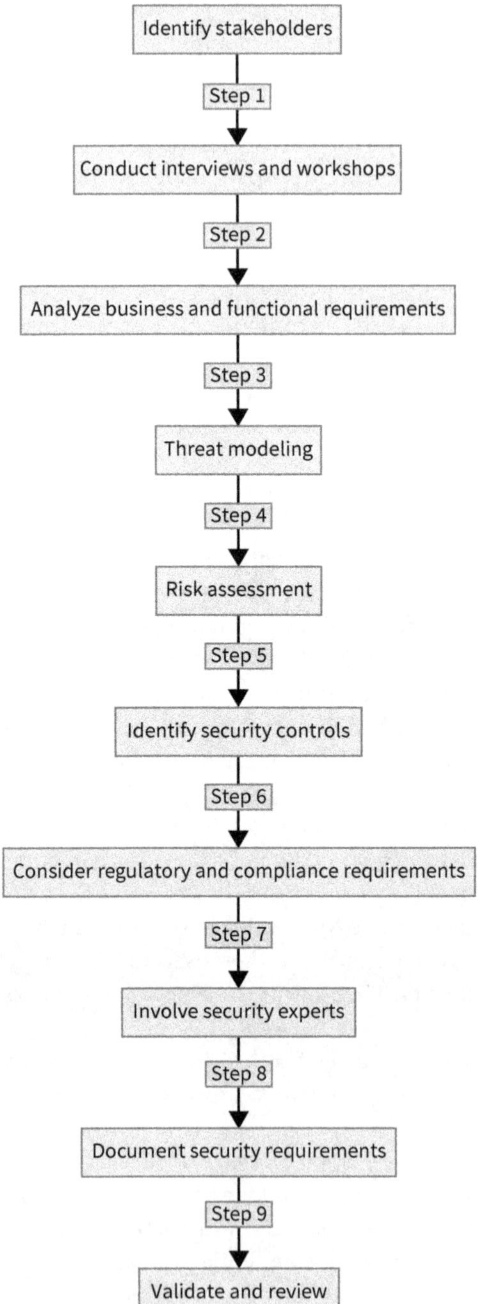

Figure 2.4 Security requirement elicitation process.

security/compliance experts. Eliciting security requirements is an iterative process that requires collaboration and ongoing communication with stakeholders throughout the development life cycle. It helps in establishing a solid foundation for designing and implementing security measures that adequately protect the system and its assets from potential threats.

2.3.2 Example of dApps Security Requirements

Let's consider the example of documenting security requirements for a decentralized lending dApp that utilizes a web interface written using React.js, server-side programming languages like JavaScript (Node.js), and smart contracts written in Solidity for Ethereum and leverages the Ethereum blockchain with its public network. Interoperability is facilitated through standardized protocols, cross-chain bridges, or APIs. Here are some security requirements that could be elicited for this dApp:

1. **Secure the Web Interfaces:**
   - **a.** Protection against common web vulnerabilities, such as cross-site scripting (XSS), cross-site request forgery (CSRF), and injection attacks. Ensure secure coding practices in React.js to prevent common web vulnerabilities like XSS and CSRF attacks.
   - **b.** Implement proper input validation and sanitization to prevent data manipulation and unauthorized access.
   - **c.** Utilize React's built-in security features, such as the Context API or React Router, to handle authentication and authorization securely.
   - **d.** Enforce secure communication protocols (HTTPS) to protect data in transit.
   - **e.** Keep dependencies up-to-date to mitigate potential security vulnerabilities in third-party libraries.
   - **f.** Implement robust user authentication mechanisms, such as multifactor authentication (MFA) or strong password requirements.
   - **g.** Enforce appropriate authorization controls to restrict user access based on roles and permissions.
   - **h.** Implement secure session management to protect user sessions from hijacking or session fixation attacks.
2. **Secure the Server Side of the dApp Application:**
   - **a.** Follow secure coding practices in Node.js, including input validation, output encoding, and parameterized queries to prevent injection attacks.
   - **b.** Implement proper error handling and logging mechanisms to detect and respond to security incidents effectively.
   - **c.** Employ security middleware, such as helmet.js, to enforce HTTP security headers and protect against known web vulnerabilities.

3. **Choose a Secure Blockchain Platform:**
   a. Choose a blockchain platform (e.g. Hyperledger Fabric, R3 Corda, Ethereum) that supports a robust consensus mechanism that provides strong security guarantees and prevents double-spending or fraudulent activities.
   b. Evaluate the consensus mechanism's security properties, such as resistance to Sybil attacks or censorship attempts, and ensure its suitability.
   c. Employ secure key management practices to protect private keys and prevent unauthorized access or theft. Implement secure key management practices for handling private keys and ensure they are stored in secure environments.
   d. Regularly monitor the blockchain network for suspicious activity, such as potential attacks or abnormal transaction patterns.
   e. Ensure the use of the latest stable versions of the Ethereum client software to mitigate known security vulnerabilities.
4. **Secure Data While in Transit and in Storage:**
   a. Employ secure communication protocols, such as TLS, to encrypt data transmitted between blockchain networks.
   b. Implement message validation and integrity checks to prevent tampering or manipulation of data during transit.
   c. Protect sensitive data, such as private keys or credentials, using encryption and secure storage practices.
   d. Apply encryption techniques to protect sensitive user data, such as personally identifiable information (PII) or financial information.
   e. Ensure compliance with data protection regulations, such as General Data Protection Regulation (GDPR) or California Consumer Privacy Act (CCPA), by implementing appropriate data privacy measures.
   f. Use secure storage mechanisms for sensitive data, both on-chain and off-chain.
5. **Secure the Integration Layers:**
   a. Implement secure communication protocols and authentication mechanisms for interoperability with external systems or APIs. Apply secure communication protocols (e.g. HTTPS) and enforce authentication mechanisms when integrating with external systems or APIs.
   b. Validate and sanitize inputs received from external systems to prevent data manipulation or injection attacks.
   c. Conduct security assessments and due diligence when utilizing cross-chain bridges or interacting with other blockchain networks.
   d. Use standardized and well-audited cross-chain bridge solutions or APIs to ensure secure interoperability between Ethereum and other blockchain networks.

   e. Implement secure authentication mechanisms to verify the identity of participants on both sides of the bridge.
   f. Utilize cryptographic techniques such as digital signatures to ensure the authenticity and integrity of messages transmitted across the bridge.
   g. Enforce authorization controls to restrict access to bridge functionalities and prevent unauthorized operations.
   h. Ensure that the bridge protocol is designed to be resistant to known attacks, such as replay attacks or man-in-the-middle attacks.
   i. Provide tools or interfaces for users to independently verify the bridge's functionality, such as displaying proof of reserves or transaction proofs.
6. **Security Audits Smart Contracts:**
   a. Conduct thorough code reviews and audits of the smart contracts to identify potential vulnerabilities and ensure adherence to best practices.
   b. Implement access control mechanisms within the smart contracts to restrict unauthorized access to sensitive functions or data.
   c. Utilize secure coding practices to prevent common vulnerabilities, such as reentrancy attacks or integer overflow/underflow. Follow best practices and security guidelines provided by the Solidity documentation and the Ethereum community.
   d. Follow secure deployment practices, including verifying contract addresses and source code on etherscan.io or other trusted platforms.
   e. Implement a proper upgrade mechanism, considering the security risks associated with contract upgrades and preserving data integrity during the process.
   f. Conduct extensive testing, including security audits, before deploying smart contracts to the live Ethereum network.
7. **Design and Implement Secure Auditing and Logging Controls:**
   a. Maintain comprehensive audit logs of user actions, transactions, and system events to support traceability and accountability. Implement mechanisms to enable transparency and auditability.
   b. Comply with applicable regulatory requirements and industry standards related to lending, financial transactions, and data protection.
   c. Conduct periodic security assessments, penetration testing, and code reviews, including security audits and simulations.
   d. Identify and address emerging security risks.
8. **Implement Emergency Response and Recovery Procedures:**
   a. Define contingency plans and procedures for handling security incidents, such as bridge failures, network disruptions, or compromised nodes.
   b. Implement mechanisms for pausing or halting operations in the event of suspicious activities or security breaches to mitigate potential risks.

   c. Regularly back up and securely store critical data to facilitate recovery in case of data loss or system failures.
9. **Institute Secure Patching Plan for Vulnerable Servers/Hosts:**
   a. Establish a robust governance model for the bridge to ensure proper decision-making, consensus, and oversight of upgrades and modifications.
   b. Implement secure mechanisms for upgrading software or protocols, considering potential security risks associated with upgrades.

These are just a few examples of security requirements that could be elicited for the given decentralized lending dApp. The actual security requirements should be tailored to the specific needs and risks associated with the dApp, taking into account the technology stack, blockchain platform, interoperability mechanisms, and the overall threat landscape. It's important to adapt these requirements based on the specific implementation and the characteristics of the blockchain networks. Regular security assessments and audits should be conducted to address emerging threats and vulnerabilities in the blockchain and dApp ecosystem.

2.4 Securing dApps

The design of secure dApps is crucial in ensuring the integrity and confidentiality of blockchain platforms. This summary provides an overview of the key principles, components, and best practices for securing dApps.

To begin with, adopting a security by design approach is of utmost importance. This means incorporating security considerations from the early stages of the blockchain platform design. By doing so, potential vulnerabilities can be identified and mitigated proactively, enhancing the overall security of the system. The secure components of a dApp include various elements such as smart contracts, UIs, storage systems, and consensus algorithms. Each component presents its own set of vulnerabilities, and it is vital to implement best practices for securing them. Examples of such practices may include code audits, regular updates and patches, input validation, and secure storage mechanisms.

Accessing blockchain platforms through APIs is a common practice. However, this introduces potential vulnerabilities related to API access. It is crucial to follow best practices for securing API access, such as implementing authentication and authorization mechanisms, using encryption for data transmission, and monitoring API activity for suspicious behavior.

Protecting confidential data and transactions within dApps is paramount. Potential vulnerabilities may arise in areas such as data storage, encryption, and transaction validation. Implementing best practices like data encryption, secure key management, and transaction validation mechanisms can significantly enhance the security of confidential data and transactions.

Consensus algorithms play a critical role in securing blockchain platforms. It is important to select and implement secure consensus algorithms to prevent vulnerabilities. Best practices include thoroughly evaluating the security properties of different consensus algorithms, considering the network's specific requirements, and regularly monitoring and updating the algorithm implementation.

Another vital aspect of securing dApps is protecting secrets and keys. Secrets and keys are used for various purposes, such as user authentication and encryption. Potential vulnerabilities can arise in their management, storage, and usage. Implementing best practices, such as using hardware-based key storage, enforcing strong access controls, and regularly rotating keys, can help safeguard secrets and keys effectively.

Finally, token-based transactions are a fundamental part of dApps, representing assets or access rights. Various token standards, like ERC-20 [60] and ERC-721 [61], are commonly used. Ensuring the security of token-based transactions requires identifying potential vulnerabilities and applying best practices specific to fungible tokens and NFTs. These practices may involve robust token transfer mechanisms, secure custody solutions, and countermeasures against attacks like double-spending, front-running, and reentrancy attacks.

As a security architect, it is crucial to recognize that secure dApps require a security by design approach and the implementation of best practices across multiple domains. By incorporating these principles and considerations, developers and organizations can significantly bolster the security and reliability of blockchain platforms and their associated applications.

2.4.1 Principles of Secure Blockchain Platform Design

Designing secure blockchain platforms is of utmost importance to ensure the creation of resilient and trustworthy systems. Following a security by design approach is key when designing dApps. As a security architect, it is essential to engage with stakeholders involved in the design and implementation of dApps and communicate the security by design principles that should be adhered to.

Secure blockchain platform design involves the implementation of measures to protect the integrity, confidentiality, and availability of the platform. It encompasses various aspects such as network architecture, consensus mechanisms, smart contract development, and access controls. By considering these principles holistically, developers can create a solid foundation for secure blockchain platforms. By prioritizing security from the outset, developers can proactively identify and address potential vulnerabilities, bolstering the overall security posture of the dApp. Incorporating security by design principles involves integrating security considerations into every stage of the design process, from conceptualization to implementation and ongoing maintenance.

Engaging with stakeholders allows for effective collaboration and a shared understanding of the security goals and requirements. This collaboration helps ensure that all parties involved are aligned with the security objectives and can contribute to the implementation of appropriate security measures.

2.4.1.1 Overview of Security Architecture Principles

Generic security by design principles provide a framework for incorporating security considerations into the design and development of various systems, including blockchain platforms. These security principles are well-documented across various authoritative sources, including academic research, industry standards, and security best practices. For instance, foundational concepts such as least privilege and fail-safe defaults were first articulated by Saltzer and Schroeder in their seminal paper, *The Protection of Information in Computer Systems* [62]. Similarly, the NIST Cybersecurity Framework [63] emphasizes principles like defense in depth, auditability, and continual monitoring to ensure resilient security postures. The Open Web Application Security Project (OWASP) foundation provides OWASP Security Principles [64] such as least privilege, secure defaults, and auditability in software development. Industry standards such as ISO/IEC 27002 [65] further reinforce these practices by providing guidelines for implementing separation of duties and secure communications. Collectively, these principles serve as a universal framework for building secure systems, regardless of the underlying technology or specific implementation.

2.4.1.2 Security Architecture Principles for dApps Design

Adapting security by design principles to the target application, in this case dApps and the specific threat environment and scenarios that are inherent to the use of dApps, it is crucial to ensure that the security principles are not abstracted but rather tailored to address the unique security challenges associated with dApp usage.

By considering the specific threat scenarios that dApps may encounter, security by design principles can be customized to provide effective protection. This involves identifying the potential threats and vulnerabilities specific to the dApp ecosystem, such as smart contract vulnerabilities, privacy concerns, or attacks on the underlying blockchain infrastructure.

To enhance the security posture of dApps, it is important to incorporate measures that directly address the identified threats. This may include implementing secure coding practices, conducting thorough code audits, adopting strong authentication and access controls, and integrating privacy-enhancing techniques. By aligning security by design principles with the specific threat environment of dApps, developers and architects can establish robust security measures that effectively mitigate risks. This approach ensures that the security measures implemented are directly relevant and tailored to the unique characteristics and challenges posed by the dApp ecosystem.

In today's evolving threat landscape, ensuring the security of blockchain applications (dApps) is of paramount importance. By following secure design principles and leveraging application secure design patterns and frameworks, developers can enhance the resilience of dApps against cyber threats. These principles provide a foundation for building secure, attack-resilient blockchain applications that protect sensitive data and maintain the trust of users and stakeholders. One effective approach to incorporating security into the design process is through threat modeling. The Process for Attack Simulation and Threat Analysis (PASTA) [66] is a widely recognized threat modeling methodology that can be applied to blockchain applications. PASTA threat modeling takes a risk-centric threat-driven approach, allowing security architects and engineers to identify, prioritize, and mitigate potential threats based on their relevance and impact. Incorporating these principles, along with utilizing the PASTA threat modeling methodology, enables security architects and engineers to proactively identify and address potential threats and vulnerabilities throughout the development life cycle of blockchain applications and dApps.

Substantiating threat modeling-driven security by design and architecture principles is essential for shaping the target architecture of dApps or blockchain applications. It's critical to contextualize security by design from the perspective of designing security controls that mitigate threats by reducing their likelihood and impact on both the dApp and its individual components. This highlights why threat modeling is a cornerstone of security by design.

Applying security by design principles should always be done in the context of mitigating identified threats, translating them into actionable security requirements for the dApp we intend to build. For example:

1. **Design for Defense in Depth:** The principle of "defense in depth" emphasizes the importance of implementing multiple layers of security controls at different tiers within the design of dApps and blockchain-based applications. By employing multiple layers of protection, this principle aims to mitigate the potential impact of a security breach by ensuring that even if one layer is compromised, other layers will provide an additional line of defense. Implementing multiple layers of security controls at different tiers ensures that even if one layer is breached, other layers provide an additional line of defense. When applying the principle of defense in depth to the design of dApps and blockchain-based applications, several key considerations should be taken into account:
   - **Implement network segmentation** to divide the dApp into different security zones or segments. Each segment should have its own set

of security controls and access restrictions. This helps contain potential breaches and prevents lateral movement of attackers within the network. Segment the dApp's network and components into smaller, isolated segments to limit lateral movement and minimize the potential impact of a compromised component.

- **Adopt secure by design patterns,** such as separation of concerns and model-view-controller (MVC), can help enhance security.
- **Segment and segregate access to the data** by implementing proper access controls, encryption, and integrity checks to protect sensitive data.
- **Deploy secure infrastructure components,** such as firewalls, IDSs/intrusion prevention systems (IPSs), and virtual private networks (VPNs), to protect the underlying network infrastructure supporting the dApp.

2. **Design for Zero Trust:** Designing dApps by applying the "zero trust" principle involves adopting an approach where trust is not automatically granted based on network location or user identity. Instead, each interaction and access request is verified and authenticated, regardless of the source. Here are some key considerations when designing dApps with the zero trust principle:
   - **Consider adopting a zero trust network architecture (ZTNA)** for the dApp, where all network traffic and access requests are treated as untrusted by default. Implement technologies such as software-defined perimeters (SDP) and secure access service edge (SASE) to enforce strict access controls, inspect traffic, and apply security policies regardless of the network location.
   - **Implement robust I&AM mechanisms** for users, devices, and services within the dApp.
   - **Enforce authentication and authorization when crossing the trust boundaries** across the various components of the dApp architecture.
   - **Implement network segmentation controls such as firewalls, VPNs, and network access control lists (NACLs)** to enforce strict trust boundaries by applying access management policies across different components and services.
   - **Utilize I&AM** solutions to enforce fine-grained access controls and policies across different cloud services and APIs used by the dApp.
   - **Implement robust monitoring and threat detection capabilities** within the dApp's infrastructure.
3. **Enforce Least Privileges:** Assigning specific roles and permissions to users and components limits access to only what is necessary, reducing the risk of privilege escalation and unauthorized actions. In the context of dApps design, this principle emphasizes the importance of limiting access rights to only what is necessary for users and components within the application. By assigning specific roles and permissions, the principle aims to reduce the risk

of privilege escalation, unauthorized actions, and potential misuse of system functionalities.

- **Implement RBAC frameworks,** such as AWS I&AM roles or Azure RBAC, to ensure that only authorized users and components have access to sensitive resources within the cloud environment.
- **Adhere to the least privilege principle** by granting users and components the minimum level of access required to perform their tasks. Separate privileged operations or administrative tasks from regular user activities within the dApp. By isolating and tightly controlling access to administrative functions, such as system configuration or critical data management, the risk of unauthorized access or malicious actions can be mitigated.
- **Secure the supply chain.** When integrating third-party components or libraries into the dApp, carefully assess their security posture and ensure that they follow secure coding practices and principles of least privilege.
- **Apply secure API design patterns,** such as the principle of least privilege and strict input validation, to prevent unauthorized access and mitigate common web vulnerabilities like injection attacks.
- **Leverage cloud-native security services,** like AWS Security Groups or Azure Network Security Groups, to enforce network-level access controls and restrict communication between different components of the dApp.

4. **Design dApps with Strong Authentication and Encryption:** This principle focuses on ensuring robust authentication mechanisms and encryption techniques to protect the confidentiality, integrity, and authenticity of user interactions and data within a dApp. This principle acknowledges the importance of verifying the identity of users, preventing unauthorized access, and safeguarding sensitive information from unauthorized disclosure or tampering. When designing a dApp with this principle in mind, several key considerations and practices come into play:
   - **Implement strong authentication mechanisms** to verify the identity of users before granting them access to the dApp. This may involve MFA techniques, such as combining passwords with biometrics or one-time password (OTP) tokens, to enhance the security of user logins.
   - **Implement secure key management practices** to protect encryption keys and cryptographic credentials used within the dApp. Keys should be stored securely in a vault, with restricted access, and regularly rotated to mitigate the risk of unauthorized key access or misuse.
   - **Implement API security frameworks,** such as OAuth 2.0 and OpenID Connect, to enable secure authentication and authorization for API interactions within the dApp.
   - **Consider implementing end-to-end encryption,** particularly for sensitive interactions or data exchanges within the dApp.

- **Utilize well-established standard encryption libraries and frameworks,** such as OpenSSL to secure sensitive data at rest and in transit within the dApp.
- **Implement secure communication protocols,** such as TLS, for encrypting data exchanged between different components and external APIs.
- **Leverage authentication frameworks,** such as JSON Web Tokens (JWT) or SAML, to enable secure and scalable user authentication within the dApp.

5. **Protect Confidential Data in Storage and Transit:** The principle of "protecting confidential data in storage and transit" emphasizes the importance of safeguarding sensitive information throughout its life cycle within a dApp. This principle recognizes that confidential data, such as PII, financial details, or proprietary business data, needs to be adequately protected to prevent unauthorized access, data breaches, or privacy violations. When designing a dApp with this principle in mind, several key aspects should be considered:
   - **Apply secure data storage patterns,** such as encryption at rest using cloud-managed encryption keys or client-side encryption, to protect sensitive data stored within cloud databases or file storage systems.
   - **Apply techniques like data anonymization or data masking** to reduce the exposure of sensitive information.
   - **Define clear policies for data** retention and secure data deletion.
   - **Implement robust access controls** to ensure that only authorized individuals or components can access and modify confidential data. *Note access controls are implemented often including access control lists (ACLs) to restrict access to approved systems only.*
   - **Implement strong encryption algorithms** to encode sensitive data both when it is stored and when it is transmitted over networks.
   - **Implement secure data transfer mechanisms,** such as secure FTP (SFTP) or HTTPS, to ensure the confidentiality and integrity of data transmitted between the dApp and external systems.
6. **Limit the Attack Surface:** Limiting the attack surface is indeed a principle aimed at reducing the opportunities for attackers to exploit vulnerabilities in exposed interfaces, including the frontend of a dApp. By minimizing the attack surface, you reduce the number of potential entry points for attackers and make it harder for them to find and exploit vulnerabilities. Examples of how security by default principle can be followed for the secure by design of dApps includes:
   - **Implement strict input validation mechanisms** on the frontend to prevent common vulnerabilities like XSS and injection attacks. Sanitize and validate user input to ensure it does not contain malicious code or unexpected data.

- **Conduct comprehensive security testing,** including penetration testing and vulnerability assessments, specifically focusing on the frontend interfaces.
- **Leverage cloud-native security services,** such as web application firewalls (WAFs) or API gateways, to filter and inspect inbound and outbound traffic to the dApp, protecting against common web-based attacks.
- **Utilize container security frameworks,** like Docker Content Trust or Kubernetes Pod Security Policies, to enforce security best practices and prevent container-based attacks within the dApp's infrastructure.
- **Configure the frontend components securely,** following best practices for security settings and access controls. This includes setting appropriate permissions, disabling unnecessary functionalities, and properly configuring security headers.

7. **"Security by Default" Throughout the Software Development Life Cycle (SDLC):** "Security by default" is a principle that emphasizes the importance of incorporating robust security measures as the default state in the design and development of blockchain applications. By following this principle, organizations ensure that security is an integral part of the application's development during the SDLC rather than an afterthought after the application is already deployed in the production environment. This approach also embraces the utilization of automated security testing as an integral part of the development process. Another aspect of "security by default" is the default configuration of dApps when they are deployed, which encompasses hardening the security of the underlying infrastructure and the components of the dApp application in the production environment. Examples of how security by default principle can be followed for the secure by design of dApps include the following:
   - **Adopt secure coding frameworks and libraries** such as OWASP Application Security Verification Standard (ASVS) or Microsoft's Security Development Lifecycle (SDL), to ensure security requirements are verified through the design, coding, and testing activities. For the design, leverage cloud provider's security best practices, such as AWS Well-Architected Framework or Azure Security Center, to guide the secure configuration of cloud resources and services used by the dApp.
   - **Leverage well-established security standard patterns** for integrating with blockchain components such as APIs and vetted software for integrating with critical components of the blockchain platform that includes APIs for basic cryptographic operations, secure access and authorization to key storages, identity validations, and transaction validations and authorizations. Utilize recognized security patterns and practices that have been proven effective in the context of dApps and blockchain applications. Some

architecture design patterns for security in the context of blockchain, specifically for use in dApp, are layered architecture, microservice architecture, and security events architecture patterns.

- **Deploy automated execution of source code audits,** incorporating static application security testing (SAST), software composition analysis (SCA), and dynamic application security testing (DAST) tools, is seamlessly integrated into the continuous integration/continuous deployment (CI/CD) pipeline when dApps are developed.
- **Secure by default all the components of dApps** that are deployed and integrated, such as the UI, such as secure defaults and configurations are set for the security controls of specific blockchain platforms, such as Ethereum, Hyperledger Fabric, or EOS, of which the dApps are benignly being implemented. Smart contracts that are deployed and integrated, such as the UI, such as secure defaults and configurations are set for the security controls of specific blockchain platform, such as Ethereum, Hyperledger Fabric, or EOS of which the dApps is benignly being implemented.
- **During deployment ensure that the underlying infrastructure of dApps and the blockchain are configured with security in mind.** This included backend services that handle the processing of transactions, communication with the blockchain network, and integration with external systems if needed. Security by default implies applying the principle of least privilege, disabling unnecessary services and ports, implementing strong authentication mechanisms, and encrypting sensitive data at rest and in transit.

8. **The Design Is as Secure as the Weakest Link:** It is an important principle in blockchain application design, emphasizing the need to identify and address vulnerabilities or design flaws that can act as weak links in the overall security posture of the system. This principle recognizes that the security of a blockchain application is only as strong as its most vulnerable component or aspect. These weak links could include insecure smart contracts, inadequate access controls, weak encryption mechanisms, or poor validation and sanitization of user inputs. Often the weakest link is constituted by a software vulnerability, and this can be prevented by conducting code reviews, penetration testing, and vulnerability scanning to uncover weaknesses that could be leveraged to compromise the integrity, confidentiality, or availability of the blockchain application.
   - **Perform regular security assessments,** vulnerability scanning, and penetration testing on the dApp's cloud infrastructure, APIs, and application components to identify and remediate any weaknesses or vulnerabilities.

- **Utilize threat modeling techniques** to identify potential attack vectors and prioritize security controls based on the likelihood and impact of threats.

9. **Design for Open Security Vetted Model:** Designing for an open security vetted model helps to mitigate risks associated with insecure designs or vulnerabilities in critical components of blockchain applications. By leveraging established security patterns and vetted software designs, organizations can benefit from the collective knowledge and experience of the security community, enhancing the overall security and resilience of the application. This approach involves leveraging the use of approved security components including security vetted and audited third-party libraries and APIs used in smart contracts. Here are some practical examples on how to follow secure by design by leveraging vetted models for security:
   - **Engage in peer reviews and independent security audits** of the design and implementation of critical components of blockchain applications and dApps as well as software architecture.
   - **Leverage existing, vetted software components** and libraries that have undergone extensive security testing and review specifically security-audited smart contract software and third-party libraries.

In conclusion, designing secure blockchain platforms and dApps is crucial to create resilient and trustworthy systems. Following a security by design approach is key in this process, which involves engaging with stakeholders, communicating security principles, and incorporating security considerations at every stage of the design process. Secure blockchain platform design involves protecting the integrity, confidentiality, and availability of the platform by considering aspects such as network architecture, consensus mechanisms, smart contract development, and access controls. By adhering to security by design principles, developers can proactively identify and address vulnerabilities, strengthening the overall security posture of the dApp.

Engaging with stakeholders enables effective collaboration and shared understanding of security goals and requirements, ensuring the implementation of appropriate security measures. Generic security by design principles, such as least privilege, defense in depth, fail-safe defaults, separation of duties, secure communication, auditability and accountability, and continual monitoring and updating, provide a foundation for secure system design across various domains, including blockchain platforms.

Adapting these principles to dApp design involves considering specific threat scenarios and customizing security measures accordingly. Identifying and addressing potential threats unique to the dApp ecosystem, such as smart contract vulnerabilities and attacks on the underlying blockchain infrastructure, enhances the security posture of dApps.

Designing attack-resilient blockchain applications and dApps requires applying threat-driven security by design principles. This involves conducting comprehensive threat modeling exercises, implementing secure architecture designs, following secure coding practices, employing strong authentication and authorization mechanisms, ensuring secure smart contract development, establishing continuous monitoring and incident response capabilities, and conducting regular security assessments.

By integrating security into the design process and following these principles, developers can build robust, resilient, and trustworthy blockchain solutions that protect sensitive data, maintain user trust, and mitigate risks in today's evolving threat landscape.

2.4.2 Securing dApps by Design

The design of secure dApps encompasses various important aspects. It involves understanding the principles of secure blockchain platform design and the underlying security architecture principles specific to dApps. These principles serve as the foundation for creating robust and secure dApps.

One crucial aspect of secure dApp design is the identification and comprehension of the key components that constitute a dApp. Each component brings its own set of potential vulnerabilities that need to be addressed. By recognizing these vulnerabilities, developers can implement best practices to ensure the security of each component and, consequently, the overall application.

In the realm of secure dApps, it is essential to delve into the components that make up a dApp and understand their significance. By examining the key components of a dApp, developers can gain insights into their functionality and potential vulnerabilities. This knowledge is crucial for implementing security measures to protect against threats and flaws. Identifying and addressing security design flaws specific to dApps is a vital aspect of secure dApp development. Understanding the potential weaknesses and vulnerabilities in the security design allows developers to proactively mitigate risks and ensure the overall integrity and reliability of the dApp.

When designing secure dApps, it is important to identify potential security design flaws. By conducting thorough assessments and evaluations, developers can pinpoint weaknesses in the dApp's security architecture and address them effectively. Identifying dApps security design flaws involves analyzing various aspects of the application, such as the overall architecture, data storage and management, authentication mechanisms, smart contracts, and communication channels. By scrutinizing these areas, developers can uncover vulnerabilities that could potentially be exploited by malicious actors.

Common security design flaws in dApps may include inadequate access controls, insufficient data encryption, improper handling of user input, insecure smart contract coding, and weak communication protocols. By recognizing these flaws, developers can take proactive measures to rectify them and enhance the overall security of the dApp.

To effectively identify security design flaws, developers can employ various techniques such as code reviews, security audits, penetration testing, and threat modeling. These methods enable a comprehensive assessment of the dApp's security posture and facilitate the identification of potential weaknesses. By diligently identifying and addressing security design flaws, developers can significantly improve the resilience and trustworthiness of dApps, ultimately providing users with a more secure and reliable experience.

Securing the components of a dApp is a critical step in building a robust and resilient application. By employing best practices and implementing security measures, developers can safeguard the individual components from potential threats and attacks. This includes implementing secure coding practices, conducting rigorous testing, and adhering to established security guidelines.

By examining dApp components, identifying security design flaws, and implementing measures to secure the components, developers can create dApps that are more resistant to attacks and provide enhanced security for users.

2.4.2.1 Identifying dApps Security Design Flaws and Vulnerabilities

In the realm of secure dApps, it is essential to delve into the components that make up a dApp and understand their significance. By examining the key components of a dApp, developers can gain insights into their functionality and potential vulnerabilities. This knowledge is crucial for implementing security measures to protect against threats and flaws. Identifying and addressing security design flaws specific to dApps is a vital aspect of secure dApp development. Understanding the potential weaknesses and vulnerabilities in the security design allows developers to proactively mitigate risks and ensure the overall integrity and reliability of the dApp.

In Section 2.2.1, we discussed the security components of dApps. These components can be associated with common secure design flaws that often emerge during the design phase of dApps. Before delving further into our discussion, it is important to define security design flaws and emphasize how they differ from vulnerabilities.

Security design flaws are essentially gaps in applying necessary security controls, such as authentication, authorization, and encryption, within the dApp's design. These flaws arise from design decisions that inadvertently compromise the overall security of the application. On the other hand, vulnerabilities primarily stem from errors in the coding or configuration of a dApp. These vulnerabilities can also be identified through security testing.

Categorizing security design flaws separately from vulnerabilities is crucial because they require distinct approaches for identification, analysis, and mitigation. While vulnerabilities pertain to specific weaknesses within individual components, security design flaws encompass broader design issues that can affect multiple components or even the entire system.

Recognizing the distinction between security design flaws and vulnerabilities enables developers to adopt a comprehensive and systematic approach to addressing potential weaknesses during the design phase of their dApps (Table 2.1). By doing so, developers can design more robust and secure applications that are better equipped to withstand potential threats and attacks. Additionally, conducting security tests plays a vital role in identifying both vulnerabilities and security

Table 2.1 Common dApps security design flaws and vulnerabilities.

dApp component	Technology stack, frameworks, and programming languages	Common design flaws	Common vulnerabilities
UI	HTML, CSS, JavaScript frameworks: React, Angular, Vue.js Mobile Apps: Swift (iOS), Java/Kotlin (Android), React Native, Flutter CLIs: Python	• Lack of input validation • Misconfigurations • Insecure data storage • Insufficient authentication and session management	• XSS • CSRF • Insecure direct object references (IDOR) • Outdated libraries with known vulnerabilities
Application logic	JavaScript (Node.js), Python (Django/Flask), Ruby (Ruby on Rails)	• Lack of secure input validation and output encoding • Inadequate access controls • Misuse of APIs	• Injection attacks • XSS • Server-side request forgery (SSRF) • Insecure deserialization
Smart contracts	Solidity, Vyper (Ethereum), Chaincode (Hyperledger Fabric), Scilla (Zilliqa)	• Improper validation of return values • Inadequate transaction ordering • Lack of authorization checks	• Reentrancy attacks • Integer overflow and underflow • DoS • Access control issues

dApp component	Technology stack, frameworks, and programming languages	Common design flaws	Common vulnerabilities
Blockchain platform	Ethereum, Binance Smart Chain (BSC), Polkadot, Hyperledger Fabric, Corda, EOSIO, Tezos, Stellar, TRON, Quorum, Avalanche	• Insecure private key storage • Weak consensus algorithms • Lack of backup and recovery mechanisms	• Vulnerabilities in authentication and key management • Single point of failure • Cross-chain communication vulnerabilities
Distributed storage	IPFS, Swarm	• Insecure file sharing • Inadequate content integrity checks • Network partitioning	• Distributed Denial of Service (DDoS) attacks • Sybil attacks • Unauthorized access to sensitive files
P2P network	Blockchain P2P network	• Design flaws exposing to network attacks • Consensus algorithm weaknesses	• DDoS attacks • Unauthorized data access • Manipulation or collusion attacks • Reentrancy and DoS attacks
I&AM	PKI (RSA, ECC, EdDSA), DID protocols: Sovrin, uPort, SelfKey	• Inadequate key management • Insufficient randomness in key generation • Weak cryptographic algorithms	• Key compromise • Identity spoofing • Privacy breaches • Man-in-the-middle attacks • Unauthorized access to personal data
Interoperability components	RESTful APIs, GraphQL, Polkadot, Cosmos, Chainlink, Cross-Chain Bridges	• Inadequate data validation • Insufficient authentication and authorization • Lack of secure communication	• Injection attacks • Transaction replay attacks • Bridge compromise • Data interception and tampering • Unauthorized access

design flaws within a dApp's implementation, further strengthening its overall security posture.

When considering the technology stack consisting of HTML, CSS, JavaScript frameworks (React, Angular, Vue.js), mobile apps built with Swift, Java/Kotlin, React Native, Flutter, and Python for CLIs, there are several common security design flaws and vulnerabilities to be aware of in the dApp UI:

1. **XSS:** XSS vulnerabilities can occur if user-provided data is not properly validated and sanitized before being rendered on the UI. This can lead to malicious code execution within the dApp, potentially compromising user data or enabling attacks on other users. XSS is primarily a security design flaw that occurs when user input is not properly validated or sanitized before being rendered on the UI. It reflects a flaw in the design of the application's input handling and output generation processes.
2. **Injection Attacks:** Injection vulnerabilities, such as SQL injection (SQLi) or command injection, can occur if user input is directly incorporated into queries or system commands without proper validation or sanitization. Attackers can exploit these vulnerabilities to execute unauthorized commands or gain access to sensitive data. Injection vulnerabilities, such as SQLi and command injection, are considered security design flaws. They arise when user input is not adequately validated or sanitized before being used in queries or commands, indicating a flaw in the design of input validation and parameterization.
3. **CSRF:** CSRF attacks can take advantage of insufficient or missing anti-CSRF mechanisms in the dApp UI. Attackers trick users into performing unintended actions on the dApp by exploiting the trust placed in the user's session or cookies. CSRF vulnerabilities are considered security design flaws when they result from a lack of appropriate anti-CSRF measures in the design of the application. Insufficient token-based or request validation mechanisms can allow attackers to manipulate user sessions and perform unauthorized actions.
4. **IDOR:** IDOR vulnerabilities can occur when the dApp UI exposes sensitive data or functionality by directly referencing internal objects or resources. Attackers can bypass authorization controls and access or manipulate unauthorized data or perform unintended actions.
5. **Insecure Data Storage:** Inadequate data storage practices, such as storing sensitive information in plain text or weakly encrypted form, can lead to data breaches and unauthorized access to user data. IDOR vulnerabilities can be categorized as security design flaws. They occur when access controls are not appropriately enforced in the design of the application, allowing unauthorized access to internal objects or resources.

6. **Insufficient Authentication and Session Management:** Weak or improperly implemented authentication and session management mechanisms can enable unauthorized access to user accounts, session hijacking, or session fixation attacks. Insufficient input validation is a security design flaw that can result in various vulnerabilities. It indicates a failure in the design of the application's input validation mechanism, which should ensure that user input meets the expected format and constraints.
7. **Insecure App Development Practices:** When developing dApps clients including mobile apps, common vulnerabilities include insecure data storage on the device, improper handling of user input, insufficient transport layer protection, and inadequate user authorization and authentication. Insufficient input validation is a security design flaw that can result in various vulnerabilities. It indicates a failure in the design of the application's input validation mechanism, which should ensure that user input meets the expected format and constraints.
8. **Lack of Input Validation:** Insufficient validation of user input can lead to various vulnerabilities, such as allowing malicious inputs that manipulate system behavior or cause unexpected errors. Insufficient input validation is a security design flaw that can result in various vulnerabilities. It indicates a failure in the design of the application's input validation mechanism, which should ensure that user input meets the expected format and constraints.
9. **Code Injection Vulnerabilities:** If user-supplied data is not properly validated or sanitized when executing dynamic code, code injection vulnerabilities can occur. This can lead to arbitrary code execution or unauthorized access to sensitive system resources.
10. **Nonsecure Configuration of Servers:** Incorrect or insecure configuration of web servers, frameworks, or dependencies can expose sensitive information, allow unauthorized access, or lead to other security risks. Misconfiguration is considered a security design flaw as it results from improper design decisions in configuring servers, frameworks, or dependencies. It reflects a failure to implement secure configurations that align with industry best practices.

It is important to include vulnerabilities in third-party libraries when considering the development of client-side code with frameworks like React, Angular, Vue.js, native languages like Swift (iOS) or Java/Kotlin (Android), cross-platform frameworks like React Native or Flutter, and Python for CLIs. Third-party libraries are commonly used to enhance the functionality and efficiency of the application development process. However, they can introduce vulnerabilities if they contain security flaws or if they are not kept up-to-date with the latest security patches. Including third-party library vulnerabilities in the assessment of the technology stack emphasizes the need to consider the security of the entire ecosystem surrounding the development process. It highlights the importance of conducting thorough security assessments and monitoring the security of the libraries used in the client-side code. Examples of third-party library vulnerabilities may include:

1. **Outdated Third-Party Libraries/Components:** Using older versions of libraries can expose the application to known security vulnerabilities that have been patched in newer releases.
2. **Known Security Vulnerabilities:** Some libraries may have known vulnerabilities that could be exploited by attackers. It is crucial to stay informed about any security advisories related to the libraries used and promptly apply patches or updates.
3. **Insecure Defaults:** Certain libraries may have default configurations that are not secure by default. It is important to review and modify these configurations to ensure proper security measures are in place. Insecure defaults can be considered a security design fault. Insecure defaults refer to the use of default configurations, settings, or options that are not secure by default, potentially leaving the application vulnerable to security risks.
4. **Code Injection Vulnerabilities:** Third-party libraries that allow for dynamic code execution, such as template engines, may be susceptible to code injection attacks if user input is not properly validated and sanitized.
5. **Insecure Cryptography:** If a library uses weak or outdated encryption algorithms or practices, it can result in insecure data storage, transmission, or authentication mechanisms. Insecure cryptography can be considered a design flaw in the context of security. Cryptographic algorithms and practices are foundational to ensuring the confidentiality, integrity, and authenticity of data and communications. When insecure cryptography is used, it indicates a failure in the design of the application's cryptographic implementation. This could involve using weak or outdated encryption algorithms, improper key management, inadequate randomness generation, or other flawed cryptographic practices.

When developing the application logic for a dApp using JavaScript (Node.js), Python (Django/Flask), or Ruby (Ruby on Rails), it is crucial for software developers to prioritize the mitigation of application layer vulnerabilities. This includes taking necessary measures to address security risks such as XSS, SQLi, OS command injection, IDOR, CSRF, and SSRF. By implementing robust security controls during the design and development stages, developers can effectively minimize these vulnerabilities and enhance the overall security of their dApp.

Smart contracts are often targeted by attackers due to their potential vulnerabilities, making them a critical focus for the security of dApps. The unique characteristics of smart contracts, such as their immutability and ability to handle financial transactions, make them an attractive target for attackers seeking to exploit vulnerabilities. These are some of the common vulnerabilities observed in smart contracts across various blockchain platforms. It's crucial for smart contract developers to follow best practices, conduct thorough testing, and perform audits to identify and address potential vulnerabilities before deploying contracts on the blockchain. While it is possible to categorize smart contract vulnerabilities as design flaws, it is important to recognize that these vulnerabilities primarily stem from coding errors rather than inherent design gaps. Smart contracts, being implemented on blockchain platforms such as Solidity, Vyper, Chaincode, or Scilla, are susceptible to a variety of vulnerabilities that can impact the security and integrity of dApps. Here's the categorization of smart contract vulnerabilities from the perspective of design flaws that can be analyzed during a design review of a dApp before starting the implementation:

1. **Reentrancy Attacks:** Since reentrancy vulnerabilities occur due to inadequate consideration of the order of execution and resource allocation within the contract, proper design considerations and implementation of mutex locks can help mitigate reentrancy vulnerabilities.
2. **Integer Overflow and Underflow:** Since designing contracts without proper checks and validations for integer arithmetic operations can lead to vulnerabilities, ensuring the use of safe math libraries or implementing explicit checks can address integer overflow and underflow issues.
3. **Unchecked External Calls:** Failing to validate return values or not using secure communication patterns when interacting with external contracts is a design flaw. Secure design practices, such as proper validation and use of secure communication patterns, can prevent unchecked external calls.
4. **Access Control Issues:** Insufficient consideration of access control mechanisms during the design phase can result in unauthorized access to contract functions or data. Designing proper access control mechanisms, such as RBAC, is essential to mitigate access control vulnerabilities.
5. **DoS Attacks:** Inadequate consideration of potential DoS attack vectors during the design phase can leave the contract vulnerable to resource exhaustion. Designing contracts with gas limits, fallback mechanisms, and complexity constraints can help prevent DoS attacks.
6. **Front-Running Attacks:** Front-running vulnerabilities occur due to inadequate consideration of transaction ordering and the potential for exploiting the order of execution. Designing contracts with commit-reveal schemes or verifiable delay functions can mitigate front-running attacks.
7. **Lack of Event Validation:** Neglecting to validate emitted events during the design phase can lead to the propagation of false or malicious information.

Designing contracts with proper event validation mechanisms helps ensure the integrity and reliability of emitted events.

8. **Information Leakage:** Careless handling of sensitive data within smart contracts can result in information leakage, which can be exploited by attackers. Implementing proper data encryption and access controls for sensitive information can help mitigate the risks associated with information leakage.

In Table 2.2, we provide a summary of the common vulnerabilities in smart contracts, their descriptions, and risk mitigations.

Table 2.2 Smart contract weaknesses and risk mitigations.

Type of weakness/ vulnerability	Description	Risk mitigation
Reentrancy attacks	This vulnerability allows an attacker to repeatedly call a contract's function before the previous invocation completes, leading to unintended state changes or loss of funds.	Use mutex locks and apply the checks-effects-interactions pattern to handle external calls securely.
Integer overflow and underflow	Arithmetic operations on integers can overflow or underflow, resulting in incorrect calculations and vulnerabilities.	Use safe math libraries or implement explicit checks and validations for all arithmetic operations.
Unchecked external calls	Failing to validate return values or using insecure communication patterns when interacting with external contracts can result in unexpected behavior or funds being sent to malicious contracts.	Validate return values and use secure communication patterns, such as the checks-effects-interactions pattern.
Access control issues	Inadequate access control can allow unauthorized access to contract functions and data.	Implement RBAC and permissioned functions to enforce access restrictions.
DoS attacks	DoS attacks can exploit contract design or execution patterns to consume excessive computational resources, leading to failures or network congestion.	Design contracts with appropriate gas limits, reduce operational complexity, and implement fallback mechanisms.

Type of weakness/ vulnerability	Description	Risk mitigation
Front-running attacks	Occurs when an attacker observes pending transactions and manipulates the order of execution for their advantage.	Use commit-reveal schemes, random delays, or verifiable delay functions to mitigate front-running risks.
Lack of event validation	Failing to validate emitted events can lead to false information being propagated, affecting other contracts or off-chain systems.	Validate emitted events to ensure their integrity and prevent potential vulnerabilities.
Information leakage	Careless handling of sensitive data can lead to information leakage, exposing it to attackers.	Use data encryption and implement strict access controls for sensitive information.

It's important to note that while design flaws contribute to vulnerabilities in smart contracts, proper implementation and adherence to secure coding practices are equally critical in addressing these design flaws and reducing the risk of successful attacks.

2.4.2.2 Securing dApps Components by Design and Implementation

It's important to note that the specific controls and techniques may vary depending on the programming language, framework, and libraries being used. Developers should consult the relevant documentation and best practices for the specific technology stack they are working with to implement appropriate security controls. Regular security assessments, code reviews, and staying updated with security patches are also crucial for maintaining a secure application.

Here are some common security measures that can be followed to prevent web application vulnerabilities in dApp UIs. These vulnerabilities can affect the security of the client-side code, user interactions, and the overall web interface. Implementing appropriate security controls and measures on the client-side can help mitigate these vulnerabilities and ensure the secure operation of dApps. In the context of dApp security, several key controls can be devised and implemented to mitigate vulnerabilities:

1. **XSS:**
   - **Input Validation and Sanitization:** Implement strict validation and sanitization of user input to ensure that it does not contain malicious scripts.

- **Output Encoding:** Properly encode user-generated content before rendering it in HTML templates to prevent script injection.
- **Content Security Policy (CSP):** Use CSP headers to restrict the types of content that can be loaded and executed on a web page.

2. **SQLi:**
   - **Prepared Statements or Parameterized Queries:** Use prepared statements or parameterized queries to ensure that user-supplied input is treated as data rather than executable code.
   - **Input Validation:** Validate and sanitize user input to reject or neutralize potential SQLi attempts.
   - **Least Privilege Principle:** Ensure that database access credentials have the least possible privileges required for the application to function.
3. **OS Command Injection:**
   - **Input Validation and Sanitization:** Validate and sanitize user input to prevent unauthorized execution of operating system commands.
   - **Use Safe APIs:** Utilize safe APIs provided by the language or framework that properly handle command execution and parameterization.
4. **IDOR:**
   - **Access Control and Authorization:** Implement proper access controls to ensure that users can only access authorized resources.
   - **Use Indirect Object References:** Instead of exposing direct references to internal objects, use indirect references that are properly validated and authorized.
5. **CSRF:**
   - **CSRF Tokens:** Generate and validate unique tokens for each user session to verify the authenticity of requests.
   - **SameSite Cookies:** Set the SameSite attribute on cookies to restrict their usage to the same origin, mitigating CSRF attacks.
6. **SSRF:**
   - **Input Validation:** Properly validate and sanitize user-controlled input to prevent unauthorized requests to internal or external resources.
   - **Whitelisting and Filtering:** Implement strict whitelisting and filtering mechanisms to only allow requests to trusted and necessary resources.

The decentralized and transparent nature of blockchain platforms also means that once a smart contract is deployed, it becomes publicly accessible, leaving little room for error. Therefore, it is essential to address potential vulnerabilities and ensure the security of smart contracts from the design and development stages to deployment and ongoing maintenance. To enhance the security of smart contracts and protect dApps, some important measures can be taken:

1. **Follow Secure Software Development Best Practices:** Smart contracts should be developed using secure coding practices and follow established guidelines and standards. This includes using validated libraries like ERC standards, conducting code reviews, and adhering to best practices specific to the programming language and platform being used.
2. **Minimize Complexity to Enhance Security and Efficiency:** Limit functionality to what is essential and avoid adding features for "just in case" scenarios or premature optimizations. In dApps, unnecessary complexity increases the risk of vulnerabilities and makes the code harder to audit and secure effectively. Simpler code not only enhances security and maintainability but also minimizes computational complexity and avoids common Turing completeness patterns such as unbounded loops as much as possible, reducing both security risk and execution costs on-chain.
3. **Thorough Security Testing and Auditing:** Smart contracts should undergo comprehensive testing to identify vulnerabilities and ensure their proper functionality. Additionally, conducting external audits by security experts helps in identifying potential issues and obtaining an independent assessment of the contract's security.
4. **Adopt Formal Verification for Smart Contracts:** Utilizing formal verification techniques can mathematically prove the correctness and security properties of smart contracts. Formal verification helps identify potential vulnerabilities and ensures that the contract behaves as intended.
5. **Enforce Access Control and Permissioning:** Implementing proper access control mechanisms ensures that only authorized parties can interact with the smart contract. This prevents unauthorized manipulation or execution of critical functions and protects sensitive data.
6. **Implement Security Events Monitoring and Incident Response:** Continuous monitoring of smart contracts and associated transactions can help detect any suspicious activity or potential attacks. Establishing an incident response plan enables swift action in case of a security incident, minimizing the impact on the dApp and its users.
7. **Security Upgrades and Patch Management:** As new vulnerabilities are discovered and platforms evolve, it is crucial to stay updated with security patches and upgrades provided by the blockchain platform. This ensures that any identified vulnerabilities are addressed promptly.

Implementing these security measures and prioritizing smart contract security can greatly enhance the overall security of dApps. By doing so, the risk of successful attacks can be reduced, safeguarding the integrity of the decentralized ecosystem.

When designing dApps, it is crucial to recognize that securing the design and implementation, specifically the coding, of smart contracts is of utmost importance. This aspect should receive significant attention during the design phase, where the business logic and functionality that the smart contract aims to automate are thoroughly reviewed. Additionally, it is essential to incorporate security testing at every stage of smart contract development, including during development, integration, quality assurance, and staging before deploying to the production environment on the targeted blockchain platform. By following these practices, the dApp can greatly enhance the security of its smart contracts and mitigate potential vulnerabilities or risks.

When designing a dApp, one crucial security consideration is undoubtedly the security of the underlying blockchain application integrated as the backend. This blockchain application serves as the foundation upon which the smart contract will execute.

In addition to addressing smart contract vulnerabilities, it is crucial to pay specific attention to various aspects when reviewing the overall design and integration of dApps with the blockchain platform. The following areas require focus during the security design review of dApps:

1. **Access Control Weaknesses:** Misconfigurations or vulnerabilities in permissioning blockchain platforms can lead to access control weaknesses. It is essential to thoroughly review and address authentication and authorization vulnerabilities to prevent unauthorized access or malicious activities. For example, dApps access the blockchain platform through various means depending on the architecture and design choices made during development. Here are some common methods (read Note) used by dApps to interact with the blockchain platform:
    a. **Web3.js Library:** dApps built on blockchain platforms like Ethereum often use the Web3.js library. It provides a JavaScript API that allows dApps to connect to the blockchain network, send transactions, and interact with smart contracts.
    b. **JSON-RPC:** dApps can communicate with the blockchain platform using JSON-RPC (Remote Procedure Call). JSON-RPC is a protocol that enables dApps to send requests to the blockchain's RPC endpoint and receive responses. It allows dApps to perform actions like submitting transactions, querying blockchain data, and invoking smart contract functions.
    c. **Native SDKs and Libraries:** Some blockchain platforms provide native software development kits (SDKs) and libraries for specific programming languages. These SDKs and libraries offer direct access to the blockchain platform's APIs, enabling dApps to interact with the blockchain natively in their preferred programming language.
    d. **API Gateways:** Some blockchain platforms provide API gateways that abstract the complexities of interacting with the blockchain. dApps can communicate with the API gateway using standard HTTP/REST

or WebSocket protocols, allowing them to access blockchain data and invoke smart contract functions without directly interacting with the underlying blockchain infrastructure.

e. **Middleware and Frameworks:** dApp development frameworks and middleware layers, such as Truffle, Embark, or Hardhat, provide abstractions and tools to simplify the integration of dApps with blockchain platforms. They offer higher-level APIs, development environments, and testing tools to facilitate interactions with the blockchain.
f. **Smart Contract Events:** dApps can listen to events emitted by smart contracts deployed on the blockchain. By subscribing to these events, dApps can react and update their UIs or trigger specific actions based on the blockchain's state changes.

Note: These techniques enable dApps to send transactions, retrieve on-chain data, execute smart contract functions, and monitor events. The exact method used depends on the chosen blockchain platform, as well as the programming languages and frameworks employed in the dApp's development.

2. **Insecure Private Key Storage and Management:** These design flaws in the blockchain platform can significantly impact the security of dApps. Examples of how these design flaws can be exploited include:
   a. **Unauthorized Access:** Private keys are crucial for signing transactions and proving ownership of assets on the blockchain. If private keys are stored or managed insecurely, they become vulnerable to unauthorized access. Malicious actors or hackers can exploit weak security practices, such as storing private keys in plain text files or using weak passwords, to gain control over users' accounts and assets.
   b. **Asset Theft:** Insecure private key storage increases the risk of asset theft. If an attacker gains access to a user's private key, they can easily transfer the user's assets to their own accounts without the user's consent. This can result in significant financial losses for individuals and can undermine the trust and adoption of dApps built on the blockchain platform.
   c. **Account Hijacking:** Inadequate private key management practices can lead to account hijacking, where an attacker takes control of a user's account by gaining access to their private key. Once the account is compromised, the attacker can manipulate the user's transactions, alter smart contract interactions, or perform malicious actions on behalf of the user.
   d. **Smart Contract Vulnerabilities:** Insecure private key storage can also impact the security of smart contracts. If a private key associated with a smart contract's administrative functions or access control mechanisms

is compromised, an attacker can gain unauthorized control over the smart contract. This can lead to financial losses, data manipulation, or disruption of the dApp's intended functionality.

e. **Irreversible Transactions:** Blockchain transactions are typically irreversible once they are confirmed on the blockchain network. If a private key is stolen or compromised due to insecure storage, it becomes challenging to recover the lost assets or reverse unauthorized transactions. This lack of recourse can have severe consequences for users and the overall reputation of dApps.

To mitigate these risks, it is crucial to follow best practices for secure private key storage and management. This includes using hardware wallets or secure hardware modules, encrypting private keys, implementing strong authentication mechanisms, and adhering to established security standards.

3. **Weak Consensus Algorithm Designs:** Weak consensus verification algorithms for transactions can introduce significant flaws that can affect the security and reliability of dApps. Here are some common flaws associated with weak consensus algorithm designs:
   a. **Weak Security Assumptions:** Weak consensus algorithm designs often rely on low-security assumptions, such as assuming that a certain percentage of honest nodes exist or that a majority of nodes are non-malicious. These assumptions can be easily exploited by adversaries, compromising the integrity and security of the dApp. Weak consensus algorithms might have inefficient or ineffective block validation mechanisms. This can allow invalid or maliciously crafted blocks to be accepted by the network, leading to data inconsistency, transaction malleability, or even network forks.
   b. **Centralization or Single Point of Failure:** Weak consensus algorithms may introduce centralization or single points of failure, where a small number of nodes or entities have excessive control over the consensus process. This centralization increases the risk of collusion, censorship, or manipulation by these privileged entities, undermining the decentralized nature of dApps.
   c. **Lack of Scalability:** Weak consensus algorithms may struggle to scale as the network grows. Inefficient consensus mechanisms can limit the throughput and transaction processing capabilities of the dApp, leading to congestion, delays, and reduced performance.
   d. **Inadequate Governance Mechanisms:** Weak consensus algorithms may lack proper governance mechanisms for decision-making, protocol upgrades, or dispute resolution. This can hinder the evolution and adaptability of the dApp, leading to governance challenges and potential conflicts within the ecosystem.

It is important for dApp developers and blockchain networks to carefully evaluate and select consensus algorithms that address these flaws and provide strong security, decentralization, scalability, and resistance against various attacks.

4. **Inadequate I&AM:** Inadequate I&AM in the design of dApps can introduce significant design flaws and pose security risks. Implementing strong authentication mechanisms is crucial to ensure that only authorized users can access the dApp. Insufficient or weak authentication mechanisms in dApps can lead to unauthorized access to the blockchain platform as a backend. Another potential design flaw in I&AM implementation stands for weak authorization. Inadequate authorization controls can result in improper access privileges and permissions within the dApp. Inadequate authorization controls can result in improper access privileges and permissions within the blockchain platform as a backend. Without granular access controls and proper role-based access mechanisms, dApps may gain unauthorized access to sensitive functionalities or perform unauthorized actions on behalf of users. Inadequate security measures during communication between dApps and the blockchain platform can result in data breaches or unauthorized interception of sensitive information. Addressing these design flaws in I&AM for dApps authenticating to the blockchain platform as a backend is crucial for ensuring the security and trustworthiness of the overall system. By implementing strong authentication mechanisms, granular authorization controls, robust identity verification processes, and secure communication channels, dApps can establish a solid foundation for protecting the integrity of transactions, safeguarding user data, and maintaining the security of the blockchain platform.
5. **Design Flaws in Interactions Between Different Chains:** Design flaws in interactions between different chains in a blockchain platform can have a significant impact on dApps that use the blockchain platform as their backend. Here's an elaboration on how these flaws can affect such dApps:
   a. **Interoperability Risks:** Design flaws in chain interactions can lead to interoperability risks, where dApps struggle to seamlessly communicate and exchange data with different chains within the blockchain platform. Lack of standardized protocols or incompatible data formats can hinder the interoperability of dApps, limiting their functionality and potential user base.
   b. **Data Consistency and Integrity:** Flaws in chain interactions can introduce challenges in maintaining data consistency and integrity across different chains. Inconsistent or conflicting data stored on different chains can lead to discrepancies and inaccuracies, affecting the reliability of dApps that rely on data from multiple chains. This can impact the accuracy of smart contract execution, transaction processing, or decision-making within the dApp.

c. **Security Vulnerabilities:** Inadequate security measures in chain interactions can expose dApps to security vulnerabilities. Weak authentication, authorization, or encryption mechanisms during cross-chain communication can make dApps susceptible to data breaches, unauthorized access, or man-in-the-middle attacks. Exploiting vulnerabilities in the interaction protocols can also result in unauthorized manipulation of data or transactions, compromising the overall security of the dApp.

d. **Scalability and Performance Issues:** Poorly designed interactions between different chains can affect the scalability and performance of dApps. Inefficient cross-chain communication protocols, excessive data transfer between chains, or delays in consensus mechanisms can result in congestion and slower transaction processing speeds. This can impact the user experience and scalability of the dApp and limit its ability to handle high volumes of transactions.

e. **Complexity and Development Challenges:** Design flaws in chain interactions can introduce complexity and development challenges for dApp developers.

Developing and maintaining cross-chain functionality requires specialized knowledge and expertise. Lack of standardized interfaces or documentation for chain interactions can make it difficult for developers to integrate their dApps with the desired chains, resulting in increased development time and potential errors. Addressing these design flaws in the interactions between different chains within a blockchain platform is crucial for ensuring the smooth functioning and security of dApps. Implementing standardized protocols for cross-chain communication, enforcing strong security measures, ensuring data consistency and integrity, optimizing scalability, and providing developer-friendly tools and documentation can help mitigate the impact of these design flaws on dApps. By addressing a review of the design for these potential design flaws, dApps can securely leverage the blockchain cross-chain functionality platform's capabilities, expand their functionality, and provide enhanced user experiences.

6. **Lack of Backup and Recovery Mechanisms:** Lack of backup and recovery mechanisms in blockchain can introduce significant design flaws and impact the security of dApps. In the absence of backup and recovery mechanisms, the lack of redundancy can create a single point of failure. If the blockchain network or nodes experience downtime or become inaccessible, dApps relying solely on that blockchain platform may become unusable. This can lead to service disruptions, financial losses, and reputational damage. Proper design of backup and recovery mechanisms is also critical in case data or smart contracts are lost or compromised due to technical failures, human errors, or cyberattacks; the consequences can be severe. It can result in permanent data

loss, financial losses, or disruption of dApp operations. If a flawed or compromised smart contract is deployed without the ability to recover or roll back its execution, it can lead to financial losses, security breaches, or disruptions in the dApp's operations.

To address these concerns, it is important to implement automated and regular backup mechanisms for critical data, smart contracts, and configuration files. It is important to design the dApp architecture to have redundancy and high availability by utilizing multiple blockchain nodes or exploring options like DLT solutions. This helps mitigate the risk of a single point of failure and ensures continuity of operations even during network disruptions, developing a comprehensive disaster recovery plan that outlines steps and procedures to be followed in the event of a major disruption. This includes backup storage locations, backup frequency, and recovery strategies to minimize downtime and quickly restore operations. By addressing these design flaws and implementing robust backup and recovery mechanisms, dApps can enhance their resilience, protect against data loss or system failures, and maintain the security and availability of the application.

7. **Lack of Multi-Sig Support for Transaction Validations/Authorizations:** Lack of multi-sig support for transactions in a blockchain can introduce design flaws that impact the security of dApps. Without multi-sig support, transactions typically require only a single signature to authorize and execute. This creates a single point of control, where a compromised private key or account can lead to unauthorized transactions or access to funds. A single point of control increases the risk of theft, fraud, or misuse of assets associated with the dApp. To mitigate this threat, integrating multi-sig functionality into the design and implementation of the dApp's transaction processing is key. This involves reviewing the blockchain platform to make sure it supports multi-sig for transactions where multiple signatures from authorized parties are required to approve and execute transactions, thereby distributing control and enhancing security.
8. **Insecure Protection of Sensitive Transactional Data in Storage and in Transmission:** Addressing the design flaws related to insecure protection of sensitive transactional data in storage and transmission is essential during the dApp security design review. It is crucial to ensure the confidentiality, integrity, and compliance of the application. To achieve this, the following considerations should be taken into account:
   a. **Strong Encryption:** Applying robust encryption techniques is vital to protect sensitive data both during storage and transmission. Encryption helps safeguard data from unauthorized access, ensuring its security and confidentiality.
   b. **Access Controls and Permissions:** Implementing stringent access controls and permission mechanisms is critical. By restricting access

to sensitive data, only authorized users or entities can interact with it, reducing the risk of unauthorized disclosure or manipulation.

c. **Comprehensive Data Flow Review:** During the design phase of the dApp, it is important to thoroughly review all data flows from the frontend to the backend. This can be done using threat modeling (refer to Chapter 3 for details). Data flow analysis ensures that sensitive data remains protected while stored and during transit, minimizing potential vulnerabilities.

d. **Protection of PII:** It is essential to prioritize the protection of PII, which includes transactional data and other confidential information. By safeguarding PII, the dApp can maintain compliance with relevant data protection regulations, such as GDPR. This involves implementing appropriate security measures, obtaining user consent, and adhering to data retention and deletion policies.

By addressing these design flaws and implementing robust security measures, the dApp can ensure the confidentiality, integrity, and compliance of sensitive transactional data. Strong encryption, access controls, and a comprehensive review of data flows contribute to a secure environment, protecting the confidentiality of user information and complying with applicable data protection regulations.

9. **Potential Single Point of Failure:** Addressing the risk of a single point of failure in blockchain design is of utmost importance during the dApp security review. A single point of failure can jeopardize the security and availability of the entire system, leading to potential breaches and loss of trust. By implementing a distributed architecture, incorporating redundancy measures, and employing continuous monitoring, the dApp can mitigate the risks associated with a single point of failure. This approach ensures that the system remains resilient, scalable, and reliable, with the ability to handle increased transaction volumes and user demands. By safeguarding against single points of failure, the dApp can maintain user confidence, protect sensitive data, and uphold the decentralized nature of the blockchain.
10. **Flaws due to Scalability and Performance:** Addressing scalability and performance flaws in blockchain design during the dApp security review is crucial. These flaws can negatively impact the dApp's security, user experience, and market competitiveness. By ensuring scalability, efficient consensus algorithms, load balancing, and performance testing, the dApp can provide a seamless user experience, mitigate security risks, and establish reliability and availability. Continuous monitoring and optimization help maintain optimal performance and enhance scalability. Overall, addressing scalability and performance flaws during the dApp security review is essential for a secure, efficient, and competitive dApp ecosystem.

Distributed Data Storage Security Risks and Mitigations: IPFS and Swarm are decentralized file storage and distribution systems that are often utilized in dApps. IPFS is a protocol and P2P network designed to create a distributed file system. It aims to replace the traditional centralized web model by enabling content addressing and P2P file sharing. In IPFS, files are given unique cryptographic hashes, and rather than being stored on specific servers, they are distributed across multiple nodes in the network. This decentralized approach allows for improved redundancy, availability, and censorship resistance. Swarm, on the other hand, is a decentralized storage and content distribution platform developed for Ethereum dApps. It provides a P2P storage infrastructure where dApp developers can store and retrieve files and data. Swarm uses a similar content addressing mechanism as IPFS, where files are identified by their content hashes. It offers features such as data replication, automatic healing, and incentivization mechanisms to encourage participation in the network.

In dApps, IPFS and Swarm are commonly used for various purposes, including:

1. **File Storage:** dApps can leverage IPFS or Swarm to store files, such as user-generated content, media files, documents, or application data. The decentralized nature of these systems ensures data availability and reduces reliance on centralized servers.
2. **Content Distribution:** IPFS and Swarm enable efficient distribution of content across the network. dApps can distribute static assets, website content, or other resources using these protocols, allowing users to access the content from various nodes, improving availability and performance.
3. **Decentralized Hosting:** dApps can use IPFS or Swarm to host their frontend components, including HTML, CSS, and JavaScript files. This enables the frontend to be served from multiple nodes, enhancing reliability and reducing reliance on a single hosting infrastructure.
4. **Data Sharing and Collaboration:** IPFS and Swarm facilitate P2P data sharing and collaboration between users. dApps can leverage these protocols to enable users to share files or collaborate on documents directly between their devices without intermediaries.

By utilizing IPFS or Swarm in dApps, developers can achieve greater decentralization, improved data availability, censorship resistance, and enhanced user experience. These protocols offer alternative approaches to traditional centralized file storage and distribution, aligning with the principles of decentralization and blockchain technology. When reviewing the architecture of dApps using IPFS or Swarm, software programmers and architects should be aware of several design flaws and vulnerabilities that can impact security. Insecure file sharing can occur if access controls and permissions are not properly implemented, leading to unauthorized access to sensitive files. Ensuring content integrity is important to prevent tampering and distribution of compromised content.

Both IPFS and Swarm allow decentralized file sharing, but if proper access controls and permissions are not implemented, sensitive or private files can be accessed by unauthorized parties. DDoS attacks pose a risk due to the distributed nature of IPFS and Swarm, necessitating measures such as load testing and mitigation strategies. Sybil attacks, where multiple identities manipulate data or disrupt operations, should be addressed with detection and prevention mechanisms. IPFS and Swarm can be susceptible to DDoS attacks due to their reliance on distributed storage and retrieval. Attackers can flood the network with requests, overwhelming nodes and causing service disruption.

Sybil attacks can affect IPFS and Swarm in several ways. By creating multiple fake identities, an attacker can manipulate the content available in the network, introduce malicious files, or overwhelm the system with false data. They can disrupt the content distribution process, compromise data integrity, and undermine the trust and reputation systems that are crucial for these decentralized file storage and distribution platforms. Preventing and detecting Sybil attacks in IPFS and Swarm is essential to maintain the security and reliability of the networks and ensure the authenticity and integrity of the shared content.

Network partitioning can affect data availability, requiring strategies like redundancy and fault tolerance. If the network experiences partitioning, where sections of the network become isolated from each other, it can impact data availability and file retrieval. Testing and prevention efforts should focus on access controls, content integrity verification, DDoS mitigation, Sybil attack detection, and network partitioning mitigation. Regular security audits, code reviews, and monitoring are essential for maintaining the security and reliability of dApps using IPFS or Swarm.

Blockchain Network Security Risks and Mitigations: A dApp can utilize the P2P network of a blockchain platform in various ways. It can store and retrieve data in a decentralized manner, process transactions by broadcasting them to the network, and rely on the network for achieving consensus on transaction validation. The P2P network also facilitates the execution of smart contracts and the broadcasting of events within the platform. Leveraging the P2P network enables the dApp to benefit from decentralization, transparency, and the reliability of consensus mechanisms. It allows the dApp to interact with other participants, securely store and access data, and rely on the network for trust and validation. However, when utilizing a P2P network for a dApp, it is important to address design flaws and vulnerabilities to ensure its security. These include network attacks, data privacy concerns, weaknesses in consensus algorithms, smart contract vulnerabilities, and the risk of DDoS attacks. During the security architecture review, measures should be taken to mitigate these risks through network security enhancements, data encryption, access controls, smart contract audits, and detection of DDoS attacks. Regular security testing and monitoring are crucial to identify and prevent vulnerabilities, ensuring the ongoing security of the dApp utilizing the P2P network.

Design flaws such as exposure to network attacks like DDoS, data privacy vulnerabilities, weaknesses in consensus algorithms, and smart contract vulnerabilities can lead to various security risks for dApps. These risks include DDoS attacks, unauthorized data access and exposure, manipulation or collusion attacks, and exploitation of smart contract flaws. To address these vulnerabilities, it is important to implement network security measures, protect data privacy through encryption and access controls, strengthen consensus algorithms, and conduct thorough smart contract audits. Additionally, measures to detect and mitigate DDoS attacks should be implemented. Regular security testing and monitoring are essential to prevent and mitigate these vulnerabilities and ensure the overall security of the dApp.

I&AM Security Risks and Mitigations: Design flaws and vulnerabilities in I&AM within dApps can stem from PKI and cryptographic protocols, as well as DID protocols. Design flaws in PKI can include inadequate key management and weak cryptographic algorithms, while vulnerabilities may involve private key compromise and algorithmic weaknesses. Similarly, design flaws in DID protocols can arise from insufficient identity claim validation and poor integration with existing frameworks, leading to vulnerabilities like identity spoofing and unauthorized access to personal data. To address these issues, dApp developers should prioritize secure cryptographic practices, conduct thorough assessments of implementation flaws, employ strong authentication mechanisms, and establish robust governance models for DID systems. Regular security audits and monitoring are vital to ensuring the integrity and protection of I&AM in dApps.

To address design flaws and vulnerabilities in I&AM from a software developer and architect's perspective, the following steps should be taken:

1. **Implement Secure Cryptographic Protocols:** Use reliable and up-to-date cryptographic protocols (e.g. RSA, ECC, EdDSA) and libraries, ensuring the proper selection of algorithms. Follow best practices for key management to prevent compromise.
2. **Conduct Comprehensive Security Assessments:** Perform code reviews, penetration testing, and vulnerability assessments to identify and address any implementation flaws in the cryptographic protocols. This helps ensure their robustness and resilience against attacks.
3. **Adopt Secure Key Storage and Management:** Follow industry best practices for key generation, storage, and distribution. Consider using hardware security modules (HSMs) or secure key storage solutions to protect private keys from unauthorized access.

4. **Use Secure Communication Protocols:** Utilize secure protocols such as TLS to safeguard sensitive data during transit. This prevents eavesdropping and unauthorized interception of communication between dApps and identity management systems.
5. **Enforce Strong Authentication Mechanisms:** Implement strong authentication methods, such as MFA, to enhance the security of I&AM. This adds an additional layer of protection against unauthorized access.
6. **Ensure Secure Identity Verifications:** Establish secure processes for identity verification, ensuring that identity claims are properly validated. Adhere to privacy and security standards to protect user data and prevent identity spoofing.
7. **Adopt a Governance, Risk, and Compliance (GRC) Practice:** Adopt GRC processes to ensure the proper management and oversight of the risks of DID systems. This includes defining roles and responsibilities, establishing trust frameworks, and enforcing compliance with privacy and security regulations.

By following these guidelines, software developers and architects can address design flaws and vulnerabilities in I&AM, enhancing the security and integrity of dApps' identity management systems.

Blockchain Interoperability Components Security Risks: Design flaws in interoperability components, such as RESTful APIs, GraphQL, Polkadot, Cosmos, or Chainlink, as well as cross-chain bridges, can have implications for dApps in the following ways:

1. **Inadequate Data Validation and Input Sanitization:** Design flaws in interoperability components may lead to insufficient data validation and input sanitization. This can result in vulnerabilities like injection attacks (e.g. SQLi, command injection) or data manipulation, impacting the integrity and security of dApps.
2. **Insufficient Authentication and Authorization:** If interoperability components lack robust authentication and authorization mechanisms, unauthorized access to sensitive data or unauthorized execution of actions may occur. This can compromise the confidentiality and integrity of dApp transactions and user information.
3. **Lack of Secure Communication:** Design flaws in interoperability components can result in insecure communication channels, potentially exposing sensitive data to eavesdropping or tampering. This can lead to data leakage or manipulation, jeopardizing the privacy and security of dApps.
4. **Incompatibility and Integration Issues:** Design flaws in interoperability components may lead to compatibility and integration challenges between different blockchain networks or systems. This can hinder the seamless

exchange of data and assets between dApps, limiting their functionality and potential.

5. **Weak Cross-Chain Bridges:** Cross-chain bridges enable interoperability between different blockchain networks. Design flaws in these bridges can introduce vulnerabilities, such as data leakage, transaction replay attacks, or manipulation of cross-chain transactions. These issues can compromise the security and trustworthiness of cross-chain interactions within dApps.

To address these design flaws, it is important to focus on the following risk mitigation measures:

- **Input Data Validation:** Implement robust data validation and input sanitization mechanisms in interoperability components to prevent injection attacks and data manipulation.
- **Authentication and Authorization:** Establish strong authentication and authorization mechanisms to ensure only authorized entities can access and execute actions within the dApp.
- **Secure End-to-End Communication:** Secure communication channels through encryption and proper security protocols to protect sensitive data in transit.
- **Through Security Testing:** Conduct thorough compatibility and integration testing to identify and address any issues between interoperability components and blockchain networks.
- **Secure Cross-Chains Interactions:** Implement secure cross-chain bridges that adhere to best practices, ensuring data integrity, authentication, and protection against potential vulnerabilities.

By addressing these design flaws, dApps can enhance their interoperability capabilities while maintaining a high level of security and trustworthiness in cross-chain interactions and data exchange.

2.4.3 Blockchain APIs

In the domain of enterprise distributed applications (dApps) that utilize BaaS and APIs, the security of APIs is of utmost importance. This section is specifically tailored to architects and software developers responsible for designing enterprise-grade dApps, offering valuable insights into the significance of securing APIs within the blockchain ecosystem. Within this section, we delve into potential design flaws and vulnerabilities related to API access and explore strategies to ensure secure access to critical blockchain functionalities such as querying transaction data, submitting transactions, and interacting with smart contracts.

When integrating with BaaS platforms through APIs, it becomes crucial to be aware of potential vulnerabilities and design flaws, both within the APIs themselves and the underlying blockchain platform. These vulnerabilities can encompass various aspects, including weak authentication and authorization mechanisms, inadequate input validation, and insufficient encryption and data protection. Identifying and addressing these design flaws and vulnerabilities should be the primary focus of security architecture reviews and comprehensive security testing prior to deploying dApps in a production environment.

To ensure secure integration of dApps with blockchain APIs, this section outlines a set of best practices that should be followed. These practices serve as guidelines for securely designing and implementing dApps that interact with blockchain platforms. Additionally, we provide a comprehensive set of security requirements that can be incorporated into the development process. By adhering to these best practices and incorporating the suggested security requirements, architects and developers can confidently integrate their dApps with blockchain APIs, mitigating potential risks and ensuring the integrity and confidentiality of sensitive data.

2.4.3.1 Securing Blockchain APIs

Building enterprise-grade dApps that leverage BaaS and blockchain APIs can be a daunting task for architects and developers tasked with designing enterprise dApps. However, there are convenient shortcuts available in the form of BaaS providers. These providers offer bundled solutions that include smart contract technology, blockchains, and network infrastructure, making the process more accessible and streamlined.

The use of APIs plays a crucial role in integrating dApps with BaaS platforms. APIs act as the bridge between the dApp and the underlying blockchain infrastructure, facilitating seamless communication and interaction. The APIs offered by BaaS platforms typically provide a standardized set of methods and endpoints, enabling developers to interact with the blockchain network securely and efficiently. These APIs abstract away the complexities of interacting with the underlying blockchain technology, providing simplified interfaces that developers can easily integrate into their dApps.

By utilizing BaaS APIs, developers can leverage the power of blockchain technology while focusing on the core functionality and features of their dApps. They can take advantage of the security, immutability, and decentralized nature of the blockchain without the need for in-depth knowledge of the underlying infrastructure. Additionally, BaaS APIs often offer functionalities beyond the core blockchain operations. They may include features like identity management, permissioned access controls, event notifications, and data encryption. These extended capabilities provide developers with additional tools and flexibility to enhance the functionality and security of their dApps.

Notable players in the BaaS market include AWS, IBM, Oracle, SAP, and VMware. These providers offer comprehensive BaaS solutions, empowering companies to leverage their expertise and infrastructure. By partnering with established BaaS providers, companies can expedite their development process, minimize operational complexities, and accelerate time-to-market for their enterprise dApps. A diverse set of APIs is available in common BaaS platforms, enabling developers and architects to interact with blockchain networks, deploy smart contracts, submit transactions, and manage various aspects of the blockchain infrastructure.

Here are examples of APIs that can be leveraged by dApp developers and architects in some common BaaS platforms:

1. **AWS (Amazon Web Services) Managed Blockchain** [67]:
   a. **Amazon Managed Blockchain APIs:** Provides capabilities to create and manage blockchain networks, interact with nodes, manage memberships, and execute transactions.
   b. **AWS Key Management Service (KMS) APIs:** Offers secure key storage and management for encrypting data and managing access to blockchain resources.
   c. **Amazon API Gateway:** Enables the creation of RESTful APIs to expose and manage dApp functionality, allowing interaction with the blockchain network.
2. **IBM Blockchain Platform** [68]:
   a. **IBM Blockchain Platform APIs:** Offers various endpoints for managing blockchain networks, including network creation, smart contract deployment, transaction processing, and network governance.
   b. **IBM Blockchain Platform SDK:** Provides libraries and tools for integrating dApps with the IBM Blockchain Platform, simplifying interaction with blockchain networks.
3. **Oracle Blockchain Platform** [69]:
   a. **Oracle Blockchain Platform REST APIs:** Provides a set of RESTful APIs to manage and interact with blockchain networks, including functionalities like network creation, smart contract deployment, and transaction submission.
   b. **Oracle Blockchain Platform SDK:** Offers SDKs for various programming languages to facilitate integration and development of dApps on the Oracle Blockchain Platform.

4. **SAP Cloud Platform Blockchain Service** [70]:
   a. **SAP Cloud Platform Blockchain Service APIs:** Allows developers to create and manage blockchain networks, deploy and interact with smart contracts, and submit transactions.
   b. **SAP Cloud Platform Integration Suite:** Offers integration capabilities to connect dApps with the SAP Cloud Platform Blockchain Service, enabling seamless interaction with blockchain networks.
5. **VMware Blockchain** [71]:
   a. **VMware Blockchain REST APIs:** Provides endpoints for managing blockchain networks, creating and deploying smart contracts, and submitting transactions.
   b. **VMware Blockchain Developer Portal:** Offers tools, documentation, and resources for developers to integrate dApps with the VMware Blockchain platform.

It's important to note that the specific APIs and functionalities may vary based on the platform and its version, so referring to the platform-specific documentation is recommended for comprehensive information. In general, APIs are built upon the foundational blockchain APIs. These specialized APIs offer a wide range of capabilities for accessing and utilizing blockchain technology. The versatility of blockchain technology allows for an unlimited number of APIs to be developed on this public platform. These APIs can be categorized based on the vast array of applications that blockchain technology enables. From financial transactions to supply chain management, identity verification to decentralized storage, the possibilities are immense.

When selecting APIs for your project, it is important to consider a set of key criteria that apply across various platforms and providers. These criteria should include at a minimum:

1. **API Functionality and Features:** Evaluate the specific functionalities and features offered by the API. Ensure that it aligns with your project requirements and provides the necessary capabilities to achieve your desired outcomes. Consider aspects such as data retrieval, data manipulation, integration possibilities, and any unique features that differentiate the API.
2. **API Documentation and Support:** Assess the quality and comprehensiveness of the API documentation. Look for clear and detailed explanations, code samples, and integration guidelines. Additionally, consider the availability of support channels, such as forums, developer communities, or direct support from the API provider. Accessible and reliable support can significantly assist in overcoming obstacles and maximizing the potential of the API.
3. **API Ease of Integration:** Consider the ease of integrating the API into your existing infrastructure or applications. Look for well-documented APIs

with clear instructions and examples that streamline the integration process. Additionally, evaluate the availability of SDKs, libraries, or developer tools that can simplify and accelerate the implementation.

4. **API Performance and Scalability:** Evaluate the performance characteristics of the API, such as response times, throughput, and latency. Consider whether the API can handle the anticipated workload and scale with your project's growth. Look for performance optimizations, caching mechanisms, or rate-limiting capabilities that can enhance the scalability and responsiveness of the API.
5. **API Security and Compliance:** Assess the security measures implemented by the API provider. Look for authentication mechanisms, encryption protocols, and data protection practices to ensure the security and confidentiality of your data. Additionally, consider compliance with relevant standards and regulations, especially if handling sensitive or PII.
6. **API Pricing and Cost:** Evaluate the pricing model of the API, considering factors such as usage-based fees, subscription plans, or transaction costs. Assess the cost structure in relation to your project's budget and the value provided by the API. Consider any potential additional charges, such as for support or data storage.

By considering these criteria, you can make an informed decision when selecting APIs for your project, regardless of the specific platforms or providers involved. This comprehensive evaluation ensures that the chosen API aligns with your requirements, provides reliable documentation and support, offers seamless integration, demonstrates strong performance and security, and fits within your budgetary constraints.

Evaluating the security of APIs requires a comprehensive assessment that includes factors such as authentication mechanisms, encryption protocols, data protection practices, vulnerability management, and response to security incidents. To assess the security of blockchain APIs, it is recommended to consider the following:

1. **Documentation and Security Measures:** Review the available documentation provided by the API blockchain provider regarding their security measures. Look for information on authentication mechanisms, encryption protocols, and data protection practices they have in place. Evaluate whether they follow industry best practices and comply with relevant security standards.
2. **Track Record:** Research the API blockchain provider track record regarding security incidents and vulnerabilities. Assess their response to security incidents, vulnerability disclosure, and how they prioritize and address security concerns. Consider any publicized audits or security assessments they have undergone.

3. **Security Testing and Audits:** Look for information on whether the blockchain API provider performs regular security testing and audits to assess the robustness of their APIs. This can include penetration testing, vulnerability assessments, and third-party audits.
4. **Compliance:** Assess whether the API blockchain provider complies with relevant regulations and standards, such as data protection and privacy laws or industry-specific compliance requirements. Compliance demonstrates their commitment to maintaining security and privacy standards.
5. **Community and Industry Reputation:** Consider feedback and reviews from the developer community and industry experts regarding the security of the blockchain APIs provided by the vendor. This can provide insights into their reputation and the confidence others have in their security practices.

It's important to note that security is a shared responsibility between the API provider and the developers using the API. It is essential to follow security best practices on your end, such as implementing secure authentication, protecting sensitive data, and keeping API keys secure.

2.4.3.2 Blockchain API Vulnerabilities

As architects and developers tasked with designing enterprise dApps that seamlessly integrate with blockchain platforms and networks, it is crucial to possess a comprehensive understanding of APIs and their role within this context. Blockchain APIs enable developers, businesses, and users to access and leverage a wide range of blockchain functionalities. These functionalities include the ability to query transaction data, submit transactions, and interact with smart contracts. When it comes to APIs, including BaaS APIs, there are several common vulnerabilities that developers should be aware of and take measures to mitigate. In general, common vulnerabilities in APIs can be prevented by following best practices and implementing appropriate security measures. Here are some of the most common vulnerabilities and their prevention techniques:

1. **Authentication and Authorization Issues:**
   - **a.** Implement strong authentication mechanisms, such as MFA, to prevent unauthorized access.
   - **b.** Use secure protocols (e.g. HTTPS) to transmit authentication credentials.
   - **c.** Employ proper authorization mechanisms, such as RBAC, to restrict API access based on user privileges.
   - **d.** Regularly review and update access control configurations to prevent misconfigurations and unauthorized access.
2. **Lack of Input Validation:**
   - **a.** Implement strict input validation by defining and enforcing proper data types, lengths, and formats.

   b. Use input validation libraries or frameworks to automate the process.
   c. Apply white-list input validation to only accept expected characters and reject any potentially malicious input.
3. **Insecure Data Transmission:**
   a. Utilize secure communication protocols like HTTPS to encrypt data transmitted over the network.
   b. Implement TLS with up-to-date cipher suites and strong encryption algorithms.
4. **Rate Limiting and DoS:**
   a. Implement rate-limiting mechanisms to restrict the number of requests from a single client or IP address.
   b. Employ load balancing and caching techniques to distribute and handle traffic efficiently.
   c. Monitor API usage patterns and set thresholds to detect and mitigate DoS attacks.
5. **Inadequate Monitoring and Logging:**
   a. Implement robust logging mechanisms to capture and analyze API activities, including access attempts, errors, and anomalies.
   b. Regularly review logs for suspicious activities and security incidents.
   c. Utilize IDSs/IPSs to monitor and identify potential threats.

By regularly updating APIs and components, conducting security audits, and performing penetration testing, you can proactively identify and address vulnerabilities, reduce the risk of successful attacks, and enhance the overall security posture of your API infrastructure. Specifically you should:

1. **Regularly Update APIs and Components:** Keep APIs and their underlying components up-to-date with the latest security patches and updates to mitigate known vulnerabilities and ensure the use of secure versions.
2. **Conduct Security Audits:** Perform regular security audits to assess the overall security of your API infrastructure, identify vulnerabilities, misconfigurations, and compliance gaps, and address them accordingly.
3. **Perform Penetration Testing:** Simulate real-world attacks through penetration testing to uncover vulnerabilities and weaknesses in your API, assess their impact, and receive recommendations for remediation.
4. **Follow Security Best Practices:** Implement security best practices such as secure authentication and authorization mechanisms, proper input validation and output encoding, secure communication protocols (e.g. HTTPS), and strong access controls.

By following these practices, you can proactively identify and address vulnerabilities, reduce the risk of successful attacks, and enhance the overall security of your API infrastructure. It is crucial to focus on preventing the exploitation of these vulnerabilities, particularly to safeguard the specific functionalities that blockchain APIs provide to dApps. Taking necessary measures to secure smart contract interaction, wallet integration, transaction management, data querying and analytics, identity/authentication, and other key functionalities will help protect dApps from potential compromises and ensure their reliable operation on the blockchain.

By prioritizing the security of these functionalities, dApp developers can maintain the integrity of user data, the proper functioning of transactions, and the trustworthiness of user interactions within their applications. Vulnerabilities in blockchain APIs can have a significant impact on these functionalities and the overall operation of dApps:

1. **Smart Contract Vulnerabilities:** If the blockchain API has vulnerabilities in its smart contract interaction capabilities, such as improper input validation or access control flaws, dApps may encounter issues when reading from or executing smart contracts, potentially leading to incorrect data or unauthorized actions.
2. **Wallet Integration Vulnerabilities:** Insecure wallet integration in the blockchain API can result in vulnerabilities, such as weak key generation or inadequate storage of private keys. This can lead to unauthorized access to user funds, compromising the functionality of dApps relying on the API's wallet management capabilities.
3. **Transaction Management Vulnerabilities:** If the blockchain API has vulnerabilities in transaction management, such as improper transaction construction or weak transaction signing mechanisms, it can impact the ability of dApps to create and submit transactions accurately. This can result in failed or erroneous transactions affecting dApp functionality.
4. **Data Querying and Analytics Vulnerabilities:** Vulnerabilities in the data querying and analytics functionalities of the blockchain API can expose sensitive blockchain data or provide inaccurate analytics results. This can affect the reliability of data retrieved by dApps and compromise the accuracy of analytics-driven functionality.
5. **Identity and Authentication Vulnerabilities:** Insecure identity and authentication services in the blockchain API can lead to unauthorized access or impersonation attacks, compromising the security of user identities within dApps. This can impact the trust and reliability of user interactions with the application.

These vulnerabilities can have a cascading effect on the functionality and security of dApps relying on the affected blockchain APIs. It is crucial for API providers and dApp developers to address and mitigate vulnerabilities to ensure the proper functioning of dApps and maintain the integrity of user data and assets on the blockchain. Regular security assessments, adherence to secure coding practices, and prompt patching of vulnerabilities are essential to mitigate risks and maintain robust API functionality for dApps.

2.4.3.3 Security Review of Blockchain API

When conducting a security architecture review of blockchain APIs, architects and developers working with enterprise dApps should adopt a proactive approach to security. Architects and developers tasked with designing enterprise dApps that leverage blockchain services and APIs must adopt a proactive approach to security. This includes a thorough security architecture review that considers best practices tailored to secure API access in enterprise settings. Key areas to focus on include implementing strong authentication and authorization mechanisms, rigorous input validation and sanitization, robust encryption and data protection, effective rate-limiting and throttling techniques, comprehensive monitoring and logging strategies, and staying up-to-date with security updates and patches. Adhering to these best practices ensures that architects and developers can build resilient and secure enterprise dApps while leveraging the potential of blockchain technology.

When evaluating the security of BaaS APIs for dApps design, specific authentication mechanisms, encryption protocols, and data protection practices play a crucial role. Here are some examples of these considerations:

1. **API Authentication and Authorization Mechanisms:**
   a. **API Keys:** Evaluate whether the BaaS API supports the generation and usage of unique API keys for authentication and authorization. Assess the strength of key generation algorithms, key management practices, and the ability to revoke or rotate keys as needed.
   b. **OAuth:** Determine if the BaaS API offers OAuth support, which allows for secure and delegated access to resources. Assess the implementation of OAuth protocols, token management, and the availability of robust authentication and authorization flows.
2. **Encryption Protocols:**
   a. **TLS:** Assess whether the BaaS API provides support for TLS, also known as SSL, to establish secure communication between the client and the API server. Evaluate the strength of TLS configurations, cipher suites, and certificate management practices.
   b. **Encryption of Sensitive Data:** Evaluate whether the BaaS API supports encryption of sensitive data, such as user credentials or sensitive payloads. Assess the encryption algorithms used (e.g. Advanced Encryption Standard [AES] or RSA), key management practices, and data decryption procedures.

3. **Data Protection Practices:**
   a. **Data Encryption at Rest:** Determine if the BaaS API provides mechanisms for data encryption at rest. Assess the strength of encryption algorithms, key management practices, and the level of protection offered for stored data.
   b. **RBAC:** Evaluate whether the BaaS API supports RBAC mechanisms to define and enforce access control policies based on roles and permissions. Assess the granularity of access control options, the ability to manage roles and permissions, and the integration of RBAC with user management systems.
   c. **Audit Logs and Monitoring:** Assess whether the BaaS API offers comprehensive logging and monitoring capabilities. Evaluate the types of events that can be logged, the availability of log aggregation and analysis tools, and the ability to monitor and detect suspicious activities or unauthorized access attempts.

During the evaluation process, consider the specific requirements of your dApp design and the sensitivity of the data and transactions involved. It's essential to consult the BaaS provider's documentation and security resources to understand the specific authentication mechanisms, encryption protocols, and data protection practices they offer. Additionally, adhere to industry best practices, standards, and compliance requirements related to data security to ensure comprehensive security in your dApp design.

There are several blockchain APIs available today that prioritize security standards and offer reliable integration options for dApps. These APIs provide secure access to blockchain networks, implement robust authentication mechanisms, and follow industry best practices to ensure the integrity and confidentiality of data. Here are a few notable examples of such blockchain APIs:

1. **Infura** [72]: Infura is an Ethereum API service that provides secure and reliable access to the Ethereum network. It offers robust infrastructure and encrypted communication and supports secure authentication and authorization mechanisms. Infura follows best practices and security standards to ensure the integrity and availability of its API.
2. **Alchemy** [73]: Alchemy is another popular Ethereum API provider known for its secure and scalable infrastructure. It offers features such as encrypted communication, secure key management, and comprehensive monitoring and logging. Alchemy follows security best practices and adheres to industry standards to provide reliable and secure integration with dApps.
3. **Chainlink** [74]: Chainlink is a decentralized oracle network that provides external data integration to smart contracts. It offers secure and reliable data

feeds through a decentralized network of nodes, ensuring data integrity and protection against manipulation. Chainlink implements security measures such as data encryption, cryptographic proof, and node reputation systems to ensure trustworthy data integration.

4. **Coinbase API** [75]: Coinbase provides a suite of APIs for various blockchain functionalities, including cryptocurrency wallets and trading. It follows robust security practices and implements industry-standard security measures like two-factor authentication (2FA), encryption of sensitive data, and secure API access controls. Coinbase API offers reliable integration options for dApps that require cryptocurrency-related functionalities.
5. **Binance API** [76]: Binance, a prominent cryptocurrency exchange, provides a secure and well-documented API for accessing its trading platform. It offers features like secure authentication, encrypted communication, and comprehensive API rate-limiting and throttling mechanisms. Binance API is widely used and considered reliable for integrating dApps with cryptocurrency trading capabilities.

When considering blockchain APIs for dApp integration, it's important to evaluate the specific security features, authentication mechanisms, encryption protocols, and data protection practices offered by the API providers. Additionally, reviewing the documentation, security resources, and reputation of the API providers can provide further insights into their adherence to security standards and reliability for dApp integration.

2.4.4 Securing dApps Confidential Data and Transactions

To protect confidential data and transactions in dApps, software developers and architects should make sure the design of the dApp incorporates security measures to ensure the confidentiality and security of data and transactions, promoting trust and reliability within the blockchain ecosystem. These measures should be implemented across various types of transactions commonly supported by dApps, including token transfers, smart contract interactions, asset exchanges, DeFi operations, voting and governance, and more.

When designing dApps, ensuring the confidentiality of data and the security of transactions requires careful consideration of the following aspects:

1. **Encryption of Sensitive Information/Data:** Implement robust encryption techniques to protect sensitive information within the dApp. This includes encrypting confidential data such as user addresses, transaction details, or private function parameters. Encryption ensures that only authorized parties with decryption keys can access and decipher the data, safeguarding its confidentiality.

2. **Integrity of Transactional Data:** Apply measures to ensure the integrity of transactional data stored on the blockchain. This is a default feature of the blockchain platform that can be leveraged such as use of cryptographic hashing and digital signatures to verify the authenticity and prevent unauthorized modifications of transactional data. Transaction signature validations and the blockchain hashing ensure verification of origination and data integrity dApp transactions.
3. **Privacy-Enhancing Technologies:** Adopt privacy-enhancing technologies to enhance data protection of user's privacy and identity verification data that is used within the dApp. Consider solutions such as zero-knowledge proofs or off-chain techniques to preserve privacy while maintaining transparency and verifiability.
4. **Secure Communication Protocols:** Implement secure communication protocols, such as TLS/SSL, to protect the transmission of data and transactions between the dApp and blockchain APIs.
5. **Follow Secure Key Management Standards:** A standard key management policy should be necessary to make sure controls are in place to protect cryptographic secret keys, adherence to access controls to prevent unauthorized access to keys, and execution of regular security audits to identify and address vulnerabilities in key storage such as digital wallets and key operations such as transaction signing.

Examining examples of how confidential data and transactional data can be protected in various types of transactions supported by dApps can provide practical insights. Here are some examples of how confidential data and transactional data can be protected in common types of transactions supported by dApps:

1. **Token Transfers Data and Transactions Protections:**
   a. Confidential data, such as the sender and receiver addresses, token amounts, and transaction details, can be protected through encryption techniques. Encryption ensures that only authorized parties with the decryption keys can access and decipher the data, maintaining its confidentiality.
   b. Transactional data, including the transaction hash, block confirmation, and timestamp, is stored on the blockchain, ensuring immutability and data integrity. Data encryption at rest can further protect this information from unauthorized access.
2. **Smart Contract Data Interactions Protections:**
   a. When interacting with smart contracts, confidential data like user inputs, private function parameters, or sensitive contract state can be encrypted to preserve its confidentiality. Encryption ensures that only authorized parties can view or modify the data.

   b. The transactional data associated with smart contract interactions, such as the function call, gas consumption, and event logs, is stored on the blockchain. As this data is public, protection mainly focuses on maintaining the integrity of the transactional information through cryptographic hashing and ensuring proper verification of the blockchain state.
3. **Digital Asset Data Exchanges Data and Transaction Protections:**
   a. For asset exchanges, the confidential data can include user order details, trade volumes, or wallet addresses. Encryption techniques can be applied to protect this information from unauthorized access, ensuring the confidentiality of user trading activities.
   b. The transactional data of asset exchanges, including trade orders, trade executions, and settlement details, is recorded on the blockchain. Data integrity is ensured through cryptographic hashes and proper verification mechanisms to prevent tampering or unauthorized modifications.
4. **DeFi Data and Transactions Protections:**
   a. In DeFi operations, confidential data such as user account balances, lending/borrowing positions, or liquidity pool shares can be encrypted to protect the privacy and confidentiality of user financial activities.
   b. Transactional data in DeFi includes lending transactions, borrowing requests, interest calculations, or liquidity provision activities. This data is stored on the blockchain with integrity protection mechanisms like hashing and digital signatures to prevent unauthorized modifications or fraud.

It's important to note that the specific protection mechanisms and implementation details may vary depending on factors such as the dApp's design, the type of blockchain platform used, permissioned or private blockchain, and the privacy and security measures implemented by the dApp developers. dApps leveraging private or permissioned blockchains provide an example of how confidentiality of data and transactions can be protected by limiting access to approved participants within specific networks, such as supply chain management, healthcare data sharing, or consortium-based financial systems.

A supply chain management dApp built on a private or permissioned blockchain can restrict access to a select group of participants, such as manufacturers, distributors, retailers, and regulators. This dApp allows authorized entities within the supply chain network to securely share and track sensitive information while maintaining data confidentiality. In this scenario, the private or permissioned blockchain ensures that only approved participants can access and participate in the network. Each participant is granted specific access rights and permissions based on their role and responsibilities within the supply chain. This restriction helps protect the confidentiality of data and transactions by limiting visibility to authorized participants, preventing unauthorized access or exposure of sensitive information.

Within the supply chain dApp, transactions can include various activities such as recording the movement of goods, verifying authenticity, tracking product origin, and managing inventory levels. These transactions are securely recorded on the private or permissioned blockchain, ensuring data integrity and maintaining a transparent and auditable history of supply chain activities. By utilizing a private or permissioned blockchain, this supply chain management dApp enables participants to share and access confidential information relevant to their role, such as shipment details, quality certifications, or compliance records. The restricted access ensures that sensitive data remains protected from unauthorized access or exposure to untrusted parties.

In a healthcare data sharing dApp built on a private or permissioned blockchain, healthcare providers, medical researchers, and other authorized entities collaborate to securely exchange and access patient data while preserving data confidentiality and privacy. The private or permissioned blockchain ensures that only authorized participants, such as healthcare institutions and approved researchers, have access to the network. Each participant is granted specific access rights and permissions based on their role and involvement in healthcare data sharing. This restriction helps protect the confidentiality of patient data and transactions by limiting visibility to trusted entities, preventing unauthorized access or exposure of sensitive healthcare information.

Within the healthcare data sharing dApp, transactions can include activities such as requesting patient data, sharing medical records, conducting research studies, or enabling interoperability between healthcare systems. These transactions are securely recorded on the private or permissioned blockchain, providing a tamperproof and auditable history of data sharing activities. For example, when a medical researcher seeks access to patient data for a research study, the dApp facilitates the request and ensures that the patient's consent and privacy preferences are properly handled. The private or permissioned blockchain securely stores the transaction, granting access only to authorized researchers while maintaining patient confidentiality.

By utilizing a private or permissioned blockchain, the healthcare data sharing dApp enables authorized participants to securely exchange and access patient data for purposes such as research, treatment coordination, or public health initiatives. The restricted access ensures that sensitive healthcare information, including medical records, diagnostic reports, or genetic data, remains protected from unauthorized access or exposure.

In a consortium-based financial system, multiple financial institutions, such as banks, credit unions, and other stakeholders, come together to collaborate and streamline their operations using blockchain technology. The dApp built on a private or permissioned blockchain facilitates secure and efficient financial transactions among the consortium members while maintaining data confidentiality. In this scenario, the private or permissioned blockchain ensures that only authorized consortium members have access to the network. Each member is granted specific privileges and permissions based on their role and participation in the consortium. This restricted access ensures that sensitive financial data, such as account balances, transaction details, and customer information, remains confidential and protected from unauthorized access. The consortium-based financial system dApp enables various transactions within the network, including fund transfers, cross-border payments, smart contract-based agreements, and trade finance operations. These transactions are securely recorded on the private or permissioned blockchain, providing transparency, immutability, and auditability of financial activities.

The dApp's smart contracts enforce business rules, automate processes, and facilitate trusted interactions among consortium members. For example, when executing a cross-border payment, the dApp can verify the sender's identity, perform necessary compliance checks, initiate the payment transfer, and record the transaction details on the blockchain. This process ensures the confidentiality of customer information and financial data throughout the transaction life cycle.

2.4.4.1 Security Requirements for the Protection of Confidential Data

Protecting confidential data and transactions in dApps is crucial for several compelling reasons. As dApps become increasingly prevalent in various industries, ensuring the confidentiality of sensitive information and the security of transactions has become paramount. This is driven by the need to preserve user privacy, foster trust and user confidence, comply with regulatory requirements, safeguard IP, prevent fraud and data tampering, and uphold the integrity of the system. Protecting confidential data and transactions in dApps is of utmost importance due to several key reasons:

1. **Data Confidentiality:** Confidentiality ensures that sensitive information remains accessible only to authorized individuals or entities. In dApps, confidential data may include personal user information, financial details, medical records, or any other sensitive data. By protecting confidentiality, dApps maintain user privacy and prevent unauthorized access or disclosure of sensitive information.
2. **Trust and User Confidence:** Confidentiality measures in dApps contribute to building trust and user confidence. When users have assurance that their data is securely handled and kept confidential, they are more likely to engage with the dApp, provide accurate information, and participate in transactions without concerns about privacy breaches.

3. **Compliance with Regulations:** Many industries, such as finance, healthcare, and data-driven services, are subject to strict regulations concerning the protection of confidential data. dApps operating in these sectors must adhere to legal requirements and industry standards to ensure compliance. Failure to protect confidential data can result in severe consequences, including legal penalties, reputation damage, and loss of user trust.
4. **IP Protection:** Confidentiality is vital for protecting IP within dApps. Whether it's proprietary algorithms, trade secrets, or innovative ideas, safeguarding confidential data ensures that valuable intellectual assets are not compromised or unlawfully disclosed, preserving the competitive advantage of the dApp and its stakeholders.
5. **Prevention of Fraud and Data Tampering:** Protecting transactions from unauthorized modifications or fraudulent activities is crucial in dApps. By ensuring the integrity and confidentiality of transactional data, dApps can prevent unauthorized changes to critical information, maintain a trustworthy record of transactions, and safeguard against tampering or malicious activities that could compromise the integrity of the system.
6. **Regulatory Compliance:** Confidentiality measures are essential for compliance with data protection regulations, such as the GDPR in the European Union or the Health Insurance Portability and Accountability Act (HIPAA) in the healthcare sector. Compliance with these regulations is not only a legal requirement but also a crucial factor in maintaining user trust and avoiding penalties associated with data breaches.

Designing a dApp that satisfies the requirements of protecting confidential data and transactions is essential for ensuring user privacy, maintaining data integrity, and building trust in the application. By implementing specific design considerations, developers can create dApps that effectively safeguard sensitive information. In this context, several examples showcase how a dApp can satisfy these requirements by design. These examples encompass various aspects such as encryption and access controls, privacy-preserving features, secure communication channels, immutable and auditable records, consent management, secure identity and authentication, and regular security audits and updates.

Here are some examples of how a dApp can satisfy the requirements of protecting confidential data and transactions by design:

1. **Encryption and Access Controls:** Implement strong encryption techniques to protect sensitive data within the dApp, such as user information or transaction details. Additionally, enforce access controls to ensure that only authorized individuals or entities can access specific data based on their roles and permissions. This design approach safeguards confidential data from unauthorized access.

2. **Privacy-Preserving Controls:** Incorporate privacy-enhancing technologies like zero-knowledge proofs or differential privacy to minimize the exposure of sensitive information while still allowing for valid transactions and data analysis. By anonymizing or obfuscating data, the dApp can protect individual privacy and maintain confidentiality.
3. **Secure Communication Channels:** Utilize secure communication protocols such as TLS/SSL to encrypt data transmission between the dApp and external systems, ensuring the confidentiality of data during transit. This prevents eavesdropping or tampering with sensitive information exchanged between parties.
4. **Immutable and Auditable Records:** Leverage the immutability and auditability of blockchain technology to ensure the integrity and traceability of transactions. By storing transactional data on the blockchain, the dApp establishes a transparent and tamper-resistant record of all interactions, providing verifiable proof of the integrity of the system.
5. **Consent Management:** Implement robust consent management mechanisms that enable users to control the sharing and usage of their data. Users should have the ability to grant or revoke consent for specific data processing activities, ensuring that their confidential information is only used as intended.
6. **Secure Identity and Authentication:** Incorporate secure identity management and authentication mechanisms to verify the identities of participants and protect against unauthorized access. This can include MFA, digital signatures, or DID solutions, ensuring that only authorized users can access and interact with the dApp.
7. **Regular Security Audits and Updates:** Conduct regular security audits to identify and address vulnerabilities within the dApp. Stay up-to-date with security patches and updates to address emerging threats and ensure the continuous protection of confidential data and transactions.

By considering privacy, encryption, access controls, secure communication, auditability, consent management, secure identity, and regular security practices, developers can design dApps that prioritize the confidentiality and security of user data and transactions.

Examples can be provided (read the following note) to illustrate how different blockchain platforms offer the necessary tools and capabilities for building dApps that meet the requirements of securing confidential data and transactions. Each platform brings unique features and advantages to the table, empowering developers to customize their dApps according to specific use cases while upholding data confidentiality and transaction security (Table 2.3).

Table 2.3 Blockchain platform, dApps, confidential data, and transactional controls.

Blockchain platform	Examples of dApps	Confidential data protection controls	Transactional data protection controls
Ethereum	DEXs (Uniswap), lending platforms (Compound), NFT marketplaces (OpenSea)	Encryption, access controls, privacy-preserving solutions	Encryption, smart contract integrity verification, auditability
Hyperledger Fabric	Supply chain management systems, healthcare data sharing platforms, consortium-based financial systems	Private channels, data collections, access controls	Immutability, auditability, secure consensus
R3 Corda	Financial systems, insurance platforms, trade finance solutions	Confidential transactions, Corda's "notary" concept, access controls	Immutability, transaction privacy, secure consensus
Stellar	Cross-border payment solutions, remittance platforms, asset tokenization platforms	Encryption within transactions, multi-sig support	Secure consensus, fast transaction confirmation
EOS	Decentralized social media platforms (Voice), blockchain-based gaming applications (EOS Knights)	Encryption, access controls, secure communication channels	Immutable ledger, fast transaction processing, access controls

Note: Please note that the data protection controls mentioned are not an exhaustive list, and the specific implementation details may vary depending on the dApp and the configurations chosen by developers. It's crucial for developers to explore the documentation and resources provided by each blockchain platform to understand the available features and best practices for protecting confidential and transactional data.

2.4.4.2 Vulnerabilities Exposing Confidential and Transactions Data

dApps have gained significant traction in recent years, offering a new paradigm of trustless and transparent interactions. However, like any software system, dApps are not immune to vulnerabilities that can compromise the confidentiality of sensitive data and transactions. Understanding and addressing these vulnerabilities is crucial to ensuring the security and integrity of decentralized systems.

There are several common vulnerabilities that can expose confidential data and transaction data in dApps. Some of these vulnerabilities include:

1. **Insecure Smart Contracts:** Smart contracts are the backbone of many dApps, and vulnerabilities in the code can lead to data exposure. Common issues include incorrect access controls, unchecked input validation, and improper handling of user data, which can result in unauthorized access to confidential information.
2. **Malicious or Compromised External Dependencies:** dApps often rely on external libraries, APIs, or oracles for functionality. If these dependencies are compromised or malicious, they can lead to data breaches or manipulation of transactions, compromising the confidentiality and integrity of the dApp's data.
3. **Insufficient Authentication and Access Controls:** Weak authentication mechanisms or inadequate access controls can allow unauthorized users to access and manipulate sensitive data within the dApp. This can result in the exposure of confidential information or unauthorized transactions.
4. **XSS and Injection Attacks:** XSS and injection attacks can occur when user input is not properly sanitized or validated. Attackers can inject malicious code or commands, leading to data leaks or unauthorized access to confidential data.
5. **Misconfigured or Insecure Data Storage:** Improper configuration or insecure storage practices can expose confidential data in the dApp. This includes storing sensitive information in plain text, weak encryption, or inadequate access controls to the data storage system.
6. **Man-in-the-Middle (MitM) Attacks:** Insecure communication channels can be susceptible to MitM attacks where an attacker intercepts and manipulates the data being exchanged. This can lead to the exposure or alteration of confidential information and transactions.
7. **Lack of Timely Security Updates and Patches:** Failure to promptly apply security updates and patches can leave dApps vulnerable to known vulnerabilities. Attackers can exploit these vulnerabilities to gain unauthorized access to data or manipulate transactions.

To mitigate these vulnerabilities, developers and architects should implement secure coding practices, conduct thorough security testing and audits, apply proper access controls and authentication mechanisms, encrypt sensitive data, validate and sanitize user input, and stay up-to-date with security updates and patches. Additionally, promoting security awareness among dApp users is important to prevent social engineering attacks and phishing attempts that can compromise confidential data.

2.4.4.3 Security Reviews to Identify Design Flaws and Vulnerabilities

Inspecting a dApp design for design flaws and vulnerabilities involves a comprehensive analysis and evaluation of the dApp's architecture, components, and implementation. Here are some steps and approaches to consider when inspecting a dApp design:

1. **Conduct Secure Architecture Reviews:** Review the overall architecture of the dApp, including the relationship between different components, data flow, and interaction patterns. Identify potential design flaws that could lead to vulnerabilities, such as improper data handling, lack of separation of concerns, or excessive trust assumptions.
2. **Conduct Threat Modeling:** Conduct a threat modeling exercise to identify potential threats and attack vectors specific to the dApp. Consider the different actors involved, potential attack scenarios, and the impact of a successful attack on data confidentiality and transaction security. This helps in understanding potential vulnerabilities and designing appropriate security controls.
3. **Conduct Manual Secure Code Reviews:** Perform a thorough code review to identify any coding flaws or vulnerabilities. Look for common coding mistakes, such as input validation issues, insufficient access controls, or insecure storage of sensitive data. Analyze smart contracts, if applicable, to ensure they are properly designed and do not contain known vulnerabilities.
4. **Follow a Standard Secure in SDLC (S-SDLC) Process:** Assess whether secure development practices have been followed throughout the dApp's development life cycle. Evaluate if proper security controls and measures have been implemented, including secure coding practices, input validation, secure data storage, secure communication protocols, and access control mechanisms.
5. **Conduct External Dependencies Vulnerability Assessments:** Evaluate any external dependencies, such as third-party libraries or APIs, for potential vulnerabilities. Check for any known security issues associated with these dependencies and assess how they are integrated into the dApp. Ensure that proper security measures, like validating inputs from external sources, are in place.
6. **Conduct Penetration Testing:** Conduct penetration testing or vulnerability scanning to simulate real-world attacks and identify potential vulnerabilities that may not be evident through static analysis. This can help uncover security weaknesses and assess the effectiveness of implemented security controls.
7. **Review Standards Compliance Requirements:** Review compliance requirements and industry best practices relevant to the dApp's domain.

Consider standards such as HIPAA, GDPR, or industry-specific guidelines. Ensure that the dApp design aligns with these standards and incorporates necessary controls to protect confidential data and transactions.

Automation security testing, specifically through SAST, SCA, DAST, and scanning of dApp components like client UI, APIs, and smart contracts, can provide valuable insights into vulnerabilities in dApps and blockchain platform development.

Automation security testing can be extended to scan specific components of dApps, including the client UI, APIs, and smart contracts, for vulnerabilities:

- **Client UI:** Automated security testing tools can analyze the client-side code of dApps, including HTML, CSS, and JavaScript, to detect potential security flaws. Vulnerability scanners specifically designed for JavaScript (Node.js), Python (Django/Flask), and Ruby (Ruby on Rails) can analyze the application's source code, dependencies, and configurations. This may involve identifying XSS vulnerabilities, insecure data handling, inadequate client-side input validation, insecure session management, or insecure direct object references.
- **APIs:** Automated tools can scan the APIs used in dApps and blockchain platforms to uncover security weaknesses, such as inadequate authentication and authorization mechanisms, injection vulnerabilities, or sensitive data exposure through API endpoints. By simulating different API requests and payloads, these tools can identify potential attack vectors and vulnerabilities.
- **Smart Contracts:** Automation can aid in analyzing the code and logic of smart contracts to identify vulnerabilities and potential attack vectors. Automated security testing tools for smart contracts can detect issues like reentrancy attacks, integer overflow/underflow, or vulnerabilities in token-related operations. By scrutinizing the code, these tools can provide insights into potential security risks before deploying the smart contracts.
- **Distributed Data Storage:** Vulnerability scanning tools for distributed data storage systems analyze the configurations and components of these technologies. They assess encryption practices, data storage permissions, and potential vulnerabilities specific to the distributed environment. The scanning process helps identify issues like insecure data encryption, exposure of sensitive information, or misconfigured access controls.
- **P2P Networks:** Vulnerability scanners for P2P networks examine the network architecture, communication protocols, and security mechanisms in place. They identify vulnerabilities related to authentication, communication, or network architecture, including Sybil attacks, eclipse attacks, or insufficient security controls. These tools help assess the overall security posture of P2P networks and identify potential weaknesses.

- **I&AM:** Vulnerability scanners for I&AM systems focus on authentication mechanisms, access controls, and user management processes. They assess the configuration settings, protocols, and implementation of I&AM systems to detect vulnerabilities like weak authentication mechanisms, misconfigured access controls, or inadequate user management practices. These scans help ensure that the I&AM system is robust and resistant to unauthorized access.
- **Interoperability Components:** Vulnerability scanners for interoperability components analyze the configuration, code, and security practices associated with APIs and cross-chain bridges. They assess input validation, encryption protocols, authentication mechanisms, and adherence to security standards. Scans help identify vulnerabilities such as injection attacks, insecure data transmission, or insufficient access controls in these components.

By following these inspection approaches, developers and architects can identify design flaws and vulnerabilities, enabling them to address these issues early in the development process and build more secure and resilient dApps. It's important to conduct regular security reviews and maintain an ongoing focus on security throughout the life cycle of the dApp.

2.4.5 Consensus Algorithms

Consensus algorithms play a crucial role in dApps and blockchain platforms/networks. They enable multiple participants or nodes in a network to agree on the state of the system and validate transactions without relying on a central authority. Consensus algorithms ensure that all nodes reach a consensus on the order and validity of transactions, ensuring the integrity and immutability of the blockchain.

Common vulnerabilities and design flaws can be found in the design and implementation of consensus algorithms, and their presence may vary across different blockchain platforms and implementations. It is of utmost importance for developers and researchers to thoroughly analyze the specific implementations and configurations of these consensus algorithms. This analysis is necessary to identify any additional vulnerabilities or risks that may be associated with the particular implementation being used. By conducting this careful examination, stakeholders can proactively address these vulnerabilities and ensure the robustness and security of the consensus algorithm in their respective blockchain systems.

All of the most commonly implemented consensus algorithms in popular blockchain platforms have inherent design flaws. For instance, PoW is vulnerable to a 51% attack, where an attacker gains control over the majority of computational power and can manipulate transactions. Similarly, PoS suffers from the "nothing at stake" problem, allowing validators to validate conflicting blocks simultaneously without incurring costs, leading to forks and consensus instability. Delegated Proof of Stake (DPoS) introduces concerns about centralization as a limited number of delegates with significant influence can compromise the principle of decentralization. Practical Byzantine Fault Tolerance (PBFT) relies on a predetermined set of trusted validators, which limits scalability and decentralization. If a majority of validators are compromised, it jeopardizes the security and availability of the blockchain. Furthermore, Proof of Authority (PoA) carries a centralization risk due to a small number of authorities. If these authorities are compromised or act maliciously, it has the potential to compromise both decentralization and security.

These examples underscore the importance of recognizing and addressing the design flaws and vulnerabilities inherent in these consensus algorithms. Developers, researchers, and stakeholders must thoroughly analyze the specific implementations and configurations of these algorithms to identify any additional risks or vulnerabilities. By doing so, they can work toward enhancing the security and integrity of blockchain platforms and mitigating the associated risks.

When designing dApps, it is crucial to review and carefully consider the available consensus algorithms offered by different blockchain platforms. It is advisable to select algorithms with well-documented designs, including the verification algorithms used for transaction validation. These consensus algorithms should also undergo rigorous testing against potential business logic attacks that attempt to forge transactions by bypassing validations. By conducting this thorough review and assessment, dApp designers can make informed decisions regarding the choice of consensus algorithms, ensuring that the selected algorithms are secure and resilient against potential vulnerabilities and attacks.

2.4.5.1 Identifying Consensus Algorithm Vulnerabilities

Ensuring the design and implementation of consensus algorithms are secure requires careful consideration and rigorous testing. Here are some practices to help mitigate security flaws and vulnerabilities:

1. **Thorough Design Security Review and Analysis:** Consensus algorithms should undergo thorough design and analysis by experienced blockchain and security experts. The algorithm's design should be well-documented, specifying its goals, assumptions, and threat models. A comprehensive security analysis should be conducted to identify potential attack vectors and vulnerabilities.
2. **Peer Review and Audits:** Consensus algorithm designs and implementations should undergo peer review by independent experts. This review process helps identify design flaws and potential security vulnerabilities and provides valuable insights to enhance the algorithm's security.

3. **Formal Verifications:** Formal methods, such as formal verification, can be employed to mathematically prove the correctness and security properties of the consensus algorithm. This rigorous approach helps identify and eliminate potential vulnerabilities before implementation.
4. **Robust Testing:** Extensive testing is crucial to identify and address security flaws. Developers should conduct thorough unit tests, integration tests, and system-level tests to validate the consensus algorithm's behavior under various scenarios, including normal operation and adversarial conditions. This testing should include edge cases, boundary values, and stress tests to uncover vulnerabilities and ensure the algorithm functions as intended.
5. **Red Team Testing:** Employing red team testing or conducting penetration testing can help simulate real-world attack scenarios and identify potential vulnerabilities. Independent security researchers or teams can attempt to exploit the consensus algorithm's weaknesses and provide valuable feedback to strengthen its security.
6. **Bug Bounty Programs:** Launching bug bounty programs encourages external security researchers and developers to scrutinize the consensus algorithm's design and implementation for vulnerabilities. Providing incentives for responsible disclosure of vulnerabilities helps identify and address potential security flaws before they can be exploited.
7. **Regular Updates and Patching:** Consensus algorithms should be actively maintained and updated to address any security vulnerabilities that may arise over time. Timely patching and updates help mitigate known vulnerabilities and ensure the security of the algorithm in the face of emerging threats.
8. **Community Engagement:** Encouraging a vibrant and engaged community around the developers of new consensus algorithms fosters active discussions, bug reporting, and knowledge sharing. Community involvement helps identify and address security concerns from diverse perspectives, enhancing the overall security of the algorithm.

Regarding documented threat models, there are well-documented threat models available for consensus algorithms. These threat models outline potential risks, attack vectors, and assumptions made when designing and implementing consensus algorithms. They provide a comprehensive understanding of the security considerations and potential vulnerabilities that consensus algorithms may face. While the specific documentation may vary based on the consensus algorithm and the associated blockchain platform, here are a few examples:

- **Bitcoin's PoW:** The original Bitcoin whitepaper by Satoshi Nakamoto outlines the threat model for PoW consensus. It discusses various potential attacks, including 51% attacks, selfish mining, double-spending, and eclipse attacks.

- **Ethereum's PoS:** The Ethereum community has published documentation and research papers that detail the threat model for PoS consensus. These documents outline risks like nothing-at-stake attacks, long-range attacks, and stake grinding and propose solutions to mitigate them.
- **Tendermint Byzantine Fault Tolerant (BFT):** Tendermint, a BFT consensus algorithm used by various blockchain platforms, provides a detailed threat model. It covers potential Byzantine faults, including validator misbehavior, network attacks, and equivocation.
- **Hyperledger Fabric's PBFT:** Hyperledger Fabric, which uses a variant of PBFT consensus, provides a comprehensive threat model in its documentation. It includes potential attacks like malicious peer behavior, DoS attacks, and collusion among peers.

These examples highlight that consensus algorithms often have associated research papers, technical documentation, or whitepapers that include threat models. These models serve as a valuable resource for understanding and addressing potential security risks and vulnerabilities within the consensus algorithm. It is recommended to refer to the specific documentation of the consensus algorithm and associated blockchain platform of interest for more detailed and specific threat models.

Formal verification tools provide developers and researchers with the means to rigorously verify the correctness of consensus algorithms and simulate different attack scenarios to assess their resilience and security. By utilizing these tools, they can identify and address potential flaws, vulnerabilities, and business logic issues, ultimately enhancing the overall reliability and trustworthiness of the consensus algorithms and the underlying blockchain system. Some examples of formal verification tools include:

- **TLA+ (Temporal Logic of Actions)** [76]: TLA+ is a formal specification language that allows developers to describe system behavior and properties. It can be used to model and verify the correctness of consensus algorithms, ensuring that they adhere to desired properties and business logic rules.
- **Coq** [77]: Coq is a proof assistant that allows for the formal verification of software and mathematical proofs. It can be utilized to verify the correctness of consensus algorithm implementations, ensuring that they satisfy specified properties and logic requirements.

Attack simulation tools provide software developers and researchers with valuable insights into potential weaknesses in the design and implementation of consensus algorithms. By leveraging these tools, they can identify not only technical vulnerabilities but also potential business logic flaws within the dApp and the underlying blockchain system. This knowledge empowers them to address these weaknesses and enhance the overall resilience, security, and trustworthiness of the system. By simulating various attack scenarios, developers and researchers can proactively identify and mitigate potential risks, ensuring that the dApp and the blockchain system are better equipped to withstand real-world threats and maintain the integrity of transactions and data.

Attack simulation tools are critical for assessing the security of consensus algorithms within blockchain platforms. These tools simulate various attack scenarios such as double-spending, Sybil attacks, forking vulnerabilities, eclipse attacks, and 51% attacks, to evaluate the resilience of the consensus mechanism. In cases where smart contracts are involved, these tools also help uncover vulnerabilities like reentrancy attacks or unauthorized access to contract functions. Rather than relying solely on theoretical models, security researchers are advised to examine the evidence from expert-conducted attack simulation tests. This empirical approach provides valuable insights into real-world performance and identifies design weaknesses that require remediation for enhanced blockchain security.

Some examples of attack simulation tools for testing the business logic and resilience to attacks of consensus algorithms include:

- **ByzSim** [78]: ByzSim is an attack simulation framework specifically designed for Byzantine fault-tolerant consensus algorithms. It enables developers to simulate various Byzantine attacks, test the resilience of consensus algorithms, and evaluate their performance under adversarial conditions.
- **TruffleHog** [79]: TruffleHog is a tool that scans source code repositories to identify sensitive information and potential vulnerabilities. It can be employed to analyze the business logic implemented within a consensus algorithm, ensuring that critical information is not inadvertently exposed.
- **Jepsen** [80]: Jepsen is a testing framework for distributed systems that can be used to evaluate the correctness and resilience of consensus algorithms. It helps simulate real-world failure scenarios, such as network partitions and node crashes, to assess the algorithm's behavior and reliability.
- **MultiChain Streams** [81]: MultiChain Streams is a tool for blockchain data stream analysis. It enables developers to simulate and analyze various attack scenarios, test the robustness of consensus algorithms, and evaluate the integrity and security of data streams in a blockchain system.

These tools provide developers and researchers with the means to rigorously verify the correctness of consensus algorithms and simulate different attack scenarios to assess their resilience and security. By utilizing these tools, they can identify and address potential flaws, vulnerabilities, and business logic issues, ultimately enhancing the overall reliability and trustworthiness of the consensus algorithms and the underlying blockchain system.

2.4.5.2 Secure Consensus Algorithm Best Practices

Selecting and implementing secure consensus algorithms involves several best practices to ensure the integrity and robustness of a blockchain system. Here are some key practices to consider:

1. **Meeting dApp Requirements:** Begin by understanding the specific requirements of the dApp and choose a blockchain platform that could fulfill them; examples include requirements of scalability, decentralization, security, and performance. This understanding will help guide the selection of an appropriate consensus algorithm that aligns with the system's goals. For example, if scalability is a crucial requirement, you may consider selecting a consensus algorithm like DPoS or PBFT that can handle a higher transaction throughput.
2. **Evaluate Security Properties:** Assess the security properties offered by different consensus algorithms. Consider factors like resistance to known attacks (e.g. 51% attacks, Sybil attacks), tolerance to Byzantine faults, and protection against various types of network disruptions or manipulations. For example, in the case of PoS, look for evidence of attack simulation tests that evaluate its resistance to known attacks like nothing-at-stake or long-range attacks.
3. **Conduct Some Research:** Look into the research papers, technical documentation, and community discussions surrounding the consensus algorithm under consideration. Evaluate the level of scrutiny and the feedback received from the research and development community to gauge the algorithm's maturity and resilience to potential vulnerabilities. For example, examine research papers and technical documentation related to the consensus algorithm, such as the original Bitcoin whitepaper for PoW. Analyze the community's feedback, discussions, and contributions to gauge the algorithm's credibility and effectiveness.
4. **Consider Implementation Complexity:** Assess the complexity of implementing and maintaining the chosen consensus algorithm. More complex algorithms may introduce additional risks and potential vulnerabilities. Opt for algorithms with clear specifications, well-defined protocols, and established reference implementations. For example, compare the complexity of different consensus algorithms. For instance, PoW has a relatively straightforward implementation, while more advanced algorithms like Tendermint's BFT consensus may require a more intricate implementation process.
5. **Evaluate Results of Rigorous Testing:** Evaluate any results of testing of the consensus algorithms done in a controlled environment before deploying it in a production setting. If required, such as in case of lack of testing reports, conduct thorough testing, including functional testing, stress testing, and security testing, to identify potential vulnerabilities and ensure the algorithm

behaves as expected. For example, before deploying in a production environment, set up a test network to evaluate the consensus algorithm's behavior. Conduct functional testing, stress testing, and security testing to identify potential vulnerabilities and ensure the algorithm performs as intended.

6. **Keep Up with Regular Updates and Security Patches:** Stay up-to-date with the latest advancements, research findings, and security vulnerabilities associated with the chosen consensus algorithm. Regularly update the algorithm implementation with security patches and enhancements to address any identified vulnerabilities or weaknesses. For example, stay informed about any security vulnerabilities or updates related to the chosen consensus algorithm. For example, Ethereum's PoS design undergoes continuous improvement and is updated based on research findings and community feedback.
7. **Engage with Security Experts:** Seek input and guidance from security experts and professionals who specialize in blockchain and consensus algorithms. Their expertise can help identify potential risks and provide recommendations for secure implementation and configuration. For example, collaborate with security experts who specialize in blockchain and consensus algorithms. They can provide insights into potential risks and offer guidance on secure implementation and configuration. Seek external security audits or reviews for an additional layer of expertise.
8. **Perform Ongoing Monitoring and Auditing:** Implement monitoring systems and perform regular audits to detect any anomalies or suspicious behavior in the consensus algorithm or the blockchain network. Continuously monitor for emerging security threats and vulnerabilities, and proactively address them. For example, implement monitoring tools that detect anomalies in the consensus algorithm or blockchain network, and use blockchain explorers or network analyzers to identify unusual behavior or potential security threats. Conduct periodic audits to assess the overall security posture of the system.
9. **Stay Up-to-Date and Informed:** Keep up with the latest research, best practices, and advancements in consensus algorithm security. Attend conferences, workshops, and forums related to blockchain security to stay informed about emerging threats and potential improvements in consensus algorithms. Stay updated with the latest research papers, conferences, and workshops focused on blockchain security. Engage in the blockchain community to remain aware of emerging threats and advancements in consensus algorithm security.

By following these best practices, developers and implementers can select and implement secure consensus algorithms that align with the specific requirements and goals of their blockchain systems. Regular evaluation, testing, and collaboration with security experts ensure that the consensus algorithm remains resilient and robust in the face of evolving security challenges.

2.4.6 Protecting Secrets

In dApps, keys and secrets play a critical role in ensuring security, identity management, and access control. Compromising private keys can lead to unauthorized access to users' funds and data, while exposing secrets can result in various security risks, such as data breaches and service abuse.

Private keys are a fundamental component of most blockchain-based dApps. They are used for cryptographic purposes, particularly in asymmetric cryptography, where a pair of keys is generated: a public key and a private key.

Specifically private keys are used to sign transactions before they are sent to the blockchain network. This signature proves the authenticity of the transaction and that it was initiated by the rightful owner of the private key. Private keys are used to access and control wallets and accounts on the blockchain. Only users who possess the correct private key can access their funds or interact with smart contracts associated with their accounts.

Secrets in dApps refer to sensitive information that must be kept confidential and secure. dApps might integrate with external services, and these services often require API keys for authentication. These API keys must be kept secret to prevent unauthorized access to the services. dApps that use OAuth or similar authentication protocols might use access tokens to grant users access to specific resources. These tokens are sensitive and should be protected from unauthorized access. Secrets used for data encryption and decryption must be kept secure to protect sensitive data stored or transmitted by the dApp. In some cases, dApps may use secrets to authenticate and authorize different components or services within the application.

The security of keys and secrets in dApps is inherently dependent on how secure the design of digital wallets is. Digital wallets serve as the primary means for users to store and manage their private keys and secrets, making them a critical target for attackers. To secure private keys, digital wallets must employ robust key management practices to protect users' private keys. A secure design should ensure that private keys are stored in encrypted formats, making them unreadable and inaccessible to unauthorized entities. Encryption ensures that even if attackers gain access to wallet data, they cannot retrieve the actual private keys without the encryption keys. Since digital wallets interact with smart contracts, a secure design should account for potential vulnerabilities in the smart contracts, preventing malicious actors from exploiting weaknesses in the code to compromise private keys or secrets. For hardware wallets, the physical security of the device is crucial. The design should consider protection against tampering and physical attacks to prevent unauthorized access to the private keys and secrets stored on the hardware.

Sensitive secrets, like API keys, access tokens, and authorization tokens, should be stored securely in a digital wallet or other secure storage solutions. The secure storage mechanism ensures that only authorized users or applications can access these secrets, protecting them from unauthorized disclosure. The process of provisioning API keys and tokens should follow security best practices. Proper authentication and authorization mechanisms should be in place to ensure that only legitimate entities receive the necessary tokens. Additionally, using short-lived tokens and implementing token rotation policies reduces the window of opportunity for attackers to misuse them.

The dApp's codebase should be robust and secure to prevent exploitation by attackers seeking to obtain secrets through code vulnerabilities. Regular code audits and adherence to secure coding practices help minimize these risks. By incorporating these security measures, dApps can strengthen the protection of secrets, including API keys and tokens. Secure storage, transmission, and provisioning practices reduce the risk of unauthorized access and misuse of these sensitive pieces of information, contributing to the overall security of the dApps.

2.4.6.1 Practical Examples of Security by Design Protection of Secrets and Keys in dApps

When designing dApps, ensuring the security of secrets and keys is of paramount importance to protect user funds and sensitive data. Implementing security by design practices helps fortify the dApp's architecture and safeguard against potential threats. In this context, we present 10 practical examples of security measures to protect secrets and keys in dApps:

1. **Secure Hardware Wallet Integration:** Integrating hardware wallets, such as Ledger [82] or Trezor [83], provides an additional layer of security by storing private keys in a secure hardware environment. This minimizes the risk of unauthorized access and key exposure, offering users a trustworthy way to interact with the dApp. Hardware wallets store private keys in a secure hardware environment, isolated from the Internet and other potential attack vectors. Hardware wallets generate private keys and sign transactions offline within the device itself. Private keys never leave the hardware wallet, preventing exposure to potential online threats. When using a dApp that integrates with hardware wallets, users interact with the application through the hardware wallet's interface. Authentication for transactions and other actions typically requires a physical button press or PIN entry on the device, ensuring that users explicitly approve each operation. Hardware wallets handle cryptographic operations related to signing transactions and generating digital signatures. These operations occur within the secure hardware environment, safeguarding private keys from being exposed during the process. Integration with hardware wallets is possible also when using software wallets such as MetaMask [84] that is a software wallet in the form of a browser extension that allows users to interact with

dApps on the Ethereum blockchain and other compatible blockchains directly from their web browsers. While MetaMask is not a hardware wallet, it can be used in conjunction with a hardware wallet for enhanced security. MetaMask allows users to connect their hardware wallets, such as Ledger Nano S [85], Ledger Nano X [86], or Trezor, to the browser extension. When a hardware wallet is connected, MetaMask serves as a bridge between the browser and the hardware wallet, enabling users to securely sign transactions and interact with dApps using the private keys stored on the hardware wallet. The advantage of using MetaMask with a hardware wallet is that it adds an extra layer of security. Private keys remain stored and protected on the hardware wallet, away from potential online threats and vulnerabilities that might affect software-only wallets. Users can initiate transactions on the browser through MetaMask, and the actual signing of the transactions happens securely within the hardware wallet, ensuring that private keys are never exposed to the Internet.

2. **Secure Key Generation and Management:** The process of generating private keys should use cryptographically secure random number generators (CSPRNGs) to ensure that keys are truly unpredictable and cannot be easily guessed by attackers. Secure key management is of utmost importance for dApps due to the critical role that keys play in ensuring the security and integrity of the entire system. In most dApps, private keys are used to control users' cryptocurrency wallets and digital assets. Secure key management ensures that these valuable assets remain protected from unauthorized access and potential theft. Private keys are often associated with users' identities on the blockchain. Securely managing these keys ensures that users' identities are safeguarded, reducing the risk of impersonation or identity theft. Private keys are used to sign transactions and smart contracts on the blockchain. Secure key management ensures that only authorized individuals can initiate transactions, preventing unauthorized changes to the blockchain's state. Implement key rotation practices periodically to change encryption keys. In case of potential key compromise, have a key revocation mechanism in place to invalidate the old keys and generate new ones. When private keys need to be stored on-chain or off-chain, encryption is crucial. Private key encryption ensures that even if the encrypted keys are somehow accessed, they remain unreadable without the appropriate decryption key. When private keys need to be stored on-chain or off-chain, encryption is crucial. Private key encryption ensures that even if the encrypted keys are somehow accessed, they remain unreadable without the appropriate decryption key. As a software developer working on a dApp that involves wallet integration and requires secure key management, some practical advice and best practices to ensure the application follows secure key management standards and uses strong encryption algorithms includes making sure you elect a reputable and well-audited wallet provider or library for wallet integration. Popular

options like MetaMask or hardware wallets (e.g. Ledger or Trezor) have undergone rigorous security assessments. It is important to Leverage Web3 libraries (e.g. web3.js or ethers.js) to interact with wallets securely. These libraries abstract away complexities and provide secure methods for transaction signing and communication with the blockchain. Avoid storing sensitive data, such as private keys or access tokens, on the client-side. Lastly make sure to implement key rotation practices to periodically change encryption keys and other sensitive credentials, when encrypting sensitive data, such as private keys or access tokens, use strong encryption algorithms like AES-256 to protect the data from unauthorized access and ensure that the dApp and its communication with the blockchain are served over HTTPS to encrypt data transmitted between the user's browser and the server.

3. **Secure Smart Contract Development:** Following secure coding practices during smart contract development mitigates potential vulnerabilities. Leveraging audited and tested code libraries, like OpenZeppelin [87], minimizes the risk of exploitable flaws that may compromise keys or secrets. Secure key storage and private key management are critical aspects of smart contract development to ensure the protection of sensitive cryptographic information and maintain the security of the entire dApp. While smart contracts are often associated with executing code based on predefined conditions and rules, they can also play a role in facilitating and authorizing transactions on the blockchain. For example, smart contracts can be used to create multi-sig wallets, where multiple parties must collectively sign a transaction for it to be executed. This adds an extra layer of security by preventing a single point of failure and ensuring that a certain number of authorized parties must agree before a transaction is approved. When a user wants to initiate a transaction that involves a smart contract, they interact with the dApp's frontend interface or use a wallet application. The dApp or wallet prompts the user to sign the transaction. To sign the transaction, the user's private key is required. Users usually store their private keys securely in a hardware wallet, software wallet, or other secure storage solutions. Once the user authorizes the transaction, their wallet application uses their private key to sign the transaction data, creating a digital signature. The transaction data, including the digital signature, is received and validated by the blockchain network. If the signature is valid and the transaction meets the conditions set by the smart contract, the transaction is executed, and the smart contract updates the blockchain state accordingly. In summary, smart contracts do not directly access private keys for signing transactions. Instead, external entities, such as users' wallets, handle the process of signing transactions using private keys. This separation of duties ensures that private keys remain under the control of users and are not exposed to smart contracts, which are public and immutable pieces of code. The use of private keys for transaction signing adds an essential layer of security and authentication to interactions with smart contracts on the blockchain.

2.4.6.2 Identification of Potential Vulnerabilities Related to Secret and Key Management with dApps

Identifying potential vulnerabilities related to secret and key management with dApps is crucial for ensuring the security and integrity of these applications. Here are some steps you can take to identify and address such vulnerabilities:

1. **Secure Code Reviews:** Start by conducting a thorough code review of the dApp's smart contracts and backend code. Look for any hardcoded secrets or private keys, as these can be easily compromised if discovered by attackers. Check the visibility of variables in the smart contract. Ensure that sensitive data and private keys are declared with appropriate visibility keywords (e.g. private or internal) to prevent unintended exposure. Check if any sensitive data is being emitted in event logs. Avoid logging private keys or sensitive information in event emissions. Best practice is to avoid storing sensitive data like private keys within the smart contract itself. Instead, use external secure key management systems or wallets. Analyze the access control mechanisms in the smart contract. Ensure that only authorized users or contracts can access sensitive functions or modify critical data. If the smart contract involves encryption or decryption processes, verify that they are implemented correctly and securely. Depending on the business criticality of the dApp being developed, engage security experts or auditors to conduct a comprehensive security audit of the dApp's smart contracts. They will be able to identify common vulnerabilities like reentrancy attacks, arithmetic overflows, and other issues that might expose sensitive information.
2. **Key Generation and Storage:** Analyze how keys (both public and private) are generated and stored within the dApp. analyzing how keys are generated and stored within the dApp involves a comprehensive review of various aspects, including software and hardware wallets. Ensure that the software wallet does not store private keys in an easily accessible location, such as plaintext files or insecure databases. Analyze how the hardware wallet interacts with the dApp. Look for secure communication channels to prevent data interception or manipulation. Ensure that proper key management practices are followed, such as using strong random number generators for key generation and securely storing private keys.
3. **Encryption:** Check if sensitive data, such as private keys or user data, are encrypted when stored or transmitted. Proper encryption ensures that even if an attacker gains access to the data, it remains unreadable without the decryption key.
4. **Secure Key Handling:** Examine how keys are used within the dApp's code. Avoid exposing private keys or sensitive information in logs, error messages, or API responses.

5. **Access Control:** Review the access control mechanisms of the dApp to ensure that only authorized users or contracts can access sensitive functions or data. Analyze the access control mechanisms in the smart contract. Ensure that only authorized users or contracts can access sensitive functions or modify critical data.
6. **Third-Party Libraries and Dependencies:** Investigate the use of third-party libraries and dependencies within the dApp. Make sure that these components are reputable and up-to-date and don't introduce any security vulnerabilities.
7. **Token Standards:** If the dApp involves the use of tokens or cryptocurrencies, ensure that the token standards (e.g. ERC-20, ERC-721) are implemented securely to prevent potential exploits.
8. **Penetration Testing:** Perform penetration testing and vulnerability scanning to identify potential weaknesses in the dApp's infrastructure, APIs, and web interfaces.
9. **Regular Updates and Patching:** Keep the dApp's software and dependencies up-to-date with the latest security patches. Vulnerabilities can emerge over time, and staying current is essential to maintain security.
10. **Security Best Practices:** Promote the adoption of security best practices among the dApp's developers and users. Educate users about the importance of safeguarding their private keys and using secure practices.

By diligently following these steps, you can significantly reduce the potential vulnerabilities related to secret and key management in dApps and enhance their overall security posture. Automated tools, such as static analysis tools and security scanners, can also assist in identifying potential vulnerabilities. However, manual review by experienced auditors is crucial as automated tools might have limitations and false positives/negatives. Always prioritize security during the development phase and consider seeking help from blockchain security experts or auditors to conduct a thorough and reliable security audit of your smart contracts. Remember that security is an ongoing process, and continuous monitoring and improvement are essential to keep up with emerging threats.

2.4.7 Securing Token-Based Transactions

Token-based transactions play a pivotal and indispensable role in the functioning of dApps within the blockchain ecosystem. These tokens serve as digital representations of real-world assets or utility, granting users access rights and enabling seamless exchanges and trades. However, as the popularity of these tokens continues to soar, ensuring the security of such transactions becomes of utmost importance to safeguard both users and valuable assets from potential vulnerabilities.

Securing token-based transactions demands robust and comprehensive security measures to uphold the integrity and trustworthiness of the underlying blockchain network. Numerous threats loom, including the looming menace of token theft, malicious attacks, and the fear of unauthorized access to critical private keys.

Different token standards, such as the widely used ERC-20 and ERC-721, come with varying security considerations due to their distinct fungible and nonfungible nature. Fungible tokens, such as ERC-20 tokens, possess interchangeability, making it vital to prevent double-spending attacks to maintain transaction accuracy and credibility. On the other hand, NFTs, like ERC-721 tokens, represent unique and singular assets, necessitating enhanced protection against fraudulent ownership transfers and reentrancy attacks, where an attacker repeatedly reenters a contract to exploit its vulnerabilities.

To ensure the security and safe custody of tokens during transfers and trades, several best practices must be diligently followed. Secure key management practices are paramount, as they form the foundation of transaction safety. Users are encouraged to utilize hardware wallets, which store private keys offline, thereby significantly reducing the risk of online attacks. Furthermore, implementing MFA adds an extra layer of defense against unauthorized access to private keys and accounts.

Standardized security protocols play a vital role in building trust within the blockchain community. Following well-established security standards can instill confidence in users and developers alike, encouraging them to participate in token-based transactions with peace of mind. Regular audits, conducted by reputable and independent security experts, further enhance security by identifying vulnerabilities and loopholes before malicious actors can exploit them.

2.4.7.1 Explanation of Token-Based Transactions

Token-based transactions play a pivotal role in the world of dApps within blockchain ecosystems. These transactions involve the exchange and transfer of digital tokens, which act as representations of various assets, access rights, or utility within the blockchain network. Tokens serve as a fundamental component of dApps, providing the means for users to interact with and participate in various blockchain-based services and applications. Beyond cryptocurrencies, tokens can represent a wide range of digital assets, including virtual goods in gaming dApps, digital certificates, IP rights, or even ownership of real-world assets like real estate or precious metals. By representing these assets as tokens, their ownership and transfer become more efficient and transparent on the blockchain.

Token can represent a wide range of assets, such as cryptocurrencies, digital certificates, virtual goods, loyalty points, and even real-world assets like real estate or commodities. The key roles of token-based transactions in dApps are as follows:

1. **Asset Representation:** Tokens act as digital representations of both digital and physical assets, enabling seamless and efficient transfer and exchange of value on the blockchain. For example, in the context of cryptocurrencies like Bitcoin or Ethereum, tokens serve as digital representations of value. When a user owns a certain amount of a particular token, it implies they possess a corresponding value of that asset.
2. **Access Rights:** Tokens can grant users access to specific functionalities or features within a dApp. For instance, in a gaming dApp, a token might provide access to premium content or exclusive in-game items. Tokens in a decentralized social media platform provide users with access to premium content, enhanced profile features, voting power in governance, and early access to new features. They enhance user engagement and offer exclusive privileges within the platform.
3. **Utility:** Tokens often serve as the primary means of payment for services and transactions within dApps. They streamline payment processes and eliminate the need for traditional intermediaries, such as banks. Utility tokens serve as the primary means of payment within dApps. An example is a decentralized e-commerce platform using utility tokens. By utilizing utility tokens, DecentralShop streamlines payment processes, eliminates the need for traditional intermediaries like banks, and provides instant, borderless, and low-cost transactions. Additionally, the blockchain's security and transparency enhance user trust, while loyalty programs and incentives further incentivize user engagement. Utility tokens offer an efficient and user-friendly payment solution, making dApps a compelling alternative to centralized platforms.
4. **Decentralized Governance:** Some dApps use tokens as a governance mechanism, allowing token holders to participate in decision-making processes related to the development and evolution of the application. For example, when a protocol upgrade proposal is made, token holders have the right to vote on whether to accept or reject the proposed changes. Token holders can cast their votes through on-chain voting mechanisms using their digital wallets. In dApps that use tokens for governance token holders can vote on protocol upgrades, platform parameters, and responses to security events; for example, if there is a vulnerability discovered in a smart contract, a governance proposal can be submitted to freeze or pause affected functions until a secure fix is implemented. This democratic approach empowers the community, fosters a sense of ownership, and ensures that the platform's development

aligns with the interests and preferences of its users, promoting transparency and community-driven growth.

5. **Crowdfunding and Fundraising:** Tokens can facilitate crowdfunding campaigns or initial coin offerings (ICOs), allowing projects to raise funds by selling their native tokens to interested investors. For example, a decentralized crowdfunding platform that uses utility tokens to enable entrepreneurs to propose projects and seek funding from interested investors. Backers can purchase these tokens to support projects, and in return, they receive various benefits and become stakeholders in the project. The successful use of utility tokens for crowdfunding fosters transparency, trust, and innovation within the dApp ecosystem.
6. **Interoperability:** Tokens can be used to bridge different blockchain networks, facilitating cross-chain transactions and interactions between various dApps. Tokens, particularly interoperability tokens such as tokens that act as the bridge between different blockchains, can play a vital role in achieving cross-chain interoperability in the blockchain space. A cross-chain protocol can enable users to wrap tokens from one blockchain into another, facilitating seamless interactions and transactions between different blockchains and dApps. This interoperability enhances liquidity provision, enables token swapping, and fosters cross-chain communication, allowing users to access services and assets across multiple blockchains effortlessly. Overall, tokens serve as bridges between blockchain networks, promoting decentralization and expanding the potential of dApps.

Overall, token-based transactions are the lifeblood of dApps, enabling efficient and secure P2P interactions within the blockchain ecosystem. They foster a trustless environment, where users can transact directly with one another without relying on central authorities, promoting transparency, efficiency, and inclusivity in various decentralized services and industries.

2.4.7.2 Secure Token Standards

Token standards are essential for the blockchain ecosystem as they provide a set of rules and guidelines for creating and managing tokens in a standardized and secure manner. The most notable token standards include ERC-20 [88] and ERC-721 [89] on the Ethereum blockchain. These standards ensure consistency, predictability, and interoperability of tokens across various applications, wallets, and exchanges. By adhering to token standards, developers can avoid common security pitfalls, benefit from external audits and reviews, and promote user trust and adoption. Token standards also foster community-driven best practices and enable seamless cross-chain interactions, enhancing the overall security and efficiency of token-based transactions within the blockchain space. Other token standards like BEP-20 [90], TRC-20 [91], NEP-5 [92], SPL Token [93], HRC-20 [94], and WRC-20 [95] serve specific blockchain networks and contribute to the diversity and innovation of the token ecosystem.

The origin of ERC-20 and ERC-721 standards for tokens can be traced back to the Ethereum blockchain and its community-driven efforts to standardize token implementations. ERC-20 stands for "Ethereum Request for Comments 20" [96]. It was proposed by Fabian Vogelsteller in November 2015 and was finalized and accepted as an Ethereum Improvement Proposal (EIP) by Vitalik Buterin, the creator of Ethereum, in September 2017. The standard outlines a set of rules and guidelines for creating and managing fungible tokens on the Ethereum blockchain.

The introduction of ERC-20 was a significant milestone in the development of Ethereum's token ecosystem. It provided a common blueprint for token creation and ensured that all ERC-20 tokens shared a standardized interface, making them easily compatible with various applications and wallets. As a result, ERC-20 tokens became widely adopted for various use cases, including ICOs, DeFi projects, and utility tokens. ERC-20 is the most widely adopted token standard on the Ethereum blockchain. These tokens are fungible, meaning each token is identical and interchangeable with any other token of the same type. ERC-20 tokens represent various cryptocurrencies, utility tokens, and digital assets within the Ethereum ecosystem. ERC-20 tokens have built-in security functions allowing developers to include additional security features in ERC-20 token contracts, such as pausing or freezing token transfers in case of suspicious activities or vulnerabilities. ERC-20 tokens also incorporate an approval mechanism, allowing token holders to authorize third-party applications to spend their tokens on their behalf. This helps prevent unauthorized access and misuse of tokens. To ensure the integrity of ERC-20 tokens and their smart contracts, projects often undergo regular security audits by external experts, identifying and fixing potential vulnerabilities.

ERC-721 stands for "Ethereum Request for Comments 721." It was proposed by William Entriken, Dieter Shirley, Jacob Evans, Nastassia Sachs, and Nathaniel Johns in October 2017. ERC-721 was also accepted as an EIP in the same month. ERC-721 introduced the concept of NFTs to the Ethereum blockchain. Unlike ERC-20 tokens, which are fungible and interchangeable, ERC-721 tokens represent unique and indivisible assets. Each ERC-721 token has its distinct value, properties, and ownership, making it ideal for representing digital art, collectibles, virtual real estate, and other unique items. The introduction of ERC-721 opened up new possibilities for asset ownership, provenance, and digital ownership in the blockchain space. NFTs gained significant attention in the art and gaming industries, as they allowed for the creation of scarce and verifiable digital assets. ERC-721 is a standard for NFTs on the Ethereum blockchain. NFTs represent unique assets, each with its distinct value, properties, and ownership. They are widely used for digital art, collectibles, virtual real estate, and other unique items.

It is worth mentioning some of the security features of ERC-721 tokens:

1. **Uniqueness and Indivisibility:** ERC-721 tokens are nonfungible, meaning each token is unique and cannot be divided into smaller units. This ensures the authenticity and exclusivity of digital assets represented by NFTs.
2. **Immutable Ownership:** NFT ownership is recorded on the blockchain through smart contracts, providing a transparent and immutable history of ownership transfers. This prevents fraudulent ownership claims and ensures the legitimacy of ownership.
3. **Secure Data Storage:** ERC-721 tokens can store metadata and additional information related to the represented asset. This metadata is tamperproof and remains associated with the token throughout its life cycle, enhancing security and provenance.
4. **Anticounterfeiting:** The uniqueness of ERC-721 tokens makes them an excellent solution for anticounterfeiting measures. Each NFT represents an authentic and verifiable asset, reducing the risk of counterfeit digital goods.
5. **Escrow and Auction Mechanisms:** NFT marketplaces often use escrow smart contracts and auction mechanisms to facilitate secure and fair asset transfers between buyers and sellers.

Overall, both ERC-20 and ERC-721 tokens incorporate various security features to protect users, assets, and data within the Ethereum blockchain ecosystem. The implementation of these security measures, coupled with regular security audits and community-driven best practices, contributes to a safer and more trustworthy decentralized environment for users and developers alike. Specifically for what concerns security, reliance on token standards is essential for the security of tokens for several reasons:

1. **Consistency and Predictability:** By adhering to standardized interfaces and functions, tokens developed under these standards are consistent and predictable. Developers and users can expect a certain level of behavior from tokens, reducing the risk of unintended vulnerabilities and ensuring better code quality.
2. **External Audits and Reviews:** These standards have undergone extensive community reviews and external security audits. Multiple eyes on the code increase the chances of identifying and fixing potential security vulnerabilities before tokens are deployed, enhancing the overall security of the token ecosystem.
3. **Interoperability:** Standardized token interfaces facilitate interoperability between different applications, wallets, and exchanges. This reduces the chances of faulty integrations and ensures that tokens can be securely used across various platforms, enhancing the overall security of token-based transactions.

4. **Preventing Common Attack Vectors:** The standards include measures to prevent common attack vectors such as double-spending, reentrancy, and unauthorized access. Implementing these features at the protocol level helps to protect tokens and the underlying applications from common security threats.
5. **Avoiding Security Pitfalls:** By following established standards, developers can avoid common security pitfalls that may arise when creating custom token contracts. Standardized implementations have already undergone rigorous security evaluations, reducing the likelihood of introducing unknown vulnerabilities.
6. **User Trust and Adoption:** Standardized tokens inspire user trust, as users are more likely to trust tokens that adhere to well-established standards. This increased trust fosters greater token adoption and usage, leading to a more robust and active token ecosystem.
7. **Community Best Practices:** ERC-20 and ERC-721 standards have paved the way for community-driven best practices in token development. Developers and auditors have developed security guidelines and best practices specific to these standards, which contribute to better security practices across the token space.
8. **Secure Smart Contract Templates:** The existence of standardized token templates and well-tested libraries allows developers to create tokens without reinventing the wheel. These templates often come with built-in security features, reducing the likelihood of introducing errors during development.

In summary, adhering to token standards provides a solid foundation for building secure tokens. By leveraging community-vetted code, established best practices, and comprehensive security reviews, token developers can mitigate risks and vulnerabilities, leading to a safer and more trustworthy token ecosystem for users and participants.

2.4.7.3 Security Considerations for Securing dApps with Token-Based Use Cases

It is crucial to use secure and audited token standard generated tokens when integrating tokens with dApps. Using well-established and audited token standards, such as ERC-20, ERC-721, BEP-20, TRC-20, etc., offers several important benefits, specifically from a security standpoint token standards undergo rigorous security audits, reducing the risk of vulnerabilities and attacks. Besides using audited and secure token standards is essential, developers should also implement additional security measures specific to their dApp's requirements. This includes secure smart contract coding practices, secure key management, regular security audits, and ongoing monitoring to identify and address potential security vulnerabilities.

Securing dApps that rely on tokens, whether they are fungible or nonfungible, and ensuring the security and reliability of token-based transactions are crucial for maintaining user trust and protecting assets. Here are the most important considerations for dApp developers to enhance the security of token-based use cases:

1. **Smart Contract Audits:** Conduct thorough security audits of smart contracts that handle token-related functionalities. Hire reputable third-party auditors to identify and address potential vulnerabilities and bugs in the code. Regularly update and improve the smart contracts based on audit findings.
2. **Token Standard Compliance:** Adhere to well-established token standards (e.g. ERC-20, ERC-721) for creating tokens. Compliance with these standards ensures compatibility, interoperability, and reduces the risk of introducing security flaws.
3. **Secure Key Management:** Implement robust key management practices to protect private keys associated with user wallets. Use hardware wallets or secure key storage solutions to prevent unauthorized access to tokens.
4. **Gas Limit and Transaction Fee Considerations:** Set appropriate gas limits and transaction fees for token transactions to avoid potential out-of-gas errors or unnecessarily high fees that may impact user experience.
5. **Preventing Double-Spending:** For fungible tokens, implement mechanisms to prevent double-spending attacks, ensuring that the same token cannot be spent multiple times.
6. **NFT Ownership:** For NFTs, ensure secure ownership transfers and enforce proper access control mechanisms to prevent fraudulent ownership claims.
7. **Escrow Mechanisms:** Use escrow smart contracts for high-value transactions to protect both buyers and sellers. These contracts can release funds or NFTs only when specific conditions are met, ensuring fair and secure exchanges.
8. **User Authentication and Authorization:** Implement MFA and authorization mechanisms to protect users' accounts and prevent unauthorized access to tokens.
9. **Regular Security Audits:** Conduct periodic security audits and reviews to assess the overall security posture of the dApp and its token-related functionalities.
10. **Emergency Response Plan:** Have a well-defined emergency response plan in place to handle security incidents and potential breaches. This plan should include steps for pausing contracts, notifying users, and mitigating the impact of the incident.
11. **Transparent Governance:** If the dApp incorporates a governance mechanism using tokens, ensure transparency in decision-making processes and involve the community in critical decisions.
12. **User Education:** Educate users about best practices for securing their wallets, avoiding phishing attacks, and protecting their assets.

By considering these security considerations and implementing best practices, dApp developers can enhance the security and reliability of token-based use cases. Ensuring a trustworthy and safe environment for users will foster greater adoption and support within the blockchain ecosystem. As practical examples, let's explore two fictional dApps utilizing fungible tokens and NFTs:

1. **TokenSwap** [96]: It is a DEX designed for trading and swapping fungible tokens. Users can effortlessly exchange their fungible tokens within this platform. TokenSwap will use a token based on the fungible token standard most commonly used on the Ethereum blockchain that is ERC-20.
2. **CryptoLend** [97]: CryptoLend is a decentralized lending platform where users can participate in cryptocurrency lending and borrowing. Through this dApp, users can lend their digital assets and borrow cryptocurrencies from other participants. CryptoLend will use the most commonly used NFT standard on the Ethereum blockchain that is ERC-721.

Considering the different use/business cases for the dApps, the types of token standards used (fungible and nonfungible), the following security considerations and requirements for secure token transactions can be documented:

1. **Token Standard Compliance:** Verify that the implemented token contracts adhere to the specific standards (e.g. ERC-20, ERC-721) to ensure compatibility with various wallets, exchanges, and applications. Noncompliance could lead to functionality issues or potential security vulnerabilities. Implement mechanisms to prevent double-spending attacks for fungible tokens (e.g. ERC-20 tokens).
2. **Ensure Secure Token Transactions:** Ensuring transaction integrity is crucial for maintaining trust and preventing loss of assets. A blockchain with consensus mechanisms like Ethereum's PoW or PoS ensures transactions are validated and recorded in a sequential manner. The implementation of the ERC-20 standard also inherently prevents double-spending as the transfer function ensures the availability of sufficient funds before processing a transaction. Double-spending attacks are not applicable to NFTs since each NFT represents a unique asset that cannot be duplicated or transferred more than once simultaneously. Each NFT has a specific ownership and cannot be replicated. For NFTs, ensure secure ownership transfers and enforce proper access control mechanisms to prevent fraudulent ownership claims or unauthorized transfer. Consider implementing token locking periods or time delays for specific transactions to prevent rapid and unauthorized token transfers.
3. **Ensure Secure Storage for Tokens:** Integrate with hardware wallets or secure key storage solutions to prevent unauthorized access to tokens. For centralized token storage in the dApp, implement strong security measures,

such as encryption and access controls, to protect tokens from unauthorized access and theft.

4. **Ensure Validations and Error Handling of Token Transactions:** Ensure that token swaps and transactions on DEXs are securely executed, with appropriate validations and error handling to prevent loss of funds or assets. Set appropriate gas limits and transaction fees for token transactions to avoid potential out-of-gas errors or unnecessarily high fees that may impact user experience.
5. **Ensure Governance and Transparency:** Ensure that the dApp incorporates a governance mechanism while using tokens, ensure transparency in decision-making processes, and involve the community in critical decisions to avoid centralized control and potential manipulation. Incorporate a governance mechanism using tokens, ensure transparency in decision-making processes, and involve the community in critical decisions to avoid centralized control and potential manipulation.
6. **Ensure Robust Testing and Auditing:** Thoroughly test smart contracts and dApp functionalities in various scenarios to identify and address potential vulnerabilities affecting token transactions. Conduct third-party security audits to verify the overall security posture of the dApp.
7. **Regular Updates and Patching:** Stay updated with the latest advancements in blockchain security and promptly apply security patches and updates to mitigate emerging threats.

In conclusion, it is paramount for developers to adhere to security best practices when building dApps utilizing fungible tokens or NFTs. Ensuring the integrity and security of these dApps and safeguarding tokens from potential threats should be a core focus throughout the development process. Emphasizing security and taking proactive measures will help minimize risks and vulnerabilities, instilling confidence in users and promoting the wider adoption of dApps.

2.4.8 Securing Cryptocurrency DEX Transactions

Prioritizing the security of DEXs is essential for dApp security and for the security of cryptocurrency transactions. This focus helps protect user assets, prevent vulnerabilities, maintain trust, comply with regulations, mitigate risks, avoid network effects, and ensure the dApp's longevity and success. When a dApp integrates with a compromised DEX, it opens the door to significant security risks that can result in financial losses for its users. DEXs facilitate the trading of cryptocurrencies and tokens in a P2P manner. However, if a DEX has security vulnerabilities or has been compromised by malicious actors, the integrity of user transactions and funds can be compromised.

Malicious actors can exploit vulnerabilities in the compromised DEX to gain unauthorized access to users' accounts. This unauthorized access can lead to the theft of users' private keys, account credentials, and other sensitive information. Once unauthorized access is gained, attackers can initiate transactions to transfer users' assets from their wallets to the attackers' wallets. This can result in the direct loss of funds from users' accounts. Compromised DEXs can manipulate the transactions initiated by users. Attackers might alter transaction details, reroute funds, or conduct unauthorized trades, causing financial harm to users. A compromised DEX can be used to launch phishing attacks on dApp users. Malicious websites or apps could mimic the dApp or DEX interface, tricking users into disclosing their private keys or login credentials.

If the smart contracts governing the DEX are compromised, attackers might manipulate trades, alter balances, or exploit vulnerabilities to drain users' funds. When a dApp integrates with a compromised DEX, users might have limited or no recourse to recover their stolen assets. Unlike centralized exchanges (CEXs), DEXs often lack the same level of customer support and insurance. From the perspective of a dApp developer, ensuring secure due diligence before integrating a DEX is of utmost importance. This is because a compromised DEX can expose users of the dApp to a range of severe security risks, leading to significant financial losses and reputational damage.

It's essential for dApp developers and architects to thoroughly assess the risks associated with integrating DEXs into their applications. This process involves starting by prioritizing the security of securing dApp integration with digital exchanges. By doing so, they can establish a solid foundation for risk management before factoring in the potential use cases. This approach ensures that the dApp's security framework is robust and well-prepared to address the specific risks posed by DEX integration, which can then be extended to mitigate risks across various use cases. Ultimately, a proactive and security-focused approach helps safeguard user assets, maintain trust, and ensure the long-term success of both the dApp and the integrated DEX.

2.4.8.1 Securing dApp Integration with Digital Exchanges

Securing dApp integration with digital exchanges involves implementing a range of design controls to mitigate risks and ensure the overall security of the integrated solution. Cryptocurrency exchanges often attract developers by presenting the idea that there is significant friction when building projects on the application layer. This notion implies that creating applications and services directly on the blockchain or decentralized platforms can be complex and challenging due to various technical and operational barriers. Exchanges, on the other hand, offer a more accessible environment for developers to access cryptocurrency markets, trade assets, and build applications that leverage the existing infrastructure they provide. This can include ready-made APIs, trading platforms, liquidity pools, and other tools that streamline the development process and enable developers to focus on building innovative solutions without having to navigate the intricacies of the underlying blockchain technology.

The idea that utilizing DEXs instead of CEXs automatically grants blockchain applications access to DEX features, such as noncustodial trading, lower security risks, and exemption from Know Your Customer (KYC) obligations, is not entirely accurate. While DEXs offer certain advantages like noncustodial trading, where users retain control of their private keys, and potentially reduced security risks due to decentralized nature, there are complexities that need to be considered. It's important to note that DEXs also have their own challenges, including potential liquidity issues, limited trading pairs, and sometimes less user-friendly interfaces compared to CEXs. Additionally, the assertion that DEXs inherently bypass KYC requirements is not universally true. Some DEXs do implement KYC processes [98] to comply with regulatory standards, especially when dealing with fiat-to-crypto transactions or certain jurisdictions. Therefore, while DEXs offer unique benefits for blockchain applications, it's crucial to thoroughly understand both the advantages and limitations before deciding to integrate them into a project.

Crypto asset tokens have the ability to symbolize a wide range of real-world items, ranging from simple items like a movie ticket or a unit of oil to more abstract concepts like someone's contributed time. These tokens essentially serve as digital representations of tangible assets or values existing in the physical realm. This process of tokenization at this level shares similarities with securitization in the traditional financial sphere, where companies can convert real assets or certain rights into tradable securities for the purpose of trading and investment. DEXs introduce a novel marketplace model that holds significant value due to its capacity for facilitating secure P2P transactions involving crypto asset tokens. DEXs provide a well-suited platform for the trading of these tokens directly among holders, fostering a secure and efficient environment for the exchange of crypto assets. Nonetheless, the situation isn't universally applicable. Trading platforms generate profits by leveraging the spread between buying and selling prices, rather than relying on transaction fees. This unique approach implies that the platform operator engages as a trading counterpart, equipped with insights into all ask and bid prices, enabling timely market reactions ahead of other participants. This advantage, though unfair, results in consistent trading profits for the central exchange operator.

It's crucial to acknowledge that while the trading of data and tickets on DEXs might appear innocuous to financial regulatory bodies, the exchange of cryptocurrencies and financial securities in a P2P manner raises serious concerns regarding money laundering, criminal funding, and terrorist activities. This concern extends beyond DEXs to the entire realm of cryptocurrencies. Due to the absence of centralized authentication, many DEXs don't mandate KYC procedures and lack a clear entity responsible for cooperating with government regulatory bodies. To facilitate broader DEX adoption, it's accurate that features such as KYC and accountability must be integrated into the code, potentially leading to a more centralized DEX operation. Additionally, DEXs provide an avenue for regulatory agencies and society to monitor suspicious transactions and operations within a significantly different framework. dApp developers must be vigilant about DEX regulatory risks. Blindly trusting the ease of integrating a dApp with DEXs through APIs and smart contracts could lead to potential regulatory challenges. It's essential to conduct thorough due diligence on the regulatory environment surrounding DEX operations. This includes assessing compliance with Anti–Money Laundering (AML) and KYC regulations, as well as understanding the responsibilities of the DEX operator in liaising with regulatory authorities. Mitigating these risks requires a comprehensive understanding of the legal landscape and careful consideration of the potential consequences of DEX integration.

From the perspective of integrating a dApp with a DEX, the risk of front-running is a significant concern. Front-running occurs when a malicious actor exploits their knowledge of pending transactions to manipulate prices or gain an unfair advantage. This can lead to several adverse outcomes:

1. **Unfair Advantage:** Front-runners can preemptively execute transactions before legitimate users, profiting from price changes caused by the impending trades. This undermines the fairness of the trading process.
2. **Market Manipulation:** Front-running can distort market prices, leading to inaccurate valuations and making it challenging for users to execute trades at fair prices.
3. **Losses for Traders:** Legitimate traders may experience losses due to price manipulation caused by front-running, affecting their trading strategies and outcomes.
4. **Erosion of Trust:** Front-running incidents can erode trust in the dApp and the DEX, discouraging users from participating in trading activities.
5. **Negative User Experience:** Front-running can cause delays in transaction confirmation, higher gas fees, and increased slippage, negatively impacting the overall user experience.

Mitigating front-running risks involves implementing measures such as using zero-knowledge proofs to obscure transaction details, exploring Layer 2 scaling solutions to reduce transaction speed differentials, and designing smart contracts that minimize the possibility of front-running by reducing the visibility of pending transactions. Additionally, educating users about front-running risks and providing clear guidelines on how to navigate these challenges can contribute to a safer trading environment within the dApp.

When considering the integration of a dApp with a DEX, several questions arise regarding the security of assets and transactions, as well as the interoperability of different networks. Here's a breakdown of the concerns that should be addressed:

1. **Asset Support and Security:** If your dApp's users are trading assets that are not widely supported on DEXs, it's important to ensure that the DEX you integrate with has robust security measures in place to protect those assets. Thorough security audits and a strong track record of handling various types of assets can indicate the DEX's commitment to asset security.
2. **Cross-Network Interoperability:** If you're planning to onboard users from one network to another or support users across multiple networks, interoperability between these networks becomes crucial. Ensure that the DEX supports cross-chain functionality securely, as cross-network transactions introduce additional complexities and potential security risks.
3. **Security of the Ecosystem:** The DEX's supported networks form an ecosystem. Assess the security protocols of this ecosystem, including the consensus mechanisms, smart contract vulnerabilities, and governance practices. A secure ecosystem contributes to the overall safety of your dApp's transactions.
4. **Integration with User Wallets:** A DEX's integration with user wallets is vital for security. Look for DEXs that provide secure and user-friendly wallet integrations, possibly through widely recognized wallet providers or DID solutions. A strong integration can prevent unauthorized access and ensure seamless transactions.
5. **Smart Contract Audits:** Evaluate whether the DEX conducts regular smart contract audits and has a history of addressing vulnerabilities promptly. This helps ensure that the transactions and assets handled by the DEX are not susceptible to hacking or exploitation.
6. **Privacy and Data Protection:** If your dApp involves sensitive user data or private transactions, inquire about the DEX's privacy and data protection measures. Encryption, data minimization, and privacy-preserving technologies can safeguard user information.
7. **Regulatory Compliance:** Depending on your dApp's jurisdiction and the DEX's supported networks, regulatory compliance may vary. Ensure that the DEX adheres to relevant regulatory standards and has mechanisms in place to prevent money laundering or other illicit activities.

The security of asset transactions, cross-network interoperability, ecosystem security, wallet integration, smart contract audits, privacy measures, regulatory compliance, and user education are all crucial factors to consider when integrating a dApp with a DEX. Thoroughly researching and assessing potential DEX partners based on these criteria will help ensure the security of your dApp's users and their transactions.

Digital exchanges pose a range of significant risks, including security vulnerabilities, regulatory uncertainty, liquidity challenges, counterparty risks, technical issues, lack of customer support, susceptibility to phishing and scams, market volatility, and the potential for exit scams. In this context, this introduction provides an overview of the key concerns associated with digital exchanges, shedding light on the multifaceted risks that users and investors should be cognizant of. Some of the key risks associated with digital exchanges include:

1. **Security Vulnerabilities:** Digital exchanges can be susceptible to hacking attacks and security breaches, resulting in the theft of users' funds and sensitive information. Even though DEXs aim to mitigate these risks by giving users control over their private keys, CEXs can still fall victim to cyberattacks.
2. **Regulatory Uncertainty:** The regulatory environment for cryptocurrencies and exchanges varies significantly from one jurisdiction to another. Digital exchanges often operate in a gray area and can face legal challenges, regulatory crackdowns, or sudden changes in regulations that impact their operations and users.
3. **Liquidity and Market Manipulation:** Some digital exchanges might have low trading volumes and liquidity for certain tokens, making them susceptible to price manipulation by a small number of participants. Illiquid markets can lead to extreme price volatility and unpredictable trading conditions.
4. **Counterparty Risk:** When users deposit funds into CEXs, they effectively trust the exchange with custody of their assets. If the exchange goes bankrupt, gets hacked, or engages in fraudulent activities, users' funds could be at risk.
5. **Technical Quality of Service Issues:** Digital exchanges can experience technical glitches, downtime, and outages, preventing users from accessing their accounts or executing trades during critical market moments.
6. **Lack of Customer Support:** Some digital exchanges might lack efficient customer support, leaving users without timely assistance in case of issues like account access problems or transaction disputes.
7. **Phishing and Scams:** Users can be targeted by phishing attacks and scams where malicious actors create fake exchange websites or communication channels to steal login credentials, private keys, or personal information.
8. **Volatility and Risky Investments:** The cryptocurrency market is known for its high volatility. Users trading on digital exchanges should be aware of the risks associated with volatile price swings and potential losses.
9. **Lack of Regulation:** DEXs, while aiming to provide greater privacy and control, can also be riskier due to the absence of regulatory oversight, making users more responsible for their security and choices.
10. **Exit Scams:** Some exchanges could engage in exit scams, where they shut down their operations suddenly and disappear with users' funds.

To mitigate these risks, individuals should conduct thorough research before using any digital exchange, prioritize security measures like 2FA and using hardware wallets, only use well-established and reputable exchanges, and stay informed about the regulatory environment in their region.

Understanding the specific use case and transaction size is essential when selecting a DEX for integration. Different DEXs are optimized for various transaction sizes, and it's important to align your choice with your dApp's needs. However, it's important to note that opting for DEXs optimized for larger transactions may indeed bear larger risks. Here's why:

1. **Market Impact:** DEXs optimized for larger transactions may have lower liquidity for smaller trades. Attempting to execute smaller transactions on these DEXs might result in slippage, where the executed price is significantly different from the expected price due to the market impact of the trade.
2. **Front-Running Risk:** Larger transactions could potentially attract the attention of front-runners who try to profit by executing trades ahead of yours. This can lead to unfavorable pricing and potentially financial losses.
3. **Smart Contract Complexity:** DEXs optimized for larger transactions might involve more complex smart contract logic to handle significant volumes, increasing the risk of vulnerabilities or bugs that malicious actors could exploit.
4. **Higher Stakes:** Larger transactions naturally involve higher amounts of assets. Any security breach, vulnerability, or unexpected behavior on the DEX could lead to substantial financial losses.
5. **Regulatory and Compliance Concerns:** Larger transactions might attract more regulatory scrutiny, and compliance requirements might differ for larger trades. Ensuring that the DEX complies with relevant regulations becomes even more crucial.
6. **Network Congestion:** Larger transactions could face network congestion, leading to delays or higher gas fees. This can impact the user experience and overall transaction costs.

When considering a DEX optimized for larger transactions, thorough due diligence is essential. Security audits, smart contract reviews, reputation checks, and understanding the DEX's track record become critical. While there might be benefits to optimizing for larger transactions, the associated risks should be carefully weighed against potential advantages. It's crucial to find a balance that aligns with your dApp's needs and ensures the security and trust of your users.

2.4.8.2 Mitigating the Risks of DEX Use Cases

Cryptocurrency transactions play a crucial role in dApps, serving as a medium of exchange and a store of value. dApps can leverage DEXs for various business cases, capitalizing on the unique features and benefits that DEXs offer.

The combination of DEX functionality with dApp features can create innovative and user-friendly experiences, enabling dApps to tap into the growing demand for decentralized and blockchain-powered services. However, digital exchanges, also known as cryptocurrency exchanges or DEXs, carry several risks due to their inherent nature and the evolving landscape of the cryptocurrency market.

Integrating a dApp with a DEX requires a strong understanding of smart contract development, blockchain interaction, UI/UX design, and security considerations. To counter these risks, dApp developers should embark on comprehensive research before engaging with any digital exchange. They should give precedence to DEX whose security vulnerabilities are tested and audited, opting for established and respected exchanges, while also keeping themselves well-informed about the regulatory landscape in their specific region.

Table 2.4 presents diverse use cases of DEXs, coupled with their associated risks and corresponding risk mitigations to look for when selecting a DEX to integrate with the dApp for the specific use case. This resource can aid in evaluating the security vulnerabilities and risks of DEXs prior to integrating them with your dApp for the specific DEX use cases that are supported.

2.4.9 Securing Digital Identities (DIDs)

As a dedicated dApp developer, understanding the concept of digital identities within the realm of Web3 is essential for creating secure and user-centric dApps. This use case explores the significance of digital identities, their role in authentication and authorization, prominent identity solutions, security considerations, and best practices to ensure the integrity of digital identities.

Digital identities form the bedrock of dApps, facilitating the verification of user authenticity and enabling personalized interactions. Unlike traditional centralized systems, Web3 digital identities empower users with ownership and control over their personal information, leading to enhanced privacy and security. In the context of dApps, digital identities serve as the key to granting access and permissions. Users can prove their identity without revealing excessive personal information, enhancing privacy. By using cryptographic methods, digital signatures, and DIDs, dApp developers can verify users' authenticity and authorize their actions within the application.

Several solutions stand out in the digital identity landscape. Examples include uPort and Civic, which offer innovative approaches to SSI management. uPort leverages Ethereum-based smart contracts to give users control over their identities, while Civic focuses on identity verification through blockchain attestation. Ensuring the security of digital identities is paramount. dApp developers must identify potential vulnerabilities like unauthorized access, data breaches,

Table 2.4 DEX use cases, associated risks, and risk mitigations.

DEX use cases	Related risks	Possible risk mitigations
Token trading and liquidity provision: DEXs serve as platforms for trading and providing liquidity for cryptocurrencies, allowing users to directly exchange assets within the realm of cryptocurrencies.	Digital exchanges can be susceptible to hacking attacks and security breaches, resulting in the theft of users' funds and sensitive information. Even though DEXs aim to mitigate these risks by giving users control over their private keys, CEXs can still fall victim to cyberattacks.	1. Choose reputable and established DEXs with a history of security measures and successful operations. 2. Use noncustodial wallets and cold storage for asset security. 3. Regularly conduct thorough security audits on DEX smart contracts and platform codebase by reputable third-party firm.
Token sales and IDOs: dApps harness DEXs to carry out token sales and IDOs, enabling transparent and decentralized fundraising approaches. These token launches through DEXs enhance user access and reduce entry barriers in comparison to traditional fundraising methods.	The regulatory environment for cryptocurrencies and exchanges varies significantly from one jurisdiction to another. Digital exchanges often operate in a gray area and can face legal challenges, regulatory crackdowns, or sudden changes in regulations that impact DEX operations and users.	1. Consult cryptocurrency and blockchain legal professionals to ensure the DEX chosen is in compliance with operating jurisdictions. 2. Confirm regular internal audits for compliance in DEX occurred. 3. Validate compliance impacts with regulations in both your dApp and DEX operating regions. 4. Understand what are your obligations, licensing, and reporting to mitigate regulatory risks. 5. Ensure robust data protection measures in your dApps for users' PII if collected or processed. Asses any scope for data privacy regulations for your dApp like GDPR.

(Continued)

Table 2.4 (Continued)

DEX use cases	Related risks	Possible risk mitigations
DeFi: Integrations of numerous cryptocurrencies play a pivotal role in DeFi, aiming to revolutionize traditional financial services through blockchain technology. DEXs are crucial elements in DeFi protocols, facilitating the smooth integration of cryptocurrencies into decentralized lending, borrowing, yield farming, and various financial functions.	In the context of integrating cryptocurrencies into decentralized financial functionalities like lending, borrowing, and yield farming, several risks should be considered. These include vulnerabilities in smart contracts, impermanent loss in liquidity provision, flash loan attacks, inaccurate oracles, fraudulent projects, collateral liquidation, volatile gas fees, user errors, phishing, and front-running manipulation.	1. Review your dApp use of protocols, conduct security audits by reputable firms. and use formal verification for smart contracts. 2. Approach participation in the DeFi ecosystem with caution, thorough research, and risk consideration. 3. Educate users to liquidity providers about impermanent loss and guide them to suitable strategies. 4. Educate users about scams, user-friendly interfaces, and zero-knowledge proofs to prevent front-running attacks. 5. Impose limits or fees on flash loans and assess risks associated with them. 6. Make sure the DEX supports a transparent liquidation process and appropriate collateralization ratios for borrower protection. 7. Explore Layer 2 scaling solutions to enhance risk mitigation.

Staking and yield generation: Users can stake tokens in DEX liquidity pools to earn rewards, providing dApps with a chance to implement yield generation mechanisms.	Staking participants risk losing a portion of their staked assets through slashing for protocol violations. Vulnerable smart contracts can be exploited by attackers, causing financial losses. Liquidity providers in yield farming face impermanent loss from token value divergence. Deceptive DeFi projects can lead to fund loss (rug pull). Borrowers in lending or yield farming risk liquidation due to collateral devaluation.	1. Follow protocol guidelines, secure validator setups, and avoid slashing triggers to protect staked assets. 2. Opt for audited platforms, interact with established projects, and diversify investments. 3. Be aware of impermanent loss, assess liquidity risks, and choose pools carefully. 4. Research projects, review teams, and opt for well-vetted DEX platforms. 5. Monitor collateral ratios, select transparent protocols, and maintain adequate collateral levels.
Token swaps within dApps: Integrating dApps with DEXs allows users to easily swap tokens within the dApp's interface, streamlining user experience and expanding token utility.	Token swaps within dApps pose risks including smart contract vulnerabilities, potential loss of funds, front-running, liquidity concerns, incorrect rates, user errors, phishing, and platform reliability issues. The integrated DEX might experience technical issues, downtime, or other disruptions, affecting the dApp's functionality.	1. Opt for DEX that adhere to protocol guidelines, secure validator setups, and avoid activities triggering slashing conditions to safeguard staked assets. 2. Be aware of impermanent loss, assess risks in liquidity provision, and choose pools carefully. 3. Thoroughly research projects, review teams, and opt for well-vetted platforms when engaging. 4. Monitor collateral ratios, select protocols with transparent liquidation mechanisms, and maintain suitable collateral levels.

(Continued)

Table 2.4 (Continued)

DEX use cases	Related risks	Possible risk mitigations
Decentralized governance: DEX governance tokens empower dApp users to participate in decision-making, fostering community involvement in shaping the platform's future.	As a dApp developer, decentralized governance presents challenges in coordinating among diverse stakeholders, possibly resulting in delays. Ensuring broad participation can be demanding, risking decisions being influenced by a limited group. There's a concern about manipulation and Sybil attacks, where fake identities or collusion disrupt the governance process. Deadlocks might arise from disagreements, obstructing crucial decisions.	1. Opt for a DEX that offers clear communication channels and facilitates discussions among stakeholders. This will help mitigate coordination issues and minimize delays in decision-making. 2. Choose a DEX that encourages active involvement by incorporating effective voting mechanisms, rewards for participation, and transparent decision-making processes. This will enhance broad participation and prevent decision-making from being dominated by a small group. 3. Look for a DEX that implements identity verification mechanisms and reputation-based voting systems or requires minimum staking thresholds for participation. These measures deter manipulation and Sybil attacks, ensuring the integrity of the governance process. 4. Prioritize a DEX that establishes clear and well-defined rules for resolving disagreements. This could involve supermajority requirements or mechanisms for third-party mediation to prevent deadlocks and enable progress in decision-making.

Cross-chain interoperability: From the perspective of a dApp developer, DEXs that enable cross-chain swaps play a vital role in facilitating seamless interaction between tokens originating from various blockchain networks. This functionality significantly enhances the overall interoperability of different blockchain ecosystem.	Interoperability risks encompass vulnerabilities in smart contracts, differing consensus mechanisms affecting transaction security, exposure of sensitive data, bugs in interoperability protocols, potential regulatory concerns, trade-offs between decentralization and centralization, custodial risks with asset locking, delays due to network congestion, standardization challenges, and incentive misalignments across chains.	1. Conduct thorough audits and adhere to standardized practices. 2. Choose solutions adaptable to varying consensus mechanisms. Ensure secure transactions between different blockchains. 3. implement privacy-preserving technologies and encryption. Rely on trusted oracles to maintain data integrity. 4. Regularly test protocols and perform audits. Foster responsible disclosure of vulnerabilities. 5. Prioritize decentralization while addressing custody risks. Minimize centralization points for enhanced security. 6. Monitor network conditions and adjust fees accordingly. 7. Adhere to industry-accepted cross-chain standards. 8. Design incentive mechanisms aligned with user behavior. Encourage participation through economic bridges and token incentives.

(Continued)

Table 2.4 (Continued)

DEX use cases	Related risks	Possible risk mitigations
Decentralized marketplaces: DEXs can be used as the trading infrastructure for decentralized marketplaces where users can buy and sell digital assets, collectibles, and other digital goods in a P2P manner.	From the perspective of a dApp developer, decentralized marketplaces integrated with DEXs face several risks. These include vulnerabilities in smart contracts allowing unauthorized actions, potential financial losses due to fraudulent listings, legal complications arising from noncompliance with regulations, issues with transparency affecting user visibility, absence of dispute resolution mechanisms causing trust concerns, and scalability challenges during periods of heightened demand.	To enhance the reliability of decentralized marketplaces on DEXs, dApp developers should focus on conducting security audits, establishing reputation systems, adhering to regulations, implementing dispute resolution methods, prioritizing transparency, and incorporating scalability solutions.
Reduced centralization: Through DEX integration, cryptocurrencies can alleviate dependence on CEXs and retain control over users' cryptocurrency holdings. This aligns with the core principles of decentralization and user empowerment within the cryptocurrency realm.	For a dApp developer, embracing reduced centralization involves managing risks such as security vulnerabilities, accountability challenges, scalability issues, regulatory uncertainties, and more. Striking the right balance between reduced centralization and addressing these risks is crucial when adopting decentralized solutions.	When mitigating the risks associated with reduced centralization, dApp developers should concentrate on enhancing liquidity management, improving user experience, ensuring robust security measures, navigating regulatory compliance, broadening asset availability, exploring privacy-enhancing solutions, and optimizing gas fee efficiency for seamless dApp integration with DEXs.

Incentive mechanisms: In the context of dApps integrating with DEXs, incentive mechanisms are strategies that encourage user engagement by offering rewards, benefits, or tokens. These mechanisms align user actions with the dApp's goals, such as providing liquidity or participating in trading competitions, enhancing the dApp's ecosystem and user experience.	The integration of incentive mechanisms in dApps using DEXs carries several risks, including token dumping, gaming the system, over-reliance on rewards, centralization of power, economic sustainability challenges, misaligned incentives, user apathy, complexity, community division, and regulatory compliance issues. To ensure success, dApp developers must carefully design and implement incentive structures that balance user engagement, dApp goals, and long-term sustainability while mitigating these risks.	To mitigate these risks, developers should carefully design incentives, monitor token behavior, assess user engagement, ensure alignment with dApp goals, manage economic sustainability, avoid undue concentration of rewards, and ensure compliance with regulations. Balancing these factors is key to successful implementation and risk mitigation.

and identity spoofing. The distributed and transparent nature of Web3 identities demands robust security measures to protect user data and prevent malicious activities.

Safeguarding digital identities in dApp implementations is paramount to ensure user privacy, security, and trust. Security considerations differ across various digital identity models. Self-sovereign identities empower users to manage their data independently, requiring strong encryption and secure key management. Centralized and federated identities demand stringent protection of central repositories and interentity communication channels to prevent unauthorized access.

Tailoring security measures to different identity models is crucial for maintaining the integrity and privacy of digital identities in dApps. Different identity models have distinct characteristics and security requirements. Security requirements vary by digital identity model. Self-sovereign identities place responsibility on users, necessitating robust encryption and secure key management. In contrast, centralized and federated identities require strong safeguards for central data stores and secure communication between entities to prevent unauthorized access.

2.4.9.1 Explanation of Digital Identities

Digital identities play a crucial role in the world of dApps, revolutionizing how users interact with online services and enhancing security, privacy, and user experience. Digital identities are digital representations of individuals, entities, or devices in the digital realm. They encompass a range of attributes, credentials, and information that uniquely define and distinguish one entity from another. These attributes can include personal information, contact details, biometric data, and more. Digital identities allow users to establish their presence and authenticate themselves in online environments.

Digital identities are the cornerstone of user interaction and security in dApps. They provide users with control over their data, enhance privacy, enable secure transactions, and contribute to building a more open and user-centric digital landscape:

1. **User Authentication and Authorization:** Digital identities serve as the foundation for user authentication and authorization in dApps. They enable users to prove their identity and gain access to dApp functionalities securely and seamlessly.
2. **Privacy and Control:** In dApps, users often have greater control over their digital identities compared to traditional centralized systems. This self-sovereign approach empowers users to manage their own data, decide what information to share, and control who has access to their identity attributes.

3. **Reduced Data Exposure:** Digital identities in dApps can minimize the amount of personal information exposed to service providers. Instead of sharing comprehensive profiles, users can share only the specific attributes necessary for a transaction or interaction.
4. **Secure Transactions:** Digital identities enable secure transactions and interactions within dApps. Users can use cryptographic methods to sign transactions, ensuring their authenticity and integrity.
5. **Interoperability:** DID systems are designed to be interoperable across various dApps and platforms. This means users can have a consistent identity across different services without creating separate accounts for each.
6. **Prevention of Identity Theft:** The use of cryptographic keys and secure authentication methods in digital identity systems helps prevent identity theft and impersonation, as it's extremely difficult for malicious actors to forge these cryptographic proofs.
7. **Trust and Reputation:** Digital identities can be tied to reputation systems within dApps. Users with established and verified identities can build trust over time, leading to better collaboration and interactions.
8. **Immutable Records:** In blockchain-based dApps, digital identity records are often stored on the blockchain, making them tamperproof and immutable. This prevents fraudulent identity claims and unauthorized alterations.
9. **Cross-Border Access:** Decentralized digital identities can facilitate cross-border access to services, as they are not tied to any specific jurisdiction or centralized authority.
10. **Decentralized Governance:** Digital identities can be used as a basis for decentralized governance systems in dApps. Participants can have proportional voting power based on their verified identities.

When considering the use cases for digital identities with dApps and blockchains, it is important to analyze the different types of identity models used to manage and verify digital identities in the digital realm. These models dictate how individuals or entities establish and authenticate their online presence. They are essential because they shape how users interact with online services, and each identity model has distinct characteristics and implications for security, privacy, and user control. Identity models define how digital identities are managed and verified in the digital realm. Choosing the right identity model is crucial to meet the specific needs of a digital application or platform and ensure a secure and user-friendly environment.

Different identity models are applied to dApps and blockchain-based systems to cater to various use cases, user preferences, and security requirements. The choice of identity model depends on factors such as the level of user control, privacy needs, interoperability, and the specific goals of the dApp.

In the context of blockchain and dApps, there are four main identity models:

1. **SSI** [99]: Users have complete control over their digital identities, owning and managing their identity attributes. SSI is decentralized, enhances privacy, and allows selective disclosure of attributes. Examples include uPort [100] and Sovrin [101]. uPort is an open-source, SSI platform built on the Ethereum blockchain. uPort follows the SSI model, giving users complete control over their identity data; it leverages Ethereum smart contracts to enable users to create and manage their identities on a decentralized network and provides users with the ability to selectively disclose specific identity attributes, enhancing privacy. Sovrin is an open-source, global, DID network designed to provide secure and self-sovereign digital identities. It uses a permissioned blockchain built specifically for identity-related use cases.
2. **Centralized Identity:** Centralized identity relies on a trusted central authority or identity provider to manage and verify user identities. It's efficient but may compromise some aspects of privacy and control. Examples include traditional email/password logins and social media logins. Centralized identity models in dApps and blockchain involve using centralized servers or authorities to manage user identities. Examples include traditional email/password logins for dApps, blockchain wallets with centralized backup services, and centralized identity verification for regulatory compliance. These models are sometimes chosen for user familiarity, convenience, or legal compliance, although they typically involve some level of centralization in contrast to fully DID models. Some dApps, especially those transitioning from traditional web applications, might initially implement centralized email/password login systems. Users register with their email addresses and create passwords to access the dApp. Certain dApps that must comply with regulatory requirements, such as KYC and AML rules, may implement centralized identity verification processes. For a discussion on the centralized identity models, especially in the context of dApps and regulatory compliance, see [102].
3. **DID:** When addressing security concerns related to digital identities, particular attention should be directed toward DID systems created to facilitate secure data exchange among parties. In such systems, users have the capability to selectively divulge precise identity details to others, all while retaining complete control over the extent of information disclosed. Unlike centralized systems where third parties (like social media platforms) control user data, DIDs give individuals sovereignty over their personal information. DIDs are designed to enable secure data exchange between parties. This means that users can share specific pieces of their identity information with others while maintaining control over what is disclosed. DID systems are built on trust frameworks, which can be established through consensus

mechanisms, cryptographic proofs, and blockchain technology. DIDs prioritize user privacy. Users have the ability to selectively disclose specific identity attributes or credentials as needed, without revealing their entire identity. DIDs are designed to be interoperable across different platforms and services; interoperability is achieved through the use of standardized protocols and DIDs, which are portable and can be used across various contexts. Trust networks, often referred to as trust frameworks, are fundamental components of DID systems. They establish trust and enable secure interactions between users and relying parties (organizations or services that require identity verification).

4. **Federated Identity:** Federated identity enables multiple identity providers to cooperate and share identity information. It promotes interoperability and often includes single sign-on (SSO) capabilities. Examples include OAuth and OpenID [103]. Connect for web SSO and enterprise solutions like Azure AD and Google Workspace dApps often implement federated identity for user convenience. They allow users to log in using their existing accounts from trusted identity providers, such as Google, Facebook, or LinkedIn, using protocols like OAuth. This simplifies the user experience and reduces the need to create new accounts for every dApp. While not fully decentralized, federated identity offers a balance between user convenience and security.

The choice of identity model for a dApp depends on factors like user control, privacy needs, interoperability, and the specific use case of the application. dApps may even blend aspects of these models to strike a balance between control, security, and user experience.

2.4.9.2 Security Considerations for Digital Identities

Securing digital identities in dApps is crucial to shield sensitive user data, uphold privacy, and foster trust in the decentralized digital landscape. When addressing security concerns related to digital identities, particular attention should be directed toward DID systems created to facilitate secure data exchange among parties. In such systems, users have the capability to selectively divulge precise identity details to others, all while retaining complete control over the extent of information disclosed. Unlike centralized systems where third parties (like social media platforms) control user data, DIDs give individuals sovereignty over their personal information.

Identifying vulnerabilities, understanding the paramount importance of security, and customizing security practices to align with various identity models are all integral components of an effective digital identity security strategy within the realm of dApps. Recognizing potential vulnerabilities related to digital identity security is a foundational step. These vulnerabilities might encompass data breaches, unauthorized access, identity theft, or phishing attacks. Vulnerabilities can manifest at various stages within dApps, including data storage, transmission, user authentication, and the management of cryptographic keys. Data breaches involve unauthorized access to sensitive user information. In the context of digital identities, this could mean an attacker gaining access to personal details, credentials, or other identity-related data. If a dApp's identity verification process is compromised, attackers may obtain enough information to impersonate users within the dApp's ecosystem. This can lead to fraudulent activities or unauthorized access to user accounts. Inadequately secured data storage can expose identity information to unauthorized parties.

As cryptography constitutes the foundational pillar of trust, integrity, and confidentiality within digital identity ecosystems, the potential risk of losing access to cryptographic keys extends its influence beyond merely decentralized systems, affecting any digital identity system that relies on such keys, including traditional centralized ones. In the event of key loss, the consequence can be the irrevocable loss of one's digital identity. Ensuring the secure protection and management of cryptographic keys is fundamental to the overall security of digital identity systems. HSMs play a crucial role in this regard, providing tamper-resistant environments for generating, storing, and managing keys. Secure key generation, utilizing trusted algorithms and processes, is essential to avoid weak or predictable keys that could be exploited by attackers.

Access controls are paramount in limiting key access to authorized personnel or systems. Robust auditing and monitoring mechanisms help track key usage, detect suspicious activities, and respond promptly to potential threats. Key recovery and backup procedures are vital to handle key loss or damage, and secure deletion and destruction processes must be in place when keys are no longer needed.

Credential forgery, or the fraudulent creation or manipulation of digital credentials, poses a significant threat to the integrity and security of all digital identity systems, since digital credentials serve as evidence of an individual's identity attributes or claims, such as their name, age, qualifications, or permissions. These credentials are used by relying parties (e.g. service providers) to verify the legitimacy of identity claims made by users. Credential forgery can lead to the presentation of false or manipulated identity assertions. An attacker might create fake diplomas, certificates, or identity cards, which can be presented as genuine during identity verification processes. If digital credentials can be easily forged or tampered with, malicious actors can gain unauthorized access to resources, services, or benefits. For example, an individual with a forged driver's license might access restricted facilities or services.

Preventing credential forgery within digital identity systems is of utmost importance for maintaining trust and security. To achieve this, several key strategies are employed. First and foremost, the utilization of digital signatures is pivotal.

These cryptographic safeguards serve to guarantee the authenticity and integrity of digital credentials, confirming that they haven't been tampered with or illicitly created. Additionally, a robust PKI plays a vital role in managing digital certificates and keys, securely associating credentials with legitimate issuers through trusted certificate authorities (CAs). Blockchain technology is another valuable tool in this endeavor. Its inherent attributes of immutability and transparency make it exceedingly challenging for malicious actors to manipulate or counterfeit credentials once they are securely recorded on the blockchain.

DIDs, specifically designed for self-sovereign identities, offer a resilient solution against forgery. DIDs exhibit persistence, global uniqueness, and cryptographic verifiability, strengthening the overall integrity of identity claims. Furthermore, adopting the W3C Verifiable Credentials standard significantly enhances security. These credentials possess tamper-evident characteristics and can be independently verified by relying parties, assuring that credentials remain unaltered. A critical component is the implementation of credential revocation mechanisms, enabling the swift invalidation and removal of compromised or obsolete credentials from circulation. Robust identity proofing processes are essential prerequisites to ascertain the legitimacy of individuals or entities before the issuance of credentials. To ensure ongoing security, continuous monitoring, diligent auditing, and effective alerting mechanisms are indispensable for the early detection of any suspicious activities or anomalies associated with the issuance and utilization of credentials.

Security considerations related to interoperability challenges in digital identity systems are crucial for ensuring the smooth interaction of these systems with various applications and services. Interoperability often involves sharing identity data between different systems. Ensuring that user consent is obtained and that sensitive information is protected during these exchanges is paramount. When using tokens (e.g. JWTs or OAuth tokens) for identity verification or authorization, securely manage these tokens and implement token validation, expiration, and revocation mechanisms to prevent misuse. Additional key aspects to follow include ensuring compliance with security standards, having an incident response plan, and performing regular audits and assessments further enhancing the security of interoperable digital identity systems, preserving data integrity and user trust.

Implementing digital identities in dApps while addressing data privacy concerns, particularly the protection of PII, is critical and will also fall in scope for user privacy regulations (i.e. GDPR, CCPA, and others). As best practice, users should be allowed to explicitly provide consent for the collection, storage, and sharing of their PII, and this consent should be granular, transparent, and revocable at any time. Moreover, a data minimization strategy should be employed, collecting and storing only the minimum necessary PII required for the dApp's intended functionality. Security measures are paramount. Encrypting and anonymizing PII both in transit and at rest is essential to protect it from unauthorized access.

Robust access controls should be in place to restrict who can access and modify PII, enhancing data security. In the context of blockchain integration, privacy-focused solutions like zero-knowledge proofs can be utilized to enable transactions on public blockchains without revealing PII. This safeguards privacy in the inherently transparent blockchain environment.

Regular privacy audits and assessments should be conducted to identify and rectify privacy vulnerabilities or noncompliance issues. Additionally, when integrating third-party services, ensure they also adhere to data privacy regulations and secure data handling practices. Seeking legal expertise is advisable, as navigating complex cross-border data privacy regulations can be challenging. Legal counsel can provide valuable guidance to ensure compliance and minimize legal risks. In summary, a comprehensive and proactive approach to data privacy is essential when implementing digital identities in dApps, especially considering the global nature of blockchain and diverse data protection regulations.

2.4.10 Securing Smart Contracts

This content is designed to provide dApp developers with a comprehensive understanding and practical guidance on enhancing the security of smart contracts, crucial for building reliable dApps. We begin with an introduction to smart contracts, outlining their operational framework within blockchain technology and discussing both their advantages such as transparency and efficiency by automating processes and eliminating intermediaries and their drawbacks, notably their immutable nature and vulnerability to programming errors.

The focus then shifts to a critical analysis of common vulnerabilities associated with smart contracts, emphasizing the importance of adopting secure coding practices. It details strategic measures such as the utilization of automated testing tools and thorough code audits, which are essential for detecting and rectifying hidden vulnerabilities that could compromise smart contract integrity. Furthermore, the narrative explores the concept of security by design, advocating for the integration of security measures early in the development process to ensure robust protection against potential threats.

This structured approach is aimed at fostering a deep and practical understanding of the security landscape in smart contract development, equipping developers with the necessary tools and methodologies to enhance the resilience and reliability of dApps.

2.4.10.1 Overview of Smart Contracts and Security Considerations

For dApp developers, a robust comprehension of smart contracts is fundamental. Smart contracts represent self-executing agreements encoded in blockchain-based code, automatically executing when predefined conditions are met.

These contracts operate within the decentralized landscape of blockchain networks, eliminating the need for intermediaries and emphasizing transparency and immutability. Understanding the programming language used for smart contract development, like Solidity for Ethereum, is paramount.

Solidity is a programming language designed specifically for creating smart contracts on the Ethereum blockchain. It has a syntax influenced by JavaScript, Python, and C++, making it accessible to developers familiar with those languages. Solidity supports various data types, custom data structures, and features like inheritance and libraries.

For developers looking to enhance their blockchain development skills, particularly in secure smart contract creation, there are several educational resources and courses available for example:

1. **Alchemy University's JavaScript Crash Course** [104]: This is a three-week intensive course designed to help newcomers become proficient in JavaScript, which is fundamental for Ethereum and other blockchain development. This course covers everything from basic programming to advanced topics such as asynchronous requests and promises, providing a solid foundation for anyone preparing to dive into Ethereum development.
2. **Online Blockchain Security Courses:** Platforms like Coursera [105], Udemy [106], and edX [107] offer a variety of courses focusing on blockchain security, especially in areas like Solidity smart contract security. These courses are often developed and taught by industry experts, providing both foundational knowledge and advanced security practices.
3. **Certified Blockchain Security Professional (CBSP) Programs** [108]: For those seeking formal certification, the CBSP credential is targeted at professionals who want to deepen their knowledge of blockchain security. It covers various aspects of blockchain technology, including advanced security mechanisms and best practices to mitigate potential vulnerabilities.

These resources provide comprehensive educational opportunities for both beginners and experienced developers aiming to specialize in the security aspects of blockchain and smart contract technology.

An essential consideration for dApp developers is that blockchain transactions and smart contract code are both transparent and immutable. This underscores the critical need for secure coding practices because once a smart contract is deployed, it becomes unchangeable. Ensuring the security and reliability of smart contract code is a top priority for dApp developers due to the unique nature of blockchain technology. The immutability of blockchain, once a smart contract is deployed, underscores the critical importance of secure coding practices. This immutability means that any vulnerabilities or bugs in the smart contract code are permanent, making it crucial to get it right from the start. One significant implication of this immutability is the potential for trust and reputation damage. Users of dApps place a high degree of trust in the security and functionality of smart contracts. If vulnerabilities are exploited, it can erode trust and harm the reputation of both the dApp and its developers.

Additionally, many dApps involve financial transactions, and vulnerable smart contracts can lead to financial losses for users, legal consequences, and damage to the dApp's credibility. Testing and auditing procedures play a pivotal role in ensuring the integrity of smart contracts. Automated testing tools and code audits are essential for identifying and addressing vulnerabilities before deployment. Considering upgradability and long-term maintenance is vital, as some blockchain platforms permit smart contract upgrades while others are immutable.

Smart contracts, the building blocks of dApps, are powerful and transformative. However, they are not immune to vulnerabilities and risks. dApp developers must be well-versed in the lexicon of common vulnerabilities such as reentrancy, integer overflow, and underflow. Understanding these vulnerabilities is akin to knowing the terrain you're about to traverse; it's fundamental. Mastery of secure coding practices is not just a feather in a developer's cap; it's a vital skill. These practices are the guiding principles that ensure your smart contracts are fortified against potential threats.

It is important to remember that security in dApp development should not follow a reactive endeavor but a proactive approach seeking to identify vulnerabilities, understanding the nuances of potential attacks, and preemptively addressing them. A single vulnerability can lead to significant financial losses or damage to a project's reputation. To mitigate these risks, developers must prioritize security throughout the development life cycle. In the intricate realm of smart contract development, where automation is the key to seamless transactions, developers assume a role of meticulous scrutiny. At the heart of their vigilance lies a profound examination of transactions and the robust security that must envelop them. This can reveal irregularities or unexpected behavior in smart contracts. A comprehensive testing strategy typically combines static and dynamic analysis, manual code reviews, audits, and custom testing scripts to ensure the security and reliability of smart contracts.

In the realm of smart contract security, the mission goes beyond identifying and mitigating common vulnerabilities. It extends into the complex terrain of interoperability, a multifaceted aspect that demands careful consideration. These risks involve dependencies on external contracts or data sources, potential data trust issues, reentrancy vulnerabilities during interactions with external contracts, reliance on oracles and data feeds that may be compromised, increased complexity leading to unintended consequences, and the presence of legacy or deprecated contracts susceptible to unpatched vulnerabilities. Addressing these risks requires thorough code audits, secure coding practices, trusted oracles, rigorous data validation, and minimizing external dependencies where possible to maintain the security of smart contracts in a decentralized environment.

In the realm of smart contract security and compliance, a holistic perspective is essential. While the technical aspects of security are paramount, legal and compliance considerations should not be underestimated. This is especially true in regulated industries, where expert legal guidance may become a necessity. The intersection of legal and compliance aspects is a pivotal concern for dApp developers, particularly when operating in regulated sectors and in cross-border contexts. In this intricate landscape, developers find themselves navigating a maze of regulations, spanning domains such as finance, securities, consumer protection, data privacy, and IP laws.

To effectively address legal risks and maintain compliance, it is important that dApp developers and dApp development managers collaborate with legal experts specializing in blockchain and smart contract law to navigate intricate regulatory landscapes effectively. Establish Service Level Agreements (SLAs) that encompass transparent terms of use and privacy policies. These SLAs not only safeguard user interests but also seamlessly align with prevailing legal requirements.

In conclusion, the significance of conducting exhaustive smart contract audits cannot be overstated. These audits represent meticulous investigations with a dual focus: ensuring strict adherence to legal and regulatory standards while also fortifying the smart contract against the ever-present threat of common vulnerabilities that could be exploited by malicious entities. Auditors meticulously assess smart contracts to verify their alignment with legal and regulatory standards. This entails a rigorous examination of the contractual terms, privacy policies, and any other legal obligations that the contract must uphold. Beyond legality, auditors probe the smart contract's code and design for security weaknesses. They scrutinize the codebase to detect and rectify common vulnerabilities that could be exploited by potential attackers. This includes vulnerabilities like reentrancy, integer overflow/underflow, and DoS susceptibility, among others.

2.4.10.2 Common Smart Contract Vulnerabilities and Associated Risks

Secure coding and defensive programming practices are paramount for smart contract developers. This involves writing code with a strong focus on security to protect against vulnerabilities and potential threats. For a dApp developer, the journey to mastering secure coding and defensive programming often begins with a valuable but unconventional approach: learning from bad examples. Understanding vulnerabilities and their implications is like navigating a treacherous terrain, and sometimes the best way to appreciate the importance of secure coding practices is by witnessing their absence.

Smart contracts, integral to blockchain-based applications, can be vulnerable to various security threats and risks. Understanding these vulnerabilities and their potential impact is crucial for developers and stakeholders in the blockchain ecosystem. In Table 2.5, we outline 18 common smart contract vulnerabilities, describe their characteristics, and assess the associated risks in terms of impact and likelihood. Each vulnerability presents a unique set of challenges, and developers must employ rigorous security practices and risk mitigation strategies to protect dApps and user assets. The risk assessments provided here serve as general estimates and can vary depending on specific implementations and circumstances.

By incorporating these vulnerabilities and mitigation strategies, developers can further enhance the security of their smart contracts and dApps. It's important to approach smart contract development with a comprehensive understanding of potential risks and implement robust security measures to protect users and their assets.

Table 2.5 offers a concise overview of prevalent smart contract vulnerabilities, accompanied by an assessment of associated risks in terms of impact and likelihood and overall risks.

Table 2.5 Common smart contract vulnerabilities and risks.

Vulnerability	Description	Risk factors levels	Overall risk (read note)
#1 Reentrancy attacks	Attackers repeatedly call a contract before previous execution is completed	Impact: high (financial loss, unauthorized access) Likelihood: high (common attack vector)	High
#2 External contract dependence	Vulnerabilities when relying on external contracts that change unexpectedly	Impact: medium (contract malfunction, data issues) Likelihood: high (depends on external contract behavior)	High
#3 Access control issues	Lack of proper access controls, allowing unauthorized execution of functions	Impact: medium (unauthorized access, data manipulation) Likelihood: high (common programming error)	High
#4 External data dependence	Relying on external data sources, like centralized oracles	Impact: medium (data reliance, data manipulation) Likelihood: high (depends on external data availability)	High

Vulnerability	Description	Risk factors levels	Overall risk (read note)
#5 MEV bots (miner extractable value)	Miners or validators extracting value from transaction ordering and execution	Impact: high (loss of profits, manipulated outcomes) Likelihood: medium (depends on network activity)	Medium
#6 Uninitialized constructors	Vulnerabilities when constructors are not initialized correctly	Impact: medium (data corruption, vulnerability) Likelihood: medium (depends on coding practices)	Medium
#7 Unvalidated inputs	Failure to validate inputs allowing malicious data to enter the contract	Impact: medium (security risks, unexpected results) Likelihood: high (common programming error)	Medium
#8 Uninitialized storage and variables	Variables or storage slots not properly initialized, vulnerable to manipulation	Impact: medium (data corruption, vulnerability) Likelihood: medium (depends on coding practices)	Medium
#9 Front-running	Exploiting the time delay between transaction submission and execution	Impact: medium (loss of expected profits, manipulated results) Likelihood: high (common in blockchain transactions)	Medium
#10 DoS	Attackers execute operations consuming excessive gas, rendering the contract unusable	Impact: medium (contract unavailability, loss of funds) Likelihood: medium (possible if not handled properly)	Medium
#11 Gas limit and loops	Excessive gas usage in loops or recursive calls leading to transaction failure	Impact: medium (transaction failure, wasted gas) Likelihood: medium (depends on gas limits and loop depth)	Medium
#12 Unchecked external calls	Improper handling of return values and exceptions from external contract calls	Impact: medium (loss of funds, contract failure) Likelihood: medium (depends on contract complexity)	Medium

(Continued)

Table 2.5 (Continued)

Vulnerability	Description	Risk factors levels	Overall risk (read note)
#13 Integer overflow/underflow	Arithmetic operations on integers can result in unexpected behavior when limits are exceeded	Impact: high (financial loss, unintended results) Likelihood: medium (depends on implementation)	Medium
#14 Timestamp dependence	Vulnerabilities due to reliance on timestamps	Impact: low (inaccurate timing, contract issues) Likelihood: low (timestamps are reliable but can be manipulated)	Low
#15 Unsafe type conversions	Incorrectly converting data types, leading to unexpected behavior	Impact: medium (unexpected behavior, security risks) Likelihood: low (depends on coding practices)	Low
#16 Predictable randomness	Vulnerabilities due to lack of true randomness, leading to predictable outcomes	Impact: low (predictable results, manipulation) Likelihood: low (unless using external sources)	Low
#17 Deprecated functions	Use of deprecated or older versions of Solidity or dependencies	Impact: low (compatibility issues, security risks) Likelihood: low (depends on developer awareness)	Low
#18 Upgradable contracts	Risks associated with implementing upgradable contracts	Impact: medium (contract modification, security risks) Likelihood: low (depends on contract design)	Low

Note: The risk levels in this table were determined using a quantitative model that evaluates both impact and likelihood on a scale from 1 to 3. For each vulnerability, the impact score reflects the potential severity of consequences (e.g. financial loss, data corruption, or operational disruption), while the likelihood score represents the probability that the vulnerability will be exploited, based on historical trends and technical feasibility. Multiplying these scores yields an overall risk score that is then mapped to a qualitative category (low, medium, or high). This method provides a structured and objective way to prioritize mitigation efforts by balancing both technical and business considerations.

Understanding how to mitigate the risk of smart contract vulnerabilities is crucial for dApp developers to ensure the security and reliability of their dApps built on blockchain platforms. In this learning process, one encounters smart contract examples riddled with vulnerabilities, each potentially exposing the application to threats. These "bad" examples act as cautionary tales, vividly illustrating the consequences of insecure coding choices. However, the path to proficiency doesn't end there. To become a proficient dApp developer, one must then pivot toward studying coding examples that showcase good defensive programming practices. These examples serve as beacons of best practices, illuminating the way toward secure and robust smart contract development. In this journey from flawed to fortified code, dApp developers gain not only an awareness of vulnerabilities but also the expertise to fortify their projects against them. It's a transformative process that equips developers with the knowledge and skills needed to build dApps that stand firm in the face of potential threats. In Chapters 3 and 4, we will explore in more detail how these vulnerabilities can be discovered and addressed through manual and automated code reviews, such as static analysis. This section will provide at high level the actionable insights for developers looking to enhance the security of their smart contracts.

2.4.10.3 Best Practices for Smart Contracts Security

In the dynamic world of smart contract development, ensuring the security and integrity of your blockchain-based applications is paramount. Smart contracts, while powerful and transformative, can be vulnerable to a range of threats and attacks if not meticulously crafted and safeguarded. This is where a comprehensive set of secure coding and defensive programming practices comes into play.

In Table 2.6 we list the key practices that are fundamental for building smart contracts that stand strong against potential threats and vulnerabilities. These practices are not just theoretical; they are the bedrock of secure blockchain development. They encompass a wide spectrum of considerations, from input validation to regulatory compliance, and from error handling to continuous monitoring.

Ensuring the security of smart contracts is paramount in the world of dApps. Adopting best practices, fortified by a multifaceted approach, can significantly enhance smart contract security. The adoption of secure coding standards, automated static analysis and dynamic analysis tools, penetration testing, secure fuzzing tests, and attack simulation tools can be woven into the fabric of secure smart contract development for example.

Secure Coding Standards: Adhering to secure coding standards forms the foundation of smart contract security and is one of the ways to help to establish a security contract key practice. Secure coding standards for dApps are the

Table 2.6 Smart contract security best practices.

Smart contact security best practice	**Details**
#1 Perform input validations	Validate and sanitize all user inputs, external data, and contract parameters to prevent malicious input, buffer overflows, and unexpected data. Use data type checks, bounds checking, and input validation libraries when necessary.
#2 Implement secure access controls	Implement robust access control mechanisms, such as RBAC or permission-based systems. Ensure that only authorized users or contracts can invoke specific functions, and apply the principle of least privilege.
#3 Identify and remediate smart contact vulnerabilities	Smart contract vulnerabilities are akin to hidden traps in the digital terrain. These risks, including reentrancy attacks and data overflows, require our diligence. Identifying and remediating vulnerabilities in smart contracts is critical for ensuring the security of blockchain applications. Use static analysis tools to examine the source code without executing it. These tools can identify potential vulnerabilities based on known patterns and weaknesses. Tools for dynamic analysis can simulate various attack vectors to see how the contract behaves. Some tools offer formal verification, which involves mathematically proving or disproving the correctness of algorithms underlying a smart contract.
#4 Secure smart contract dependencies and third-party data sources	Regularly update and audit external dependencies and smart contract libraries to address vulnerabilities and take advantage of security patches. Ensure that the dependencies you use are trusted and well-maintained. As external data sources can be manipulated, it is important to incorporate security checks to validate incoming data for authenticity and integrity. Design contracts so that they can be upgraded to fix vulnerabilities. This often involves separating state and logic into different contracts. Leverage well-tested libraries from reputable sources, like OpenZeppelin, which reduce the likelihood of introducing vulnerabilities.
#5 Code audits and testing	Regularly perform code audits, including static analysis and dynamic testing. Consider using security tools and conduct fuzz testing to uncover vulnerabilities. External audits by reputable firms or experts can provide valuable insights into contract security. Before deployment, have the contract reviewed by experts who specialize in smart contract security.

Smart contact security best practice	Details
# 6 Adopt security in the SDLC and integrate with secure development tools	Follow a secure SDLC that incorporates specific security coding standards code reviews, automated security testing and penetration testing, and security-focused testing phases. Integrating security at each stage of development enhances overall security. Critical of security in the SDLC is the integration of the software development pipelines with secure development tools, frameworks, and integrated development environments (IDEs) that follow best practices for smart contract development. These tools can help you catch vulnerabilities early in the development process and promote secure coding practices.
#7 Enable smart contract continuous monitoring	Set up continuous monitoring and alerts to detect and respond to unusual contract behavior or security incidents. Monitoring can help you identify and mitigate threats promptly, reducing the impact of potential breaches.

baseline to establish a set of rules and guidelines that developers must follow to write secure code. For developers interested in secure coding standards for dApps, there are several valuable resources and standards to consider:

1. **OWASP Smart Contract Security Verification Standard** [109]: The OWASP smart contract verification standard provides a comprehensive set of guidelines designed to enhance the security of smart contracts and dApps. This standard addresses unique security challenges inherent in smart contracts, providing actionable guidelines to mitigate vulnerabilities such as reentrancy, overflows/underflows, and gas optimizations.
2. **AuditFirst's Best Practices** [110]: AuditFirst offers insights into essential security practices for dApps. This includes managing private key storage securely, implementing robust authentication mechanisms, and securing API layers to prevent reentrancy attacks and other common vulnerabilities.
3. **DEV Community's Guide on Secure dApp Development** [111]: The DEV Community outlines best practices for building secure dApps, covering everything from smart contract auditing to frontend security measures. It discusses the use of established libraries, thorough smart contract auditing, and the importance of considering frontend to smart contract interaction security. These guidelines help developers understand the complexities of dApp security and provide strategies to address them.

By strictly adhering to these standards, developers ensure that their codebase is resilient to common vulnerabilities. Smart contracts, the self-executing contracts with the terms of the agreement directly written into code, have transformed industries by automating processes and enabling trust in dApps. However, their power comes with a crucial caveat: security vulnerabilities in smart contracts can lead to significant financial losses and pose serious risks to users and the blockchain ecosystem as a whole. Ensuring the security of smart contracts is of paramount importance. Smart contract developers must be well-versed in common security issues, secure coding practices, and the tools available to identify and prevent vulnerabilities. In this guide, we will explore the world of smart contract security, focusing on the importance of secure coding standards and the tools and practices that can help developers build robust and secure smart contracts.

From understanding the common pitfalls that smart contract developers face to mastering the art of secure coding and leveraging tools like linters and static analysis, this guide will equip you with the knowledge and skills to create smart contracts that not only execute transactions flawlessly but also stand up to the rigorous security standards required in today's blockchain landscape. Here are examples of secure coding standards and code checkers/linters that can help identify potential issues in Solidity smart contract development:

1. **Solidity Style Guide and Code Checker** [112]: The Ethereum project maintains a Solidity Style Guide that provides coding standards and best practices for writing secure and maintainable Solidity code. Ethlint [113] (formerly Solium) is a popular code checker that enforces the rules defined in the Solidity Style Guide. It can issue warnings and errors when code deviations are detected.
2. **ConsenSys Secure Smart Contract Guidance and Code Checker** [114]: ConsenSys, a blockchain technology company, offers a set of Solidity recommendations for secure smart contract development. These recommendations cover topics such as error handling, gas optimization, and access control. MythX [115] is a security analysis platform that integrates with various development environments and linters, including Truffle [116] and Hardhat [117]. It checks for vulnerabilities based on best practices and security recommendations.
3. **OpenZeppelin Contracts Library** [118]: OpenZeppelin Contracts is a widely used library of reusable smart contract components designed with security in mind. Developers can follow OpenZeppelin's coding standards to create secure contracts. While OpenZeppelin Contracts itself doesn't provide a linter, developers can integrate existing linters, such as Solhint or Solium (now Ethlint), to check for adherence to these coding standards.
4. **SWC Registry (Smart Contract Weakness Classification)** [119]: The SWC Registry provides a list of known smart contract weaknesses and

vulnerabilities along with recommended mitigation strategies. It's a valuable resource for secure coding guidelines. Some security tools, like the MythX platform, use the SWC registry to identify and report potential vulnerabilities in smart contracts based on these weaknesses.

5. **Solidity Security Blog by ConsenSys** [120]: The Solidity Security Blog maintained by ConsenSys Diligence provides in-depth articles on secure coding practices, including examples and explanations of common vulnerabilities and how to prevent them.
6. **Slither Code Checker/Linter** [121]: Slither is an open-source static analysis tool for Solidity smart contracts. It identifies security issues and code smells and provides recommendations based on a set of predefined security checks.
7. **Mythril Code Checker/Linter** [122]: Mythril is another popular open-source security analysis tool for Ethereum smart contracts. It can detect vulnerabilities and provide actionable advice for remediation.
8. **Securify Security Scanner** [123]: It is a security scanner that automatically checks Ethereum smart contracts for vulnerabilities. It provides a detailed report of issues found and suggestions for improvement.

These coding standards and tools are essential for ensuring the security of smart contracts written in Solidity. Developers should adopt a combination of these resources and tools to proactively identify and address security issues during the development process.

Software Development Frameworks: They can assist in performing input validations and implementing secure access controls in your smart contracts. Software development frameworks for smart contracts can help prevent coding vulnerabilities to some extent by promoting best practices, providing preaudited code components, and facilitating secure coding patterns, which can significantly reduce the likelihood of vulnerabilities. For example, frameworks often include standardized and audited libraries for common functionalities like access control, token standards (e.g. ERC-20, ERC-721), and arithmetic operations. Using these libraries reduces the risk of coding errors and vulnerabilities, as they have undergone thorough security assessments. Software frameworks may provide templates and guidelines for implementing secure coding patterns, such as the "checks-effects-interactions" pattern to prevent reentrancy attacks or RBAC patterns for access control. Here are some popular and well-established options for secure smart contract development:

1. **Truffle Suite:** It is one of the most widely used development frameworks for Ethereum smart contracts. It provides a development environment, testing framework, and asset pipeline for building and deploying smart contracts. Truffle offers built-in support for automated testing, migrations,

and dependency management. It integrates with tools like Ganache [124] (for local development blockchains) and supports popular testing frameworks like Mocha [125] and Chai [126]. You can use external libraries like OpenZeppelin's SafeMath [127] for input validation and arithmetic operations. Truffle allows you to write and run tests for your smart contracts. You can use this testing capability to ensure that access control mechanisms are correctly implemented in your contracts.

2. **Hardhat:** It is a development environment for Ethereum that aims to provide a comprehensive toolkit for smart contract development. It's known for its extensibility and developer-friendly features. Hardhat offers built-in tasks for compilation, testing, and deployment of smart contracts. It supports TypeScript and integrates seamlessly with popular Ethereum networks. Developers can create custom plugins to extend its functionality. Hardhat allows you to write tests for your smart contracts, including tests that validate input data and validate access control logic.
3. **Brownie** [128]: It supports Python-style smart contract development, multiple Ethereum networks, and custom network configurations, which includes using Python libraries and tools for input validation. You can write Python functions to validate input data and incorporate them into your contracts. Brownie's Python-based development environment enables you to write Python scripts to test access control logic.
4. **Embark** [129]: It focuses on dApp development and provides various tools for interacting with smart contracts. While it doesn't have built-in input validation, you can integrate external libraries or write custom JavaScript functions to validate input. Embark provides a testing environment that allows you to validate access control mechanisms. You can write tests to ensure that only authorized parties can interact with your contracts.
5. **OpenZeppelin:** It is an SDK development framework for building secure, upgradable smart contracts and dApps. It's known for its emphasis on security best practices. OpenZeppelin SDK provides tools for creating, testing, and deploying smart contracts with a focus on security. It offers upgradeable smart contracts out of the box and supports Ethereum and other EVM-compatible blockchains.
6. **Remix** [130]: It is an open-source, web-based IDE for smart contract development. It's suitable for both beginners and experienced developers and offers a range of features for writing, testing, and deploying smart contracts. Remix provides a user-friendly interface for writing and testing smart contracts directly in the browser. It supports multiple Ethereum networks and offers debugging capabilities. You can write and test input validation functions directly in Remix to ensure data integrity and write and execute tests to verify access control logic.

7. **Vyper** [131]: It is a Python-like programming language designed for Ethereum smart contracts. While not a framework itself, it's worth mentioning for developers who prefer a language that enforces strict security measures. Vyper promotes security by design with a simplified and auditable syntax. Developers can use Vyper with existing Ethereum development tools. It enforces strict rules and validation checks, making it inherently suitable for input validation. Vyper enforces strict access control rules and can help you avoid common access control vulnerabilities.

In summary, these development frameworks and tools provide the means to perform input validations and implement secure access controls in your smart contracts. While some frameworks may not provide dedicated input validation libraries, they support the integration of external libraries and encourage secure coding practices to address these aspects of smart contract security.

Automated Security Testing: In the realm of smart contract development, harnessing the power of automated security testing tools is a strategic advantage. These tools encompass secure development environments, frameworks, and IDEs designed to adhere to the highest standards of smart contract development. They play a pivotal role in early vulnerability detection and the cultivation of secure coding practices. Leveraging automated static analysis tools is akin to having a vigilant code inspector. These tools scrutinize the codebase without executing it, identifying vulnerabilities, coding errors, and security flaws. By integrating such tools into the development pipeline, developers can catch issues early in the development process. Dynamic analysis tools, in contrast, assess code behavior during execution. They identify runtime vulnerabilities and can uncover security issues that may not be apparent in static analysis. Using dynamic analysis tools in tandem with static analysis provides a comprehensive security net. While DAST tools are not commonly used for smart contracts due to the deterministic and isolated nature of blockchain execution environments, there are some approaches to dynamic testing in this context. One method involves transaction simulation, where developers and testers simulate transactions on the blockchain network that interact with the smart contract. This helps uncover issues related to unexpected contract behavior in response to specific inputs or transactions. Additionally, custom test scripts can be written to interact with the smart contract, submitting various inputs and transactions to assess its behavior and security.

Desktop IDE for Solidity development provides a convenient and efficient environment for creating and testing smart contracts on a developer's local machine, contributing to faster development cycles and improved code quality. Examples of popular desktop IDEs for blockchain and Solidity development include Remix, Truffle, and Visual Studio Code (with suitable extensions). These IDEs are widely used in the blockchain developer community and are specifically designed for blockchain development and offer several advantages, including a controlled local environment, code editing tools, smart contract compilation, local testing capabilities, debugging tools, integration with blockchain networks, version control support, and plugin ecosystems.

One effective method for identifying vulnerabilities in smart contracts involves automated vulnerability testing. Several specialized static analysis tools are designed for smart contract security, and they play a crucial role in the development process. Examples of such tools include MythX, Slither, and Ethlint. These tools operate by conducting automated static analysis of the smart contract code. Importantly, they do this without executing the code, which enhances safety during the testing process. By analyzing the code statically, they can flag potential vulnerabilities and issues within the codebase.

MythX is a security analysis platform for Ethereum smart contracts. It can help identify vulnerabilities in both your smart contract code and its dependencies. It integrates with popular development frameworks like Truffle and Hardhat. Slither is a static analysis tool for Solidity smart contracts. It can detect vulnerabilities not only in your smart contract code but also in your contract's dependencies. It provides detailed reports on identified issues. Mythril is another static analysis tool for Ethereum smart contracts. It can scan both contract code and dependencies to identify vulnerabilities.

Automated tools, like MythX and Slither, can analyze the code to detect potential reentrancy issues. They look for instances where external contract calls are made within functions that manipulate state variables. These tools can identify whether the contract's state is properly locked during external calls, preventing reentrancy attacks. Automated tools can analyze mathematical operations within the contract to detect potential integer overflow or underflow vulnerabilities. They check whether the contract uses safe math libraries or follows secure coding patterns to handle arithmetic operations safely. DoS vulnerabilities are often related to gas consumption. Automated tools can assess the code to identify excessive loops or operations that consume an unusually high amount of gas. They can also detect unbounded loops that can potentially lead to a DoS attack.

MythX and Slither are tools used in the development and auditing of Ethereum smart contracts. MythX is a cloud-based security analysis platform for Ethereum smart contracts. It performs static and dynamic analysis of smart contract code to identify potential vulnerabilities and security issues. MythX can detect a wide range of vulnerabilities, including common ones like reentrancy attacks, integer overflows/underflows, and more. It integrates with popular Ethereum development frameworks like Truffle and Hardhat. Ethlint is a linter for Solidity smart contract code. Linters analyze code for adherence to coding standards, style guidelines, and best practices. While Solhint is not a security-focused tool like MythX or Slither, it helps developers write cleaner, more readable, and potentially more secure code by enforcing coding conventions. It can also catch some common coding mistakes.

These tools are valuable assets for enhancing the security of smart contracts, and they can seamlessly integrate into the SDLC to fortify the development process. One significant integration point is within CI and CD pipelines. These tools can be programmed to run automatically at different stages of development, ensuring that security analysis becomes an integral part of the development process. For instance, they can be triggered to scan code changes in real time and provide instant feedback. Precommit hooks are another powerful integration option. They enable developers to enforce security checks using these tools before committing code to the repository. This prevents potential vulnerabilities from becoming part of the codebase and reinforces a security-first approach.

Smart contract development relies on several widely used repositories and platforms. These resources are essential for simplifying, securing, and streamlining the development process. Some of the most commonly utilized ones include GitHub, hosting various smart contract libraries and projects. You can automate security scans with tools like MythX, Slither, and Ethlint in GitHub Actions by creating a workflow file in your repository. This file defines trigger events, sets up jobs, and configures the environment for security analysis. You run the scans within the workflow's steps, and you can report results and trigger notifications based on scan findings. GitHub Actions allows you to proactively integrate security checks into your CI and CD pipelines, ensuring the early detection of vulnerabilities in your smart contract code.

Penetration Testing and Attack Simulations: Often referred to as ethical hacking, they can be used to simulate real-world attacks on smart contracts. Ethical hackers can attempt to exploit vulnerabilities in smart contracts, providing valuable insights into potential weaknesses. This approach helps identify and rectify vulnerabilities before malicious actors can exploit them. Penetration testing for smart contracts involves assessing multiple interfaces and attack vectors to evaluate the security of a contract comprehensively. These interfaces include external contract interactions, UIs, smart contract APIs, blockchain transaction interfaces, data feeds and oracles, contract state manipulation, Ether handling, access control mechanisms, economic attack vectors, and more. The objective is to identify and mitigate vulnerabilities, ensuring the contract's resilience against a wide range of potential threats and risks across these interfaces.

If the smart contract is part of a dApp, penetration testers can assess the UIs through which users interact with the contract. They can look for vulnerabilities in the dApp's frontend code, such as input validation issues, XSS vulnerabilities, and other web application security concerns. Since smart contracts often expose public functions that can be called by external entities, penetration testers can simulate interactions with these APIs to identify vulnerabilities, including reentrancy attacks, unauthorized access, and input validation issues.

A penetration test can also be used to test the security of blockchain transactions that interact with the smart contract, specifically the smart contract's ability to receive and process transactions that might include transaction ordering vulnerabilities. Testers attempt to manipulate the contract's state variables to identify vulnerabilities related to improper state changes, variable manipulation, or unexpected behavior.

One important aspect of pen testing is to emulate various attack scenarios, allowing developers to assess how a smart contract responds to potential threats. These tools provide an invaluable means of proactively identifying and addressing vulnerabilities. A pen test can also be used with dApp-specific attack vectors, such as front-running, miner extractable value (MEV) attacks, and other economic incentives that could be exploited to the detriment of contract users. Attack simulation tools are a crucial component of a smart contract security toolkit, allowing developers to proactively assess how their contracts respond to various attack scenarios and vulnerabilities. These tools simulate real-world threats, helping developers identify and address weaknesses in their contracts. Before conducting attack simulations, pen testers can create a threat model tailored to the smart contract. Manticore, for example, is a symbolic execution tool that can be used for testing Ethereum smart contracts. It allows developers to simulate various scenarios to identify vulnerabilities.

Attack simulation involves identifying and assessing potential threats, attack vectors, and vulnerabilities that could affect the contract's security. Common vulnerabilities such as reentrancy attacks, integer overflow/underflow, and access control issues are considered. Attack simulation tools then execute these custom scenarios against the smart contract. They simulate actions an attacker might take, such as invoking specific functions, manipulating contract state, or exploiting vulnerabilities. The simulations are instrumental in detecting vulnerabilities that align with the threat model. For example, if the threat model identifies an access control issue as a potential risk, the simulation aims to determine if unauthorized access can indeed occur and whether countermeasures are effective.

Finally, a pen test can also help analyze the contract's input validations and access control mechanisms, including RBAC or permission systems. They verify that only authorized users or contracts can invoke specific functions. A pen test might also include secure fuzzing tests, such as security tests involving feeding unexpected, often malformed data into a smart contract to observe its behavior. This process helps identify vulnerabilities related to data handling and input validation. Automated fuzzing tests can efficiently explore a wide range of inputs, uncovering potential weaknesses.

Security Event Monitoring and Alerts: Setting up continuous monitoring and alerts for your smart contracts is essential for detecting and responding to unusual behavior or security incidents promptly. These are some of the steps that can be followed:

1. **Define the Metrics:** Determine the critical metrics, events, and behaviors that indicate the normal functioning of your smart contract. This can include transaction volume, gas usage, contract balance, user interactions, and specific function calls.
2. **Choose the Monitoring Platform:** Choose appropriate monitoring tools and platforms. Some popular choices for blockchain and smart contract monitoring include services like Etherscan, Chainlink, Amberdata, and custom solutions using blockchain APIs.
3. **Integration:** Integrate with monitoring services. For public blockchains like Ethereum, you can often integrate directly with blockchain explorers and monitoring services. These platforms provide APIs that allow you to access blockchain data in real time. Set up accounts on these platforms and create API keys or tokens for authentication.
4. **Establish Thresholds:** Establish alert thresholds for the identified metrics and events. These thresholds should reflect normal behavior and deviations from it. For example, you might set an alert if the contract balance drops below a certain level or if there is a sudden spike in gas consumption.
5. **Configure Alerts:** Use the monitoring platform's alerting system to configure alerts based on your predefined thresholds. For example, you can specify the alert conditions, such as "if gas usage exceeds X," "if contract balance falls below Y," or "if a specific function is called."
6. **Select Channels:** Select alert channels for receiving alerts. Common channels include email notifications, SMS, instant messaging platforms like Slack, and custom webhooks. Consider setting up multiple alert channels to ensure that critical alerts are noticed promptly.
7. **Test Alerts:** Before deploying your smart contract, thoroughly test the alerting system to ensure that alerts are triggered correctly and delivered to the intended recipients. Simulate different types of incidents and ensure that corresponding alerts are generated.

Once security monitoring and alerts for smart contract security events are implemented, it is important to develop a well-defined incident response plan that outlines the steps to be taken when an alert is triggered. Include roles and responsibilities for team members. It is important to determine the severity levels of alerts and the appropriate responses for each level. For example, a critical alert might require immediate action, while a lower-severity alert may require further investigation. Whenever possible, it is important to be able to automate responses to alerts. For example, you can configure automated scripts or processes to perform actions like pausing the contract, freezing funds, or triggering emergency measures. Continuous monitoring and alerting are critical components of a proactive security strategy for smart contracts. By promptly identifying and responding to security incidents, you can reduce the impact of potential breaches and protect your contract and its users.

2.5 Conclusions for This Chapter

In this chapter, we delved into the realm of dApps and their significance in the blockchain ecosystem. We compared dApps to traditional centralized applications, highlighting the unique advantages and security challenges they present. By identifying the specific security requirements for dApps, we laid the foundation for designing secure dApps.

We explored the principles of secure blockchain platform design and outlined security architecture principles for dApps. The proactive approach of securing dApps by design was emphasized, with a focus on identifying security design flaws and vulnerabilities. This included securing various components, such as blockchain APIs, dApp's confidential data and transactions, consensus algorithms, secrets and keys, token-based transactions, cryptocurrency transactions, and digital identities. We also discussed the secure development and integration of dApps with blockchain APIs. We examined the vulnerabilities that can compromise the integrity of dApps and provide insights into conducting security reviews to identify design flaws and vulnerabilities effectively. We did a deep dive into consensus algorithms and their usage in dApps and blockchain platforms was provided. We recognized potential vulnerabilities associated with consensus algorithms and recommended best practices for selecting and implementing secure consensus algorithms. The importance of safeguarding secrets and keys in dApps was underscored through practical examples. We outlined potential vulnerabilities related to secret and key management and offered guidance on securing these crucial components.

Token-based transactions were explained, along with secure token standards. We explored security considerations for implementing dApps with token-based use cases, providing a robust framework for protecting these transactions. dApps' integration with DEXs was explored, along with strategies for mitigating the risks associated with DEX use cases. This section provided a comprehensive view of securing cryptocurrency transactions within dApps. The concept of digital identities was introduced, accompanied by critical security considerations. We emphasized the importance of securing digital identities within dApps, reflecting the growing significance of identity management in the blockchain landscape.

Finally, we turned our attention to the core of many dApps: smart contracts. We provided an overview of smart contracts, highlighted associated vulnerabilities and risks, and presented best practices for secure coding and defensive programming. This section equipped developers with the knowledge needed to fortify their smart contracts against potential threats.

In conclusion, this chapter serves as a comprehensive guide to designing secure dApps. By understanding the unique security requirements of dApps and applying the principles and best practices outlined herein, developers and organizations can build resilient, trustworthy, and secure dApps in an ever-evolving blockchain landscape. The proactive approach to security by design and continuous monitoring ensures that dApps remain resilient against emerging threats, providing a safe environment for users and their assets. In Chapter 3, we will offer practical examples of how to apply security by design principles to effectively identify and mitigate the risks associated with dApp vulnerabilities through risk-centric threat modeling exercises.

2.5.1 Future Trends in Blockchain Technology and Security

The field of secure blockchain application design, development, and testing is dynamic and ever-evolving. As blockchain technology continues to mature and gain adoption across various industries, several future directions and trends are emerging:

- **AI and Machine Learning in Security:** AI and machine learning will play a more significant role in blockchain security. These technologies will be used to improve the analysis of threats and their impact, predict vulnerabilities in smart contracts, and enhance threat detection and response.
- **AI-Augmented Smart Contract Auditing Tools:** The demand for smart contract auditing tools and services will grow as the complexity of dApps increases. Automated analysis tools will become more sophisticated with augmentation of AI in identifying vulnerabilities and security flaws.
- **Enhanced Security Testing and Attack Simulation:** As blockchain applications become more complex, testing and simulation tools will evolve to provide more comprehensive coverage of vulnerabilities and attack scenarios. Attack simulation platforms will become integral to security assessments.
- **Stronger Identity Verifications and Privacy-Preserving Technologies:** DID solutions will gain prominence, providing users with more control over their personal information. Secure authentication mechanisms and identity management will be crucial for these systems. Future blockchain applications will likely incorporate stronger privacy features to protect user data and transaction details.

- **Expansion of DeFi:** DeFi has exploded in popularity, and the trend is expected to continue. Secure DeFi applications will require robust smart contract development and security practices to protect users' assets from vulnerabilities and exploits.
- **Enhanced Regulatory Compliance:** Blockchain applications are facing increasing regulatory scrutiny due to the need to prevent financial crimes. As regulations evolve, more DeFi applications may start to implement KYC, especially those seeking to operate legally in multiple jurisdictions.
- **Emerging Quantum Computing Threats:** While still in the early stages, quantum computing poses a potential threat to blockchain security. Postquantum cryptography research is ongoing to develop cryptographic algorithms that are resilient to quantum attacks.

In conclusion, the future of secure blockchain application design, development, and testing is marked by innovation, integration with new emerging technologies like AI, increased security measures, and broader adoption of dApps. As blockchain technology matures, staying updated with these trends and integrating them into development practices will be crucial for building resilient and secure blockchain applications.

3

Mitigating Blockchain Vulnerabilities

Every security flaw is an engineering lesson waiting to be learned.

—Ross Anderson

3.1 Introduction

Blockchain technology has emerged as a transformative force across various industries, providing unparalleled transparency, trust, and decentralization. As blockchain applications, including decentralized applications (dApps), continue to proliferate, the need to ensure their security becomes paramount. In this chapter, we explore the practical aspects of securing blockchain-based applications from a software development perspective. Our focus extends to software and application security, where we delve into topics such as identifying software vulnerability bugs. We'll cover the utilization of specific tools and technologies to detect these vulnerabilities and assess their potential impact. Additionally, we will discuss effective strategies for remediating these issues, prioritizing actions based on the severity and criticality of the associated threats.

This chapter is crafted as an extensive guide, providing actionable insights on mitigating the risks posed by vulnerabilities, with a keen focus on enhancing the security posture of blockchain-based applications, particularly dApps. Our exploration will traverse several critical areas to deepen understanding and enhance the application of security measures.

First, we will delve into the common vulnerabilities and risks associated with dApps, elucidating the most prevalent issues such as smart contract bugs, consensus attacks, and wallet vulnerabilities. By highlighting their relevance to dApps and showcasing real-world examples, we aim to demonstrate the tangible impact

Blockchain Application Security: How to Design Secure and Attack Resilient Blockchain Applications, First Edition. Marco Morana, Harpreet Singh and Francesco Piccoli.
© 2026 John Wiley & Sons, Inc. Published 2026 by John Wiley & Sons, Inc.

these vulnerabilities can have on dApps. This section aims not only to inform but also to illustrate the critical nature of these vulnerabilities in real-world settings.

Following this, the chapter will guide you through various vulnerability identification techniques. You will gain valuable insights into the methodologies, techniques, and tools essential for pinpointing vulnerabilities within dApps. This section includes hands-on learning with coding examples that will equip you to effectively recognize and address smart contract vulnerabilities. The focus here is on practical skills that can be immediately applied to enhance the security of your applications.

Next, we introduce the concept of risk-centric threat modeling, an essential process for identifying threats, mapping these threats to vulnerabilities, and conducting attack modeling (AM) and simulation exercises. We will provide a real-case example on how to conduct threat modeling on a dApp. This detailed walkthrough will help you understand how to manage risks effectively by making security by design decisions early in the software development life cycle (SDLC). This approach ensures that security considerations are integrated from the onset, paving the way for safer, more secure applications.

Equipped with a profound comprehension of dApp vulnerabilities and the associated identification methods, we will next delve into potent strategies for vulnerability risk management. This part of the chapter is designed to fortify your understanding and strategic approach to mitigating risks effectively, ensuring that you can apply these insights in a manner that significantly reduces the potential impact of security threats.

Finally, the chapter addresses the auditing process for compliance, which plays a pivotal role in providing assurance that compliance risks are thoroughly assessed. Auditing serves as a vital means to identify and rectify control gaps, ensuring alignment with established standards and regulations. This section will emphasize the importance of regular audits and how they can be leveraged to maintain a robust security framework, thereby safeguarding dApps against both current and emerging threats.

3.1.1 Focused dApp Application Security

Blockchain application security is of paramount importance for the development of secure dApps. dApps, as the cornerstone of blockchain ecosystems, rely on the principles of transparency, decentralization, and immutability to provide users with trust and reliability. Understanding the significance of blockchain application security in the context of dApps involves several key aspects:

1. **Trust and User Confidence:** Trust is a fundamental component of dApp adoption. Users need to trust that their data, transactions, and assets are

secure within the dApp's ecosystem. When news about security incidents and data breaches involving specific blockchain applications become public, it erodes trust, leading to a loss of users and credibility. On the other hand, visibility of robust security measures being implemented and audited can be leveraged to inspire user confidence, encouraging adoption and usage of dApp products and services.

2. **Protection of Digital Assets:** Many dApps involve the management of digital assets, including cryptocurrencies, tokens, and nonfungible tokens (NFTs), and are used for various types of digital assets for value exchanges. Security vulnerabilities in dApps can lead to compromise of these digital assets and financial losses for customers.
3. **Data Integrity:** dApps often handle sensitive data, such as personal information or proprietary business data. A breach in data security can have serious consequences, including legal ramifications and reputational damage. Blockchain application security is essential to maintain the integrity and confidentiality of sensitive data such as transactional data.
4. **Preventing Vulnerability Exploitation:** dApps are built on smart contracts, which are self-executing contracts with predefined rules. Vulnerabilities in smart contracts can be exploited by malicious actors to manipulate the dApp's behavior, leading to unintended consequences. Robust security practices such as security by design and source code vulnerability scanning are crucial to identify these vulnerabilities early during the product life cycle by fixing them before production releases.
5. **Continuity and Reliability:** dApps are designed to be decentralized, which means they rely on a network of nodes to function. Security vulnerabilities can disrupt the dApp's operation, leading to possible denial of service (DoS) or inconsistent performance.

Blockchain application security could serve as the cornerstone of a resilient and trustworthy dApp ecosystem. As blockchain technology continues to evolve and gain adoption across industries from finance and supply chain to healthcare and gaming, the need for robust security practices becomes more critical than ever.

Unlike traditional applications, dApps operate in an environment where transactions are irreversible, smart contracts execute autonomously, and data is publicly accessible on distributed ledgers. These characteristics create unique security challenges, such as smart contract vulnerabilities (e.g. reentrancy attacks, integer overflows), private key mismanagement, consensus mechanism exploits, and bridge attacks that compromise cross-chain interactions. We explored real-world examples of these design flaws and vulnerabilities in Chapter 2, highlighting how such weaknesses can arise in blockchain architectures, Application Programming Interfaces (APIs), and smart contract implementations.

To build a secure and sustainable dApp ecosystem, developers and stakeholders – including product owners, security teams, and platform providers – must integrate security from the ground up. This includes embedding secure development practices, implementing rigorous smart contract auditing, and continuously monitoring blockchain networks for emerging threats. Security by design principles, such as least privilege access control, threat modeling tailored to blockchain architectures, and proactive vulnerability scanning, should be standard practices in blockchain application development.

In this chapter, we will delve into the details of identifying and remediating vulnerabilities in blockchain applications. We will begin by examining real-world security incidents that have impacted blockchain applications and components, extracting key lessons from these cases. Next, we will focus on proactive vulnerability identification during the design phase, utilizing techniques such as threat modeling.

Following that, we will explore strategies for securing critical components by implementing essential security controls. A thorough approach to vulnerability testing, including smart contract auditing, will be emphasized to detect and mitigate potential exploits. We will analyze various types of vulnerabilities that attackers can exploit and discuss effective prevention methods.

All of this will be presented within the broader context of managing dApp security risks both from a technical standpoint and through business impact risk management to ensure a resilient and trustworthy decentralized ecosystem.

3.1.2 dApp Vulnerabilities Risks

In the rapidly evolving digital landscape, dApps and blockchain-based technologies are heralding a new era of transformation, characterized by enhanced transparency, security, and trust. These innovations promise to reshape industries ranging from finance and supply chain management to data privacy and more. However, beneath the innovative veneer, these technologies are beset by a complex array of security vulnerabilities and risks that continually evolve.

For software developers and architects of dApps, it's crucial to have an in-depth understanding of the various attack vectors (AVs) and vulnerabilities that could undermine the security of these applications. Such vulnerabilities include, but are not limited to, the compromise of private keys and the exploitation of smart contracts, posing significant threats to users and developers alike, and potentially undermining the foundational principles of trust and decentralization inherent to blockchain technology.

Our discussion will explore the range of vulnerabilities that could impact dApps and blockchain applications, starting with private keys which grant access to blockchain assets and are thus prime targets for attack. For example, attackers might use phishing schemes to deceive users into disclosing their secret recovery phrases or pose as legitimate services to steal these phrases. Once obtained, attackers can gain access to a user's wallet and assets. Moreover, the insecure storage of recovery phrases, such as in plain text files or unencrypted cloud storage, poses a substantial risk, leaving them vulnerable to theft.

Another critical concern is malware-based attacks targeting dApps, such as malware and keyloggers that stealthily record passwords and secrets that are inputted directly by the user. Client-side malware-based threats typically exploit browser-based vulnerabilities for man-in-the-browser (MitB) and with stolen cookies and secrets including spoofed credentials also perform man-in-the-middle (MitM) attacks aiming to steal sensitive information such as transaction details during transmission with the aim to commit fraudulent transactions.

Further, the direct exploitation of smart contract vulnerabilities through targeted attacks remains probably the most significant threat for dApps. This is primarily because a single vulnerability in a smart contract can be instantly and irreversibly exploited, leading to substantial financial loss, disruption of services, or compromise of user trust. Moreover, the transparent nature of blockchain code allows attackers to analyze deployed contracts publicly, making it easier to discover and exploit flaws. High-profile incidents, such as the Decentralized Autonomous Organization (DAO) hack or the Wormhole bridge exploit, underscore how even minor coding errors can result in multimillion-dollar losses. Unlike traditional applications, where patches and hotfixes can be applied post deployment, smart contracts typically require complex upgrade mechanisms, if allowed at all making proactive security measures and rigorous auditing essential. For these reasons, securing smart contracts against direct exploitation is critical to the safety and sustainability of the entire dApp ecosystem.

Examples of these attacks include reentrancy attacks, where attackers repeatedly call a vulnerable contract to execute malicious code, and front-running, where attackers manipulate transaction orders on a blockchain to their benefit, often resulting in financial losses for other users. Moreover, unchecked external calls in smart contracts pose risks when contracts interacting with others fail to adequately validate input data, allowing malicious contracts to exploit these oversights.

To effectively mitigate these vulnerabilities, dApp developers and users must prioritize identifying and addressing potential security issues in dApps software and components and smart contracts prior to production deployment. Conducting regular vulnerability assessments and smart contract audits are essential for ensuring the ongoing security and operational viability for dApps.

3.1.3 Security Incidents: Lessons Learned for Future Resilience

In the rapidly evolving landscape of blockchain-based applications, the occurrence of security breaches has provided the industry with invaluable insights and lessons. These incidents have not only exposed vulnerabilities in the technology but also highlighted the critical need for proactive security measures and user awareness. Blockchain, often lauded for its security features, is unfortunately not immune to security breaches impacting customers' financial assets with fraud and theft of digital assets. As seen in numerous incidents, both dApps and smart contracts can be susceptible to a wide range of very critical types of vulnerabilities. Understanding these security breaches and their implications is essential for strengthening the blockchain ecosystem and fortifying its resilience against threats.

3.1.3.1 Smart Contract Exploits: Confronting a Critical Threat

Security incidents involving smart contracts have underscored the significant risks posed by vulnerabilities within them. Smart contracts, like any software, can harbor bugs and security flaws that necessitate comprehensive code audits and rigorous testing to ensure their integrity. This is a compilation of some of the most significant smart contract security incidents that have occurred in the past:

1. **The DAO Incident** [132]: The DAO was an Ethereum-based crowdfunding venture aiming to operate as a venture capital fund for dApps. Unfortunately, it became infamous due to a critical vulnerability in its smart contract that permitted an integer underflow. The attacker exploited this by making recursive calls to the DAO's split function, enabling them to withdraw Ether repeatedly without depleting their DAO token balance. This vulnerability led to the theft of approximately 3.6 million Ether, worth tens of millions of dollars at the time. The severity of the attack prompted the Ethereum community to execute a controversial hard fork, dividing Ethereum into two separate chains: Ethereum (ETH) and Ethereum Classic (ETC).
2. **Bancor Front-Running Attack** [133]: Bancor, a decentralized liquidity protocol, fell victim to a sophisticated front-running attack that exploited discrepancies in its automated market-making mechanism, which adjusts token prices based on supply and demand. The attacker manipulated the conversion process between different tokens leading to significant financial discrepancies and user losses.
3. **FOMO3D Attack** [134]: FOMO3D, a blockchain-based gaming platform, had smart contract vulnerabilities, notably in the random number generation process and external calls. These vulnerabilities were exploited and leveraged predictable outcomes and poor security practices within smart contract design to siphon funds or disrupt the game operations.
4. **BatchOverflow Attack** [135]: The BatchOverflow was due to a vulnerability in multiple Ethereum-based token contracts that allowed attackers to

generate vast numbers of tokens by exploiting an integer overflow. This led to unresponsive smart contracts and exchanges due to the overwhelming volume of tokens, effectively creating a DoS within the ecosystem.

5. **DeversiFi Incident** [136]: DeversiFi incident entailed an erroneous payment of a $23.7 million transaction fee due to a software bug in a part of the system that constructs transactions. The core issue stemmed from a bug in the automated transaction-forming software, which incorrectly calculated and assigned the transaction fee.
6. **bZx Incidents** [137]: The bZx platform was targeted through a series of clever exploits that leveraged both design flaws and operational weaknesses in the smart contract system. The attackers were able to manipulate token prices and execute unauthorized loans. These incidents served as a critical wake-up call to the decentralized finance (DeFi) community about the potential risks and the need for more rigorous security measures, better risk management practices, and the importance of continuous monitoring and updating of smart contract protocols.
7. **DeFi Liquidity Pool Manipulation Attacks**: In the DeFi sector, liquidity pools allow users to provide liquidity and earn rewards. These pools, however, are vulnerable to Miner Extractable Value (MEV) bots [138] where the transaction order within a block can be manipulated for financial gain. An attacker used an MEV bot to alter transaction orders, manipulating asset prices within a liquidity pool to obtain favorable trade rates, resulting in considerable profits at the expense of regular users who faced unexpected financial losses.
8. **Poly Network's Cross-Chain Vulnerabilities Exploits** [139]: The attackers exploited a vulnerability in the smart contract code that governs the Poly Network's cross-chain asset transfers. Specifically, they found a flaw in the function responsible for verifying transaction data between different blockchains. Using this vulnerability, the attackers were able to forge transaction requests that appeared legitimate and bypass the security mechanisms in place. This allowed them to transfer a vast sum of digital assets out of the Poly Network's control across various blockchains including Ethereum, Binance Smart Chain (BSC), and the Polygon network.
9. **BSC Attacks** [140]: The BSC suffered a significant security breach where attackers exploited vulnerabilities in its cross-chain bridge, the BSC Token Hub. This attack, occurring on 6 October 2022, led to the unauthorized creation and drainage of approximately $570 million worth of BNB tokens. Binance responded by halting the blockchain's operations and freezing affected funds to mitigate the damage.
10. **Token Bridge Flash Loan Attacks** [141]: Nomad, a token bridge that allows users to send and receive tokens between blockchains was exploited due to

a security flaw in its contract logic, allowing attackers to bypass verification steps and repeatedly withdraw funds. This incident resulted in the loss of nearly $200 million [8] and highlighted the vulnerabilities associated with cross-chain bridges.

11. **The Beanstalk Farms Flash Loan Attack** [142]: The Beanstalk Farms Flash Loan Attack was a flash loan attack targeted at Beanstalk Farms, a stablecoin protocol. The attackers exploited Beanstalk Farms, a decentralized credit-based stablecoin platform, using a flash loan, a type of loan where borrowing and repayment occur within the same transaction block on the blockchain. The attackers borrowed a large amount of funds through a flash loan from another DeFi protocol. They then used these funds to gain temporary majority control over Beanstalk Farms' governance system, as Beanstalk uses a governance model where voting power can be purchased.
12. **SushiSwap $3.3 Million Compromise** [143]: The exploit was triggered by a bug that allowed an attacker to manipulate token approvals within the SushiSwap protocol. This flaw enabled unauthorized access to funds by bypassing normal security checks. The exploit was triggered by a bug that allowed an attacker to manipulate token approvals within the SushiSwap protocol. This flaw enabled unauthorized access to funds by bypassing normal security checks. The attacker was able to manipulate transaction routing paths within the platform, resulting in unauthorized token swaps and the extraction of funds worth around $45 million. This incident emphasized the need for thorough testing and auditing of all contract functionalities, especially those handling routing and token swaps.
13. **ByBit Billion-Dollar Hack** [144]: In February 2025, ByBit, a prominent cryptocurrency exchange based in Dubai, experienced a significant security breach resulting in the theft of approximately $1.5 billion worth of Ethereum, marking one of the largest cryptocurrency thefts to date. The attack occurred during a routine transfer from cold storage to a warm wallet, where attackers exploited the transfer process by redirecting funds to an unauthorized address. Hackers exploited this process by manipulating the transfer mechanism, redirecting 401,000 ETH to an unauthorized address. The breach highlighted a critical flaw: reliance on frontend security and contract address verification alone is insufficient. The attackers compromised the user interface (UI; e.g. via an S3 bucket tampering attack), which still displayed the correct contract address, matching the one verified through hardware wallets and multi-signature (multi-sig) controls. However, the UI manipulated transaction parameters, enabling malicious interactions with the legitimate contract, exploiting trust in the UI while bypassing backend safeguards. This incident highlights the importance of enforcing strong on-chain security measures

rather than relying solely on frontend verification. Specifically, delegate calls within smart contracts must be carefully managed to prevent unauthorized upgrades or manipulation of contract logic. This can be achieved by whitelisting trusted upgrade paths, requiring multiple layers of approval for delegate calls, and restricting sensitive operations directly within the contract code. By implementing these controls on-chain, blockchain applications are better equipped to withstand compromised UIs and safeguard critical transactions from tampering.

These incidents illustrate the imperative need for developers to prioritize security during the design, development, and maintenance of smart contracts. The numerous security incidents in DeFi highlight the critical need for the blockchain community to adopt rigorous, comprehensive smart contract audits and regular security testing. These measures are vital to identify and mitigate vulnerabilities before they are exploited, as evidenced by incidents ranging from the DAO to SushiSwap. Reducing the complexity of smart contract code and minimizing dependencies can also diminish risks, as the complexity often introduces security vulnerabilities.

Moreover, the vulnerabilities in governance models exposed by attacks such as the one on Beanstalk Farms necessitate the development of robust governance frameworks. These frameworks should prevent manipulation through financial leverage, such as that seen with flash loans. Similarly, the unique vulnerabilities associated with cross-chain functionalities, demonstrated by attacks on Poly Network and Finance Smart Chain, call for specific security measures to ensure the integrity of cross-chain transactions.

Quick and effective incident response strategies also play a crucial role in mitigating damage. For instance, Binance's rapid response to freeze transactions following the BSC exploit helped limit further losses. This underscores the importance of preparedness and the ability to act swiftly in the face of security breaches.

Educating end users about the risks involved in DeFi and the best practices for security is equally important. Users must understand the mechanisms of the platforms they use and verify contract interactions to safeguard their investments. Additionally, the evolving nature of DeFi and smart contract exploits necessitates ongoing updates and improvements to security protocols and practices.

As DeFi continues to grow, there is also an increasing need for clear regulatory frameworks that guide the development of secure, transparent, and equitable financial technologies. This collective shift toward higher security standards will involve better risk management practices, enhanced security protocols, and greater community involvement in security audits and governance decisions. These steps are essential to fostering a safer and more resilient DeFi ecosystem.

3.1.3.2 Digital Wallet Design Flaws: Mitigating Emerging Threats

Besides smart contract vulnerabilities, another critical area of focus for the security of dApps, particularly DeFi, involves securing private keys, which are frequently targeted in hacks. Effective management and storage of these keys are essential, typically handled through the use of digital wallets. These wallets play a pivotal role in safeguarding private keys from unauthorized access and are a fundamental aspect of maintaining the overall security integrity of blockchain applications.

This is a compilation of significant security incidents in digital wallets used in blockchain-based applications that can provide several valuable lessons for the industry:

- **BtcTurk Hot Wallet Hack** [145]: BtcTurk reportedly lost $55 million due to compromised hot wallets. This incident underlines the vulnerabilities that can occur in digital wallet structures and the significant financial implications of such breaches. This incident serves as a stark reminder of the vulnerabilities inherent in digital wallet structures, particularly hot wallets which are connected to the Internet and thus more susceptible to attacks. The breach at BtcTurk highlights the significant financial implications and operational disruptions that can result from such security lapses. It underscores the critical need for enhanced security protocols, regular security audits, and stringent access controls to mitigate risks associated with digital wallets, especially those handling large volumes of transactions.
- **Parity Multi-Sig Wallet Incident** [146]: Parity Technologies, aiming to enhance Ethereum network security, developed a multi-sig wallet that used a smart contract to enable shared ownership and management of Ether. Designed to require multiple approvals for transactions, this wallet was considered a more secure option compared to single-signature wallets. However, it harbored a significant vulnerability related to unchecked external calls within its smart contract. This flaw was exploited when an attacker manipulated the contract's constructor function to assign ownership to themselves during the initialization of a new multi-sig wallet. This breach allowed the attacker to execute a "kill-switch" feature in the smart contract, rendering all connected multi-sig wallets inoperable and permanently locking users out of their funds. The incident affected numerous Ethereum projects and resulted in the loss of tens of millions of dollars, highlighting the critical need for comprehensive audits and the potential risks of complex smart contract functionalities.
- **Mist Wallet Security Flaw** [147]: Mist, developed as a user-friendly Ethereum browser wallet, facilitated interactions with smart contracts on the Ethereum blockchain. It aimed to provide a secure and intuitive platform

for managing cryptocurrency transactions. However, a critical design flaw was exposed when attackers began deploying deceptive smart contracts. These contracts, often disguised with attractive offers like free tokens, contained hidden malicious code. Users, deceived by these offers and trusting Mist's interface, interacted with these contracts, unwittingly triggering the malicious actions embedded within. This vulnerability allowed attackers to perform unauthorized actions, such as transferring users' funds to their own accounts. The exploit not only led to financial losses for users but also raised significant concerns about the safety of interacting with smart contracts through browser wallets.

The incidents involving BtcTurk, Parity Multi-Sig Wallet, and Mist Wallet offer crucial lessons for the design, development, and operation of digital wallets. Here are the key takeaways:

1. **Enhanced Security Protocols:** The need for stronger security measures for hot wallets, which are always online and more vulnerable to attacks, is critical. Implementing multifactor authentication (MFA), using hardware security modules (HSMs), and employing advanced encryption methods can help secure these wallets.
2. **Regular Security Audits:** Regularly scheduled audits and penetration testing can help identify and mitigate vulnerabilities before they can be exploited by attackers. As smart contracts can contain complex and critical vulnerabilities, comprehensive security audits are necessary to identify issues before deployment. Using both automated tools and manual inspection by experts can help uncover potential risks.
3. **Stringent Access Controls:** Limiting who can access wallet infrastructure and under what conditions can significantly reduce risk. This includes both physical and digital access controls.
4. **Isolation of Contract Interactions:** Implementing features that isolate or sandbox smart contract interactions within the wallet can prevent malicious contracts from accessing wallet funds directly.
5. **Focus on Phishing and Security Awareness:** Wallet interfaces should be designed to make clear what is a genuine offer or alert and what might be a phishing attack. This includes clear warnings about how users are expected to interact and the information that should be requested. Users should be educated about the inherent risks of interacting with unknown or unverified smart contracts and the importance of verifying the authenticity of dApp interactions such as deceptive pop-ups or transaction requests that mimic legitimate services. Educating users on the dangers of keyloggers, clipboard hijackers, and other malicious software that targets private keys or transaction data is also critical.

6. **Emergency Response Plan:** Organizations should establish a comprehensive plan to respond swiftly to smart contract failures or security exploits in order to minimize impact. This plan may include mechanisms to pause or restrict contract functionality (e.g. circuit breakers), clearly defined procedures for contract upgrades, and, where appropriate, a well-secured kill switch that can disable critical functions under predefined and tightly controlled conditions. The plan should also outline roles, responsibilities, and communication protocols to ensure coordinated incident response and rapid recovery, while maintaining transparency and trust with stakeholders.

By integrating these lessons learned into the development and management of digital wallets, organizations can enhance their security and resilience against increasingly sophisticated cyber threats.

3.1.3.3 Proactive Security Recommendations from Blockchain Breaches

We outline herein some general recommendations for proactively implementing security measures as these have been learned from data breaches and fraudulent activities in blockchain-based applications, particularly those involving decentralized transactions with crypto assets. Special attention is given to vulnerabilities, exploits, and security flaws associated with smart contracts and digital wallets.

3.1.3.3.1 Conduct Security Audits of Smart Contracts

The necessity of rigorous security audits cannot be overstated in the blockchain domain. Many security breaches in blockchain applications trace back directly to vulnerabilities within the code itself. Conducting regular and thorough security audits is essential to detect and address these vulnerabilities before they manifest into security breaches.

Security audits provide a comprehensive review of smart contracts and blockchain applications, uncovering potential vulnerabilities and weaknesses that could be exploited. These audits cover a wide range of potential AVs, including reentrancy attacks, integer overflow/underflow issues, the proper handling of external contract calls, unsafe type conversions, and ensuring the correct initialization of storage and variables. Additionally, audits assess defenses against front-running attacks, strategies to counteract MEV bots, access control measures, optimization of gas usage in loops, secure external contract dependencies, and the robustness of upgradable contracts.

An illustrative case highlighting the consequences of inadequate security auditing involves the DeFi platform, YAM. In August 2020, YAM was launched without a comprehensive security audit and soon encountered a critical bug in its rebase mechanism – a key feature of the protocol. This oversight led to a significant security incident where the YAM token, along with its ecosystem, nearly collapsed, resulting in substantial financial losses for users. A thorough audit prior to launch could have identified this flaw, potentially averting the crisis and safeguarding user assets.

This example underscores the importance of security audits in the blockchain and DeFi sectors, where vulnerabilities threaten not only security but also user confidence and financial stability. Engaging in detailed security audits before deploying any smart contract is crucial, especially in environments where user funds are extensively involved.

3.1.3.3.2 Require Responsible Disclosure of Vulnerabilities

Transparency and open-source practices are foundational to building trust within the blockchain community. It is vital for users and investors to have access to smart contract code and understand its functionality. DeFi projects that openly share their smart contract code and operational practices typically garner greater trust from the community. This transparency not only fosters user confidence but also enhances security by allowing independent verification of the code.

An illustrative negative example of what happens when transparency and open-source principles are not adhered to is the case of ForkDelta, a decentralized exchange (DEX) that was initially built on the EtherDelta smart contract. ForkDelta aimed to provide a community-driven, decentralized trading platform for Ethereum-based tokens. However, the project suffered from a lack of transparency in its development and governance processes. This opaqueness led to inadequate maintenance and updates of the smart contract code, coupled with minimal oversight from the community.

In 2018, this lack of transparency resulted in a security breach when an attacker exploited a vulnerability in the ForkDelta smart contract, executing a fraudulent transaction that led to substantial financial losses. Several thousand dollars' worth of cryptocurrency was stolen as the attacker withdrew funds from users' accounts. This incident underlined the critical need for openness and community involvement in projects, especially DEXs where users' funds are at risk. Transparent practices enable community members to review and identify potential security vulnerabilities, thereby helping prevent fraudulent transactions and data losses.

This lesson emphasizes that maintaining an open-source nature and ensuring project transparency are not just ethical practices but also crucial for the security and reliability of blockchain applications.

3.1.3.3.3 Comply with Regulatory Requirements

The regulatory landscape for blockchain and cryptocurrencies is constantly evolving, posing a significant challenge for businesses operating in this space. It is crucial for companies to stay abreast of and comply with relevant laws and regulations to avoid legal repercussions. The consequences of noncompliance can be severe, ranging from fines and penalties to operational restrictions and reputational damage.

Binance's Regulatory Challenges: Binance, one of the largest cryptocurrency exchanges globally, faced intense scrutiny from regulatory authorities across multiple countries due to its operational practices and lack of adherence to local financial regulations, especially concerning Know Your Customer (KYC) and Anti–Money Laundering (AML) standards. This lack of compliance led to various legal actions, including fines, penalties, and restrictions on its services, severely impacting its global operations and tarnishing its reputation. This situation underscores the critical need for cryptocurrency exchanges to ensure they are in compliance with evolving regulatory requirements to maintain their operations and avoid punitive measures.

The BitMEX Case: Similarly, BitMEX, a prominent cryptocurrency derivatives exchange, encountered significant legal issues in October 2020 when the US Commodity Futures Trading Commission (CFTC) charged the platform and its executives with facilitating unregistered trading and violating the Commodity Exchange Act. These charges were primarily due to BitMEX's failure to implement adequate KYC and AML procedures, reflecting a broader noncompliance with US regulations. The repercussions were substantial, leading to the resignation of BitMEX's founders from their management roles and the exchange agreeing to pay a hefty fine. Furthermore, in response to these legal challenges, BitMEX introduced stringent KYC requirements for its users. This incident highlights the critical importance of regulatory compliance for cryptocurrency exchanges, particularly those operating in jurisdictions with stringent financial regulations like the United States.

These examples emphasize that for cryptocurrency businesses, understanding and adhering to regulatory requirements is not optional but a necessity. Compliance helps in safeguarding against legal risks, ensuring the continuity of operations, and maintaining trust among users and stakeholders. Noncompliance, on the other hand, can result in severe consequences, affecting an organization's ability to operate and compete in the global market. It's important to note that regulatory compliance is typically determined by where the business is legally registered, where it operates, and where its users are located. Since cryptocurrencies are borderless by nature, organizations must often comply with multiple regulatory frameworks across jurisdictions such as General Data Protection Regulation (GDPR) in the European Union (EU), FinCEN regulations in the United States, or Financial Action Task Force (FATF) guidelines internationally. Identifying the relevant legal jurisdictions early is critical to determining which compliance obligations apply. We will explore these aspects in greater detail later in this chapter, in a dedicated Section 3.3 titled "Auditing Blockchain Applications for Compliance."

3.1.3.3.4 Address Scalability and Network Congestion Issues

As blockchain networks grow and experience increased transaction volumes, they can face significant scalability and congestion challenges. These issues not only affect transaction processing times and costs but also increase the network's vulnerability to various types of cyberattacks. Implementing scalability solutions and maintaining vigilant network monitoring are crucial to managing these risks effectively.

Impact of CryptoKitties on Ethereum: In late 2017, the Ethereum network experienced severe congestion due to the unexpected popularity of CryptoKitties, a blockchain game that allowed players to breed and trade digital cats as ERC-721 tokens. This surge in activity led to slower transaction times and higher costs, severely impacting the usability of the Ethereum network. This incident underscored the urgent need for scalable blockchain solutions capable of handling sudden increases in demand without compromising network performance.

Vulnerabilities During Network Congestion: DeFi protocols, in particular, have shown susceptibility to attacks during periods of high congestion. Attackers have exploited such times to conduct flash loan attacks, utilizing the slow network response to manipulate DeFi protocols for financial gain. For example, an attacker could use a flash loan to manipulate a price oracle (a service that feeds price data to smart contracts), tricking a DeFi platform into selling assets far below market value, then quickly buying them and making a massive profit before repaying the flash loan. These attacks often involve exploiting price discrepancies and smart contract vulnerabilities, exacerbated by delayed transaction confirmations.

Moreover, network congestion can amplify the risk of reentrancy attacks, where an attacker makes repeated calls to a contract, manipulating its state before previous transactions are confirmed. This can result in significant financial losses, as seen in various incidents across the DeFi landscape.

Strategies to Mitigate Risks: To combat these challenges, it is essential for blockchain developers and network engineers to focus on:

- **Scalability Solutions:** Implementing technologies such as sharding, Layer 2 solutions (like rollups and sidechains), and state channels to increase throughput and reduce latency.
- **Optimizing Gas Fees:** Adjusting gas fees based on network demand can help manage congestion and deter spam transactions.
- **Enhanced Monitoring:** Continuous monitoring of network activity to anticipate and react to periods of high congestion.

Addressing scalability and network congestion is not merely about enhancing performance but also about bolstering security. As blockchain technology continues to evolve, prioritizing these aspects will be crucial for maintaining robust, secure, and efficient networks capable of supporting a wide range of applications.

3.1.3.3.5 Strengthen Incident Response Process

Being well-prepared with a robust security incident response and recovery plan is crucial for effectively managing and mitigating the impacts of security breaches in blockchain environments. When security incidents occur, the ability to respond swiftly and decisively is key to minimizing damage and protecting user interests. A clear communication plan should also be an integral part of any incident response strategy. Effective communication, both internally among teams and externally with stakeholders, is essential for managing incidents efficiently, maintaining trust, and ensuring timely, coordinated responses when security events occur.

Effective incident response plans are crucial across different segments of the blockchain ecosystem, from exchanges to dApps. Here's how a robust incident response strategy plays a critical role in managing security breaches and vulnerabilities in the context of various types of security incidents scenarios:

- Managing Potential Exploits of Smart Contract Vulnerabilities
  - **Security Incident Possible Scenario:** Smart contracts, especially those integral to DeFi applications, are prone to sophisticated exploits that can lead to significant financial losses.
  - **Incident Response Strategy:** The response strategy for managing smart contract vulnerabilities encompasses a comprehensive approach that starts with regularly scanning and updating smart contracts using automated tools to identify and fix vulnerabilities. It includes deploying real-time monitoring systems capable of detecting signs of exploitation, such as unusual transaction rates or unexpected contract interactions. Swift incident response measures are crucial, involving quick isolation and mitigation of affected contracts to prevent further damage, which may include pausing or migrating contracts. Conducting forensic analysis is essential to understand how vulnerabilities were exploited, providing insights to develop more robust smart contracts in the future. Maintaining open communication with all stakeholders, including users, investors, and regulatory bodies, ensures transparency throughout the incident management process. Legal preparedness to handle potential consequences arising from vulnerabilities is also critical. The strategy also involves continuous education for development teams on the latest security practices, regular security audits by independent third parties, and simulation of attacks to test and refine response strategies. Creating a feedback loop after resolving incidents helps learn from exploits and improves both the smart contracts and incident response strategies. Integrating these targeted strategies within broader incident response plans enables organizations to enhance their ability to effectively respond to incidents, protect user assets, and maintain

trust in their blockchain applications, crucial in the rapidly evolving blockchain technology landscape where new threats can emerge swiftly.

- Cryptocurrency Exchange Security Breaches: Example of Cryptopia
  - **Security Incident Possible Scenario:** Cryptocurrency exchanges can suffer a major security breach resulting in significant losses.
  - **Incident Response Strategy:** When a security breach occurs at a cryptocurrency exchange, an effective immediate response includes halting all trading to prevent further unauthorized transactions. Simultaneously, it's crucial to secure user assets by transferring funds to cold storage or implementing other robust security measures. Promptly notifying affected users and the public about the breach and the expected next steps is essential for maintaining trust and transparency. To mitigate losses, partnerships with forensic specialists can be invaluable in tracking stolen funds, and exploring legal avenues for recovery can also be effective. Following the immediate response, a thorough postincident analysis is necessary to identify the causes of the breach and any system vulnerabilities. This analysis should lead to a comprehensive strengthening of security protocols to prevent future incidents.
- DeFi Protocol Vulnerabilities
  - **Security Incident Possible Scenario:** Various DeFi protocols might face exploits, especially during network congestion.
  - **Incident Response Strategy:** When an exploit is detected, swift actions include immediately disabling any vulnerable functions or contracts to halt further exploits. Alongside this, freezing transactions temporarily helps prevent additional exploitation of the vulnerability. To maintain trust and address financial losses, establishing a compensation scheme for affected users is crucial. Further, conducting a thorough incident analysis to understand the mechanism of the exploit is essential. This analysis should guide the implementation of necessary upgrades or fixes to the protocols, ensuring that such vulnerabilities do not recur.
- DoS and Spam Attacks on Blockchain Networks
  - **Security Incident Possible Scenario:** Blockchain networks are susceptible to DoS and spam attacks that can destabilize operations.
  - **Incident Response Strategy:** The response strategy for managing network threats involves employing advanced monitoring tools to detect and manage unusual traffic patterns effectively. Rapid mitigation techniques such as rate limiting, rerouting, and other network management strategies are essential to quickly mitigate attacks. Implementing recovery procedures helps in promptly restoring services and ensuring network stability. Additionally, continuously adapting and improving defensive strategies based on observed attack patterns and trends is crucial to enhancing network security over time.

- Phishing and Wallet Compromises
  - **Security Incident Possible Scenario:** Users are vulnerable to phishing attacks and wallet compromises.
  - **Incident Response Strategy:** The response strategy to enhance user security includes conducting regular educational sessions and issuing alerts to inform users about the latest phishing tactics and wallet security practices. Providing immediate support to users who report compromises is crucial, helping them secure their accounts and recover their assets. Strengthening authentication processes and introducing additional security measures such as MFA are vital steps. Additionally, collaborating with law enforcement and cybersecurity firms to track attackers and recover lost funds is essential to address legal and recovery aspects effectively.

By tailoring the incident response plans to specific threats and vulnerabilities within the blockchain and cryptocurrency landscapes, organizations can not only mitigate the immediate impacts of security incidents but also enhance their resilience against future threats. When conducting root cause analysis of security incidents, it is also possible to leverage threat intelligence not only to understand the specifics of the incident but also to gain insights into what has gone wrong with similar products or systems in the industry. By analyzing breaches and vulnerabilities in other applications, you can identify common failure points and patterns and then apply this knowledge proactively to assess whether your own product is susceptible to the same or related issues. This approach ensures that lessons learned from external incidents are integrated into your product's security posture, reducing the risk of similar vulnerabilities and improving overall resilience.

3.2 Enhancing Blockchain Security: Mitigating Vulnerabilities and Design Flaws

As a security architect in charge of the design of dApps, performing security architecture reviews is crucial to identifying necessary design changes that mitigate potential risks. This responsibility involves integrating a security by design strategy right from the outset of dApp development. By proactively incorporating security considerations during the design, implementation, and deployment stages of dApps, security architects can significantly reduce vulnerabilities and design flaws. Equally important is the incorporation of privacy by design principles, particularly given the transparent and immutable nature of blockchain technology. This means proactively identifying how personal or sensitive data may be handled, minimizing data exposure, ensuring user consent, and applying techniques such as data minimization, anonymization, or selective disclosure where appropriate. Together, these approaches ensure that both security and privacy are foundational elements of dApp architecture, fostering trust, compliance, and resilience.

Given the unique challenges and high stakes associated with blockchain technology, where dApps often handle sensitive transactions and store substantial value, establishing a robust security framework from the very beginning is essential.

Threat modeling plays a pivotal role in this strategy. It is a structured approach that helps identify potential security issues before they are coded into the dApp. By simulating various AVs and analyzing possible threats during the design phase, security architects can foresee and mitigate a wide range of vulnerabilities. This proactive approach not only helps in designing more secure systems but also reduces the need for costly fixes after deployment.

Incorporating threat modeling within the security by design strategy ensures that each component of the dApp is scrutinized for security weaknesses. This includes evaluating the smart contracts for common vulnerabilities such as reentrancy attacks, improper access controls, and other exploit vectors that could be used maliciously.

By establishing security by design as a fundamental philosophy in dApp development, backed by rigorous threat modeling, security architects create more resilient and trustworthy applications. This strategy not only protects the applications and their users but also enhances the overall reliability and adoption of blockchain technology.

A security by design strategy has to include basic pillars of technology/tools, people resources and skills/training, and processes/standards and guidelines. This strategy should include a capability and maturity assessment of the organization's current security posture, aligning with business objectives, customer expectations, and available resources. A robust strategy will also define metrics and key performance indicators (KPIs) to measure progress and effectiveness over time. As part of this, the maturity of threat modeling activities must be assessed and continuously improved. This can be achieved by evaluating the organization's current threat modeling capabilities against established frameworks such as Open Web Application Security Project (OWASP) SAMM (Software Assurance Maturity Model) [148], which helps determine maturity levels in secure design and risk assessment practices. Alternatively, peer benchmarking through BSIMM (Building Security in Maturity Model) [149] can offer insights into industry-standard practices and highlight areas for improvement.

By aligning threat modeling execution with assessed maturity levels, organizations can tailor their approach ranging from basic ad hoc activities to fully integrated and automated threat modeling across development pipelines ensuring the strategy remains actionable, scalable, and effective as the organization evolves.

3.2.1 Introduction to Threat Modeling

As organizations increasingly turn toward dApps and blockchain technologies to drive new business models and enhance operational efficiencies, the imperative for robust security measures grows exponentially. The unique architecture of dApps, coupled with the complex nature of blockchain networks, introduces distinct security challenges that demand a sophisticated, risk-focused approach to security by design.

Threat modeling is a crucial methodology for developing secure by design dApps and blockchain-based applications and should be considered as a foundational practice essential for ensuring robust, compliant, and secure blockchain applications. According to National Institute of Standards and Technology (NIST)'s "Guide to Threat and Risk Assessment" (Special Publication 800-30 [150]), threat modeling is defined as: "A process for identifying and evaluating security threats to a system. It involves characterizing system components, potential attackers, and possible threats to determine potential vulnerabilities and devise strategies to mitigate risks."

As the chief information security officer (CISO) of a DeFi start-up focused on dApps, the integration of threat modeling into our security and development life cycle is not just a best practice, it's a crucial pillar of applications and products security posture. In this dynamic environment, where the pace of innovation and the stakes of security are both exceedingly high, aligning the organization's security program with top-tier standards and control frameworks is one way to get a threat modeling business case for designing, implementing, and testing security controls. There are standard control frameworks that explicitly include threat modeling as activity such as the Security and Privacy Controls NIST 800-53 [151] and NIST Special Publication 800-154 [152], "Guide to Data-Centric System Threat Modeling," that focuses on threat modeling for data-centric systems whose principles are highly relevant to dApps, which are inherently data-centric in nature. Additional standards that explicitly include or focus on threat modeling are the OWASP Application Security Verification Standard (ASVS) [153] and the Center for Internet Security (CIS) Critical Security Controls [154]. Adopting any of these standards allows you to frame the requirement of threat modeling assessment as a foundational element of security controls risk assessment for the risk of your developed dApps.

Besides the alignment with industry-endorsed risk management and control standards, threat modeling enables your organization to proactively identify, assess, and mitigate potential threats right from the earliest stages of system development. This approach shifts your strategy from reactive (i.e. fix vulnerabilities as these are identified during testing) to proactive (i.e. fix them by implementing security controls as soon as the design phase). By embedding threat modeling into your software development processes such as a part of SecDevOps, you can ensure that your security measures are designed in parallel with secure product development, not as an afterthought. The importance of threat modeling in this context can be summarized as a list of benefits that can be articulated for your different types of business case such as:

1. **Aligning Security with Business Objectives:** Your focus is on identifying threats that could have the most significant impact on our business and tailor our security controls to mitigate these risks effectively as these carry business impact. Threat modeling aims to present identified risks and proposed mitigations to decision-makers, enabling them to allocate budget and resources efficiently and effectively in alignment with business goals. This alignment is critical in ensuring that our operational priorities and product security evolve in unison.
2. **Ensuring Compliance:** The DeFi industry is becoming increasingly regulated. Effective threat modeling supports compliance with various regulatory frameworks that mandate rigorous risk assessments as early as the design phase. Moreover, a threat model serves as an efficient and effective way to document such risk assessments, providing clear evidence of due diligence in identifying, evaluating, and addressing potential security and privacy risks.
3. **Enhancing Risk Management:** Early identification of potential threats through structured threat modeling significantly lowers the likelihood of security breaches by exploitation of vulnerabilities by these threats with a focus primarily on identifying and analyzing risks but do not extend to handling or treating those risks.
4. **Promoting Security by Design:** By integrating security considerations into the application architecture from the onset, we minimize the need for costly security retrofits later in the product life cycle and enhance overall control over the security posture of your product releases.

In the evolving landscape of cybersecurity, threat modeling stands as a cornerstone practice, essential for identifying and mitigating potential threats early in the system development life cycle. Threat modeling frameworks provide structured methodologies for identifying and mitigating security threats.

These frameworks can be categorized based on their focus. We provide here a brief introduction to various threat modeling methods that you can apply when designing and securing a dApp. Each method offers distinct benefits and focuses, allowing you to choose the approach that aligns best with your goals, development style, and risk profile:

1. **STRIDE** [155]: STRIDE (Spoofing, Tampering, Repudiation, Information Disclosure, DoS, Elevation of Privilege) is a well-established, system-centric framework created by Microsoft. It systematically walks you through different categories of threats based on security properties, helping you identify

potential vulnerabilities in your dApp's smart contracts, data flows, and infrastructure. If you're looking for a straightforward way to categorize and address common threats in a structured manner, STRIDE provides a balanced starting point.

2. **PASTA** [156]: PASTA (Process for Attack Simulation and Threat Analysis) takes a risk-driven, attacker-centric approach to threat modeling, integrating business impact analysis (BIA) with detailed simulation of potential attacks. For a dApp handling high-value transactions or tokenized assets, PASTA can help you focus on attacks that pose the greatest financial or reputational risk. This methodology is particularly useful if you need to align security decisions with broader business objectives and ensure that resources are allocated based on actual threat impact.
3. **LINDDUN** [157]: LINDDUN (Linkability, Identifiability, Nonrepudiation, Detectability, Disclosure of information, Unawareness, Noncompliance) is primarily focused on privacy-related threats, making it useful if your dApp deals heavily with sensitive user data or regulatory requirements around anonymity. By analyzing data flows and identifying privacy weaknesses, LINDDUN ensures compliance with data protection standards (such as GDPR) and helps foster user trust in your decentralized solution.
4. **VAST** [158]: VAST (Visual, Agile, and Simple Threat Modeling) is designed to be scalable, visual, and easily integrated into agile or DevSecOps environments. If your dApp is evolving quickly or you're continually deploying new features and smart contracts, VAST can keep pace by providing clear diagrams and rapid threat identification. Its visual approach helps both technical and nontechnical stakeholders grasp potential risks, making it a good fit for teams that prioritize collaboration and iterative development.
5. **Trike** [159]: Trike integrates threat modeling with risk management and auditing processes, translating security properties into measurable risks. It can be helpful in dApp environments where governance and auditability are critical, such as when working with regulators or third-party auditors. By combining detailed threat identification with explicit risk assessment, Trike supports transparent decision-making and can demonstrate robust security practices to stakeholders and users.
6. **ATASM** [160]: ATASM (Attack Tree-Based Agile Security Methodology) is a lightweight, agile-friendly threat modeling approach developed by Daniel Shoenfield. It uses attack trees to visually map out potential threats, breaking down attack objectives into smaller, actionable steps. Designed for agile development environments, ATASM supports iterative, continuous threat modeling throughout the software life cycle. Unlike heavier methodologies, ATASM emphasizes flexibility, risk prioritization, and ease of integration into

development workflows, making it ideal for teams that need practical and scalable threat modeling without extensive overhead.

Selecting the right framework depends on your dApp's complexity, regulatory requirements, and the nature of assets or data it handles. A risk-driven approach like PASTA may suit projects where financial impact and attacker motivations are paramount, while VAST or STRIDE might be more appropriate if you seek a simpler, more iterative model that aligns with agile development. Remember that many teams adopt a hybrid strategy, combining elements from different frameworks to cover all angles of a rapidly evolving dApp ecosystem.

Among the myriad methodologies available, the PASTA distinguishes itself as a particularly effective framework to mitigate technical and business risks. Unlike many traditional threat modeling methodologies, PASTA is explicitly risk-centric, offering a systematic approach that aligns closely with both technical security needs and overarching business objectives. The PASTA is a comprehensive threat modeling methodology that is particularly well-suited to the needs of dApps security. Unlike traditional threat modeling methods, which often focus primarily on identifying vulnerabilities without a direct link to business impact, PASTA is inherently risk-centric, offering a structured framework that aligns closely with your strategic business objectives and technical requirements.

As a security architect specializing in enterprise-level blockchain solutions for financial services and DeFi, you will find the PASTA methodology particularly well-suited for conducting thorough and effective threat modeling in dApps with substantial risk exposure. When weaknesses and vulnerabilities can lead to significant data compromises and fraud, potentially resulting in millions of dollars' worth of misappropriated digital assets, PASTA's risk-driven framework is precisely what high-stakes security environments demand. Beyond identifying architectural risks, it facilitates critical design revisions and validates the effectiveness of security measures. Here's why I strongly advocate adopting PASTA in your security by design practices, especially for high-risk dApps and DeFi use cases:

- **Alignment with Business and Technical Goals:** Starting with a clear understanding of our business objectives and the technical scope, PASTA ensures that our security measures are not just technically sound but also strategically aligned with our business priorities. This alignment is crucial as it ensures that our security efforts support our core business goals and are not just peripheral checks.
- **Detailed Threat and Vulnerability Analysis:** PASTA's methodical seven-step process allows for a deep dive into the complex architecture of our dApps and blockchain infrastructure. It helps us dissect the system comprehensively, identifying and prioritizing potential threats from smart contract

flaws to consensus mechanism vulnerabilities. This meticulous analysis is key to ensuring that no stone is left unturned and all components are secured.

- **Risk-Centric Prioritization:** By integrating risk analysis right from the initial stages, PASTA helps us focus our efforts on the most critical threats – those that could have the most significant impact on our business. This risk-centric approach ensures efficient resource allocation and targets our security efforts where they are needed most, enhancing our overall security posture.
- **Proactive Simulations and Countermeasures:** PASTA goes beyond mere identification of potential threats. It actively involves simulating attacks and proposing strategic countermeasures. This proactive stance is crucial for us to fortify our dApps against both known and emerging threats, significantly reducing the potential for exploitation and ensuring that our applications are robust against adversarial actions.
- **Support for Regulatory Compliance and Due Diligence:** In an environment where regulatory demands are constantly evolving, PASTA's comprehensive documentation and risk management capabilities are invaluable. They support our compliance efforts and simplify due diligence processes, ensuring that our dApps meet all necessary standards before they go live.

Using PASTA, we are able to effectively identify risks in our architecture, drive necessary design changes, and conduct security tests to validate that our countermeasures effectively mitigate risks before our products reach production. This thorough preparation not only secures our applications but also aligns our security practices with the strategic goals of our organization, ensuring that our security architecture is both comprehensive and business-oriented. Applying the PASTA methodology to a Web3 dApp involves a structured approach tailored to the unique architectural components and security concerns of blockchain-based systems.

The PASTA methodology for threat modeling systematically executes activities across several stages to provide a thorough approach to securing dApps. It effectively combines threat identification with proactive mitigation strategies, ensuring dApps are safeguarded against potential risks. This structured approach not only protects dApps from existing and emergent threats but also ensures that the security strategies are seamlessly integrated with the business objectives of the organization.

3.2.2 PASTA Threat Modeling

Below, I present a cheat sheet for PASTA-based threat modeling tailored for dApps, which can serve as a reference for any dApp. PASTA consists of seven stages, each with specific activities and outcomes. In this cheat sheet, we provide a high-level overview of these stages, outlining key actions and expected results to guide the threat modeling process. Additionally, this chapter will include an example of applying PASTA threat modeling to a DeFi dApp, demonstrating its execution in practice.

3.2.2.1 Definition of Business Objectives

The primary goal in this stage is to pinpoint the critical business assets that require protection, ensuring that security measures are effectively aligned with the overarching business priorities of the dApp. This alignment is crucial for safeguarding the dApp's operational integrity and maintaining user trust.

Focus Activities: To achieve this goal, it's essential to answer several key questions that will guide the threat modeling assessment during this stage:

- **What are the primary business goals of the dApp?** For example, is the dApp focused on DeFi, gaming, or operating as an NFT marketplace? Understanding the business goals helps in tailoring the security measures to the specific needs and challenges of the dApp.
- **What are the critical assets that need to be protected?** Identifying assets such as private keys, smart contracts, user funds, and transaction data is crucial. These assets represent the core components of the dApp that must be protected against threats.
- **What would be the impact in case of a security breach?** Evaluating potential consequences such as financial loss, reputational damage, intellectual property damages, and regulatory penalties helps in prioritizing security efforts and understanding the broader implications of security incidents. A security breach impacts both users and the organization, but in different ways. Users may face loss of assets, privacy violations, or trust issues, while organizations risk financial loss, reputational harm, and regulatory penalties.

Outputs: The expected outcomes of this stage might include:

- **A Clear Understanding of the Business Impact and Security Priorities for the dApp:** This understanding is fundamental for developing a targeted and effective security strategy that supports the dApp's business objectives.
- **Alignment with Compliance Requirements:** It ensures that the security measures adhere to relevant legal and regulatory standards depending on the scope, such as GDPR [161] and California Consumer Privacy Act (CCPA) [162] for data protection of Personally Identifiable Information (PII) and AML and KYC [163] for compliance and operations with digital assets, blockchain, and DeFi. Organizations must not only adhere to data protection and financial compliance regulations but also align with technical cybersecurity standards like NIST Cybersecurity Framework (CSF) [63] and NIST SP 800-53 [151] or product-focused legislation like the EU Cyber Resilience Act

(CRA) [164], depending on the application's use case, infrastructure role, and geographical market.

Tools and Techniques: Several tools and frameworks can be utilized during this stage to facilitate the risk analysis, business impacts, and compliance risks and derive the necessary security and risk profile for the dApp:

- **Risk Frameworks for Deriving the dApp Inherent Risk Profile**
  - **Likelihood and Impact Analysis:** The NIST Guidance for Conducting Risk Assessments [165] methodology assesses the likelihood of a threat event occurring and its potential impact to quantify and prioritize risks. At this stage, risk evaluation is based on the inherent characteristics of the dApp, including its role in storing and handling sensitive data and transaction-signing secrets, as well as its exposure to the Internet. The likelihood assessment provides a high-level estimate of both technical risks (e.g. vulnerability exposure) and business risks (e.g. operational criticality).
  - **Cybersecurity Risk Management Frameworks (RMFs):** If your organization does not yet have a risk management standard in place, you can consider adopting established methodologies such as the NIST CSF [166], OCTAVE (Operationally Critical Threat, Asset, and Vulnerability Evaluation) [167], or FAIR (Factor Analysis of Information Risk) [168] to develop a comprehensive risk profile for the dApp in scope. These frameworks provide structured approaches for assessing and managing security risks effectively such as managing corrective action plans for control gaps being identified as well as risk acceptances and exceptions.
- **BIA Templates and Standards**
  - **BIA Template:** NIST SP 800-34 Rev. 1 [169] provides a free BIA template designed to assist organizations in analyzing the impact of disruptions to their information systems. It includes sections for identifying critical functions, potential impacts, recovery time objectives (RTOs), and recovery point objectives (RPOs).
  - **ISO 22301/Business Continuity Management** [170]: This international standard provides a robust framework for establishing, implementing, and maintaining an effective business continuity management system, which includes conducting detailed impact analyses.
- **Blockchain-Specific Governance, Risk, and Compliance (GRC) Tools**
  - A combination of security, compliance, and blockchain-specific tools [171] can be utilized to help organizations adhere to legal and regulatory requirements, including jurisdiction-specific blockchain regulations. Once the risk profile of the dApp has been documented, the requirement for relevant compliance processes such as architecture governance and vulnerability

risk management, including threat modeling, will be documented into the risk asset management tool. This risk profile will also serve as a key factor in determining whether the dApp falls within the scope of specific audit and compliance requirements.

By leveraging these tools and focusing on key security activities, organizations can develop a well-informed strategy that safeguards critical assets while aligning with business objectives and compliance requirements. This strategic approach strengthens the security of the dApp, recognizing it as a critical asset that the organization must own, manage, and protect from a risk management perspective, aligning with Stage I of the PASTA threat modeling methodology, which emphasizes defining business objectives and valuing assets in context.

As the threat modeling process advances, the business logic and architectural components are primarily examined during Stages II through V of PASTA, with a specific focus on identifying how business processes, workflows, and application logic could be exploited. These stages enable a detailed understanding of how technical elements support key business functions, allowing organizations to uncover potential flaws such as issues in authentication flows or transaction handling that could lead to security breaches or business disruption if not addressed.

3.2.2.2 Definition of the Technical Scope

The primary goal of defining the technical scope is to clearly delineate the technical boundaries of the dApp. This involves a comprehensive assessment of all pertinent hardware, software, and network components that constitute the application/system whose threats, vulnerabilities, and impacts on the data and transactional assets need to be further analyzed.

Focus Activities: To accomplish this, it is essential to conduct a comprehensive review of architectural diagrams and technical design documents within the established architecture governance processes. This review will serve as the foundation for a detailed analysis of threats impacting the technical components within the defined scope.

- **Review the Scope of the Design/Architecture:** Conduct a detailed analysis of the key components of the dApp that should be included in the scope of the threat modeling assessment. The focus can be on the end-to-end (E2E) dApp architecture or specific critical components (e.g. smart contract interactions, API gateways, Kubernetes services), depending on the assessment objectives.

Outputs: From these activities, the outputs should include:

- **Identification of Application Components:** The DTS focuses on identifying and documenting the architectural components including the technical

stack and any data assets that need to be analyzed as exposure to threats as part of the threat modeling process. This stage aims to develop a comprehensive understanding of the application's architecture, its components, and the technical stack used to design and build those components. Each component is associated with data assets, both in storage and in transit, as well as the processes that operate within them. A thorough scoping process also involves analyzing these data assets and processes in detail during this phase of the analysis.
- **Updated Architecture Diagrams:** Detailed architecture diagrams are a critical artifact in designing, analyzing, and understanding blockchain-based systems. It provides a visual representation of the system's structure, highlighting its key components, their interactions, and the flow of data. Such diagrams ensure clarity for stakeholders and serve as a foundation for activities like architecture risk analysis/threat modeling, continuity of business and performance reviews, and system scalability assessments.

Templates and Tools: To assist in defining the technical scope, you can utilize the following artefacts and tools:

- **dApp Architecture Templates/Blueprints** [172]: These templates or blueprints are instrumental in highlighting the components within the scope (e.g. wallets, off-chain APIs, blockchain network and nodes, smart contracts) versus those that are out of scope. Using these templates ensures that all relevant aspects are considered and adequately secured. Draw.io can be used to create architecture diagrams. In addition to templates, reference architectures for developing specific static hosted web applications that interact with blockchain networks can be useful such as Web3 dApp Amazon Web Services (AWS) [173] reference architectures.
- **Architecture Diagramming Tools:** To complement the use of templates and blueprints, architecture diagramming tools are indispensable for visualizing the structure and workflow of a dApp. These tools facilitate a deeper understanding and communication of how various components of the dApp interact, which is crucial for both development and security auditing purposes. Tools like Draw.io [174] allow users to visually map out the entire dApp architecture, providing a clear representation of how different components like smart contracts, wallets, and off-chain APIs interact with each other.

By clearly defining the technical scope and leveraging the appropriate tools, you can ensure a thorough assessment of the dApp's architecture throughout the threat modeling process. This foundational effort is crucial for embedding security by design, providing a strong framework for in-depth security control analysis.

It is essential to recognize that the scope of the threat modeling exercise may evolve as new components are introduced in updated versions or releases. Consequently, the initial threat modeling baseline must be continuously updated to reflect architectural changes and account for emerging threats that these modifications may introduce.

3.2.2.3 Application Decomposition and Analysis

The primary goal of Application Decomposition and Analysis (ADA) is to systematically break down the dApp to facilitate a detailed examination of its components and their interactions, including server-to-server and application-to-application communication with data and users. This thorough analysis provides a deep understanding of the dApp's architecture and workflows, which is essential for defining security domains and identifying potential entry points for threats that will be further analyzed in subsequent assessment stages.

Focus Activities: A dApp is built on interconnected components enabling secure, decentralized functionality. During this stage, we will thoroughly dissect and analyze the critical components of the dApp, examine how these components interact with each other, trace the data flows that occur between them, and identify the trust boundaries that separate different domains of control. This analysis helps reveal potential vulnerabilities, misconfigurations, or implicit assumptions that could be exploited by adversaries:

- **Deconstruct Key Components:** Starting with the architectural design of the application for the components identified as in scope, we aim to break these down further to the level of detail required to capture interactions between users and components, as well as component-to-component interactions. For example (*note this is not a comprehensive list, it is provided as an example*):
  - **Frontend Application:** Assess how the frontend interacts with users and integrates with blockchain wallets, ensuring seamless communication and secure transactions.
  - **Middleware:** Expand on **JSON-RPC** and other communication protocols that facilitate interaction between the frontend and blockchain nodes.
  - **Blockchain Layer:** Identify the underlying blockchain platform (e.g. Ethereum, Solana, Polygon) and evaluate its consensus mechanism, security features, and transaction processing capabilities.
  - **Smart Contracts:** Analyze the functionality, storage mechanisms, and deployment environment of smart contracts, ensuring they align with security best practices and operational requirements.
  - **Off-Chain Components:** Examine APIs, centralized servers, oracles, and external data feeds that interact with the blockchain. Understanding these

off-chain elements and their integration with on-chain components is critical for a comprehensive architectural assessment.

- **Perform an In-Depth Component-Level Interaction Analysis:** After detailing the key components to the required level for analyzing data interactions between architectural components, users, processes, and data storage, the next step involves representing these interactions in data flow diagrams (DFDs) tailored to the specific use case. The next step is to capture them in DFDs for the specific use case:
  - **User Case:** Blockchain platform-specific dApp functionalities that are considered business critical and need to be analyzed in depth.
  - **Data Flows:** For the specific use case, analyze how components interact from the user to the frontend and to the backend blockchain platform and with the type of functionality that is supported and the type of data that is processed.

Outputs: Detailed DFDs that map data interactions within the dApp and identify components, endpoints, and smart contract functions.

- **Detailed DFDs:** These diagrams illustrate the flow of data through the dApp, aiding in the identification of interactions and data exchange points.
- **Identification of Trust Boundaries:** Identify and document all trust boundaries within the system, such as the interfaces between the user wallet and the blockchain. This facilitates the analysis of trust boundary threats, vulnerabilities, and potential attacks in the subsequent stages of the risk and threat modeling process.

Tools and Techniques: For ADA, tools are employed to create detailed diagrams that visualize data flows and architecture, facilitate code-level diagramming, and enhance the understanding of component interactions and security implications within the dApp.

- **DFD Tools:** Use tools like Draw.io and Lucidchart [175] to manually create detailed DFDs that visualize data movement and interactions within the dApp. Additionally, prebuilt DFD stencils are available in tools like OWASP Threat Dragon [176] and Microsoft Threat Modeling Tool [177], both of which support STRIDE-based threat mapping by component, enabling structured security analysis.
- **DFD Automatic Generation Tools:** In the case of cloud deployments, there are tools that can be used for automatic generation of architecture diagrams, allowing network flow analysis. CloudMapper [178] facilitates the extraction of AWS cloud infrastructure diagrams including AWS security group data flows, providing visibility into network interactions. Commercial threat

modeling tools like IriusRisk [179] enable scalable, repeatable threat modeling with automated DFD generation, reducing manual effort, minimizing errors, and ensuring consistent modeling across projects. Open-source toolkits like Threagile [180] can automate DFD generation by using YAML-based architecture definitions. Users can define technical assets, data flows, and trust boundaries as code, and Threagile automatically generates DFDs and risk reports from this input. Products like Avocado Systems Reveal [181] have the ability to automatically generate DFDs at runtime, mapping process components and their interactions without manual input by feeding into threat modeling tools by offering real-time visibility into application behavior, helping validate models, uncover unmodeled threats, and prioritize mitigations based on actual risks.

By rigorously applying the ADA process, the architecture and interactions within a dApp are fully mapped and analyzed, allowing for a thorough security review. This ensures that all potential security risks are identified and addressed early in the development process, setting the stage for a secure, efficient, and reliable dApp deployment. This structured approach is crucial for safeguarding against cybersecurity threats and ensuring the integrity of the dApp in a blockchain environment.

3.2.2.4 Threat Analysis

The primary objective of threat analysis (TA) is to identify and assess relevant threats to the dApp, based on the detailed architectural and DFDs developed in previous stages. This process includes analyzing key components, their interactions, entry and exit points, and the trust boundaries they traverse. By leveraging structured threat libraries and threat intelligence, this stage systematically evaluates potential risks, ensuring alignment with the application's specific scope and context.

Focus Activities: The focus activities for TA of a dApp should comprehensively address its unique architecture, dependencies, and operational use cases, such as those found in DeFi applications. Key activities include:

- **Architectural Review:** Analyze the dApp's components, including the UI, middleware (e.g. JSON-RPC), blockchain node, distributed ledger, smart contracts, and oracles. Identify trust boundaries and entry points, such as APIs, external data feeds, and interactions with wallets.
- **Data Flow Analysis:** Map data flows between components, capturing sensitive data such as private keys, transaction payloads, and user identity credentials. Highlight points of interaction that could be vulnerable to interception, tampering, or unauthorized access.

- **Threat Enumeration:** Utilize frameworks such as STRIDE to systematically enumerate threats, including spoofing, tampering, and information disclosure. Additionally, tailor the TA to the dApp's technology stack by identifying risks specific to smart contracts, API configurations, and dependency management, ensuring a contextualized approach to threat assessment.
- **Attack Surface Analysis:** Identify all possible AVs, including user inputs, communication channels, and third-party integrations (e.g. wallet software or external oracles). Focus on both direct interactions (e.g. user-to-dApp) and indirect interactions (e.g. oracle to blockchain). *Please note that the analysis of expected attackers in relation to the attack surface primarily takes place in Stage IV: Threat Analysis, building upon the technical scope and attack surface defined in Stage II.*
- **Threat Intelligence Integration:** Utilize real-world threat intelligence to identify emerging attack patterns and vulnerabilities observed in past security incidents involving blockchain-based systems. Apply these insights to refine threat scenarios and enhance the accuracy of attack simulations, ensuring a proactive approach to security assessment.
- **Mitigation Strategy Development:** Identify and implement security controls, also known as countermeasures, to mitigate threats. These may include encryption for data protection, input validations, access controls such as authentication and authorization, and transaction validation mechanisms to ensure integrity and prevent unauthorized modifications.

Outputs: Involves documenting and associating identified threats with specific components and interactions within the dApp.

- **DFDs:** Draw DFDs to visualize the flow of data within a system. A DFD that identifies the processes, data stores, external entities, and communication channels involved in the system provides a foundation for analyzing potential security risks and threats to the data in transit and in storage.
- **Threats and Countermeasures Lists**: Documenting and mapping identified threats to specific components and assets (e.g. data and processes) within the dApp at the architectural and DFD levels enhances the understanding of potential AVs and the identification of countermeasures to mitigate the risks of these threats. These vectors target components by exploiting vulnerabilities, potentially causing impacts to the associated assets and components.
- **Threat Models Diagrams:** Threat model diagrams can be based on a previously analyzed DFD to incorporate security-related elements. These might highlight potential threats, vulnerabilities, and trust boundaries within the system's data flow and architecture. By overlaying threat information onto

the DFD, the diagram becomes a powerful tool for analyzing risks and identifying security controls/countermeasures.

Tools and Techniques: Diagramming tools like Draw.io are highly effective for drafting DFDs and can also be used to manually create threat models. Dedicated threat modeling tools, while not strictly necessary, offer significant advantages during this stage by leveraging built-in threat libraries tailored to the components of a dApp. Threat modeling tools, such as OWASP Threat Dragon, Microsoft Threat Modeling Tool, ThreatModeler [182], and IriusRisk, streamline the process by contextualizing threats based on the application architecture in scope, often using built-in threat libraries (e.g. STRIDE, CAPEC, or custom rulesets). While these tools are generally built around standard or generic architectural patterns, they can be customized or extended to address dApp architectures, even if dedicated threat libraries for dApps are not provided by default.

Streamline the process by contextualizing threats specific to the generic application architecture in scope even if there are build-in threat libraries that are specific to a dApp. However, effective TA can still be conducted manually by applying frameworks like STRIDE to enumerate threats at the component/architecture level. This approach systematically identifies potential data assets that could be impacted, the AVs that threat actors might use to exploit vulnerabilities, and the potential impact on those assets. In addition to threat modeling tools, threat intelligence platforms can enrich the analysis by providing insights into blockchain-specific vulnerabilities and real-world threat scenarios. This integration enhances the understanding of threats and supports the development of countermeasures that address both architectural weaknesses and emerging risks in the dApp ecosystem.

By comprehensively analyzing threats using these focus activities and tools, the TA stage ensures a robust understanding of the security threat and vulnerability landscape surrounding the dApp. This detailed threat knowledge enables developers and security professionals to design and implement effective security measures to mitigate the risk of relevant threats that the dApp might be exposed to, safeguarding the dApp against potential exploits and vulnerabilities by the threats.

3.2.2.5 Vulnerability Analysis

In the context of risk-based threat modeling, vulnerability analysis (VA) focuses on identifying and mapping potential threats to specific vulnerabilities within the dApp. This stage provides a detailed understanding of weaknesses that adversaries could exploit, enabling developers and security teams to implement precise security measures. The objective of this stage is not to conduct vulnerability assessments directly, such as scanning source code or performing penetration testing, but rather to leverage the findings from such assessments. For example, vulnerabilities identified through static analysis scans or audits of smart contracts – such as reentrancy vulnerabilities, integer overflows/underflows, or improper access controls – are evaluated and correlated with the threats previously analyzed during the TA phase. By mapping these vulnerabilities to specific components, such as smart contracts, the process ensures that the identified risks are contextualized and addressed effectively within the dApp's architecture.

Focus Activities: VA in the context of risk-based threat modeling for a dApp would include the following activities:

- **Mapping Threats to Identified Vulnerabilities:** Correlate threats identified in the dApp's components (e.g. dApp client, dApp frontend, middleware, backend blockchain, off-chain components, digital wallets, distributed storage, etc.) by vulnerability assessments (e.g. source code scanning, pen tests, etc.) to the threats listed documented as part of the TA. Example is the use of threat lists and frameworks to link previously analyzed threats (e.g. spoofing, tampering) against components with identified vulnerabilities or design flaws that facilitate these threats.
- **Vulnerability Risk Prioritization:** List vulnerabilities and security design flaws identified for each of the components that are in scope for the threat modeling assessment and determine the risk by considering the severity, exploitability, and potential impact on critical assets for the identified threats. For determining the exposure to threat agents, analyze the DFD threat model and for severity/criticality of the vulnerabilities use metrics like Common Vulnerability Scoring System (CVSS) [183] risk scores to quantify risk and prioritize remediation efforts. However, it's important to note that CVSS base scores must be reevaluated in the context of the specific dApp architecture and the organization's threat environment. Without this contextualization, the scores may not accurately reflect the actual risk posed to the business. Vulnerability management prioritization takes place in the last stage, Stage VII of the PASTA threat modeling methodology, where risk assessment methodologies are applied to incorporate business-specific factors such as the value of the affected asset, potential compliance implications, and the downstream impact on trust, operations, and organizational objectives.

Outputs: A comprehensive identification and analysis of vulnerabilities, assessing their potential exposure to exploitation by previously identified threats. Vulnerabilities are detailed and ranked based on severity and threat exposure across the dApp, categorized by affected components. This approach enables further analysis of their exploitability through AVs that could be used to realize such threats.

- **Mapped Threats to Weaknesses and Vulnerabilities:** Each vulnerability is identified through vulnerability scanning tools such as Static Application Security Testing (SAST), Dynamic Application Security Testing (DAST), Infrastructure as Code (IaC) scanning, and container security scanning, which are typically integrated into SecDevOps workflows [184]. The scans from these tools are typically triggered by code events such as source code merges or builds, ensuring continuous security assessment throughout the development life cycle. When vulnerabilities are detected, they are categorized using Common Vulnerabilities and Exposures (CVEs) [185] to classify the issue type, while the CVSS assigns a risk score based on severity. Beyond vulnerabilities, security design flaws and weaknesses are identified through architecture and design reviews. These flaws, classified under Common Weakness Enumerations (CWEs) [186], require appropriate countermeasures to mitigate potential threats. Examples of such countermeasures include MFA, distributed denial-of-service (DDoS) defense mechanisms, Web Application Firewalls (WAFs), and API gateways. These security measures are mapped back to the threats previously analyzed, ensuring a structured approach to vulnerability mitigation and reinforcing security controls identified during the TA stage. The presence of such vulnerability or security control gap or weaknesses for the specific threat is highlighted as a risk that factors the severity of the vulnerability as severity calculated as CVSS helps to quantify their impact and exploitability. This risk is assessed by considering the severity of the vulnerability, the likelihood of its exploitation (exposure to the threat), and the potential impact on the targeted asset. The evaluation of exposure and impact of these weaknesses and vulnerabilities is informed by the threat model and the DFD diagrams, which provide insights into the system's architecture and data flows.
- **Risk Prioritized Threat–Vulnerability List:** A detailed compilation of vulnerabilities and weaknesses ranked by the threat risk as high, medium, and low to be prioritized for AM and simulation based upon the threat risk. This prioritized list of threat–vulnerability risks serves as the foundation for simulating attack scenarios to test effectiveness of security controls. These attack-driven tests/simulations leverage the knowledge of the possible AVs for identified vulnerabilities to derive a list of simulation attack tests in Stage VI of the threat modeling exercise that is AM whose goal is to determine the risk exposure and impact to the assets for the threats, vulnerabilities, and attacks being analyzed.

Tools: The tools for VA support identifying issues during development, detecting runtime vulnerabilities, monitoring production environments, assessing third-party dependencies, and enriching threat intelligence for comprehensive risk understanding. A list of these tools might include:

- **Static Analysis Tools:** In addition to Software Composition Analysis (SCA), SAST, IaC scanning, and container security scanning tools [187], specialized tools are required when the target of the scans is smart contracts. Tools such as MythX [188], Slither [189], and Remix IDE [190] play a crucial role in analyzing and scanning smart contract code during development. These tools help identify vulnerabilities early in the development life cycle, mitigating the risk of deploying flawed or insecure smart contracts.
- **Dynamic Analysis Tools:** Tools like OWASP ZAP [191] are used for scanning frontend applications and APIs to detect runtime vulnerabilities such as injection attacks or insecure authentication flows.
- **Production Monitoring Tools: Defender OpenZeppelin** [192] is ideal for managing smart contract vulnerabilities postdeployment, providing continuous monitoring and protection in production environments.
- **Dependency Scanning, SCA Tools:** Use libraries like **npm audit** [193], **Snyk** [194], or similar tools to evaluate third-party libraries and components for known vulnerabilities.

VA bridges the gap between theoretical threats and practical risks by connecting identified threats to real vulnerabilities within the dApp. This stage emphasizes a detailed evaluation of the dApp's architecture, workflows, and dependencies, uncovering exploitable weaknesses across all layers. By leveraging a combination of static, dynamic, and dependency analysis tools, the process ensures a thorough understanding of the dApp's security posture. The resulting prioritized vulnerability list and mitigation recommendations enable the development of a robust and secure application, safeguarding both the dApp and its users from potential attacks.

3.2.2.6 Attack Modeling (AM)

The purpose of AM is to simulate threat scenarios through targeted security tests, building on the insights gathered in earlier stages of the risk-based threat modeling process. This includes the analysis of threats, identification of vulnerabilities and weaknesses, evaluation of AVs, mapping of trust boundaries, and understanding of data flows. The primary objective is to assess the effectiveness of the dApp's security measures in addressing identified vulnerabilities and design weaknesses, highlighting areas of risk exposure for the evaluated threat scenarios and recommending appropriate mitigation strategies.

AM delves into how adversaries could exploit vulnerabilities in critical components by following the most likely paths or sequences of events to compromise dApp assets. It leverages methodologies such as attack trees to model the attacker's probable paths to achieve their objectives and the attack kill chain to analyze the sequence of events leading to the exploit.

By conducting attack analysis and simulating attack scenarios, this stage provides actionable insights into mitigating risks associated with vulnerabilities and exposure to threats. The recommended risk mitigations are contextualized as countermeasures tailored to the specific threat scenarios. Ultimately, AM strengthens the dApp's defenses by guiding the implementation of effective security controls and countermeasures.

Focus Activities: AM entails analyzing threats, crafting attack scenarios, and simulating adversarial tactics to identify potential exploit paths and evaluate the dApp's resilience to these scenarios. This process highlights risks and proposes targeted mitigations such as remediating vulnerabilities in the source code or modifying the dApp's design to address security control gaps to reduce both the likelihood and impact of potential attacks.

Effective AM also involves assessing the attacker's skills, resources, and motivations, which helps in understanding the plausibility and sophistication of each threat scenario. Factoring in these elements improves risk prioritization and guides the development of proportionate, context-aware defenses. In practice, this involves the following activities:

- **Analyzing Threat Realization:** Assess how security control weaknesses and vulnerabilities identified during the VA phase could be exploited by specific types of attacks. Focus on realistic scenarios aligned with the dApp's architecture, such as bypassing authentication mechanisms or manipulating transaction validation.
- **Developing Attack Scenarios:** Developing attack scenarios involves contextualizing potential exploits within the dApp architecture, focusing on components that can be targeted at the trust boundaries identified during the TA phase. These scenarios should reflect the threats previously analyzed and detail how adversaries might attempt to exploit vulnerabilities at these critical interaction points. Examples of these attack-targeted dApp components include:
  - **Digital Wallet Attacks:** Simulate scenarios like phishing for private keys, malware-based key extraction, transaction tampering, and session hijacking to explore wallet vulnerabilities.
  - **Smart Contract Attacks:** Develop attack trees for issues like reentrancy exploitation, authentication bypass, and vulnerabilities in transaction validation mechanisms, such as consensus manipulation.

- **Transaction Validation Layer Attacks:** Model scenarios targeting transaction validation processes, including double-spending exploits, oracle data manipulation, or temporary blockchain forking.
- **Simulating Adversarial Tactics:** Leverage blockchain-specific intelligence tools to emulate real-world attacker behaviors, allowing to identify potential weaknesses in the dApp's ecosystem. Simulate advanced adversarial techniques such as front-running exploits, smart contract exploits, network attacks, consensus manipulation, identity and access exploits, and oracle and flash loan attacks using frameworks like MITRE ATT&CK for blockchain [195].

Outputs: The outputs of AM provide actionable insights into potential attack paths, key vulnerabilities, and risk prioritization to inform mitigation strategies and enhance the dApp's security posture.

- **Visualized Attack Scenarios:** Graphical representations, such as attack trees, map out exploit paths for vulnerabilities, showing the logical steps an attacker might take. These scenarios provide a clear understanding of potential attack surfaces and their impact on the dApp's security.
- **Attack Simulation Tests:** Document a list of attack-driven tests for the attack scenarios that are most likely to occur using specific AVs identified during the TA and weaknesses and vulnerabilities identified during the VA for conducting attack-driven tests. Resources like MITRE ATT&CK for blockchain allow to structure adversarial simulations and analyze tactics, techniques, and procedures (TTPs) specific to blockchain ecosystems. Execution of attack simulation tests should be prioritized based on the likelihood and potential impact of specific threat scenarios, as part of an overall risk assessment for the dApp. These simulations replicate real-world exploitation techniques to validate the dApp's exposure to threats and assess its resilience under targeted attack conditions. While baseline penetration testing is essential for all Internet-facing dApps, attack simulation that involves more sophisticated, scenario-driven testing should be performed for dApps classified as high risk or business critical.
- **Prioritized Attack Risks and Mitigation Recommendations:** A comprehensive list of attack scenarios is developed to represent how threats can be realized through the exploitation of identified vulnerabilities, leading to an impact on targeted assets. These scenarios are validated through attack-driven testing and prioritized based on their probability of occurrence and potential impact, focusing on the highest-risk scenarios. To ensure effectiveness, it can be highly efficient to first execute the attack to confirm the vulnerability, then reexecute the same scenario after implementing the mitigation to verify the correct technical implementation of the control. This approach

provides strong assurance that the risk has been properly addressed and that the dApp is resilient against reexploitation. Recommendations include implementing tailored countermeasures to address control gaps and weaknesses, such as enhancing authentication mechanisms, securing transaction validation processes, and remediating critical vulnerabilities. This prioritization ensures effective management of critical risks and supports the mitigation of high-priority threats.

Tools: Tools for AM help model attack paths, simulate adversarial tactics, and enrich scenarios with real-world threat intelligence for a comprehensive security evaluation.

- **Attack Tree Modeling Tools** [196]: Use modeling tools and templates to create detailed attack trees, visualizing the flow of potential exploits and highlighting weak points in the dApp's architecture.
- **Breach and Attack Simulation (BAS) Tools** [197]: BAS tools automate the process of simulating cyberattacks, enabling continuous assessment of security defenses. These tools replicate various AVs to uncover weaknesses in security controls. Many of these BAS tools offer E2E visualization across the MITRE ATT&CK framework, enabling organizations to map attack techniques, assess security control effectiveness, and enhance their defense strategies against real-world cyber threats.

In summary, AM bridges the gap between theoretical vulnerability identification and practical security assessment by simulating realistic attack scenarios. It enables teams to identify exploit paths, determine AVs, and conduct targeted testing to validate the dApp's security measures and recommend risk mitigations. By focusing on vulnerabilities in key areas like digital wallets, smart contracts, and transaction validation, AM provides the insights necessary to prioritize risks and implement effective mitigations, ensuring a resilient and secure blockchain application.

3.2.2.7 Risk Assessment and Mitigation

The risk assessment and mitigation (RAM) stage serves as the final phase of the PASTA threat modeling process, where the risks associated with previously simulated attack scenarios are systematically documented, analyzed, and prioritized based on their likelihood and potential impact on dApp assets. The goal of this stage is to provide a holistic understanding of the risks to the dApp by integrating analyzed threats, identified design weaknesses, component vulnerabilities, and results from attack simulation tests. These results offer evidence of how threats can be realized, along with an assessment of their likelihood and impact on the dApp's assets. This comprehensive risk assessment supports informed decision-making by dApp risk owners, enabling them to implement effective risk management strategies. The ultimate objective is to minimize risk exposure to the business while ensuring that the dApp's security measures align with organizational risk management practices. Additionally, this process helps fulfill contractual obligations with customers by providing evidence of security by design principles, addressing audit requirements, and ensuring compliance with applicable regulations and standards.

Focus Activities: The focus activities of the RAM stage are designed to systematically evaluate and address risks to the dApp by leveraging insights from earlier stages of the PASTA threat modeling process. These activities include documenting and analyzing risks based on threats, vulnerabilities, and attack simulation results; prioritizing risks by their likelihood and impact; and developing targeted mitigation strategies for example:

- **Risk Documentation and Analysis:** Compile and document risks derived from previously simulated attack scenarios. Analyze risks by integrating identified threats, design weaknesses, component vulnerabilities, and attack simulation results. Assess the likelihood of threat realization and the potential impact on dApp assets.
- **Risk Assessment and Prioritization:** Use attack simulation test results as evidence of risk realization to validate the need for specific mitigations. Prioritize risks based on a combination of likelihood and impact assessments. Use structured risk frameworks such as FAIR or OCTAVE to quantify and rank risks systematically, as these methodologies are better suited than CVSS for capturing business context and aligning risk decisions with organizational priorities. Focus on high-risk vulnerabilities and critical components that present the greatest threat to dApp security and business continuity.
- **Alignment with Organizational Risk Management:** Ensure that proposed mitigation strategies align with the organization's overarching RMF. Integrate findings into broader organizational practices for continuous monitoring and risk management.
- **Stakeholder Communication:** Provide comprehensive RAM recommendations to dApp risk owners and stakeholders. Clearly articulate the implications of identified risks and the effectiveness of proposed mitigations.
- **Compliance and Audit Readiness:** Collaborate with development and security teams to implement the proposed countermeasures effectively. Monitor mitigation efforts to ensure risks are addressed appropriately and residual risks are within acceptable limits.
- **Final Risk Mitigation Implementation:** Collaborate with development and security teams to implement the proposed countermeasures effectively. Monitor mitigation efforts to ensure risks are addressed appropriately and residual risks are within acceptable limits.

Outputs: These outputs collectively ensure that risks to the dApp are effectively identified, addressed, and monitored. They provide a structured approach to documenting and tracking risk management efforts, aligning with business goals, and demonstrating accountability and compliance. For example:

- **Comprehensive Risk Register/Repository:** A detailed record of identified risks, ranked by severity and likelihood, linked to associated threats, vulnerabilities, and impacted components. Includes detailed descriptions of risks, supporting evidence from attack simulations, and the estimated likelihood and impact for each. Serves as a central reference to guide mitigation strategies and monitor risk management progress over time.
- **Risk Mitigation Plan:** An actionable roadmap outlining technical and procedural measures for addressing identified risks. Includes specific timelines, milestones, resource allocations, and responsibilities for implementing controls. Prioritizes high-risk vulnerabilities and aligns security measures with organizational objectives and compliance requirements.
- **Threat–Vulnerability Mapping (from VA Stage):** A documented map linking threats to vulnerabilities and affected components, along with associated risk levels. Provides actionable insights for addressing specific risks and informs security control prioritization.
- **Attack Simulation and Testing Results (from AM Stage):** A record of attack scenarios tested during the AM phase, including identified weaknesses and the effectiveness of existing countermeasures. Supports evidence-based risk assessments and demonstrates the validity of mitigation strategies.
- **Compliance and Audit Readiness Documentation:** Evidence of TA and modeling conducted in compliance/alignment with security by design standards/best practices, control implementation, and risk management aligned with regulatory and contractual requirements.
- **Stakeholder Risk Communication Report:** A summary report for stakeholders, including risk owners, developers, and business leaders, outlining key risks, mitigation strategies, and their impact on organizational objectives. Ensures transparency and alignment in risk management initiatives.

Tools: Tools in the RAM stage enable systematic risk evaluation, visualization, and mitigation planning while ensuring compliance with regulatory requirements.

- **GRC Tools** [198]: They play a crucial role in managing and visualizing risks across components within an organization. Enterprise-grade platforms like Archer (RSA) provide robust capabilities for comprehensive risk management, enabling teams to track, analyze, and address risks effectively. Similarly, RiskWatch offers an extensive platform designed for monitoring and analyzing risks, while OneTrust simplifies compliance management and documentation processes. For simpler applications, tools like Excel or Google Sheets offer customizable templates that are highly effective for creating and maintaining risk registers.

- **Issue Tracking and Management Platforms** [199]: They, such as Jira and Confluence, are invaluable for organizing mitigation strategies, setting timelines, and monitoring milestones. For more detailed project planning and resource allocation, Microsoft Project provides specialized tools to support these efforts.
- **Risk Reporting and Documentation Tools:** They are equally important for effective communication. Microsoft Word and Google Docs allow teams to draft detailed risk and mitigation reports, while PowerPoint and Google Slides are ideal for presenting findings and recommendations to stakeholders. These tools ensure that threat modeling insights are communicated clearly, aligning security initiatives with business objectives and fostering informed decision-making. By streamlining documentation and presentation, these platforms enhance collaboration and facilitate the implementation of risk mitigation strategies.

In conclusion, the execution of the risk-based threat modeling exercise following the PASTA framework provides actionable outcomes that significantly enhance the security posture of the dApp. By systematically analyzing threats, vulnerabilities, and attack scenarios, this process generates a comprehensive understanding of the risks to critical components and assets. The actionable insights derived from the exercise empowered dApp engineering managers and security architects to design and implement targeted security measures, ensuring the dApp's architecture is resilient to identified threats.

For security architects and heads of security architecture, the exercise provides a structured methodology to align security controls with the unique requirements of the dApp's architecture and use cases. It enables informed decision-making regarding resource allocation, security investments, and prioritization of mitigation efforts.

dApp product business owners benefit from the exercise by gaining confidence in the platform's ability to safeguard user trust and meet compliance requirements, enhancing the dApp's market value and competitive edge.

For information security officers (CISOs) and risk managers, the process delivers a risk register and mitigation roadmap that facilitates ongoing risk management and compliance tracking. The outputs of the exercise demonstrate adherence to security by design principles, ensuring contractual and regulatory obligations are met. By linking technical security measures to business objectives, this approach enables stakeholders to balance operational efficiency with robust risk management, ensuring the dApp's long-term success and security.

For software developers, the threat modeling exercise translates complex security concerns into clear, actionable tasks within the development workflow. It provides concrete guidance on secure coding practices, defensive design patterns, and the implementation of specific controls such as input validation, access control, and secure key handling based on identified threats and attack paths. Developers not only gain a deeper understanding of the security implications of their code but also become active participants in building a security-aware development culture, contributing directly to the overall resilience of the dApp.

Ultimately, the PASTA-based threat modeling exercise fosters collaboration across technical, security, and business teams, driving a unified approach to securing the dApp ecosystem.

Table 3.1 consolidates the key stages, focus activities, outcomes, and tools in a summarized format, highlighting the comprehensive nature of the PASTA risk-based threat modeling process.

3.2.3 Threat Modeling Example: DeFi Lending and Borrowing dApp

In Chapter 1, we examined the "Applications of Blockchain within DLT Ecosystems," highlighting DeFi as a pivotal use case for dApps. A DeFi dApp is a blockchain-based application that delivers financial services without depending on traditional intermediaries such as banks or brokers. DeFi stands out as an ideal use case for blockchain technology, leveraging its core features – decentralization, transparency, and immutability to transform financial services.

We explored how DeFi dApps utilize smart contracts, self-executing programs on the blockchain, to automate financial operations in a trustless and transparent way. These applications aim to reimagine traditional financial systems by enabling users to engage in activities such as lending, borrowing, trading, and earning interest on assets directly through their cryptocurrency wallets.

DeFi dApps can be developed by integrating with permissionless blockchain networks like Ethereum [200], BSC [201], Solana [202], and others, allowing anyone with Internet access and a compatible wallet to participate without the need for a central authority or gatekeeper. Prominent examples include Uniswap [203] (DEXs), Aave [204] (lending and borrowing), and MakerDAO [205] (stablecoin issuance and governance). These platforms have revolutionized the financial landscape by offering open, borderless, and innovative alternatives to traditional finance systems.

In the context of DeFi, we examined token transactions, which facilitate activities like exchanges, lending, borrowing, and other financial operations conducted on decentralized platforms using smart contracts. These transactions employ various token types, such as stablecoins, governance tokens, and utility tokens,

Table 3.1 PASTA-based threat modeling stages.

PASTA stage	Description	Focus activities	Outputs/outcomes	Tools
Definition of business objectives (DBOs)	Identifies critical business assets to align security with the dApp's operational priorities	• Define primary business goals and objectives. • Identify critical assets (e.g. private keys, user funds, smart contracts). • Evaluate potential impacts of security breaches (e.g. financial loss, reputational damage).	• Clear understanding of dApp business impact and security priorities • Compliance alignment with regulations and contractual obligations • BIA	• BIA tools (e.g. ISO 22301) • Risk frameworks (e.g. NIST, FAIR, OCTAVE) • BIA toolkits for structured analysis • Blockchain-specific GRC tools
DTS	Defines technical boundaries by analyzing hardware, software, and network components in scope	• Review dApp architecture, identifying frontend, wallets, blockchain layer, off-chain components, and smart contracts. • Map technical stack components and their interactions. • Analyze data assets and flows in the context of their processes.	• Comprehensive architectural component list • Detailed architecture diagram	• Architecture templates and blueprints for scoping • Diagramming tools (e.g. Draw.io, Lucidchart) for architecture visualization
ADA	Deconstructs the dApp to analyze interactions between components, users, processes, and data flows	• Break down key components (e.g. UI, wallets, middleware, blockchain platform). • Conduct data flow and trust boundary analysis. • Generate DFDs capturing component interactions and critical processes.	• Detailed DFDs highlighting data interactions and trust boundaries • Identification of processes and data assets at risk	• DFD tools (e.g. Draw.io, Lucidchart) • DFD automatic extraction tools (e.g. CloudMapper, IriusRisk. Threagile, Avocado Systems)

TA	Identifies potential threats to the dApp using threat libraries and contextual analysis	• Analyze architecture and data flows for trust boundaries. • Enumerate threats using frameworks like STRIDE. • Identify attack surfaces and interaction vulnerabilities (e.g. smart contracts, APIs, wallets). • Integrate real-world threat intelligence to refine threat scenarios.	• Comprehensive threat–vulnerability mapping • Threat model diagrams for identified risks • Lists of threats and countermeasures for targeted mitigation	• Threat modeling tools (e.g. OWASP Threat Dragon, Microsoft Threat Modeling Tool, IriusRisk, ThreatModeler, SD Elements) • Threat intelligence platforms for contextual insights
VA	Maps vulnerabilities identified during assessments to threats, prioritizing based on risk	• Correlate threats to identified vulnerabilities (e.g. SAST, DAST findings). • Rank vulnerabilities by severity, exposure, and impact using metrics like CVSS. • Map vulnerabilities to threat model and DFD insights.	• Risk-prioritized threat–vulnerability list • Detailed analysis of vulnerabilities mapped to threats • Recommendations for prioritized remediation of risks	• Static and dynamic analysis tools (e.g. Checkmarx, Snyk, OpenText Fortify, Synopsis, Veracode, MythX, Slither, Remix IDE, OWASP ZAP) • Dependency scanning tools (e.g. Snyk) • Production monitoring tools (e.g. Defender OpenZeppelin)

(Continued)

Table 3.1 (Continued)

PASTA stage	Description	Focus activities	Outputs/outcomes	Tools
AM	Simulates adversarial tactics to evaluate security control effectiveness and identify risk exposure	• Develop attack scenarios for analyzed threats and vulnerabilities. • Model likely attack paths using attack trees and kill chains. • Conduct attack-driven testing of high-risk vulnerabilities.	• Visualized attack scenarios (e.g. attack trees) • Attack-driven test results validating risk exposure • Recommendations for prioritized mitigations and security control enhancements	• Attack tree tools for modeling • Simulation frameworks (e.g. MITRE ATT&CK for blockchain) • BAS tools
RAM	Final phase documenting, analyzing, and prioritizing risks to the dApp, integrating insights from all previous stages	• Compile and prioritize risks based on likelihood and impact. • Develop mitigation strategies aligned with business goals and compliance. • Create actionable risk mitigation plans with timelines and milestones.	• Comprehensive risk register • Risk mitigation roadmap • Stakeholder communication reports aligning security measures with business objectives	• GRC tools (e.g. Archer, RiskWatch) • Issue tracking tools (e.g. Jira) • Documentation tools (e.g. Microsoft Word, Google Docs)

typically executed on DEXs or other platforms via automated market makers (AMMs) [206] or order books. Blockchain technology ensures transparency and immutability for these transactions. A notable advantage of token usage in DeFi is interoperability, which allows users to seamlessly move assets between protocols, for instance, borrowing one token from a DeFi platform and using it to provide liquidity in another. In Chapter 1 we discussed security risks associated with digital tokens in DeFi, particularly vulnerabilities in token transactions. These risks include flash loan exploits, oracle manipulation, and fraudulent projects, which are significant concerns in distributed applications like DeFi.

In Chapter 2, we explored the security challenges unique to DeFi, focusing on vulnerabilities in smart contracts and flaws in business logic. Notable threats include flash loan exploits, oracle and market manipulation, and the risks of impermanent loss faced by liquidity providers. Operational challenges such as volatile gas fees, collateral liquidations, phishing, and user errors further compound the risks. We also examined front-running attacks, where transaction ordering is exploited for financial gain, and rug pulls, where malicious developers lure liquidity before absconding with funds.

Many of these vulnerabilities arise from insufficient audits or unforeseen edge cases in smart contract implementation. Flash loans, a common tool in sophisticated exploits, allow attackers to manipulate large sums without collateral. Oracle manipulation also poses a significant threat, with inaccurate data feeds potentially disrupting protocols reliant on precise pricing. Rug pulls represent another critical risk, where fraudulent developers exploit user trust, attract liquidity, and withdraw all funds, leaving participants with losses.

The DeFi ecosystem has seen several high-profile breaches that underscore these risks. The Poly Network experienced a $600 million breach [207] due to a cross-chain bridge vulnerability, marking one of the largest hacks in the space. Wormhole suffered a $325 million loss [208] from a smart contract exploit that enabled unauthorized minting of wrapped tokens. The collapse of Terra's UST stablecoin destabilized the market, causing billions in losses [209]. Similarly, the Ronin Network faced a $620 million exploit [210] targeting validator vulnerabilities in its cross-chain bridge.

These incidents underscore the urgent need for robust security frameworks in the DeFi ecosystem to protect platforms and users from escalating risks. The adoption of comprehensive security measures must become a foundational aspect of DeFi application development. We emphasized the critical importance of conducting thorough audits and testing of smart contracts to identify and address vulnerabilities before deployment. Smart contracts, as the backbone of DeFi, require rigorous validation to ensure they operate securely and reliably under both expected and edge-case scenarios.

In addition to safeguarding code integrity, the protection of data privacy is essential in DeFi operations. Confidential information, including user account balances, lending/borrowing positions, and liquidity pool shares, must be encrypted to shield sensitive financial activities from unauthorized access. Ensuring privacy fosters user confidence and aligns DeFi practices with growing regulatory demands for data protection.

We further explored the significance of transactional data protection within DeFi ecosystems. Transactions such as lending activities, borrowing requests, interest calculations, and liquidity provisioning are recorded on the blockchain, which inherently provides transparency. However, additional safeguards, including hashing algorithms and digital signatures, are necessary to guarantee the integrity and authenticity of this data. These mechanisms prevent tampering and fraud, ensuring that transactional records remain immutable and verifiable.

To build trust and ensure the sustainability of DeFi platforms, security must extend beyond technical safeguards. Developers and platform architects are encouraged to adopt a risk-centric approach, including threat modeling, to identify potential AVs and implement targeted mitigations. Moreover, regular monitoring and incident response planning should be integral to the operational workflow, enabling platforms to swiftly address vulnerabilities and recover from security breaches. By prioritizing these practices, DeFi can continue to innovate while minimizing risks to users and the broader ecosystem.

In Chapter 2, we also provided examples of security requirements and documented security by design principles for the design, development, and security testing phases of dApp creation. Furthermore, we conducted a comprehensive review of security by design requirements at the component level, identifying common design flaws and vulnerabilities. We outlined security controls and assessment methodologies to address these issues, ensuring that dApps are secure before they are deployed into production.

One key process discussed was threat modeling, where we emphasized the need for a comprehensive threat modeling exercise. This process involves identifying potential AVs and threats specific to blockchain applications by analyzing risks, understanding attacker motivations, and evaluating the impact of threats on application security. By addressing these risks proactively, dApp developers can establish a robust security posture and safeguard their applications against emerging threats.

In this chapter, we introduced risk-centric threat modeling, outlining the various stages of the methodology, its goals, key focus activities, the outputs/artifacts generated at each stage, and the tools that can be utilized to produce them. This serves as general guidance in the context of a dApp. Additionally, we aim to provide a deeper understanding of this methodology by presenting a working example of risk-centric threat modeling for a DeFi dApp. This example leverages a commonly used Web 3.0 technology stack and associated technologies to illustrate the practical application of the methodology.

In this chapter, we will demonstrate how these challenges can be addressed through a risk and attack-centric threat modeling approach, ensuring a security by design methodology. This approach proactively identifies threats, vulnerabilities, and AVs specific to DeFi lending and borrowing platforms. By systematically analyzing risks and simulating real-world attack scenarios, we will outline actionable countermeasures to mitigate these threats before deployment, strengthening application security and enhancing user protection.

3.2.3.1 Stage 0 – Setting the Stages for PASTA Threat Modeling

For this PASTA threat modeling exercise, we selected a DeFi application focused on lending and borrowing services as the application in scope. This class of dApp has become one of the most widely adopted and business-critical applications of the DeFi ecosystem, enabling users to lend cryptocurrency assets to earn interest or borrow against their holdings all without relying on traditional financial intermediaries like banks.

For the use case example in this threat model, we selected a DeFi application focused on lending and borrowing platforms, one of the most popular and widely adopted use cases in the DeFi ecosystem. These platforms allow users to lend their cryptocurrency assets to earn interest or borrow against their holdings, eliminating the need for traditional financial intermediaries such as banks.

DeFi lending and borrowing platforms operate by facilitating financial interactions directly between users through blockchain-based protocols. On the lending side, users deposit their cryptocurrency into the protocol, where the funds are pooled and made available for borrowers. In return, lenders earn interest on their deposits, which is paid by the borrowers. Borrowing involves locking cryptocurrency as collateral within the protocol to secure a loan. Typically, these loans are overcollateralized, requiring users to deposit assets worth more than the loan amount to minimize risk. Borrowers then pay interest on the borrowed funds, which goes to the lenders. The entire process is managed by smart contracts, which enforce loan terms, calculate interest rates, and manage collateral automatically and transparently.

Collateralization is a cornerstone of DeFi lending and borrowing. Borrowers must lock cryptocurrency as collateral, with collateral ratios, such as 150%, ensuring a buffer against market volatility. If the collateral value drops below a specified threshold, the protocol automatically liquidates the assets to repay the loan and protect lenders. Interest rates are determined algorithmically, responding dynamically to the supply and demand of the lending pool. Higher borrowing demand raises interest rates, attracting more deposits from lenders.

Flash loans, another innovative feature, are unsecured loans that must be repaid within the same transaction. These are often used for arbitrage, collateral swaps, or liquidation opportunities. Liquidation occurs when a borrower's collateral value falls below the required threshold, and the protocol sells off the collateral to cover the loan and maintain the pool's stability.

DeFi lending and borrowing platforms offer numerous benefits. They eliminate intermediaries, allowing users to transact directly via smart contracts. This makes them accessible to anyone with cryptocurrency and an Internet connection, without the need for credit checks or traditional banking infrastructure. Transparency is inherent, as all transactions and loan conditions are recorded on the blockchain. These platforms also provide flexibility, enabling users to lend or borrow a wide range of cryptocurrencies without restrictions on how borrowed funds are used. Additionally, lenders can earn passive income on idle cryptocurrency, and borrowers can use the funds programmatically within other DeFi protocols for activities like yield farming or trading.

However, these platforms face challenges. Overcollateralization can limit access for users who lack significant holdings, while market volatility can lead to rapid collateral liquidation and losses for borrowers. Smart contracts, while efficient, are not immune to vulnerabilities that could be exploited by hackers. Price oracles, which feed data into the protocols, are also susceptible to manipulation, potentially disrupting loan conditions. Lastly, regulatory uncertainties add an additional layer of risk, as changes in laws could impact the operation of DeFi platforms. Despite these challenges, DeFi lending and borrowing continue to drive innovation in the financial ecosystem.

This use case was chosen due to its high risk profile, significant financial impact, and known exposure to targeted threats, including flash loan attacks, oracle manipulation, collateral price manipulation, and liquidity pool draining. Many of these incidents involve fraudulent transaction patterns aimed at misappropriating tokens for monetary gain, effectively constituting a type of DeFi-specific financial fraud. These attacks exploit flaws in smart contracts and business logic to manipulate asset flows, often without technically violating the code as written.

Given this context, lending and borrowing platforms are an ideal subject for a risk- and attack-centric threat modeling exercise. By applying the PASTA methodology, we aim to assess the platform's architecture through the lens of business impact, technical exposure, and adversarial tactics, ensuring that appropriate mitigations can be identified and prioritized to reduce both the likelihood and impact of potential exploitation.

3.2.3.2 Stage I – Definition of Business Objectives

In Stage I of the risk-centric threat modeling methodology, defining business objectives for the dApp, we begin by analyzing the use case in greater depth to establish the application's inherent risk profile at a high level. This stage sets the foundation for the threat modeling exercise, enabling a security architect to thoroughly explore the dApp's design. Through this process, architects can identify potential threats, determine vulnerabilities in the dApp's components, and propose countermeasures to mitigate the risks posed by possible attacks.

Starting with the lending and borrowing use case for the dApp is an effective approach, as the core functionality of the application provides a foundation for identifying potential abuse or malicious use cases. This initial analysis helps document, at a high level, the potential risks associated with the dApp's operations and interactions.

DeFi lending and borrowing platforms operate through a streamlined, decentralized process enabled by blockchain technology. Users looking to earn passive income can deposit their cryptocurrency into a lending protocol. These deposits are pooled together, creating a fund that is made available to borrowers. Lenders earn interest on their contributions, with the interest payments funded by borrowers who utilize the protocol.

For borrowers, the process involves locking cryptocurrency as collateral to secure a loan. Typically, these loans are overcollateralized, meaning that borrowers must deposit assets worth more than the loan amount to reduce the risk for lenders. For example, a user might provide $150 worth of collateral to borrow $100. Borrowers are charged interest on the loan, which is automatically distributed to the lenders minus a small percentage retained by the dApp as a protocol fee or platform revenue. This fee structure supports the sustainability and continued operation of the platform. This overcollateralization model is standard across DeFi lending platforms to manage the risk of borrower default due to crypto market volatility. The interest paid by borrowers is distributed to lenders as a return on their deposits; however, a small percentage is typically retained by the dApp as a protocol fee to fund platform maintenance and development. This fee structure helps ensure the sustainability and continued operation of the lending ecosystem.

The entire operation is governed by smart contracts, which act as self-executing agreements on the blockchain. These smart contracts handle critical functions such as enforcing loan terms, calculating interest rates, and managing collateral. By automating these processes, smart contracts eliminate the need for intermediaries, ensuring efficiency, transparency, and trust throughout the lending and borrowing life cycle.

DeFi lending and borrowing platforms offer a decentralized solution for financial services, but their inherent functionality introduces specific risks that must be carefully assessed to establish a risk profile and devise mitigation strategies. By understanding how these platforms operate, we can identify key assets at risk, the potential AVs, and the strategies needed to mitigate these threats effectively.

The process begins with users depositing cryptocurrency into the lending protocol. These deposits are pooled together and made available for borrowers, forming the basis of the platform's functionality. Borrowers secure loans by locking their cryptocurrency as collateral, often providing more collateral than the loan amount to minimize default risks. For example, a borrower might deposit $150 worth of assets to obtain a $100 loan. Interest is charged on the borrowed funds, which is paid to the lenders as a return on their deposits.

The entire operation is automated and managed by smart contracts, which enforce loan terms, calculate interest, and handle collateral management. These smart contracts are critical assets as they underpin the platform's reliability and security. If compromised, they can lead to loss of funds, unauthorized transactions, or incorrect execution of loan terms.

During this stage of threat modeling, we focus on analyzing the business cases and the associated critical assets (e.g. critical data), along with their roles and functionalities within the dApp. These assets form the foundation for a high-level BIA, enabling us to identify potential vulnerabilities, evaluate how they could be exploited, and assess their potential impact:

- **User's PII Assets:** User's PII is a critical data asset in DeFi lending and borrowing dApps. PII includes sensitive information such as names, addresses, government-issued identification, contact details, and financial information, often collected to comply with KYC/AML regulations. This data is essential for verifying user identities, ensuring regulatory compliance, and facilitating trust between users and the platform. However, improper handling, storage, or exposure of PII can lead to severe privacy breaches, regulatory penalties, and loss of user confidence. Centralized storage, insecure APIs, or inadequate access controls can increase the risk of unauthorized access or data leaks. For example, attackers exploiting vulnerabilities in data storage systems or APIs can steal PII, leading to identity theft, fraud, or misuse of user credentials.
- **Possible Business Impacts**
  - **Privacy Breaches and User Trust Erosion:** If PII is exposed, users are at risk of identity theft, phishing, and other forms of fraud. A privacy breach undermines user confidence in the platform's ability to protect their sensitive information, leading to mass user attrition. For instance, a breach exposing government IDs or addresses can result in significant harm to users, deterring them from future interactions with the platform.
  - **Operational Disruptions:** A data breach can force the platform to halt operations to assess the extent of the compromise and implement remediations. Resources must be allocated to investigating the breach, patching vulnerabilities, and enhancing security controls, delaying regular platform activities and development initiatives.

  - **Regulatory Fines and Legal Actions:** Noncompliance with privacy laws depending on the jurisdiction where the dApp operates such as GDPR in EU; CCPA in California, United States; or Personal Data Protection Act (PDPA) in Singapore [211] can result in hefty fines and legal liabilities. For example, GDPR violations can incur penalties of up to €20 million or 4% of global annual revenue. Additionally, the platform may face lawsuits from affected users, further straining financial resources and damaging its reputation. Regulatory bodies may impose stricter oversight on the platform following a PII breach. This could result in higher compliance costs, more frequent audits, and stricter operational requirements, adding to the platform's financial and administrative burdens.
  - **Reputational Damage:** News of PII leaks spreads rapidly within the DeFi community and mainstream media, harming the platform's reputation. Affected users may publicize their grievances, discouraging potential users and investors from engaging with the platform. Competitors with stronger privacy protections may capitalize on this damage, eroding the platform's market share.
- **User Deposits Assets:** User deposits, made by lenders, form the liquidity pool that powers the lending and borrowing functionalities of a DeFi platform. These funds are used to issue loans to borrowers, with lenders earning interest as compensation. Bugs or vulnerabilities in the underlying smart contracts can allow attackers to drain liquidity pools. Reentrancy attacks or unchecked permissions can be leveraged to withdraw funds without proper authorization. Rogue actors exploiting governance mechanisms might redirect funds or execute unauthorized transactions. Exploits that manipulate the valuation of collateral can result in undercollateralized loans, leading to a depletion of the deposit pool.
- **Potential Business Impacts**
  - **Immediate Loss of Funds for Lenders:** The funds deposited by lenders form the liquidity backbone of DeFi lending platforms. An exploit targeting these deposits can lead to catastrophic financial losses. Exploits such as smart contract vulnerabilities, unauthorized withdrawals, or liquidity pool drains can result in lenders losing their assets entirely.
  - **Impact on Liquidity:** With user deposits depleted, the platform may become unable to issue loans, disrupting its core functionality. Borrowers reliant on these loans for other operations may also suffer cascading financial consequences.
  - **Reputation Damage:** Even a single high-profile exploit can tarnish the platform's reputation, discouraging existing users from continuing to use it and deterring new users from joining. News of a breach can spread rapidly across social media and forums and trigger a wave of withdrawals as users

panic. A breach can cause a sharp decline in token value, further eroding the platform's financial standing and user confidence.

- **Operational Challenges:** The platform may need to divert significant resources to address the breach, such as compensating users, auditing code, and rebuilding infrastructure. This can strain finances and delay future development.
- **Regulatory Scrutiny:** High-profile losses can attract attention from regulators, leading to potential fines, stricter oversight, or even shutdowns in certain jurisdictions.
- **Closure:** In extreme cases, the platform may be forced to cease operations entirely, leaving remaining users and investors at a loss.

- **Collateralized Assets:** Borrowers must lock cryptocurrency assets as collateral to secure loans. These collateralized assets serve as security for lenders, ensuring that loans are repaid or recovered through liquidation if necessary. Attackers can exploit price oracles to undervalue or overvalue collateral, leading to premature liquidations or excessive borrowing. Flaws in liquidation mechanisms can result in collateral being sold at unfair prices or seized unnecessarily. Collateralized assets moved between chains (via token bridges) can be targeted, as seen in high-profile bridge exploits (e.g. Ronin Bridge hack). If collateralized assets are moved across chains using token bridges, an attack on the bridge can result in the theft or loss of collateral. Attackers can manipulate price oracles to artificially undervalue the collateral assets. For example, if the price of a borrower's collateral asset is manipulated to appear lower than its actual market value, the platform may trigger a liquidation process, selling off the collateral at a discount. Vulnerabilities in the smart contract governing liquidations can cause collateral to be liquidated unnecessarily or at an undervalued price, depriving borrowers of their assets.
- **Potential Business Impacts**
  - **Loss of Collateral for Borrowers:** Borrowers lose assets they locked in good faith, leading to financial hardship.
  - **Cascading Loan Defaults:** A single manipulation event, such as an oracle attack, can impact multiple borrowers simultaneously, leading to widespread undercollateralization and a liquidity crisis. Exploits that overvalue borrowed assets or undervalue collateral can result in loans that are not fully covered by locked assets.
  - **Platform Insolvency:** If attackers successfully steal or manipulate collateral, the platform may face significant financial losses, leaving it unable to repay lenders. The inability to recover sufficient collateral results in financial losses that could lead to insolvency.
  - **Reputation Damage:** Perceived unfairness in the process can discourage borrowers from using the platform in the future.

- **Smart Contracts Assets:** Smart contracts automate the core functionalities of the dApp, including deposits, loans, interest calculations, and liquidations. They are the operational backbone, removing the need for intermediaries. Smart contracts are more than just pieces of code, they are critical enablers of a business's operations, financial stability, and competitive advantage. As such, they are undeniably a business asset, demanding careful design, rigorous testing, and ongoing management to protect and enhance their value. Errors in contract code, such as faulty interest rate calculations or improper access controls, can disrupt functionality or enable unauthorized actions. Malicious or poorly reviewed code in smart contracts can allow attackers or insiders to siphon funds exploiting backdoors. Sophisticated attacks like reentrancy, integer overflows, or front-running can result in funds being stolen or functionality being altered.
- **Possible Business Impacts**
  - **Financial Losses from Stolen Funds or Miscalculated Transactions:** Smart contracts in lending and borrowing dApps manage critical financial processes such as deposits, loans, interest calculations, and liquidations. Exploits like reentrancy attacks, logic flaws, or unauthorized access can lead to draining of liquidity pools, leaving the platform unable to meet withdrawal requests. Borrowers might underpay their loans due to incorrect calculations, impacting the platform's revenue. Incorrect or missing interest payments directly impact the platform's profitability. Theft of collateralized assets could render the platform undercollateralized and financially unstable. A single attack can result in the loss of millions in user funds or collateral, destabilizing the platform's liquidity and solvency.
  - **Platform Downtime as the Compromised Contract Is Fixed or Redeployed:** The platform may freeze lending, borrowing, and withdrawal activities to prevent further exploitation. While services are suspended, the platform generates no transaction fees or interest revenue, impacting its financial performance. Fixing and redeploying a compromised contract is time-consuming and may require rigorous auditing to prevent repeated exploitation. Migration of funds and data to new contracts can be operationally complex and risky. Addressing the issue requires additional resources, such as security consultants, auditors, and legal advisors, increasing operational expenses. Inaccessible funds or interrupted services can frustrate users, leading to complaints, negative publicity, and potential user attrition.
  - **Long-Term Damage to the Platform's Reputation:** Reputation is a cornerstone of success in DeFi, where users place trust in the platform to securely manage their funds and provide reliable services. A smart contract exploit can severely damage this trust. A compromised smart contract

undermines the platform's perceived security and technical competence, discouraging existing users and deterring new ones. Exploits often attract significant attention on social media, industry forums, and news outlets, amplifying the reputational damage.

- **Transaction Data Assets:** Transaction data includes records of lending, borrowing, repayments, and interest calculations. This data is essential for ensuring transparency, accountability, and auditability of platform operations. Insecure storage or manipulation of transaction data can result in discrepancies in user balances or loan terms. Attackers targeting APIs, indexing services (e.g. the Graph), or off-chain data sources can alter transaction history, leading to disputes. Off-chain data storage without robust tamperproof mechanisms may be vulnerable to insider threats or unauthorized modifications. Compromise of transaction data through inaccuracies, manipulation, or tampering can have far-reaching consequences.
- **Possible Business Impacts**
  - **Inaccurate Records Leading to Disputes, Financial Losses, or Legal Challenges:** Borrowers may challenge incorrect repayment records, claiming they have repaid more than what the platform records. Lenders may dispute discrepancies in their earnings from interest or withdrawals. Incorrect interest calculations can lead to under- or overpayment, impacting the platform's revenue streams. Manipulated records could allow malicious actors to claim refunds or withdraw funds fraudulently. Disputes over transaction data accuracy may escalate into lawsuits, especially if large sums are involved or if institutional participants are affected. Regulators may scrutinize platforms that fail to maintain accurate records, potentially imposing fines or operational restrictions.
  - **Erosion of Trust If Users Cannot Verify the Accuracy of Their Transaction Data:** Trust is paramount in DeFi platforms, where users expect transparency and immutability. Compromised transaction data undermines this trust. Users rely on transaction data to verify the accuracy of their balances, repayments, and earnings. If they suspect manipulation or inaccuracies, their confidence in the platform diminishes. Dissatisfied users are likely to withdraw their funds and migrate to competing platforms, resulting in a loss of liquidity and a shrinking user base.
  - **Difficulty in Conducting Audits or Recovering from Malicious Activity:** Accurate transaction data is essential for internal audits, forensic investigations, and regulatory compliance. When this data is compromised. Internal teams and external auditors may face difficulties verifying financial operations, interest calculations, and compliance with governance policies. In the event of an exploit, compromised transaction data makes it harder to trace the source of the attack, quantify losses, or identify affected users.

Recovery efforts may be delayed or insufficient, exacerbating the damage to the platform and its users.

- **Price Oracles Assets:** Price oracles are critical data assets in DeFi lending and borrowing dApps. These external systems provide real-time asset price data that the dApp uses to calculate collateral values, determine loan-to-value (LTV) ratios, and trigger liquidation events. Oracles ensure borrowers' collateral remains sufficient to back their loans, dynamically adjusting collateral requirements based on asset price fluctuations. When collateral value drops below a defined threshold, oracles provide the price data necessary to initiate liquidations. Oracles ensure borrowers and lenders receive equitable treatment by providing accurate, up-to-date market valuations of assets. Attackers may manipulate price feeds to exploit the dApp's reliance on inaccurate data. Flash loans can temporarily manipulate prices on DEXs, causing oracles to report inflated or deflated asset values. If the oracle goes offline, the dApp may be unable to update prices, freezing operations or triggering unnecessary liquidations. Smart contracts used to fetch, aggregate, and validate oracle data can be exploited; errors in contract logic may process invalid or malicious data, leading to platform misbehavior.
- **Possible Business Impacts**
  - **Financial Losses from Liquidations and Borrowing Abuse:** Manipulated oracle data [212] can force borrowers into unnecessary liquidations, causing significant financial loss and eroding borrower trust. Attackers may exploit inflated collateral values to borrow excessively without proper backing, leaving the platform insolvent. In 2020, the bZx protocol suffered from multiple oracle manipulation attacks [213], resulting in millions of dollars in losses. In October 2021, Cream Finance, a decentralized lending protocol, suffered a significant flash loan attack [214], resulting in the loss of approximately $130 million.
  - **Operational Disruptions:** If oracles fail to provide data, the platform may suspend lending, borrowing, and liquidations to prevent further harm, impacting revenue. Users may withdraw funds en masse during periods of uncertainty, depleting liquidity and straining the platform's operations.
  - **Reputational Damage:** News of compromised price data spreads quickly in the DeFi community, reducing user confidence and trust in the platform. Disillusioned users may move their funds to competitors with more robust oracle systems, leading to market share losses.
  - **Systemic Risks:** DeFi platforms are interconnected; manipulated oracle data on one platform can propagate incorrect valuations to others, causing widespread disruptions. Platforms issuing governance or utility tokens may see a decline in token value, further impacting financial stability.

- **Private Keys Assets:** Private keys are critical functional assets in DeFi lending and borrowing dApps. These cryptographic keys are used to sign transactions, manage collateral, authorize fund transfers, and execute governance actions. Private keys ensure the integrity and security of operations by verifying the identity of users and the validity of transactions. They play a vital role in protecting the platform's funds, maintaining user accounts, and enabling secure execution of platform logic. If private keys are compromised, attackers can gain unauthorized access to funds, governance mechanisms, and operational controls, potentially causing catastrophic damage. Single points of failure, insecure storage, or poor key management practices increase the risk of private key theft or misuse. For example, malicious actors can exploit exposed keys to drain user accounts, manipulate governance votes, or disrupt platform operations.
- **Possible Business Impacts**
  - **Financial Losses from Unauthorized Transactions:** If private keys are compromised, attackers can perform unauthorized withdrawals or manipulate critical platform functions. For instance, stolen admin keys can allow an attacker to drain liquidity pools, move collateralized assets, or approve undercollateralized loans. This can lead to significant financial losses for users and the platform, as seen in incidents like the KuCoin hack [215], where private key exposure resulted in a major asset theft.
  - **Operational Disruptions:** Compromised private keys can force the platform to pause operations to prevent further exploitation. This may include freezing lending, borrowing, and liquidation processes. Redeploying contracts or migrating funds to secure wallets is often time-consuming, causing operational delays and user frustration. Additionally, resources are diverted to investigate the breach and implement remediations, increasing operational expenses.
  - **Reputational Damage:** A private key compromise undermines user trust and confidence in the platform's security practices. News of such incidents spreads rapidly within the DeFi community and can result in mass withdrawals, eroding liquidity and market share. Existing users may abandon the platform, and potential users or investors may be deterred by the perceived risks.
  - **Regulatory Fines:** Incidents involving private key compromise can attract regulatory attention, especially if users incur significant financial losses. This scrutiny may lead to stricter compliance requirements, fines, or even platform shutdowns in certain jurisdictions.

Each of these assets plays a pivotal role in ensuring the platform operates smoothly and securely. The corresponding attack scenarios provide a critical foundation for assessing potential vulnerabilities and prioritizing security measures effectively. By integrating these assets and scenarios into the threat modeling process during Stage II, the technical scope can be comprehensively defined to mitigate risks. The corresponding attack scenarios provide essential insights for defining the technical scope during Stage II of threat modeling, Definition of Technical Scope, which focuses on the definition of the system's technical boundaries. These inputs help assess potential AVs and prioritize security measures effectively.

User's PII is a critical data asset in DeFi lending and borrowing dApps. PII includes sensitive user details such as names, addresses, government-issued IDs, email addresses, and financial information. This data is often collected to comply with KYC/AML regulations and is vital for verifying user identities and ensuring compliance with legal and regulatory requirements. PII forms the foundation for building trust between users and the platform, as well as enabling secure and lawful operations. However, improper handling, storage, or exposure of PII can lead to privacy breaches, identity theft, and legal complications. Attackers targeting insecure APIs, poorly secured databases, or insider threats may steal or misuse this information, resulting in significant harm to both users and the platform. To mitigate the risks associated with user PII in DeFi lending and borrowing dApps, platforms should implement robust encryption to protect data both at rest and in transit, coupled with role-based access control (RBAC) to restrict data access based on user roles and enforce least privilege. Adopting decentralized identity (DID) solutions, such as Civic or uPort, can minimize centralized storage of PII and enable users to retain control over their data, sharing only necessary attributes. Data minimization principles should be followed, collecting only what is required for regulatory compliance and operational purposes.

APIs and interfaces should be secured with authentication mechanisms, input validation, and rate limiting to prevent unauthorized access or data exposure. Storing PII in decentralized or tamperproof systems further reduces the risk of unauthorized modifications or deletions. Continuous monitoring and logging of data access, along with regular reviews of access logs, can help detect and address suspicious activities promptly.

Platforms must adhere to privacy regulations like GDPR and CCPA, incorporating user consent mechanisms, data erasure policies, and breach notification procedures. Regular security assessments, including penetration testing and third-party audits, are essential to identify and resolve vulnerabilities. Educating users on best practices for securing their accounts and recognizing phishing attempts also plays a key role in enhancing overall security. These measures collectively strengthen the platform's ability to safeguard PII, ensuring trust and compliance.

User deposits, the foundation of the platform's liquidity, are susceptible to risks like smart contract vulnerabilities, unauthorized withdrawals, and collateral manipulation. To safeguard these assets, rigorous auditing of contracts, governance safeguards, and real-time collateral monitoring are crucial. By including deposit-related functions and mechanisms within the threat modeling scope, the platform can identify vulnerabilities that might lead to depletion of funds or systemic failures.

Collateralized assets, which serve as security for loans, face threats such as oracle manipulation, flawed liquidation mechanisms, and cross-chain vulnerabilities. Mitigating these risks involves adopting decentralized oracles, designing resilient liquidation logic, and securing token bridges with redundancy and enhanced validation. During the DTS, interactions involving collateral, liquidation mechanisms, and cross-chain operations should be analyzed to anticipate and address exploitation scenarios.

Smart contracts automate the platform's core functionalities, making them critical assets for operational integrity. These contracts, however, can be exploited through logic flaws, backdoors, or advanced techniques like reentrancy. Comprehensive testing, modular architectures for upgradability, and fail-safe mechanisms are essential to reduce risks. Defining the technical scope to include smart contracts managing loans, deposits, and interest calculations allows for a detailed analysis of potential vulnerabilities, ensuring robust security measures.

Transaction data, encompassing records of lending, borrowing, and repayments, is vital for transparency and accountability. Risks such as inaccurate records, API attacks, or tampered indexing services can undermine data integrity. Mitigations include tamperproof logging, secure APIs, and decentralized indexing solutions. By scoping APIs, storage systems, and indexing operations during threat modeling, the platform can safeguard against discrepancies and ensure the reliability of its data.

Price oracles provide the real-time asset valuations that underpin the platform's functionality. Vulnerabilities in oracles, such as manipulation of price feeds or aggregation logic flaws, can lead to unnecessary liquidations or borrowing abuse. Mitigation strategies include employing multiple oracle providers, implementing anomaly detection safeguards, and opting for decentralized oracles. Including oracle interactions and external data integrations within the technical scope ensures that scenarios like oracle manipulation and flash loan exploits are adequately addressed.

Private keys are critical functional assets in DeFi lending and borrowing dApps, serving as the backbone of secure platform operations. These cryptographic keys are used to sign transactions, manage collateral, authorize fund transfers, and execute governance actions, ensuring the integrity of the platform by verifying user identities and transaction validity. Private keys are indispensable for protecting platform funds, maintaining user accounts, and enabling the secure execution of smart contract logic. However, their compromise can lead to unauthorized access, catastrophic financial losses, and operational disruptions. Risks arise from single points of failure, insecure storage, and inadequate key management practices, which may be exploited by malicious actors to drain accounts, manipulate governance, or disrupt platform functionality.

To mitigate the risks associated with private keys in DeFi lending and borrowing dApps, platforms should implement robust security measures and best practices for key management. Using HSMs ensures keys are stored in tamper-resistant, secure hardware, while multi-sig wallets distribute control among multiple parties to eliminate single points of failure. Cold storage solutions should be used for long-term security of significant funds, complemented by regular key rotation policies to limit exposure from prolonged use. Strict access controls and authentication mechanisms are necessary to restrict unauthorized use, and continuous monitoring systems with real-time alerts help detect and respond to suspicious activity. Regular audits of key management practices are essential for identifying and addressing vulnerabilities, while staff and user training on secure key handling further reduces risks. Collectively, these measures provide a comprehensive framework for safeguarding private keys and ensuring the platform's security and operational resilience.

Table 3.2 provides a comprehensive risk profile for dApp DeFi lending and borrowing, highlighting key risks to protected assets and business objectives, along with mitigation strategies to safeguard application assets.

In this risk assessment, risk is calculated using the formula *Risk* = *Likelihood* × *Impact*, a widely adopted approach in both security and risk management. Likelihood refers to the probability that a threat will be realized, while impact measures the severity of consequences if the threat does occur. Each factor is scored on a scale from 1 (low) to 3 (high), resulting in a total risk score ranging from 1 to 9. This score is then mapped to qualitative categories: low (1–2), medium (3–4), medium to high (5–6), and high (7–9). This structured method supports clear prioritization of security efforts based on both technical and business relevance.

Documenting the key risks associated with protected assets establishes the business case for threat modeling and a high-level risk mitigation strategy. The outcomes of this stage directly inform the analysis of the technical scope where these risks must be managed. By proactively addressing these risks, the platform can enhance operational resilience, protect user trust, and ensure long-term sustainability in the competitive DeFi ecosystem.

In addition to addressing security risks as outlined in the high-level requirements for the lending and borrowing dApp, another critical area for preliminary risk analysis involves compliance with applicable regulations. This includes adherence to laws governing the protection of users' personal identities, such as those related to PII, as well as compliance with KYC and AML regulations. Failure to meet these requirements poses significant legal and operational risks.

Table 3.2 dApp assets, key risks and business risk mitigation objectives for risk mitigation.

Asset type	Impact (score)	Likelihood (score)	Risk level (score)	Key risks	Business objectives for risk mitigation
User deposits	High (3): A compromise of user deposits can result in direct financial loss, disruption of lending operations, and severe reputational damage, eroding user trust and potentially triggering regulatory and legal consequences.	High (3): User deposits are a primary target in DeFi lending and borrowing platforms, making them extremely attractive to attackers due to their immediate monetary value.	High (9): User deposits represent the financial backbone of a DeFi lending and borrowing platform. A risk score of 9 (high) reflects the convergence of two critical factors: severe impact if compromised, and a high likelihood of such compromise due to historical patterns and technical exposure.	Exploits draining liquidity pools, unauthorized withdrawals, governance abuses, or manipulation of collateral values	Rigorous smart contract audits, real-time anomaly detection tools, secure governance frameworks, and multi-sig transaction systems.

Collateral	High (3): A compromise of collateralized assets can cause loan defaults, platform insolvency, and loss of borrower funds, leading to user distrust, capital flight, and reputational damage, especially when tied to oracle manipulation or cross-protocol dependencies.	High (3): Collateral mechanisms are commonly targeted through oracle manipulation, flash loan attacks, and cross-bridge vulnerabilities. These AVs are well-documented and have been used in high-profile DeFi attacks (e.g. bZx, Alpha Homora, and Cream Finance).	High (9): Collateralized assets are cryptocurrency tokens that borrowers lock into smart contracts to secure loans. These assets are central to the functioning and solvency of DeFi lending platforms. A risk score of 9 (high) reflects the critical business and security importance of these assets, combined with real-world exploitability seen across the DeFi ecosystem.	Oracle manipulation, liquidation flaws, cross-chain vulnerabilities, and undercollateralization of loans	Use decentralized oracles with fallback mechanisms, implement robust liquidation logic, and secure token bridge infrastructure.

(Continued)

Table 3.2 (Continued)

Asset type	Impact (score)	Likelihood (score)	Risk level (score)	Key risks	Business objectives for risk mitigation
Smart contracts	High (3): Severe financial losses (i.e. total or partial loss of funds because of drained pools), protocol manipulation (e.g. governance takeovers or payout miscalculations), permanent locking of funds due to logic bugs or failed fallback paths, reputational and legal repercussions, especially when funds cannot be recovered.	High (3): Smart contracts are highly attack-prone due to their public accessibility, complex logic, and critical financial functions. Common flaws, like reentrancy, logic errors, and oracle misuse have led to major DeFi exploits (e.g. DAO, bZx, Cream), making them a frequent and likely target.	High (3): Smart contracts are rated high risk (9) due to their critical role, frequent exploitation, and irreversible impact when compromised. Their public exposure and complex logic make them a consistent attack target, requiring strong preventive and monitoring controls.	Reentrancy attacks, backdoor vulnerabilities, logic flaws in calculations, and improper access controls	Conduct continuous audits, adopt secure coding practices, implement automated testing, and use monitoring for anomalies.

Transaction data	High (3): Compromise of transaction data can lead to inaccurate user balances, repayment disputes, regulatory exposure, and loss of trust, especially if the data is relied upon for audits, compliance, or financial integrity. These issues directly affect the platform's credibility and user confidence.	Medium (2): While transaction data is typically recorded on-chain and harder to alter, risks remain through API vulnerabilities, indexing errors, or off-chain data handling flaws. These are less frequently exploited than smart contract or key-related issues but still present a realistic threat.	Medium to high risk (6): Due to its role in maintaining financial integrity and user trust. While not as frequently targeted as assets or contracts, manipulation or tampering can lead to incorrect balances, legal disputes, or audit failures. The risk is elevated by API vulnerabilities or insecure data handling, requiring tamperproof storage, secure APIs, and regular integrity checks.	Manipulated transaction histories, tampering with repayment records, or API vulnerabilities leading to data integrity issues	Use tamperproof data storage, secure APIs, conduct regular audits, and establish robust indexing and verification processes.

(Continued)

Table 3.2 (Continued)

Asset type	Impact (score)	Likelihood (score)	Risk level (score)	Key risks	Business objectives for risk mitigation
Price oracles	High (3): Price oracles are critical to determining collateral valuations, triggering liquidations, and setting borrowing limits. Manipulated or unreliable oracle data can lead to massive financial losses, market manipulation, and cascading effects across multiple protocols, especially in interconnected DeFi ecosystems.	High (3): Oracle manipulation is a well-known and repeatedly exploited AV in DeFi. Techniques like flash loan attacks, data feed delays, or exploiting centralized oracle sources have been used in real-world exploits, making this a highly probable and well-documented threat.	High (3): Given their central role and high exposure, price oracles represent a critical risk surface in DeFi lending platforms. The combination of severe impact and frequent real-world attacks warrants robust mitigations, such as decentralized oracles, data source diversity, and fallback mechanisms.	Oracle manipulation, flash loan exploits, data feed outages, and errors in price aggregation or validation logic	Integrate decentralized oracles, diversify data sources, implement fallback strategies, and use robust error-handling in smart contracts.

User's PII	High (3): Exposure of PII can lead to privacy breaches, identity theft, regulatory noncompliance, and loss of user trust, with potential for significant legal and financial consequences.	Medium (2): While DeFi platforms typically minimize PII collection, risks still exist through off-chain integrations, misconfigured APIs, or KYC/AML providers, making PII exposure plausible but less frequent than smart contract or financial asset attacks.	Medium to high (6): The combination of high impact and moderate likelihood places PII at a medium to high risk level, requiring strong encryption, access controls, and regular compliance audits to protect user data and maintain regulatory alignment.	Exposure or theft of PII leading to privacy breaches, identity theft, and noncompliance with regulations	Encrypt PII at rest and in transit, enforce access controls, adopt DID solutions, and conduct regular compliance audits.
Private keys	High (3): Compromise of private keys can result in unauthorized access to funds, governance manipulation, and complete loss of control over key platform functions, potentially leading to severe financial, operational, and reputational damage.	High (3): Private keys are a high-value target for attackers and can be compromised through phishing, malware, poor key management, or insecure storage. Given their sensitivity and history of real-world breaches, the likelihood of compromise is high without strong controls.	High (9): Due to their critical function and high exposure, private keys represent a top-tier risk requiring robust protections like HSMs, multi-sig wallets, access controls, and key rotation policies.	Compromise of private keys enabling unauthorized transactions, governance manipulation, or disruption of platform operations	Store keys in HSMs, use multi-sig wallets, implement key rotation policies, enforce strict access controls, and monitor for anomalies.

Noncompliance with privacy and AML/KYC regulations presents significant risks for dApps focused on lending and borrowing cryptocurrencies. Legally, noncompliance can lead to severe penalties, including fines, operational shutdowns, and personal liability for platform operators. Financially, it can result in loss of access to banking services, frozen assets, high remediation costs, and compensation claims from affected users. Reputationally, the dApp risks losing user trust, attracting negative media attention, and deterring investors. Operationally, it may face increased regulatory scrutiny, fragmented compliance requirements across jurisdictions, and restrictions on market access.

Privacy risks include data breaches, unauthorized data sharing, and failure to honor user rights, which could lead to regulatory penalties and diminished user confidence. Moreover, without proper AML/KYC controls, the dApp becomes vulnerable to criminal exploitation, such as money laundering, sanctions violations, and fraud. This damages not only the platform but also the broader DeFi ecosystem, inviting stricter regulations.

Failure to address these compliance issues could lead to delisting from exchanges, market bans, and reduced user participation. To mitigate these risks, dApps must prioritize robust compliance measures, align with legal requirements, and implement privacy-focused, secure designs. Key risks for noncompliant dApps include regulatory penalties, such as fines imposed by authorities like the US CFTC [216] for operating unregistered financial services. Reputational fallout can result from involvement in illicit activities, as seen with platforms like Tornado Cash, leading to operational bans and loss of user trust. Additionally, noncompliance with licensing requirements in jurisdictions such as the EU, United States, or Singapore can restrict market access and user participation, significantly impacting the platform's operations and growth potential.

When designing a lending and borrowing dApp, several key considerations are essential. Regulatory requirements, such as AML/KYC compliance, often mandate the collection of PII to meet jurisdictional laws with respect to AML such as the EU's AMLD [217], and in the United States, FinCEN requires companies that deal with cryptocurrencies on a frequent basis to be registered as money service businesses (MSBs) [218].

If PII is collected, data protection laws like GDPR, CCPA, or PDPA require secure storage, user consent, and data minimization. The dApp's design plays a critical role in compliance, offering options like fully decentralized architectures that avoid PII storage or hybrid models that process PII off-chain through regulated entities. Additionally, user trust and adoption depend on minimizing or eliminating PII processing to protect privacy, aligning with the growing demand for secure and privacy-preserving platforms.

Based on the regulatory scope applicable to the dApp, design-phase decisions related to privacy preservation and compliance should consider integrating privacy-focused and compliance-auditing tools. Privacy engineering tools such as Civic enable DID verification for KYC compliance, while uPort provides DID management that aligns with GDPR and other data privacy laws, ensuring secure handling of PII.

Additionally, the dApp's compliance strategy can benefit from integrating tools that facilitate compliance audits. For example, blockchain governance tools like OpenZeppelin Defender [219] help implement secure governance and compliance controls directly in smart contracts. Regulatory assessment tools such as Chainalysis [220] provide transaction monitoring and compliance insights to identify illicit activities and maintain adherence to financial regulations. These integrations collectively enable the dApp to achieve privacy preservation, regulatory compliance, and operational transparency. Engaging legal advisors is essential to navigate complex regulatory landscapes and avoid noncompliance. Regular audits should be conducted to review compliance measures, including transaction monitoring and reporting.

As a recap, this chapter introduced Stage I of the risk-centric threat modeling methodology, focusing on defining business objectives for the dApp. We explored the core lending and borrowing use case to establish the application's high-level risk profile and identified critical functional assets, including user deposits, collateral, smart contracts, transaction data, price oracles, user PII, and private keys. Each asset was examined in terms of its operational role, exposure to potential threats, and the business impact if compromised. Real-world attacks were referenced to contextualize risks and align them with business concerns. To support prioritization, we applied a risk scoring model based on the combination of likelihood and impact, using a 1–3 scale. We also highlighted key compliance requirements, such as KYC, AML, and data protection laws like GDPR and CCPA, that influence architectural decisions around identity, data handling, and governance. The insights and outputs from this stage form the baseline for identifying technical boundaries and attack surfaces in Stage II.

3.2.3.3 Stage II – Definition of Technical Scope

In Stage II of the risk-centric threat modeling methodology, defining technical scope for the dApp involves delineating the technical boundaries of the dApp. This step focuses on clearly identifying the components of the architecture that fall within the scope of the design, as well as the technical stack and services included in the assessment. This ensures a well-defined scope for the threat modeling process, targeting relevant aspects of the dApp's architecture and functionality.

From the initial technical scope assessment, a comprehensive baseline for threat modeling is established, encompassing all primary components of the system. This approach ensures that every element of the architecture, from core functionalities to external integrations, is thoroughly evaluated for potential risks. This baseline assessment is a foundational step in securing the overall system, as it captures the architecture's intended design and operational flow before deployment.

This methodology contrasts sharply with on-project threat modeling, which focuses exclusively on changes or updates made to the system after deployment. For the sake of this example, we will focus on the baseline threat model, which includes all components of the architecture from the user-facing elements to the backend infrastructure. This baseline threat model serves as the reference for the application's threat modeling analysis, forming the foundation for assessing risks across the system. By comprehensively examining the entire architecture in its intended operational context, this baseline ensures a consistent and accurate framework for identifying vulnerabilities and threats that might impact the application throughout its life cycle.

dApps for lending and borrowing cryptocurrencies rely on a comprehensive architecture that integrates multiple components. Each element serves a specific purpose, ensuring functionality, scalability, and security while delivering a seamless user experience. For each component, the choice between open-source and commercial solutions depends on factors like budget, scalability needs, and security maturity. Open-source solutions provide transparency and flexibility, while commercial offerings deliver enhanced support, service level agreements (SLAs), and enterprise-grade security. By applying robust controls and carefully selecting technology solutions, a lending and borrowing dApp can ensure a secure, scalable, and user-friendly platform. Each component in a lending and borrowing dApp architecture plays a critical role in ensuring functionality, security, and scalability. By carefully selecting open-source or commercial solutions based on specific needs, platforms can achieve a balance between flexibility, support, and operational security.

Each component of a lending and borrowing dApp architecture can be targeted for different types of threats, whose exposed weaknesses and vulnerabilities need to be further analyzed in a comprehensive risks-centric threat modeling exercise. The following outlines the key technical components of the DeFi lending and borrowing dApp, which should be formally scoped and documented within the application's architecture design:

1. **UI or Mobile Interface:** The UI or mobile application is the primary interaction layer for users, enabling access to lending and borrowing functionalities. Developers can use frameworks like React Native [221] and Flutter [222],

offering flexibility and innovation. Additional libraries such as Web3 Modal [223] for wallet integration, TailwindCSS [224] for styling, and Formik [225] for secure input validation are commonly employed. Open-source frameworks like React Native and Flutter offer flexibility and innovation but rely on developers to implement robust controls, such as secure session management, input sanitization, and MFA.

Note: In this dApp taxonomy, the UI refers to the user-facing interface delivered via browser or mobile app.

2. **Frontend Application:** The frontend application connects the UI to blockchain wallets and smart contracts, facilitating user input processing and API communication. Popular frameworks like Vue.js, AngularJS, or React are used for developing responsive frontends. Libraries such as Web3.js or ethers.js enable blockchain interaction.

Note: In this dApp taxonomy, the frontend includes the application logic and interface components. The interface server or application backend handles off-chain processing and communication between the UI and decentralized infrastructure.

3. **Identity Solutions:** DID solutions manage user authentication. Open-source frameworks like uPort [226] allow flexibility in implementing DID solutions. Commercial solutions like Civic [227] provide SLA-backed identity verification, key management, and compliance tools to address identity risks effectively.
4. **Distributed Storage:** Distributed storage solutions such as InterPlanetary File System (IPFS) [228] or Arweave [229] handle transaction data, loan records, and user activity. Open-source systems like IPFS offer flexibility but depend on additional security layers, such as encryption and token-based access control, to protect against threats. Developers use libraries like OrbitDB [230] for peer-to-peer (P2P) databases and Textile [231] for enhanced IPFS management. Ceramic Network [232] supports dynamic data management. Commercial platforms like AWS Blockchain [233] offer SLA-backed reliability, integrated encryption, and regular updates to protect against these risks.
5. **JSON-RPC Provider Middleware:** Middleware acts as a bridge between the frontend and blockchain nodes, enabling smart contract interactions. Open-source providers like Infura [234] and Alchemy [235] offer robust middleware but require developers to implement additional layers of monitoring and protection. Commercial providers such as QuickNode [236] offer SLA-backed services, encrypted communications, and proactive patching to address these middleware-related risks effectively. Libraries like web3-provider-engine [237] and rpc-websockets [238] facilitate robust middleware implementation, enabling secure and efficient RPC interactions.

6. **Digital Wallets:** Digital wallets manage private keys for signing transactions and authenticating users. Open-source wallets like Trust Wallet [239] provide transparency and broad adoption but rely on users for key security. Commercial platforms like Fireblocks [240] implement delegated signing, MFA, and SLA-backed updates, ensuring comprehensive wallet security.
7. **Blockchain Layer:** The blockchain layer forms the foundation of the dApp, hosting smart contracts and enabling decentralized operations. Open-source blockchains like Ethereum, Solana, and Polygon provide flexibility but require rigorous node management to mitigate risks. Managed blockchain services like AWS Managed Blockchain [241] offer enterprise-grade security, regular patching, and SLA-backed performance monitoring, reducing risks associated with network vulnerabilities and scalability constraints.
8. **Smart Contracts:** Smart contracts automate core lending and borrowing functionalities. Open-source libraries like OpenZeppelin provide secure contract templates but require developers to ensure thorough testing and compliance. Solidity-Parser [242] is used for programmatic code analysis and Foundry [243] for testing. Auditing tools like MythX and Slither and commercial services like Trail of Bits ensure secure deployment and ongoing protection.
9. **Oracles and External Data Sources:** Oracles provide off-chain data like asset prices to smart contracts. Open-source options like Chainlink supported by libraries such as Chainlink-SDK [244] are widely trusted for their decentralization but require careful implementation to mitigate manipulation risks. Commercial solutions like API3 [245] offer SLA-backed assurances, timely patching, and robust validation frameworks to ensure data accuracy and integrity.
10. **Indexing Applications:** Indexing applications like the Graph [246] enable efficient retrieval of blockchain data for user transactions and balances. Developers rely on graphql-request [247] for querying indexed data and databases like Dgraph [248] for performance optimization. Open-source indexing frameworks provide flexibility but require additional security mechanisms, while commercial indexing services offer SLAs, proactive monitoring, and regular updates to address these risks comprehensively.
11. **Scaling Applications:** Scaling solutions such as zkSync [249] or Optimism [250] supported by SDKs such as zkSync SDK and libraries like PlasmaJS enhance dApp scalability by offloading computations to sidechains or Layer 2 networks. Open-source frameworks provide customization flexibility but require stringent testing to address these vulnerabilities. Commercial solutions like ConsenSys Quorum [251] offer SLA-backed reliability, proactive patching, and built-in safeguards to ensure secure synchronization and seamless scalability.

Transaction Monitoring Tools: Transaction monitoring ensures that all on-chain activities are compliant, preventing malicious actions such as money laundering or illicit fund transfers. Open-source solutions like Ethereum Transaction Explorer (ETX) [252] enable basic transaction analysis [253], while commercial platforms like Chainalysis or Elliptic provide robust monitoring tools with features such as risk scoring, real-time alerts, and SLA-backed updates to their data models. Developers can use these tools to identify and flag high-risk transactions in real time while maintaining compliance with global regulatory standards.

Table 3.3 provides a summary of the assets and the associated tech stack included in the scope for establishing the threat modeling baseline.

Table 3.3 Summary of assets, functionalities, and tech stack for dApp high-level design.

Asset	Description	Tech stack
UI or mobile interface	The primary interaction layer for users, enabling access to lending and borrowing functionalities; ensures secure session management, input sanitization, and MFA	React Native, Flutter, Ionic Enterprise; Web3Modal (wallet integration), TailwindCSS (styling), Formik (input validation)
Frontend application	Connects the UI to blockchain wallets and smart contracts, processing user inputs and facilitating API communication; provides encryption, secure API gateways, and authentication	Vue.js, AngularJS, React; Web3.js, ethers.js (blockchain interaction); Apollo Client (API data fetching); Okta (API security)
Identity solutions	Manages user authentication and PII securely using decentralized identifiers and cryptographic protections; supports compliance with regulations	uPort, Civic; DID-JS, Veramo (decentralized identifiers and cryptographic proofs)
Distributed storage	Stores transaction data, loan records, and user activity on decentralized networks; requires encryption, token-based access control, and redundancy	IPFS, Arweave; OrbitDB (P2P databases), Textile (IPFS management), Ceramic Network (dynamic data), AWS Blockchain Templates
JSON-RPC provider middleware	Bridges the frontend and blockchain nodes, enabling smart contract interactions; ensures secure API calls and efficient RPC communication	Infura, Alchemy; QuickNode (commercial middleware); web3-provider-engine, rpc-websockets (middleware libraries)

(Continued)

Table 3.3 (Continued)

Asset	Description	Tech stack
Digital wallets	Manages private keys for signing transactions and authenticating users; implements institutional-grade protections, delegated signing, and MFA	Trust Wallet, MetaMask; Fireblocks (commercial); WalletConnect SDK (multiwallet support), BIP39 (mnemonic management)
Blockchain layer	Hosts smart contracts and facilitates decentralized operations; requires rigorous node management and monitoring	Ethereum, Solana, Polygon; Ganache (testing), Hardhat (development); Azure Blockchain (managed services)
Smart contracts	Automates lending and borrowing functionalities; requires secure coding, testing, and auditing for reliability and compliance	OpenZeppelin (secure templates), Solidity-Parser (code analysis), Foundry (testing); MythX, Slither, Trail of Bits (auditing)
Oracles and external data sources	Provides off-chain data such as asset prices to smart contracts; ensures data validation and accuracy	Chainlink, Chainlink-SDK (integration), API3 (SLA-backed oracles)
Indexing applications	Retrieves and indexes blockchain data for user transactions and balances; ensures data consistency and cryptographic validation	The Graph, graphql-request (querying), Dgraph (performance optimization)
Scaling applications	Enhances scalability by offloading computations to sidechains or Layer 2 networks; ensures secure synchronization and proactive patching	zkSync, Optimism; zkSync SDK, PlasmaJS (scaling integrations); ConsenSys Quorum (commercial scaling)
Transaction monitoring tools	Tracks and analyzes on-chain activities for compliance and security; provides real-time alerts and risk scoring for suspicious transactions	ETX; Chainalysis, Elliptic (commercial tools)

After the critical assets are identified, from an architectural design perspective, the next step is to capture these assets in scope in an architecture diagram that is where all components of the architecture are shown to interact to each other through interfaces such as APIs and communication protocols as well as with the user. The architectural diagram helps define the attack surface by outlining exposed components, such as APIs, smart contracts, data stores, or communication channels with the blockchain. By understanding how these components interact and depend on each other, potential vulnerabilities in their connections can be analyzed in later stages of the threat modeling assessment.

In the case of a DeFi dApp for lending and borrowing, the architectural diagram needs to illustrate the components of the architecture involved in lending and borrowing functionalities, highlighting their component-to-component interactions and integration points. Each component within the system collaborates to ensure secure, scalable, and decentralized financial services while maintaining a strong focus on user experience and compliance. Figure 3.1 depicts these interactions and connections using tools such as Draw.io.

At the forefront, the UI or mobile interface acts as the primary interface for user interactions, enabling access to features such as initiating loans, depositing collateral, or tracking account balances. This interface communicates directly with the dApp frontend application, which processes user inputs and facilitates secure interactions with other components of dApp such as digital wallets, the distributed identity system, the distributed file storage, and the various middleware and backend components (i.e. blockchain platform, smart contracts). The communication relies on robust protocols like JSON-RPC for connecting to blockchain nodes and APIs.

Identity solutions play a critical role in managing user authentication and safeguarding personal information. These systems leverage DID protocols like

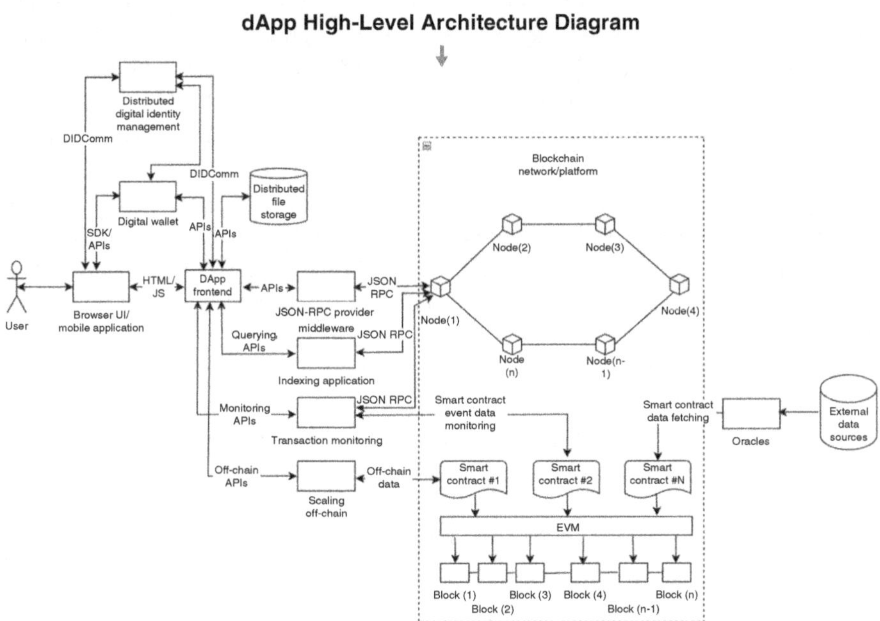

Figure 3.1 dApp high-level architecture diagram.

DIDComm, enabling secure, P2P communication for identity verification without exposing sensitive data. By integrating identity management with OAuth 2.0 or OpenID Connect where needed, the system ensures a seamless authentication experience while adhering to regulatory requirements.

Beneath the user-facing layers, the blockchain serves as the backbone of the system, hosting smart contracts that automate core business logic. These contracts govern lending and borrowing operations, including interest accrual, collateral management, and liquidation processes. To make informed decisions, the smart contracts rely on data fetched from oracles such as Chainlink or Band Protocol, which provide accurate and tamperproof external information, such as asset prices. These external inputs are essential for maintaining fair and efficient operations within the dApp.

The architecture also incorporates distributed storage systems to handle user data, transaction metadata, and other application-critical information in a decentralized and reliable manner. By storing only essential references or hashes on the blockchain, these systems achieve scalability without compromising data integrity.

To enhance performance and scalability, off-chain solutions, such as rollups or sidechains, manage a significant portion of computational and transactional workload. Transactions are aggregated and validated off-chain, with only the final proofs or aggregated data being submitted to the blockchain. This reduces congestion and lowers costs while ensuring security through cryptographic proofs like zk-SNARKs. Off-chain computation frameworks further optimize complex operations, allowing the blockchain layer to remain lightweight and efficient.

For querying and monitoring, indexing applications play a pivotal role. They organize blockchain data for efficient retrieval, ensuring that UIs can quickly display information such as loan statuses or transaction histories. Meanwhile, transaction monitoring tools provide compliance and security oversight by analyzing smart contract events and user activity. These tools work in tandem with event-driven mechanisms, listening to smart contract updates or flagged anomalies for real-time insights.

The entire system is underpinned by digital wallets that serve as secure repositories for user private keys and as interfaces for blockchain interactions. They empower users to sign transactions, authorize smart contract calls, and manage their funds. Wallet APIs ensure seamless integration with both mobile and desktop platforms.

As a subsequent step in the threat modeling process, DFDs will provide a more granular view of component data interactions and capture the detailed flow of data between external entities, architectural components, running processes, and data stores. DFD analysis helps to analyze how the data is being transmitted, processed, or stored and exposes areas of risk, especially when this data crosses trust boundaries. These diagrams are particularly valuable for identifying entry points in the system where potential attacks could be initiated.

3.2.3.4 Stage III – Application Decomposition and Analysis

ADA, the third stage of the PASTA threat modeling methodology, focuses on deconstructing the dApp's architecture to gain a detailed understanding of its components, workflows, and interactions. By analyzing these elements through DFDs and system interaction mappings, ADA lays the foundation for identifying security-critical areas, defining trust boundaries, and pinpointing potential entry points for threats. This process evaluates the system's attack surface as a whole, exposing vulnerabilities and AVs that could compromise the application. By systematically analyzing the architecture, workflows, and interactions, ADA ensures the dApp's decentralized and blockchain-based design is fully accounted for.

In the context of a DeFi dApp for lending and borrowing, ADA begins by identifying the technical domains and their interactions. The architecture includes user-facing interfaces such as browser-based and mobile applications, which handle user authentication, input, and interaction. These interfaces connect to the dApp's frontend services, which facilitate data flows between users, the blockchain network, and external systems.

The architecture of the lending and borrowing dApp consists of several critical components, each integral to maintaining the system's functionality, security, and scalability. Decomposing the application into its architectural components highlights these key elements, necessitating a deeper analysis of their interactions and data flows, which can be effectively mapped and examined using DFDs. At a high level, these components were identified during the technical scoping phase of threat modeling in Stage II. This identification was based on the documentation of a high-level architecture diagram of the DeFi dApp, which was also created and recorded as part of the process.

To someone new to modeling, it's helpful to understand that an architecture diagram provides a high-level view of a system's structure. It shows the main components such as frontends, APIs, smart contracts, databases, and third-party services, and how they interact or are deployed. It answers the question: what are the parts of the system, and how are they connected?

Once you've depicted that high-level architecture and defined the scope, you can then derive a DFD. A DFD focuses on how data moves through the system – between users, processes, data stores, and external entities. It helps you understand the paths that data takes, what kind of data is being exchanged, and where trust boundaries exist. This is particularly useful for identifying security concerns, data exposure, and potential threats.

For example, in a DeFi application, the architecture diagram might show a Web3 frontend, backend API, smart contracts on a blockchain, a price oracle, and decentralized storage. From that, you can derive a DFD that maps how user loan requests are sent to the backend, how those requests are validated and passed to the smart contract, how price data flows in from the oracle, and how lending records are stored or accessed. The DFD makes it easier to reason about where validation, authentication, or encryption should happen, and where data might be vulnerable.

In short, architecture diagrams help you understand the structure of a system, while DFDs help you understand the behavior of that system through the movement of data. Typically, you start with the architecture to establish context, and then use the DFD to drill into how the system functions and where security controls should be applied.

During the ADA phase, we aim to gain a deep understanding of the functionalities of the components as functional assets, the types of processes they host or invoke, and the nature of their data interactions, including both app-to-app and user-to-app interactions. This analysis lays the groundwork for documenting these details through the creation of detailed DFDs.

Before proceeding to draw the DFDs, we will first delve into the components identified as within the scope of the analysis. This involves examining their core functionalities, the processes they support, and how they interact with other components through data interfaces. This exploration will provide the necessary foundation for accurately mapping the system's data flows and interactions.

1. **Users:** At its core, the dApp interacts with users, including borrowers, lenders, and administrators, each with distinct roles and permissions. Borrowers and lenders engage with the dApp to perform specific financial transactions related to lending and borrowing, such as submitting loan requests, approving loan offers, providing collateral, repaying loans, or earning interest on lent funds. Administrators oversee the system, managing operational functions like user onboarding, monitoring transactions, updating configurations, and ensuring security compliance. User credentials, roles, and access rights play a crucial role in maintaining the integrity of the system. Borrowers and lenders access functionalities tailored to their roles, such as initiating loan requests, approving loans, and viewing loan histories. Administrators have higher-level access for managing the platform, such as adjusting lending parameters, reviewing logs, and managing smart contract updates. These permissions ensure that users can only perform actions aligned with their assigned roles, maintaining the security and functionality of the DeFi lending and borrowing ecosystem.
2. **UI Browser-Based or Mobile Application-Based:** The UI serves as the primary interaction point for users, providing an accessible and intuitive platform for engaging with the DeFi lending and borrowing dApp. Available as either a browser-based platform or a mobile application, the UI caters to

diverse user preferences, ensuring usability and convenience across devices. Through the UI, users can perform essential actions such as initiating loan requests, approving or declining loan offers, managing repayments, and handling collateral submissions or withdrawals. Additionally, the UI facilitates viewing transaction histories, monitoring loan statuses, and accessing personalized dashboards for real-time updates on borrowing or lending activities. The UI acts as a gateway between the user and the dApp's frontend, seamlessly connecting to the blockchain-powered backend via secure APIs. It validates user inputs, manages session authentication, and relays transaction data to the dApp's frontend for further processing. This connection ensures that all user-initiated actions are securely transmitted to the underlying blockchain, enabling smart contract execution and state updates while providing feedback to the user through notifications or status changes. Moreover, the UI incorporates user-centric security features, such as two-factor authentication (2FA), biometric verification (for mobile apps), and RBACs, ensuring that only authorized users can access specific functionalities.

3. **Frontend dApp Component:** The frontend dApp component is a critical hub within this architecture. It interfaces directly with the UI and facilitates interactions with the distributed digital identity solution (such as uPort). This integration supports credential verification, authentication, and user authorization, ensuring that only legitimate users can access the dApp's functionality. Additionally, the frontend integrates with digital wallets, including hardware wallets, allowing users to securely store private keys, sign transactions, and manage blockchain interactions seamlessly. These interactions are further secured by robust cryptographic protocols.
4. **APIs:** In a dApp the frontend component interacts with various APIs that enable integration with external systems. For example, it utilizes JSON-RPC APIs to communicate with blockchain nodes, ensuring that transactions are submitted, verified, and recorded accurately. The frontend of a dApp interacts with APIs using various authentication mechanisms to ensure secure and authorized communication with the backend. When interacting with services like Infura or Alchemy, which are popular providers of blockchain JSON-RPC endpoints, the frontend often relies on API key authentication. In these cases, a unique API key is included in the request headers to verify that the request originates from an authorized source. For APIs requiring more granular access control, the frontend may use OAuth 2.0, obtaining access tokens after user authentication. These tokens are included in API requests to platforms such as indexing services or transaction monitoring tools, granting access based on assigned permissions. Some APIs, particularly those related to blockchain interactions like oracles or Layer 2 solutions, require requests to be signed with a private key. This ensures that only the key owner can

send valid requests, with the signature being verified by the backend using the corresponding public key. In scenarios involving user-specific resources, the frontend often employs token-based authentication. Distributed identity solutions or wallets authenticate the user and issue a token, such as a JSON Web Token (JWT), which is then used for subsequent API calls. For highly secure interactions, mutual TLS (mTLS) may be employed, requiring both the client and server to present and verify TLS certificates, ensuring a secure connection.

5. **Decentralized (or Distributed) Digital Identity Systems:** The distributed digital identity system ensures that users are authenticated securely before accessing the application. By leveraging technologies like DIDs and verifiable credentials (VCs), the identity system integrates seamlessly with the dApp's frontend and the UI. This not only reduces the risk of identity spoofing but also enhances user privacy by minimizing the exposure of sensitive data. Solutions such as IPFS, Filecoin, and Arweave are commonly used, offering P2P data sharing, financial incentives for storage, and permanent data retention.
6. **Distributed File Storage System:** The distributed file storage system plays a critical role in dApps by securely storing user data and application-related metadata in a decentralized manner, promoting resilience, transparency, and censorship resistance. This system is typically accessed via HTTPS or APIs, allowing the dApp's frontend and other components to interact seamlessly. Distributed file storage solutions like IPFS, Filecoin, and Arweave are commonly used for dApps. These platforms ensure that data is distributed across multiple nodes, reducing the risk of centralized points of failure and enhancing the system's overall robustness. IPFS is one of the most popular frameworks for decentralized file storage. It uses a P2P network to store and share files, identifying data via unique cryptographic hashes. This ensures that data integrity is preserved and users always retrieve the correct version of a file. Building on IPFS, Filecoin adds a financial incentive layer, allowing users to rent storage space in exchange for cryptocurrency, making it suitable for applications requiring scalable and incentivized storage. Arweave, another distributed storage framework, focuses on permanent data storage, often chosen for applications where data immutability is critical.
7. **Digital Wallets:** Another vital component is the digital wallet, which interacts with users through the UI and integrates with the frontend to enable key management and transaction signing. Secure grade wallets, such as hardware wallets, provide a secure environment for managing private keys and executing cryptographic operations, ensuring the integrity of user-initiated transactions. Some hardware wallets and HSMs used in dApps are FIPS 140-2 compliant [254], providing robust cryptographic security. For instance, ledger devices, such as the Nano X [255] and S+ [256], incorporate a secure element

(ST33J2M0) that meets FIPS 140-2 certification and can integrate with dApps via platforms like MetaMask or WalletConnect. Similarly, the YubiHSM 2 [257], a FIPS 140-2 Level 3 certified HSM, is designed to manage cryptographic operations, including blockchain transaction signing. Enterprise-grade solutions like the Thales Luna HSM offer industrial-strength security and FIPS compliance, making them suitable for managing private keys and blockchain integrations. While consumer-grade wallets focus on usability, enterprise HSMs are ideal for scenarios requiring stringent security measures.

8. **JSON-RPC Blockchain Provider (dApp Middleware):** The dApp middleware serves as a vital bridge between the dApp's frontend and the blockchain platform, enabling seamless interactions such as smart contract executions, transaction submissions, and data retrieval. Middleware translates JSON-RPC calls from the frontend into actionable requests that the blockchain nodes can process, ensuring smooth communication and operation within the dApp's architecture. To establish secure connectivity between the frontend and the middleware, secure communication protocols such as HTTPS or WebSocket Secure (WSS) are employed. These protocols encrypt data in transit, safeguarding sensitive information such as user credentials, transaction details, and API keys from interception or tampering. In addition to encryption, mTLS can be implemented for enhanced security, ensuring both the frontend and middleware authenticate each other before establishing a connection. On the backend, the middleware connects to the blockchain network nodes using similarly robust security measures. Encrypted RPC communication ensures that requests and responses between the middleware and blockchain nodes remain confidential and tamperproof. For sensitive operations such as smart contract executions or transaction verifications, additional authentication mechanisms, like signing requests with cryptographic keys, can provide an extra layer of security. Middleware providers like Infura, Alchemy, and QuickNode offer built-in security features, including encrypted communication channels and access control mechanisms, to further protect these interactions. For developers building custom middleware solutions, libraries such as web3-provider-engine or rpc-websockets enable the creation of secure and efficient connections by supporting advanced features like request validation, caching, and real-time updates via WebSockets.
9. **Blockchain Platform and Network:** The blockchain platform and network form the backbone of a dApp, providing the essential infrastructure for secure, transparent, and immutable operations. The backend of the dApp is closely tied to this platform, consisting of blockchain nodes that manage critical tasks such as transaction submissions, verification, and state updates. These nodes are responsible for maintaining the decentralized ledger and ensuring the integrity of the blockchain. The nodes interface with the frontend of

the dApp through JSON-RPC middleware, which acts as a communication layer. This middleware enables the frontend to interact with blockchain functionalities, such as submitting signed transactions, querying the ledger for specific data, and retrieving smart contract states. For example, when a user initiates a loan request, the frontend communicates with the blockchain node via JSON-RPC to validate the transaction, update the ledger, and execute the associated smart contract logic. Popular blockchain platforms widely used for building dApps include Ethereum, BSC, Solana, Polygon, and Avalanche.

10. **Smart Contracts:** The blockchain platform also hosts the dApp's smart contracts, which are responsible for automating core operations, including lending, borrowing, collateral management, and transaction monitoring. Hosted on blockchain platforms, these self-executing programs ensure trustless, transparent, and secure workflows. Popular programming languages for smart contracts include Solidity, widely used on Ethereum and compatible chains; Vyper, for security-focused contracts; Rust, favored on high-performance platforms like Solana; and Move, designed for resource management on emerging blockchains like Aptos. Frameworks such as Hardhat, Truffle, Remix, and Anchor simplify development, testing, and deployment processes, while libraries like OpenZeppelin provide secure, reusable contract components. Blockchain platforms like Ethereum, BSC, Polygon, Solana, and Avalanche host these contracts, offering unique features such as scalability, low transaction costs, and high performance. Smart contracts interact with the dApp frontend via libraries like web3.js or ethers.js, enabling seamless execution of decentralized workflows and ensuring secure, efficient operations within the dApp ecosystem.
11. **Indexing Applications:** Indexing applications play a critical role in ensuring the dApp can query and display blockchain data efficiently. These applications interface with the blockchain via APIs and indexing nodes, allowing users to access real-time data such as transaction histories, loan statuses, and contract interactions. Indexing applications are essential for ensuring that a dApp can efficiently query and display blockchain data in a user-friendly and accessible manner. They address the inherent limitations of querying data directly from blockchain networks, which can be slow and resource-intensive due to the decentralized and immutable nature of blockchains. By creating structured and searchable data repositories, indexing applications provide dApps with real-time access to critical information such as transaction histories, loan statuses, and smart contract interactions. These applications work by interfacing with the blockchain through APIs and specialized indexing nodes. Indexing nodes continuously monitor the blockchain for new transactions and events, extracting relevant data and organizing it into a structured format. This data is then made available to the dApp through APIs, enabling efficient queries

and retrieval. For example, when a user of a lending and borrowing dApp wants to view their loan status, the frontend queries the indexing application, which returns the relevant information almost instantly, eliminating the need for the dApp to parse the blockchain directly. The Graph is one of the most popular indexing solutions for blockchain-based applications. It uses a GraphQL-based interface to allow dApps to query specific data points from the blockchain, such as account balances, transaction details, or contract states. Developers define subgraphs, which are custom schemas that specify the types of data the application needs to index. These subgraphs make it easier to retrieve targeted information, ensuring that the dApp remains responsive and efficient.

12. **Scaling Applications:** Similarly, the architecture includes a scaling application that improves the dApp's performance by offloading computationally intensive operations to off-chain platforms. These scaling solutions integrate with smart contracts and the blockchain to synchronize states and ensure seamless operation. These solutions enable dApps to handle larger user bases and higher transaction volumes while reducing latency and costs, all without compromising the decentralized nature of the system. Popular scaling techniques include rollups, state channels, and sidechains, each providing unique approaches to improving efficiency and scalability. Rollups are a widely used method that bundles multiple transactions off-chain into a single transaction, which is then committed to the blockchain. Optimistic rollups operate on the assumption that transactions are valid unless proven otherwise, while ZK-rollups use cryptographic proofs to verify transaction validity securely and efficiently. State channels offer another solution by allowing participants to conduct multiple transactions off-chain while only recording the final state on-chain. This approach minimizes the number of blockchain interactions, reducing fees and improving transaction speed. Developer tools like Alchemy SDK, QuickNode, and Infura provide the necessary infrastructure for connecting dApps to scaling solutions, ensuring seamless integration and reliable operation. By incorporating these technologies, dApps achieve greater scalability, deliver lower transaction costs, and enhance the user experience while maintaining the transparency and security of decentralized systems.
13. **Transaction Monitoring Applications:** Transaction monitoring applications are vital for enhancing transparency, security, and user confidence in dApps by providing real-time tracking of transactions. These applications interface with the dApp's frontend via APIs, monitoring blockchain interactions such as smart contract executions and transaction confirmations. They notify users about transaction statuses, such as pending, confirmed, or failed, and help ensure informed decision-making. For example, in a lending

and borrowing dApp, a transaction monitoring application might track the progress of a loan request. When a user submits a transaction, the application monitors its status on the blockchain, notifying the user in real time when the transaction is mined, confirmed, or rejected. Similarly, it can provide updates on collateral liquidation events or repayment confirmations, ensuring users are always aware of their financial status. Security is a significant aspect of transaction monitoring applications. These systems are designed to detect anomalies or suspicious activities, such as failed transactions, unusually high gas fees, or unexpected contract interactions. By identifying these issues promptly, the application can alert users and, in some cases, prevent further actions that might result in losses or errors. Transaction monitoring applications also play a role in compliance and auditability. In use cases requiring adherence to regulatory frameworks, these tools can provide detailed logs of transaction histories, user actions, and smart contract interactions. This is particularly important for financial dApps, where maintaining a transparent and auditable record is essential. Popular tools like Etherscan and Tenderly integrate seamlessly to track and debug transactions, while subscription models like webhooks and WebSocket connections enable real-time updates. These applications also enhance security by detecting anomalies, support compliance by maintaining auditable logs, and contribute to the overall reliability and usability of the dApp, ensuring smooth and trustworthy operations within the decentralized ecosystem.

14. **Oracles:** The architecture integrates oracles to connect the blockchain with external data sources, enabling dApps to access essential off-chain information. Oracles act as secure bridges, fetching, verifying, and delivering data such as asset prices, interest rates, and market conditions. This functionality is critical for lending and borrowing dApps, where real-time data directly impacts financial transactions and risk management. For example, oracles provide price feeds to monitor collateral values, triggering liquidation events when necessary, or dynamically updating interest rates based on market changes. Oracles work through smart contracts that receive and verify data before executing predefined operations. For instance, a price oracle might supply Ethereum's current value to a smart contract, which recalculates collateral ratios and determines whether additional collateral is required or liquidation should occur. Decentralized oracle networks like Chainlink and Band Protocol enhance reliability by aggregating data from multiple sources, achieving consensus to ensure accuracy and mitigate manipulation risks. To secure the connection, oracles use cryptographic techniques such as signed data packages and secure communication protocols like HTTPS or mTLS, ensuring data integrity and confidentiality. Advanced methods

like TLSNotary further authenticate data retrieved from HTTPS endpoints. First-party oracles, such as API3, offer direct data feeds from providers, reducing latency and improving reliability for high-frequency updates like real-time market monitoring. Developers rely on frameworks like Hardhat, Truffle, and Foundry to build and test oracle integrations within smart contracts. Platforms like Chainlink provide tools such as prebuilt data feeds and verifiable random functions (VRFs) to simplify integration. These technologies ensure seamless and secure interactions between oracles and smart contracts, enabling the dApp to operate dynamically and reliably in response to external market conditions.

The next step, after identifying and analyzing the critical components of the dApp application, involves creating a detailed map of workflows and data flows within the system using DFDs. This process focuses on understanding how information moves between components and pinpointing where sensitive data, such as private keys and transaction details, is stored or processed. The DFD analysis is critical for identifying critical trust boundaries, such as those between users and the application, the application and blockchain nodes, and the dApp and external APIs. Understanding these boundaries is essential for assessing how threats might propagate through the system and prioritizing areas requiring robust security controls.

For the DeFi dApp, ADA also emphasizes the identification of critical components, including smart contracts managing financial transactions, digital wallets safeguarding private keys, and APIs enabling external integrations. These components, due to their functional and security-critical roles, are prioritized for in-depth analysis and rigorous testing.

Creating DFDs: Before documenting the DFD, it is important to understand what DFDs represent and how they can be created from a use case (Table 3.4). A DFD helps analyze data flows and interactions between architecture components, represented as process circles, and entities such as users, systems, application components, and data stores. In the context of threat modeling, DFD analysis focuses on identifying trust boundaries crossed by data interactions. These boundaries often represent entry points for potential AVs that threat agents could exploit, leading to impacts such as data compromise, information disclosure, spoofing, tampering, repudiation, privilege escalation, or DoS. A DFD is a powerful tool for visually representing the movement of data within a system. It provides a clear picture of how data flows between processes, data stores, and external entities, offering an overview of the system's structure and interactions. DFDs are particularly valuable in system design and threat modeling, where they help analyze workflows, identify potential risks, and optimize secure and efficient operations.

Table 3.4 DFD legend – symbols and components.

Symbol	Component	Description	Example
Rectangle	External entity	Represents an external entity, which is any system, person, organization, or device outside the scope of the system being modeled that sends data to or receives data from the system	User or customer, a third-party payment processor, an external KYC service, another system outside the dApp's trust boundary
Circle	Process	Represents an activity or function where data is processed or transformed	Authenticate user, validate loan
Parallel lines	Data store	Represents a location where data is stored persistently within the system	Credential database, blockchain node
Arrow	Data flow	Indicates the flow of data between entities, processes, and data stores, showing the direction of flow *Note: DFD arrows are typically unidirectional, and bidirectional flows should be shown as two separate, clearly labeled arrows to ensure clarity, traceability, and accurate threat modeling.*	Credentials flow, transaction data flow
Dashed line	Trust boundary	Separates trusted components from untrusted components within the system	User outside versus backend processes
Blue label/ text	Entry point	Marks where external entities interact with the system, providing inputs or initiating requests	User submitting credentials

In a DFD, key components are represented by specific symbols, each serving a unique purpose. Processes are depicted as circles or ovals and represent functions or activities where data is processed, transformed, or utilized. For instance, in a dApp, a process might include actions like "Authenticate User" or "Sign Transaction." Data flows, shown as arrows, indicate the movement of data between components in the system. These flows point in the direction of data transfer, such as from a user to a process or from a process to a data store. External entities, represented as rectangles, illustrate sources or destinations of data that exist outside the system, like users, third-party APIs, or external services such as oracles or identity providers. Data stores, symbolized by two parallel horizontal lines or an open rectangle, represent locations where data is stored, such as blockchain nodes or off-chain databases. In the context of blockchain systems, this can refer to on-chain storage within blockchain nodes or off-chain storage solutions such as external databases. It's important to note that while these symbols are standard in diagrammatic representations like DFDs, the term "process" or "store" may have different meanings depending on the audience; for example, developers may think of multithreading or memory structures, while business users may interpret it as steps in a workflow. Therefore, it's helpful to clarify that in this context, "Data Store" specifically refers to persistent storage locations, not computational processes or business steps.

These symbols work together to create a comprehensive view of a system's data interactions. For example, a user (external entity) might submit credentials to a process like "Authenticate User." This process would retrieve stored credentials from a data store to validate the input and then send the validation outcome back to the user or forward it to another process for further action.

In the context of a lending and borrowing dApp, DFDs might illustrate the following interactions:

1. A user submits a loan request through the UI, which flows into a process such as "Validate Loan Request."
2. This process interacts with a data store, like a blockchain node, to fetch the user's balance and collateral data.
3. Once validated, the data is sent to another process, such as "Calculate Loan Terms," which interacts with an oracle (external entity) to fetch the current asset price.
4. The results of this calculation are then stored in a data store, such as the smart contract state, and communicated back to the user.

This flow provides a comprehensive DFD structure to model how data moves through the lending and borrowing dApp, highlighting key interactions and dependencies between system components.

In Figure 3.2, we present the high-level DFD for the DeFi lending and borrowing dApp highlighting basic business logic and interactions among external entities like the user and the oracle processes to validate loan requests and calculate loan terms and data storages like the one located within the blockchain node and the smart contracts. To create DFD diagrams like the one depicted you can utilize drawing tools like Draw.io, utilizing the tools DFD stencils to visualize the system's data flows and various interactions among components.

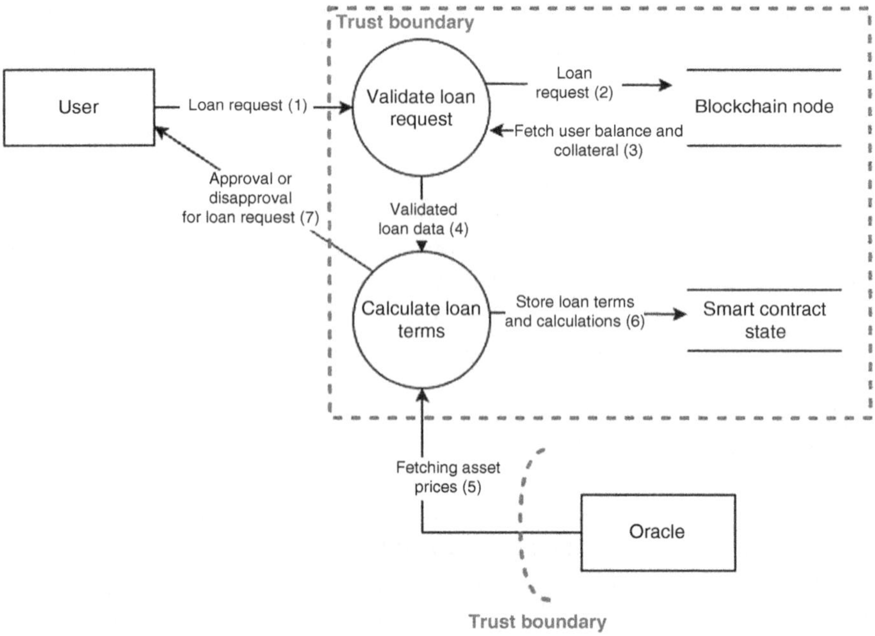

Figure 3.2 DFD of the dApp loan and borrowing use case.

Conducting a DFD analysis for a dApp focused on DeFi lending and borrowing is essential for understanding how data flows through the system, identifying potential security risks, and optimizing system integration. As a security architect tasked with this responsibility, here's how you can approach the analysis and structure key use cases for the DFD capturing in scope the key assets previously identified:

1. **User Interaction Flows:** The user interaction flow focuses on how users engage with the dApp's components, particularly through the UI. These flows are specific to the functionality of the dApp. For example, when a user interacts with the dApp UI to create a transaction, such as submitting a loan request, the process is abstracted from the user. After the user's identity is verified, they do not directly interact with the background components. Instead, the dApp calls an API to sign the transaction using the private keys securely stored in the digital wallet. Once the transaction is signed, it is submitted to the blockchain network for validation. Following initial validation by signature and by consensus, the transaction is executed via a smart contract to process the loan request. The smart contract evaluates the loan request based on programmatic rules and relevant data, such as the value of

the asset being borrowed and the collateral provided. For instance, the smart contract ensures that the collateral meets predefined criteria to allow the requested asset to be borrowed. If the conditions are satisfied, the transaction is approved, resulting in the loan being issued. This seamless interaction flow ensures traceability of the user interactions with front components like the UI that is the dApp client and by application to application components (e.g. frontend, middleware, and blockchain-platform/network as backend) as well as all the other entities and components of the dApp (e.g. digital wallet, credential identity validation systems, oracles, indexing, scaling, and transaction monitoring components).

2. **APIs Interaction Flows:** The API flow in a dApp is pivotal, particularly for the dApp frontend, which relies extensively on APIs to interact with key components such as the digital wallet, middleware, DID management systems, and other backend services. Given the central role of APIs in facilitating these interactions, they represent a critical area for optimization and security enhancement. This flow analyzes the interaction between the dApp's frontend and its backend components, focusing on how APIs enable communication with the blockchain and other systems. APIs handle crucial tasks such as submitting blockchain transactions, querying blockchain data, and retrieving user-specific information, including balances, loan statuses, and transaction histories. The analysis also examines how the frontend communicates with DID management APIs to verify user credentials and with wallet APIs to facilitate transaction signing and private key management. A robust DFD should capture these API interactions, emphasizing the role of the frontend in coordinating requests and responses across the dApp's ecosystem. For example, the DFD should illustrate how the frontend submits signed transactions to the middleware via authenticated API calls, enabling them to be processed by the blockchain. Similarly, it should detail how the frontend queries real-time transaction updates, fetches market prices and interest rates through oracle APIs, and dynamically feeds this data into smart contracts for decision-making.
3. **Blockchain Interaction Flows:** The blockchain interaction flows detail how smart contracts and blockchain nodes are utilized to execute lending and borrowing operations, manage collateral, and verify transactions. Key components include the blockchain nodes, which process transactions and maintain the decentralized ledger, and the smart contracts that define and enforce lending and borrowing logic. This flow examines how the frontend communicates with blockchain nodes through middleware for submitting transactions, invoking smart contracts, and retrieving state updates. It also illustrates how event logs emitted by smart contracts are captured by indexing applications and fed back to the frontend to provide users with real-time data about their loan statuses or collateral positions.

4. **Oracle Integration Flow:** The oracle integration can be analyzed in a DFD to examine how the dApp retrieves and integrates external data into its blockchain ecosystem. Oracles like Chainlink or API3 are key components, serving as intermediaries that fetch data such as asset prices and interest rates from external sources. Smart contracts then use this data to dynamically update collateral ratios, calculate interest rates, or trigger liquidations. This flow visualizes how external data moves from APIs to oracles, then to blockchain nodes, and finally into smart contracts. The interaction between oracles and the frontend is also highlighted, showing how data fetched through oracles is displayed to users for real-time insights. The DFD analysis ensures the reliability and security of these data flows, critical for maintaining accurate and trustworthy lending and borrowing operations.

Additionally if implemented as part of the dApp design:

5. **Layer 2 Data Flows:** Scaling in dApps consists of the ability to handle high transaction volumes efficiently by leveraging off-chain and on-chain solutions. Key components include Layer 2 scaling solutions like zkSync or Polygon, which bundle transactions off-chain, reducing the load on the main blockchain, and the blockchain nodes that synchronize state updates between the two layers. This flow explores how the frontend interacts with scaling APIs to offload computationally intensive tasks while ensuring data consistency. It also examines the communication between smart contracts and scaling solutions to validate and commit off-chain operations back to the main blockchain, ensuring seamless integration and improved performance.
6. **Transaction Monitoring Flows:** The transaction monitoring flow analyzes how the dApp tracks and displays the status of blockchain transactions. Key components include transaction monitoring applications, JSON-RPC middleware, and blockchain nodes. This flow visualizes how transaction monitoring applications retrieve real-time updates from the blockchain and communicate these updates to the frontend dApp component. It ensures users can track their transactions, such as loan issuances or repayments, from initiation to completion. Additionally, the flow highlights how monitoring tools verify transaction statuses using event logs and data from smart contracts, providing users with transparency and confidence in the dApp's operations.

Once you become familiar with basic DFD diagramming, you can start practicing diagramming DFDs to depict the various types of dApp processes, the interactions between users and components, as well as the data flows that is the data that is sent and received between processes and entities.

As an example we created in Figure 3.3 a DFD for the transaction signing use case of the lending and borrowing dApp. This DFD illustrates the

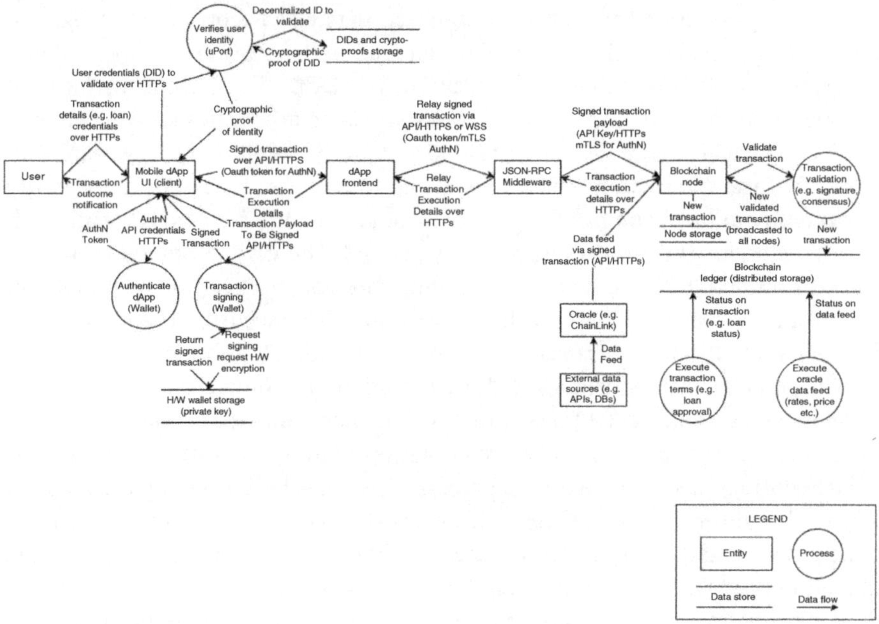

Figure 3.3 dApp lending and borrowing transaction DFD.

interactions with the primary components of the architecture, including processes, entities, data stores, and data flows, represented using standard DFD notation. Additionally, we have incorporated trust boundaries and identified the entry points, which serve as key inputs for a comprehensive threat modeling exercise. These elements enable a more detailed analysis of potential threats at the architectural component level.

The flow begins with the user, represented as an external entity, interacting with the mobile dApp UI, which is depicted as a separate component. Through the dApp client, such as a mobile dApp and its UI, the user provides transaction details, such as the loan amount they wish to borrow, along with digital credentials for authentication. This user-to-dApp client interaction ensures secure data transmission using HTTPS.

The mobile dApp UI transmits the user's credentials, formatted as a decentralized identifier, to a process within the distributed digital identity component (e.g. uPort) for validation. Depicted as a process in the DFD, the identity validation mechanism verifies the DID by retrieving its stored value and issuing a cryptographic proof of validation. This proof is returned to the mobile dApp UI, confirming the user's identity and granting access to the blockchain dApp.

Following identity verification, the transaction payload is prepared for signing. The signing process involves a digital wallet with both a software component and a hardware wallet for enhanced security. In the DFD, the software wallet handles transaction signing, with the payload sent via an authenticated API call over HTTPS. The wallet accesses the private key securely stored in the hardware wallet to sign the transaction. Once signed, the transaction is sent from the dApp client entity to the dApp frontend entity through an authenticated API call, such as one using an OAuth token secured by HTTPS. The dApp frontend relays the signed transaction to the JSON-RPC middleware via another authenticated API call, using mechanisms like OAuth tokens or mTLS for security. The middleware then forwards the signed transaction to the blockchain platform for processing.

Within the blockchain platform, the transaction first undergoes signature verification, depicted in the DFD as a process. It is then validated through the blockchain's consensus mechanism before being recorded in the distributed ledger. A smart contract, also depicted as a process, executes the business logic for loan approval, running as a program within the blockchain platform (e.g. on the Ethereum Virtual Machine). Upon execution, the smart contract updates its state, changing the loan status from "pending" to "active." The final state of the validated transaction, including the updated loan status, is recorded in the distributed ledger. A copy of the updated blockchain, including the new transaction, is broadcast to all participating nodes in the network, including the node that originated the transaction.

When the blockchain platform integrates with oracles to fetch real-time market data, such as interest rates or cryptocurrency prices, the oracle is depicted in the DFD as an external entity. The oracle interacts with external data feeds and the blockchain platform through a signed transaction. Once validated by the blockchain platform, the oracle's signed transaction triggers a smart contract to update the data feed. The updated feed status is recorded in the distributed ledger. After all processes are complete, the final transaction status, such as loan approval and activation, is relayed back to the user as a notification. This notification follows the reverse pathway through the JSON-RPC middleware, dApp frontend, and mobile dApp UI.

The DFD also highlights trust boundaries (indicated by red dotted lines) that are crossed by data flows among various components of the dApp. These trust boundaries mark points of interaction between components with different levels of trust, such as between external users and the system or between the frontend application and the blockchain backend. These boundaries are crucial for identifying potential vulnerabilities, as they represent the system's primary exposure to external threats. For example, trust boundaries in this dApp include user interactions with the dApp client, software and hardware wallets, distributed digital identity systems, JSON-RPC middleware, the blockchain platform, and external oracles. Within the blockchain platform, the local node and smart contracts are considered distinct trust boundaries due to their specific security and operational roles.

Entry points, such as UIs, APIs, and protocols (e.g. DIDConnect, WSS), enable data to flow into and out of the dApp components, including wallets, distributed identity systems, middleware, oracles, blockchain nodes, and smart contracts. While these entry points are essential for functionality, they also present potential attack surfaces where vulnerabilities may be exploited. By systematically analyzing the data flows that cross these trust boundaries, we can identify threats, vulnerabilities, and AVs specific to each boundary. This analysis will form the foundation for the next stages of the threat modeling exercise.

It should be noted that in addition to DFDs, other diagram types such as use case diagrams and sequence diagrams can also be employed to analyze the transaction signature process used by the dApp. Use case diagrams help capture the functional interactions between users and system components, while sequence diagrams detail the step-by-step flow of messages and actions during transaction signing. Together, these visualizations provide a comprehensive view of the process, helping to identify potential vulnerabilities and ensuring that each stage of the transaction life cycle is properly secured.

A final note on using diagramming tools for creating DFDs, such as Draw.io, MS Visio, or threat modeling tools with built-in DFD capabilities (e.g. OWASP Threat Dragon): it is important to emphasize that creating DFDs is a manually intensive process. While DFDs are highly recommended as a starting point for a baseline threat model, they may not be ideal for ongoing, real-time changes to the architecture. When new components are added to the system, their processes, entities, and data interactions require updating the DFD to maintain an accurate and comprehensive view. This can be a time-consuming task, particularly in dynamic projects with frequent architectural revisions. Manual generation of DFDs can become a bottleneck in rapidly evolving systems where frequent updates are necessary. One solution is to automate DFD generation. For example, a YAML-based modeling tool can integrate with your architecture repository to update diagrams automatically. Additionally, runtime tools can extract traffic data from the deployed architecture, dynamically updating the DFD to reflect current operational flows. This dual approach ensures that DFDs remain current and relevant, eliminating delays in risk assessment and decision-making processes.

In the on-project phase, threat modeling could adopt automation to extract real-time DFDs of the deployed architecture allowing the identification and mapping processes, data stores, and communication flows in their actual run-time state. By doing so, DFD automated extraction can provide an up-to-date representation of the system's components and their interactions, reflecting changes that may not have been part of the initial design. These real-time DFDs serve as an input for advanced threat modeling tools, which integrate with the architecture's dynamic environment. These tools analyze the extracted information to identify processes that are actively running, new components introduced during development, or configuration changes that may introduce vulnerabilities. The analysis is then enriched with updated threat intelligence, allowing for the identification of new risks associated with these components or changes.

This dynamic threat modeling process is particularly critical in the QA staging environment, which is designed to closely replicate the production setup. By mirroring the production environment, the QA stage provides a realistic context for evaluating how new components or changes might interact with the existing system. Automated threat modeling tools in this environment can proactively identify security gaps or risks that could manifest in production, enabling teams to address these issues before deployment.

3.2.3.5 Stage IV – Threat Analysis

The focus of this stage is TA, which involves examining the relationships and interactions between system components such as entities, processes, data flows, and data stores – rather than evaluating components in isolation. According to NIST Special Publication 800-30, TA identifies and evaluates threat sources, such as external actors, internal issues, or environmental factors, along with system vulnerabilities that could be exploited. It also includes impact analysis to assess how realized threats could affect system components, data flows, and processes.

Incorporating TA as a dedicated step in the threat modeling process ensures a thorough evaluation of potential threats by proactively identifying, prioritizing, and mitigating them in a way that aligns with the real-world operational environment. This step is particularly critical for dApps, which operate in decentralized ecosystems where the attack surface extends beyond the external-facing nodes of blockchain networks. It also encompasses entities outside the direct control of the dApp, such as oracles, digital exchanges, and other third-party components. These external dependencies introduce additional risks, as they can be vulnerable to manipulation or exploitation, making a robust and comprehensive TA essential for safeguarding the entire system.

For this stage, the goal is to comprehensively evaluate threats to both custom-developed dApp code and third-party components and the supporting infrastructure ensuring the entire attack surface of the application is thoroughly assessed. This holistic approach considers all layers of the application, including the frontend, middle-tier, and backend components, to ensure no part of the system is overlooked. A strong starting point for TA is to focus on the specific threats faced by the architectural components identified as in scope for the threat modeling process. The following is a detailed overview of the key technical components of the DeFi dApp for lending and borrowing. Each component is essential to the system's architecture and introduces unique vulnerabilities that must be addressed.

The **UI or mobile interface** serves as the primary point of interaction for users, enabling access to lending and borrowing functionalities. This layer faces risks such as phishing attacks, session hijacking, and user input manipulation, which could result in unauthorized access, data breaches, or fraudulent transactions. These vulnerabilities often arise from insufficient authentication mechanisms, insecure session handling, and a lack of input validation. Robust measures, such as implementing MFA, validating user inputs, and securely managing session tokens, are essential to mitigate these risks.

The **frontend application** bridges the UI with blockchain wallets and smart contracts, handling user input and managing API communications. Threats to this layer include API abuse, MitM attacks, and data leakage, which could expose sensitive user data or enable unauthorized actions. Common vulnerabilities stem from weak API authentication, unencrypted communications, and improper data handling. To address these risks, developers must enforce strict API gateway authentication, encrypt data in transit, and implement access control measures. Popular frameworks such as Vue.js, AngularJS, or React offer tools for building responsive frontends but require developers to implement robust security practices to safeguard data flows.

DID solutions manage user authentication and protect PII. These systems face risks like identity fraud, PII breaches, and vulnerabilities related to centralization. Weak integration practices or poor cryptographic controls could lead to data exposure or identity theft. Open-source frameworks like uPort, which use decentralized identifiers and zero-knowledge proofs (ZKPs), offer flexibility but require careful implementation to ensure secure key storage and cryptographic protections. Commercial solutions like uPort and Civic provide SLA-backed identity verification and compliance tools, adding a layer of trust and reliability.

Distributed storage solutions, such as IPFS, store sensitive data like transaction records, loan details, and user activity logs. These systems are vulnerable to data tampering, unauthorized access, and DDoS attacks targeting storage nodes. Risks often arise from unencrypted data storage, insufficient access controls, or single points of failure within distributed nodes. Encrypting data at rest, implementing token-based access controls, and ensuring redundancy within storage nodes are critical measures to secure these systems against threats.

The **JSON-RPC provider middleware** acts as an intermediary between the frontend and blockchain nodes, enabling smart contract interactions and facilitating API communications. Risks associated with middleware include API abuse, data interception, and insufficient access controls, which could allow attackers to intercept or manipulate sensitive data. Vulnerabilities often arise from inadequate API authentication, unencrypted communications, and lack of rate limiting. Providers like Infura and Alchemy offer robust middleware but require developers to implement additional layers of protection, such as monitoring and encrypted communications.

Digital wallets are critical for managing private keys, signing transactions, and authenticating users. They face threats such as phishing attacks, malware infections, and private key exposure, often due to poor user practices or insufficient wallet encryption. Open-source wallets provide transparency and broad adoption but place the responsibility for key security on users. In contrast, commercial platforms offer institutional-grade protections, including MFA, delegated signing, and SLA-backed updates, reducing the risk of private key theft.

The **blockchain layer** forms the backbone of the dApp, hosting smart contracts and enabling decentralized operations. It faces threats such as 51% attacks, node compromise, and scalability challenges. These vulnerabilities can stem from weak consensus mechanisms, outdated node configurations, or network congestion. Developers must ensure rigorous node management and keep nodes updated with the latest patches. Tools like Ganache for testing and Hardhat for development can help developers build secure and scalable blockchain applications. Managed services like Azure Blockchain offer enterprise-grade security, proactive patching, and performance monitoring.

Smart contracts, which automate core functionalities such as lending and borrowing, face risks like reentrancy attacks, logic errors, and unauthorized access. These vulnerabilities often arise from poor code quality, insufficient validation, and inadequate testing. Open-source libraries like OpenZeppelin provide secure contract templates, while tools such as MythX, Slither, and Foundry facilitate testing and auditing. Engaging commercial auditing services like Trail of Bits ensures that smart contracts are rigorously tested before deployment.

Oracles and external data sources provide off-chain data such as asset prices to smart contracts, but they introduce risks like oracle manipulation, inaccurate data, and latency issues. These vulnerabilities often result from reliance on a single data source or insufficient validation mechanisms. Open-source solutions like Chainlink are widely trusted for their decentralization but require careful configuration to mitigate manipulation risks. Commercial solutions like API3 offer SLA-backed assurances, timely patching, and robust validation frameworks to ensure data accuracy and integrity.

Indexing applications, such as the Graph, enable efficient retrieval of blockchain data for user transactions and balances. These applications face threats like data tampering, API exploitation, and inconsistencies between indexed and on-chain data. Vulnerabilities often arise from weak cryptographic validation or insufficient API security. Developers can mitigate these risks by implementing additional cryptographic mechanisms and robust API authentication, while commercial indexing services provide proactive monitoring and SLA-backed reliability.

Scaling applications, such as zkSync or Optimism, enhanced dApp scalability by offloading computations to sidechains or Layer 2 networks. These solutions face risks like cross-chain replay attacks, synchronization failures, and data inconsistencies. Open-source frameworks offer flexibility and customization but require extensive testing to address these vulnerabilities. Commercial solutions like ConsenSys Quorum provide SLA-backed reliability, proactive patching, and safeguards to ensure secure and seamless scaling.

Note: Enhancing scalability through sidechains or Layer 2 means that a blockchain can offload some of its transactions or computations onto secondary networks, which operate alongside the main chain. These additional layers process transactions more quickly and at lower cost, then settle the results back on the main chain, thereby increasing overall throughput without compromising security or decentralization.

Finally, **transaction monitoring tools** ensure that all on-chain activities comply with regulatory standards, preventing malicious actions like money laundering or illicit fund transfers. Risks include failing to detect suspicious activity, which could result in reputational damage or legal penalties. Open-source tools like ETX offer basic analysis, while commercial platforms like Chainalysis provide robust risk scoring, real-time alerts, and SLA-backed updates, enabling comprehensive compliance and security.

By analyzing each component and its unique risks, the architecture design document can guide the development of robust security controls, ensuring a resilient and secure DeFi dApp for lending and borrowing.

Table 3.5 provides a summary of the dApp components, their functionalities, specific threats, and potential security controls or countermeasures to mitigate these threats.

Table 3.5 dApp component level high-level threats and countermeasures.

dApp component	Functionality	Threat exposure	Security controls/ countermeasures
UI or mobile interface	Enables user interaction for lending and borrowing	Phishing attacks, session hijacking, input manipulation	MFA, secure session handling, input validation, anti-phishing measures
Frontend application	Connects UI to wallets and smart contracts	API abuse, MitM attacks, data leakage	API authentication, encrypted communications (HTTPS), rate limiting, secure API gateway
Identity solutions	Manages user authentication and DID	Identity fraud, PII breaches, cryptographic vulnerabilities	ZKPs, secure key storage, SLA-backed identity verification, cryptographic safeguards

(Continued)

Table 3.5 (Continued)

dApp component	Functionality	Threat exposure	Security controls/ countermeasures
Distributed storage	Stores transaction data, loan records, and user activity	Data tampering, unauthorized access, DDoS attacks	Encryption for data at rest, token-based access control, redundancy within distributed nodes
JSON-RPC provider middleware	Bridges frontend with blockchain nodes	API abuse, data interception, insufficient access controls	Monitoring, encrypted communication channels (e.g. HTTPS, mTLS), API rate limiting
Digital wallets	Manages private keys for transaction signing	Phishing, malware, private key exposure	Hardware wallets, delegated signing, MFA, robust wallet encryption
Blockchain layer	Hosts smart contracts and decentralized operations	51% attacks, node compromise, scalability challenges, node software vulnerabilities, local storage risks	Rigorous node management, consensus mechanism security, regular software updates, encryption for local storage, redundant backups
Smart contracts	Automates core lending and borrowing functionalities	Reentrancy attacks, logic errors, unauthorized access	Code audits, secure contract templates (e.g. OpenZeppelin), thorough testing, tools like MythX and Slither
Oracles and external data sources	Provides off-chain data to smart contracts	Oracle manipulation, inaccurate data, latency issues	Decentralized oracle networks, data source validation, secure configuration of oracle nodes
Indexing applications	Retrieves blockchain data for user transactions	Data tampering, API exploitation, inconsistencies between indexed and on-chain data	Cryptographic validation of data, secure API authentication, proactive monitoring of indexed data
Scaling applications	Enhances scalability through sidechains or Layer 2 (see the note)	Cross-chain replay attacks, synchronization failures, data inconsistencies	Secure cross-chain communication protocols, extensive testing, SLA-backed scaling solutions
Transaction monitoring tools	Monitors on-chain activity for compliance	Failure to detect suspicious activity, legal penalties	Risk scoring, real-time alerts, SLA-backed monitoring tools, adherence to global regulatory standards

This initial mapping helps pinpoint security controls that should be integrated into the application's design to mitigate potential risks. However, in the context of a dApp, achieving a truly comprehensive TA requires going beyond this static evaluation. To fully understand the risk landscape, it is crucial to assess threats specific to the application's environment. This includes analyzing the development environment, such as the risks associated with using open-source repositories compared to on-premises infrastructure, as well as the operational environment, which might involve multitenant cloud deployments, cloud-based virtual private networks, virtualized platforms, or legacy servers.

The initial TA focuses on identifying and documenting a list of threats and mitigations, using the trust boundaries defined in the DFD diagram as a framework. These trust boundaries represent critical points where a threat actor may attempt to bypass security controls or exploit vulnerabilities to gain unauthorized access to servers, services, or data assets.

To prevent such breaches, security controls like authentication, authorization, input validation, and session management and audit trail are implemented to enforce the required level of trust for crossing these boundaries. For instance, at entry points such as UI and mobile interfaces, these controls ensure that only legitimate users are granted access, thereby mitigating risks and maintaining the integrity of the system. This approach provides a structured basis for analyzing and addressing potential vulnerabilities at each point of interaction within the application.

The outcome of the TA at this stage can be achieved either manually or with the assistance of threat modeling tools. Tools can automatically associate relevant threats with DFD components based on the specific type of component being analyzed, streamlining the process.

When conducting the analysis manually, leveraging threat frameworks such as STRIDE, which addresses Spoofing, Tampering, Repudiation, Information Disclosure, DoS, and Elevation of Privilege, provides a good enough initial list of threats to be further evaluated. STRIDE helps systematically map potential threats to the components identified in DFDs, ensuring a thorough assessment.

This process, commonly referred to as "threat enumeration" in threat modeling practices, involves systematically identifying and cataloging potential threats to a system, application, or environment. It is a structured approach designed to ensure all possible threats are considered, taking into account the system's architecture, components, data flows, and operational context.

To comprehensively identify threats against the trust boundaries, we analyze each trust boundary individually and enumerate the threats that a potential threat actor might exploit. This begins with the trust boundary between the user and the client dApp and continues systematically through all trust boundaries documented in the DFD, as shown in Figure 3.4, ensuring a complete E2E analysis of potential vulnerabilities.

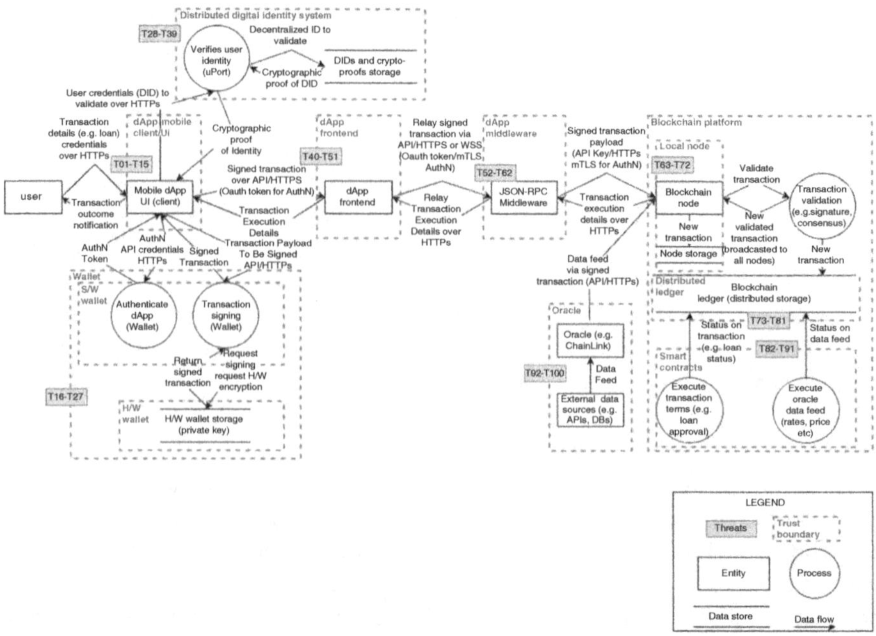

Figure 3.4 dApp lending and borrowing threat model.

For example, to cross a trust boundary, a threat actor or malicious user may initiate specific threat events. A threat event refers to the potential occurrence of a threat where a threat actor targets a specific component. This could involve an external or internal entity depicted in the DFD, a particular process, inbound data in transit to the component, or sensitive data or assets stored within the system.

Each of these threat events highlights the possibility to be realized for impact by AVs targeting the various assets. An AV in threat modeling is a specific pathway or mechanism through which a threat actor can exploit vulnerabilities to compromise a system, application, or component. In the context of threat modeling and its use for attack simulation testing, AVs serve as actionable descriptions that identify how an attack might be initiated, propagated, and executed. They provide the groundwork for simulating real-world attack scenarios to validate the system's defenses. These AVs will be further analyzed in depth during Stage VI: Attack Modeling.

In the analysis of various threats, it is essential to consider the threat actor and the assets impacted. This approach enables a clear definition of risk within the context of risk-centric threat modeling. A threat actor (also known as an adversary) is an individual, group, or entity that is responsible for carrying out malicious actions that pose risks to an organization's systems, networks, data, or applications.

Threat actors can have various motivations, capabilities, and resources, ranging from financial gain to political or ideological goals. These factors in the TA allow the analysis of the risks that can be assessed as the likelihood of a threat actor exploiting a vulnerability or security control gap to cause an impact on the targeted assets.

Table 3.6 categorizes STRIDE threat events, each representing a distinct instance of a threat along with their impacts, AVs, and recommended countermeasures to enhance security at the trust boundary between the user and the mobile UI.

The next TA considers component-specific including technology stack-specific threats; these are security risks at dApp component level that arise due to gaps or misconfigurations of security controls (i.e. security design flaws) as well as vulnerabilities, misconfigurations, or in the specific technologies, frameworks, tools, and libraries used to build the dApp component (Table 3.7). These threats are tied to the unique aspects of the software stack employed in an application, such as the mobile UI, and include dependencies, protocols, APIs, SDKs, and storage mechanisms.

The next list of enumerated threats covers the analysis of dApp DeFi lending and borrowing context business functionality-specific threats. These are security risks that target the unique business processes, rules, and operations inherent in a DeFi platform for lending and borrowing. These threats arise from vulnerabilities in the design, implementation, and operation of the dApp's functionalities, which facilitate user interactions with smart contracts, digital wallets, and blockchain networks (Table 3.8).

Appendix A: Threat Modeling Matrix provides a comprehensive TA of threat events, encompassing threats at the trust boundaries, component/tech stack level, and business context level for the targeted dApp component functionalities, categorized into nine primary threat scenarios.

Table 3.9 provides a consolidated summary of threat scenarios targeting the various components and interactions within a DeFi dApp architecture. Each threat scenario is characterized by a range of specific threat events, the components impacted, and the associated weaknesses or vulnerabilities that can be exploited by malicious actors. This structured representation links threat scenarios to targeted components and highlights key vulnerabilities, offering a clear framework for further risk analysis, mitigation planning, and prioritization of security measures. The table also organizes threats into distinct scenarios, making it easier to understand the scope and potential impact of each scenario within the dApp environment.

This analysis identifies approximately 100 enumerated threat events, with their relevance to the dApp DFD trust boundaries. The updated DFD in Figure 3.4 illustrates the scope and potential impact of these threats on assets at the defined trust boundaries, helping to visualize how changes affect threat exposure and control points.

Table 3.6 STRIDE enumerated threat events and analysis.

Enumerated STRIDE threats at the dApp trust boundary: external entity user to UI dApp component

Threat ID	Threat event	Description	AVs	Assets targeted	Potential impact	Countermeasures
T1	Spoofing	A malicious user impersonates a legitimate user to access the mobile UI	Phishing, session hijacking	User credentials, session tokens	Unauthorized access, data theft, or account compromise	MFA, secure login mechanisms, user behavior monitoring
T2	Tampering	Malicious actor alters data inputs sent from the user to the mobile UI	Payload manipulation, injection attacks	Transaction payloads, input data	Corrupted data, injection attacks, or unexpected behavior	Input validation, sanitization, and secure communication protocols (e.g. HTTPS)
T3	Information disclosure	Eavesdropping on data transmitted between the user and the mobile UI	Network sniffing via MitM	Data in transit (e.g. user information, transaction details)	Data breaches, exposure of sensitive information	Encryption for data in transit (e.g. TLS), secure session management
T4	DoS	Overwhelming the mobile UI with excessive requests to disrupt service	Flooding API endpoints, automated bot attacks	Application resources, backend services	Service unavailability	Rate limiting, CAPTCHA, and IP-based throttling

T5	Elevation of privilege	Exploiting flaws in the mobile UI to gain higher access privileges than authorized	Exploiting privilege escalation vulnerabilities, weak role definitions	Access control settings, sensitive application functionality	Unauthorized access to sensitive functionality or data	RBACs, privilege auditing, secure coding practices
T6	Repudiation	A malicious user denies performing an action on the mobile UI, such as initiating a sensitive request	Exploiting insufficient logging or lack of nonrepudiation mechanisms	System logs, audit trails	Disruption of accountability mechanisms	Implementing secure logging, nonrepudiation techniques, and user activity tracking

Table 3.7 Analysis of threats against the mobile UI component exploiting vulnerabilities in the tech stack (e.g. software components/libraries and misconfigurations).

Enumerated threat by component for dApp UI						
Threat ID	**Threat event**	**Description**	**AVs**	**Assets targeted**	**Potential impacts**	**Countermeasures**
T7	Exploitation of authentication flaws	An attacker bypasses weak or improperly implemented authentication mechanisms to impersonate a user.	Brute force attacks, session hijacking, or stolen credentials	User accounts, authentication tokens	Unauthorized access to user accounts, theft of sensitive data, and financial or reputational damage	Implement strong authentication mechanisms, such as MFA and cryptographic proof.
T8	Exploitation of insecure data storage	Sensitive data, such as private keys or credentials, is stored insecurely in local storage without encryption.	Exploiting unencrypted local storage or accessing insecure directories	Local storage, sensitive data (e.g. private keys, intellectual property, PII)	Exposure of sensitive user data, leading to data breaches and potential regulatory fines	Use secure storage solutions like Keychain or Keystore for sensitive data, and apply encryption for local storage.

T9	Exploitation of insufficient encryption	Data transmitted between the mobile UI and backend systems lacks strong encryption protocols, making it vulnerable to interception.	MitM attacks, packet sniffing, or weak encryption protocols	Data in transit between mobile UI and backend	Loss of confidentiality and potential data breaches due to intercepted communications	Encrypt all data in transit using strong protocols like TLS 1.2 or 1.3 to prevent interception.
T10	Code tampering and reverse engineering	Malicious actors reverse engineer the mobile app to extract sensitive information or tamper with app logic.	Reverse engineering with debugging tools or modifying app binaries	Application logic, sensitive algorithms	Compromise of application logic, enabling attackers to bypass security controls or launch further attacks	Apply code obfuscation and tamper detection to protect app logic and sensitive information.
T11	Exploitation of vulnerable APIs	Attackers exploit weak or unprotected APIs to perform unauthorized actions, such as accessing sensitive data or modifying user accounts.	Exploiting insecure endpoints or bypassing weak authentication mechanisms	API endpoints, user data, application integrity	Unauthorized actions, such as data theft or account manipulation, compromising the system's integrity	Regularly audit and secure APIs with proper access controls and authentication mechanisms.

(Continued)

Table 3.7 (Continued)

Enumerated threat by component for dApp UI						
Threat ID	**Threat event**	**Description**	**AVs**	**Assets targeted**	**Potential impacts**	**Countermeasures**
T12	Exploitation of sensitive data logging	Sensitive information, such as private keys or credentials, is inadvertently exposed in debug or application logs.	Analyzing debug logs for exposed sensitive data or credentials	Debug logs, sensitive credentials, private keys	Unintended exposure of critical data, increasing the risk of exploitation and user distrust	Sanitize logs to ensure sensitive information is excluded, and disable logging in production environments.
T13	Exploitation of outdated third-party libraries/ dependencies	Libraries or frameworks used in the mobile app are outdated, containing known vulnerabilities that can be exploited by attackers.	Exploiting known vulnerabilities in outdated libraries or frameworks	Vulnera application libraries, software dependencies	Exploitation of known vulnerabilities to compromise the app, resulting in system downtime or data breaches	Regularly update libraries and dependencies to address known vulnerabilities and ensure security patches are applied.
T14	Exploitation of lack of RBACs (i.e. authorization flaws)	The absence of RBAC allows attackers to escalate privileges and perform unauthorized actions.	Bypassing privilege checks or exploiting misconfigured RBAC policies	Access control policies, administrative privileges	Unauthorized access to privileged functionalities, leading to significant disruption and potential data loss	Implement RBAC to restrict privileges based on user roles and enforce least privilege principles.

Table 3.8 Mobile UI business logic contextual-specific threats with dApp DeFi lending and borrowing.

Enumerated business context-specific threats for dApp UI						
Threat ID	**Threat event**	**Description**	**AVs**	**Assets targeted**	**Potential impact**	**Countermeasures**
T15	Unauthorized loan creation	Attackers exploit weak input validation or authentication mechanisms to create fraudulent loan requests	Sending unauthorized loan requests or bypassing input validation	Loan request records, user authentication mechanisms	Financial losses due to fraudulent loans and reputational damage	Implement robust input validation and MFA for loan requests.
T16	Manipulated collateral valuation	Using compromised oracle feeds to alter the valuation of collateral, enabling users to borrow more than permitted	Manipulating oracle data feeds or sending spoofed data to oracles	Collateral valuation data, oracle data feeds	Incorrect collateral valuation leading to financial imbalances and platform instability	Secure oracle feeds with cryptographic proofs and regular audits of data sources.
T17	Fraudulent repayment transactions	Replay attacks on signed repayment transactions to falsely settle loans multiple times	Replaying valid transaction payloads using network sniffing tools	Repayment records, transaction logs	Multiple repayments falsely credited, causing financial inaccuracies and trust issues	Use nonce-based mechanisms and timestamp validation to prevent replay attacks.

(Continued)

Table 3.8 (Continued)

Enumerated business context-specific threats for dApp UI						
Threat ID	**Threat event**	**Description**	**AVs**	**Assets targeted**	**Potential impact**	**Countermeasures**
T18	Exploitation of smart contract vulnerabilities to cause fraud	Manipulating loan or repayment logic in smart contracts to bypass business rules or increase financial exposure	Exploiting smart contract vulnerabilities and logic flaws through custom transactions	Smart contracts, business rules logic	Smart contract exploitation leading to significant financial losses and system vulnerabilities	Conduct rigorous audits of smart contracts and apply runtime monitoring for anomalies.
T19	Abuse of loan terms	Exploiting API vulnerabilities to alter loan terms, such as repayment periods or interest rates	Modifying API requests to alter loan parameters	API endpoints, loan term configurations	Unauthorized changes in loan terms causing unfair or fraudulent lending/ borrowing practices	Implement strict API authentication, rate limiting, and parameter validation.
T20	Exploitation of data integrity for transaction fraud	Injecting or tampering with user-submitted data (e.g. loan amounts or collateral details) to create inconsistencies in records	Injecting tampered data into the system via compromised inputs	User-submitted data, loan and collateral records	Data inconsistencies undermining trust in the platform and leading to operational failures	Validate all user-submitted data and maintain integrity checks for loan and collateral records.

Table 3.9 Threat scenario, targeted components, and weaknesses.

Threat scenario and description[a]	Targeted components	Targeted dApp weaknesses/ vulnerabilities
TS1: Malicious actors targeting critical processes and assets in a dApp's mobile UI and associated components	Mobile or browser client/ UI and associated components	Authentication flaws, weak access controls, insecure local storage, software vulnerabilities, outdated libraries, insufficient encryption
TS2: Malicious actors targeting critical processes and assets in a dApp's interaction with digital wallets and associated components	Digital wallet and associated S/W and H/W components	Weak authentication mechanisms, insecure APIs, hardware and software vulnerabilities in wallets
TS3: Malicious actors targeting critical processes and assets in the interaction between the dApp and the distributed digital identity system (uPort)	Distributed digital identity system (uPort) and components	Misconfigured trust settings, insecure APIs, compromised nodes, insecure credentials storage
TS4: Malicious actors targeting critical processes and assets in the interaction between the dApp UI and the dApp frontend	dApp frontend and components	Session management flaws, misconfigured and insecure E2E communications, insecure APIs, web application vulnerabilities, software vulnerabilities and outdated dependencies
TS5: Malicious actors targeting the interaction between the dApp frontend and dApp middleware (JSON-RPC)	dApp middleware (JSON-RPC)	Unencrypted communication, outdated dependencies, weak access controls, middleware misconfigurations
TS6: Malicious actors targeting the interaction between the dApp middleware and blockchain local node	Blockchain node	Node software vulnerabilities, outdated libraries and dependencies, weak access controls, API vulnerabilities, misconfiguration flaws, local storage vulnerabilities
TS7: Malicious actors targeting the interaction between the blockchain node and the distributed ledger	Blockchain platform/network and distributed ledger	Misconfigurations and logical flaws in consensus algorithms, weak access controls for nodes to access the distributed ledger/ blockchain network, unsecured local node to distributed ledger communication channels

(Continued)

Table 3.9 (Continued)

Threat scenario and description[a]	Targeted components	Targeted dApp weaknesses/ vulnerabilities
TS8: Malicious actors targeting smart contracts for DeFi dApp loan and borrowing processes	Smart contracts	Smart contract vulnerabilities, outdated and vulnerable software third-party frameworks/libraries, weak/insecure smart contract state and management, flawed transaction validation logic
TS9: Malicious actors target the oracle and its data feed/ sources	Oracle platform and external data sources	Oracle to blockchain platform integration flaws, insecure communication, outdated and vulnerable oracle software, smart contracts and dependencies, compromised oracle external data feed/sources

a. Details of these threat events instances, including descriptions of threat events, AVs, targeted assets, potential impacts, and corresponding countermeasures, are provided in the Appendix A: Threat Modeling Matrix.

These countermeasures outline security by design principles and strategies for mitigating risks, reducing the likelihood of successful attacks, and minimizing their potential impact on the dApp.

In analyzing threats to the dApp, we employed threat frameworks such as STRIDE and contextualized threats based on the inherent risks of the dApp components and the business context, recognizing that threats are often designed to create business impacts that must be effectively risk-managed.

Factoring Threat Intelligence in the TA of the dApp: In addition to its application in risk-based threat modeling, TA plays a crucial role in examining threats identified through data gathered via threat intelligence. Threat intelligence involves collecting, analyzing, and contextualizing information about potential threats and adversaries targeting systems, applications, and organizations. It includes indicators of compromise (IoCs), TTPs of attackers, as well as insights into vulnerabilities and exploits relevant to specific contexts.

In the context of a DeFi dApp for lending and borrowing, threat intelligence enhances preparedness for attacks, for example, by understanding attacker motivations and highlighting the specific AVs used by the attacker that could target DeFi dApps. By providing real-world examples of attacks, threat intelligence helps to refine threat scenarios used in threat modeling and simulating known threats to validate the effectiveness of implemented security controls. Studying threat intelligence during threat modeling exercises helps to contextualize threat scenarios with real-world examples, improving the relevance and accuracy of the threat modeling process.

Threat intelligence, when applied in an operational environment, is instrumental in identifying, detecting, and responding to potential or ongoing attacks targeting an application. In the context of security incident preparedness and response, threat intelligence leverages external and internal sources, such as IoCs, to assess the current threat landscape and monitor for suspicious activities. For instance, analyzing IoCs like IP addresses, file hashes, or malicious behaviors linked to active threats against similar applications enables the review of system, application, and network logs for attack patterns, such as unusual API usage or anomalies in smart contract transactions. Additionally, threat intelligence collected during an attack provides actionable insights to guide incident response efforts. For example, identifying anomalies in blockchain node communications that align with known IoCs for malware targeting DeFi platforms can inform a targeted response strategy, ensuring threats are mitigated effectively in real time.

In the context of threat intelligence, TTPs describe the methods adversaries use to achieve their objectives. Tactics represent the high-level goals, such as gaining initial access or exfiltrating data. Techniques are the specific ways attackers accomplish these goals, like phishing or exploiting software vulnerabilities. Procedures provide detailed, implementation-level descriptions of how techniques are executed, including the tools and steps used in an attack. Understanding TTPs is essential for detecting threats, enhancing defenses, guiding incident response, and attributing attacks to specific threat actors. Frameworks like MITRE ATT&CK provide a structured approach to categorizing and analyzing TTPs.

Table 3.10 highlights the key differences between TA for application threat modeling and TA based on threat intelligence, while emphasizing their complementary nature. These two approaches mutually reinforce each other to enhance the security of DeFi dApps.

These approaches complement each other by leveraging threat intelligence to simulate real-world attacks during testing, which helps validate the application's resilience and the effectiveness of implemented countermeasures. Additionally, real-world threat intelligence can be applied to refine threat scenarios, ensuring they accurately reflect actual attack trends and patterns. By combining proactive measures, such as threat modeling, with reactive measures derived from threat intelligence, organizations can achieve a more robust and comprehensive security posture across both the development and operational phases.

A valuable approach to designing and building attack-resilient dApps is to leverage insights of security incidents that occurred for similar dApp and DeFi use cases. For instance, in the case of DeFi lending and borrowing platforms, numerous publicly disclosed incidents have revealed critical vulnerabilities and risks inherent to these protocols. These examples provide essential lessons for improving the security and robustness of DeFi systems:

1. **Lendf.me and Uniswap Exploits** [258]: In April 2020, the Lendf.me and Uniswap platforms experienced significant hacks resulting in the theft of approximately $25 million. Attackers exploited a known vulnerability, leading to substantial financial losses.
2. **Aave Protocol Incident** [259]: In November 2022, the Aave v2 lending platform on Ethereum faced a significant liquidation event. A user borrowed a substantial amount of Curve DAO Token (CRV) using USD Coin (USDC) as collateral, leading to notable financial implications for the protocol. The impact of this incident is approximately $10 million in protocol debt (estimated based on public reports).
3. **Onyx Protocol Exploit** [260]: In November 2023, the Onyx Protocol suffered an attack resulting in the loss of approximately $2.1 million (1,164.53 ETH). The methods used were similar to previous exploits, indicating recurring vulnerabilities in DeFi lending protocols.

Table 3.11 also highlights IoCs and TTPs derived from notable publicly disclosed security incidents involving DeFi dApps for lending and borrowing.

In addition to leveraging open-source intelligence (OSINT), integrating specialized threat intelligence tools into a dApp can greatly enhance VA. Tools like Inter OWL provide blockchain-specific threat intelligence, offering a deeper, contextual understanding of risks. These tools analyze real-world threats and IoCs targeting blockchain systems, such as vulnerabilities in smart contracts, dApps, and consensus mechanisms. By identifying patterns of known exploits – like reentrancy attacks, oracle manipulation, or transaction replay vulnerabilities – Inter OWL maps threats to specific blockchain components, enabling a more focused approach to vulnerability prioritization and remediation. Furthermore, these tools track threat trends across blockchain networks, identifying risks from misconfigurations, outdated dependencies, and emerging AVs, ensuring a proactive defense strategy.

In addition to Inter OWL [261], several other tools can supplement VA by incorporating blockchain-specific threat intelligence, offering a deeper understanding of risks. For instance, Chainalysis provides compliance and investigative software to analyze blockchain transactions, enabling organizations to track virtual currencies and monitor suspicious activities. This tool has been widely used by entities, including law enforcement agencies, to enhance security and compliance efforts. Another notable tool is BLOCKEYE [262], a real-time attack detection system tailored for DeFi projects on the Ethereum blockchain helps in identifying

Table 3.10 Threat modeling versus threat intelligence.

Aspect	TA for application threat modeling	TA based on threat intelligence
Purpose	To design security controls and mitigate risks during development	To detect and respond to active threats in an operational environment
Scope	Architecture, components, and data flow vulnerabilities	Monitoring, IoCs, and attack patterns
Focus	Prevention and secure design	Detection and response
Sources of input	DFDs, architectural review, application-specific risks	Threat intelligence feeds, logs, operational activity
Output	Threats, weaknesses, countermeasures, and security recommendations	Indicators of attacks, alerts, and incident response actions
Example	Identifying insecure API endpoints during dApp development	Detecting a replay attack based on transaction patterns in logs

potentially vulnerable DeFi projects and monitoring transactions to detect security attacks, thereby strengthening the overall security framework of blockchain applications. BAD [263] (Blockchain Anomaly Detection) is also an effective solution that leverages blockchain metadata to identify potentially malicious activities within the network.

Together, these tools, along with Inter OWL, enrich VA by integrating threat intelligence specific to blockchain ecosystems. They provide actionable insights into potential vulnerabilities, such as smart contract weaknesses, anomalous transaction patterns, and risks posed by misconfigured blockchain components. By leveraging these tools, organizations can adopt a proactive approach to securing blockchain-based systems, ensuring they remain resilient against evolving threats.

3.2.3.6 Stage V – Vulnerability Analysis

In the context of risk-based threat modeling, VA focuses on evaluating vulnerabilities identified and mapped to the threats modeled and analyzed during Stage IV of the PASTA methodology, Threat Analysis. This stage extends the TA by examining identified dApp vulnerabilities and weaknesses that could become the potential targets for threat agents to exploit and cause an impact to the dApp assets. dApp vulnerabilities might include vulnerabilities inherent to the technology stack used to build and deploy dApp components, including software and third-party

Table 3.11 Threat intelligence from dApp security incidents.

Incident and date	IoCs	TTPs
Lendf.me and Uniswap (April 2020)	• Unusual repeated transactions involving ERC-777 tokens • Drained contract balances • Interaction with imBTC	• Reentrancy attack: Exploited state changes by calling functions before completion • ERC-777 token vulnerability: Leveraged reentrant calls within the token standard
Aave Protocol (November 2022)	• Large-scale borrowing of CRV tokens • Collateral imbalance with disproportionate USDC deposits • Triggered liquidations	• Price manipulation: Leveraged oracles to feed false price data • Liquidation exploit: Forced collateral liquidation by manipulating asset prices
Onyx Protocol (November 2023)	• Flash loan transactions in large volumes • Sudden asset price anomalies • Rapid fund movement across wallets and platforms	• Flash loan attack: Utilized flash loans to manipulate pricing and lending logic • Collateral exploit: Exploited price-sensitive operations via manipulated asset values

dependencies, as well as security weaknesses such as design flaws and control gaps such as the absence or inadequacy of security controls and countermeasures.

A VA stage is necessary in risk-centric threat modeling to ensure that identified vulnerabilities are accounted for in factoring the risk of the threats that represent a possibility to be realized as an attack. For vulnerabilities that have been identified and not yet remediated, threat modeling and VA provide a detailed understanding first of the possible threat scenarios and the possible exploitable vulnerability and weaknesses, enabling developers and security teams to understand the likelihood of exploitation in the context of the dApp application, the component/assets that could be targeted, and the possible impacts.

However, the purpose of this stage is not to perform vulnerability assessments directly to validate the exploitability of a threat, such as targeted code scanning or penetration testing.

Instead, threat modeling and VA leverage the findings from previously executed application vulnerability assessments. For example, for threats targeting smart contract vulnerabilities to cause an impact on business functionality such as alter the business logic for fraudulent transactions and for data or digital asset (e.g. cryptocurrency) exfiltration, the scope of the vulnerabilities to be analyzed includes review of vulnerability assessment reports such as the one produced by the use of static code analysis tools to scan smart contract code as part of audits of smart contracts as part of the SecDevOps practice to review issues identified during testing. This approach ensures that vulnerabilities are not only understood in isolation but also addressed effectively in relation to the broader threat land architecture landscape, reinforcing the security posture of the dApp.

The expected outcome of the VA stage in the PASTA risk-centric threat modeling framework is to map the identified dApp vulnerabilities to specific components and threats. This process contextualizes the severity of these vulnerabilities in relation to the threat scenarios and threat events previously analyzed during the TA stage. While the foundational concepts of vulnerabilities and their analysis are discussed in Chapter 2, this chapter focuses on the unique context of the DeFi dApp example. It highlights the evaluation of vulnerabilities in specific dApp components and examines their exposure to threat agents targeting the dApp's architecture, as defined during Stage II of the threat modeling process. This analysis provides a structured mapping of threat scenarios to the identified vulnerabilities, illustrating how these vulnerabilities are risk-scored using frameworks such as CWE, CVEs, and CVSS.

The CWE is a comprehensive, standardized catalog of software and hardware weaknesses designed to help identify, categorize, and address vulnerabilities that can lead to security issues. Maintained by MITRE, CWE provides detailed descriptions, examples, and mitigation strategies for each weakness, making it a valuable resource for developers, security professionals, and organizations. Each weakness is assigned a unique identifier (e.g. CWE-79 for cross-site scripting [XSS]), which is widely used in security assessments and integrated into various tools and methodologies. For threat modeling, CWE is particularly relevant because it offers a structured framework to identify and prioritize potential weaknesses during the design and development phases. By mapping identified threats to specific CWE entries, security practitioners can derive targeted mitigations and develop attack-driven test cases to validate the effectiveness of their defenses. CWE also enhances communication between teams by providing a common language to discuss and address security concerns, ultimately supporting the creation of more resilient systems. Additionally, the CWE framework is organized in a tree-like structure with categories and subcategories, which aligns well with the principles of threat modeling. Similar to an attack tree, addressing a high-level CWE (e.g. CWE-398) can mitigate numerous underlying weaknesses (15–20 related CWEs), thereby enhancing the overall security posture through a more efficient, holistic approach.

CVE is a system for identifying and naming publicly disclosed cybersecurity vulnerabilities in software and hardware. Each CVE entry has a unique ID (e.g. CVE-2023-12345), a brief description of the vulnerability, and references for additional details or fixes. CVEs are widely used to standardize vulnerability tracking, enhance communication, and prioritize risk mitigation efforts. Managed by MITRE, CVEs are often linked to severity scores like CVSS and integrated into security tools and databases like NVD. In the context of VA within risk-centric threat modeling, CVEs are commonly reported for vulnerabilities identified in frameworks and libraries used to develop dApp components, for example CVE-2024-54134 is a vulnerability of publish-access account for solana/web3.js that when compromised allows an attacker to publish unauthorized and malicious packages. For a complete listing, please refer to MITRE's CVEs [185]. This resource provides detailed descriptions and classifications of vulnerabilities, serving as an essential reference for identifying and mitigating security threats. MITRE's CWE [186] provides detailed descriptions of various weaknesses, including their causes, potential impacts, and recommended mitigations, making it an essential tool for anyone looking to understand and address security vulnerabilities.

Both CWEs and CVEs can be assessed for criticality using a standardized framework like CVSS, which quantifies the severity of software and hardware vulnerabilities. CVSS provides a numerical score (ranging from 0.0 to 10.0) and a qualitative severity rating (e.g. low, medium, high, or critical) to help organizations prioritize and manage risks. In addition to the base score that reflects the intrinsic properties of a vulnerability, CVSS includes temporal and environmental metrics that further adjust the score based on factors such as exploitability over time and the specific impact within an organization. For more details consult FIRST's CVSS [183].

By mapping threats to weaknesses (CWEs) and vulnerabilities (CVEs) and scoring them for risk, organizations can prioritize the threat modeling of those threats that are most likely to exploit critical vulnerabilities. However, this constitutes only an initial risk prioritization based solely on the vulnerability's severity; a comprehensive risk evaluation must also consider factors such as the likelihood of exploitation and the potential impact on the targeted asset.

This is why AM is essential. By modeling attacks, it becomes possible to evaluate all the factors necessary for a vulnerability to be exploited, simulating real-world scenarios. These scenarios should closely align with past attacks or those that are highly likely to occur, enabling more accurate RAM strategies.

Framing the vulnerability risk in terms of the threat–vulnerability combination ensures that vulnerabilities are identified and prioritized not only based upon criticality/severity but also in the context of the exposure to threats being analyzed as part of the threat modeling process.

The threat–vulnerability mapping is a process designed to identify potential weaknesses that a threat agent could exploit within the analyzed threat scenario.

In Step 1 of this threat to vulnerabilities/weaknesses mapping process, all relevant potential weaknesses (CWEs) are documented and mapped to each threat scenario. These are the CWEs that correspond to the weaknesses and vulnerabilities identified for the threat scenarios analyzed during Stage IV of the TA. (Refer to Appendix B: Mapping of Threat Scenarios to Targeted Weaknesses and Asset Impacted for a detailed enumeration of the threat events associated with the targeted assets and weaknesses/vulnerabilities.)

The initial mapping of threats to weaknesses and vulnerabilities is presented in Table 3.12.

In Step 2 of the threat-to-vulnerability/weakness mapping process, CWEs are used as a structured framework to analyze vulnerabilities reported for the dApp and identify matches based on the dApp's vulnerability assessments. These assessments may include findings from a security architecture review, where weaknesses in security controls are identified, or results from testing and vulnerability scanning methods. Such methods include analyzing dApp source code, third-party libraries, APIs, web interfaces, and any software components developed or integrated into the system. The process involves mapping the weaknesses (CWEs) identified during TA to those vulnerabilities uncovered through various security testing methods. These may include DAST, SAST, SCA, manual penetration testing, IaC, and container security scanning (for cloud-based deployments). This mapping includes vulnerabilities identified in testing environments (e.g. staging or QA) prior to production deployment.

The goal of this step is to establish a clear connection between the potential weaknesses targeted by threat events in specific threat scenarios and the CWEs and CVEs detected during security testing. Before performing this mapping, it is essential to validate all identified vulnerabilities, including CWEs and CVEs, to eliminate false positives. Remediation efforts should then be prioritized according to the organization's vulnerability management policies (e.g. critical vulnerabilities to be addressed within 15 days, high within 30 days, medium within 60 days, and low severity within 90 days). It's important to clarify that this process is not a component of the threat modeling exercise itself; rather, it is part of the overall risk handling process. The same process is used to manage any issues identified through security activities such as SAST, SCA, and other vulnerability scanning efforts.

The scope of vulnerabilities includes CWEs identified through secure architecture reviews, CWEs and CVEs reported by security testing tools related to the source code developed for dApp components owned by the development team and vulnerabilities originating from third-party libraries, frameworks, or packaged software integrated into the dApp.

Any CWE identified and mapped as a candidate weakness in the context of threat exploitation represents a specific vulnerability that could be leveraged by a threat actor to compromise a targeted asset. This mapping process correlates threats, targeted assets, and candidate weaknesses (CWEs), thereby enabling a systematic identification and mitigation of security risks that can have significant business impacts.

Table 3.12 Mapping of threats to weaknesses and vulnerabilities.

Threat scenario ID	Threat scenario description	Targeted assets	Candidate weaknesses (CWEs) for threat exploitation
TS1	Malicious actors targeting critical processes and assets in a dApp's mobile UI and associated components	Mobile or browser client/UI	Authentication flaws (CWE-287, CWE-304, CWE-613, CWE-269) Weak access controls (CWE-284, CWE-200, CWE-732, CWE-276) Insecure local storage (CWE-922, CWE-312, CWE-922) Software vulnerabilities (CWE-20, CWE-119, CWE-787, CWE-434) Outdated libraries (CWE-1104, CWE-937, CWE-114) Insufficient encryption (CWE-326, CWE-327, CWE-331, CWE-311)
TS2	Malicious actors targeting critical processes and assets in a dApp's interaction with digital wallets and associated components	Digital wallet (e.g. S/W and H/W) smart contracts, on-chain, off-chain components	Weak authentication mechanisms (CWE-287, CWE-304, CWE-613, CWE-294, CWE-798, CWE-288, CWE-308) Insecure APIs (CWE-200, CWE-287, CWE-285, CWE-352, CWE-601, CWE-918, CWE-611, CWE-829) Smart contract vulnerabilities specific to digital wallets (CWE-20, CWE-284, CWE-285, CWE-287, CWE-288, CWE-346, CWE-703, CWE-532, CWE-400, CWE-125, CWE-129, CWE-190, CWE-191) Hardware and software vulnerabilities in digital wallets (CWE-922, CWE-256, CWE-330, CWE-326, CWE-327, CWE-706, CWE-285, CWE-798, CWE-610) Key management and cryptography (CWE-320, CWE-331, CWE-312, CWE-324, CWE-255) On-chain components (CWE-502, CWE-611, CWE-295, CWE-345, CWE-400, CWE-284, CWE-916) Off-chain components (CWE-200, CWE-311, CWE-522, CWE-284, CWE-255, CWE-319, CWE-352, CWE-918, CWE-312)

TS3	Malicious actors targeting critical processes and assets in the interaction between the dApp and the distributed digital identity system (uPort)	Distributed digital identity system, smart contracts, on-chain and off-chain, bridge components	Misconfigured trust settings (CWE-295, CWE-296, CWE-346, CWE-703) Insecure APIs (CWE-200, CWE-287, CWE-285, CWE-352, CWE-918, CWE-601, CWE-611) Compromised nodes (CWE-322, CWE-327, CWE-284, CWE-703) Insecure credentials storage (CWE-922, CWE-256, CWE-798, CWE-759, CWE-311) Data leakage and exposure (CWE-200, CWE-359, CWE-312, CWE-209, CWE-203) Smart contract vulnerabilities (*)(CWE-120, CWE-400, CWE-787, CWE-840, CWE-284, CWE-703, CWE-352, CWE-641) Identity wallets (CWE-200, CWE-522, CWE-311, CWE-359, CWE-640, CWE-918) Verifiable Credentials (CWE-345, CWE-295, CWE-209, CWE-290) Off-chain storage systems (CWE-284, CWE-200, CWE-611, CWE-917, CWE-918) Authentication and communication protocols (CWE-319, CWE-352, CWE-287, CWE-522, CWE-326) Issuers, verifiers, and backend systems (CWE-601, CWE-862, CWE-287, CWE-502, CWE-434) Key management and cryptography (CWE-320, CWE-331, CWE-312, CWE-324, CWE-255) Bridge off-chain to on-chain components (CWE-200, CWE-311, CWE-522, CWE-284, CWE-255, CWE-319, CWE-352, CWE-918, CWE-312)

(Continued)

Table 3.12 (Continued)

Threat scenario ID	Threat scenario description	Targeted assets	Candidate weaknesses (CWEs) for threat exploitation
TS4	Malicious actors targeting critical processes and assets in the interaction between the dApp UI and the dApp frontend	dApp frontend and components	Session management flaws (CWE-287, CWE-613, CWE-299, CWE-384, CWE-305) Misconfigured and insecure E2E communications (CWE-295, CWE-296, CWE-319, CWE-323, CWE-326, CWE-327) Insecure APIs (CWE-200, CWE-285, CWE-287, CWE-352, CWE-918, CWE-601, CWE-611) Web application vulnerabilities (CWE-79, CWE-89, CWE-601, CWE-918, CWE-352, CWE-209) Software vulnerabilities and outdated dependencies (CWE-937, CWE-1104, CWE-20, CWE-119, CWE-787, CWE-434) Resource exhaustion and consumption (CWE-400, CWE-770, CWE-404, CWE-701, CWE-776, CWE-835, CWE-409)
TS5	Malicious actors targeting the interaction between the dApp frontend and dApp middleware (JSON-RPC)	dApp middleware (JSON-RPC)	Unencrypted communication (CWE-319, CWE-311, CWE-326, CWE-327, CWE-308) Outdated dependencies (CWE-937, CWE-1104, CWE-829, CWE-20) Weak access controls (CWE-284, CWE-287, CWE-285, CWE-798, CWE-613, CWE-522) Middleware misconfigurations (CWE-16, CWE-346, CWE-703, CWE-290, CWE-295, CWE-601, CWE-918) Resource exhaustion and consumption (CWE-400, CWE-770, CWE-404, CWE-701, CWE-776, CWE-835, CWE-409)

TS6	Malicious actors targeting the interaction between the dApp middleware and blockchain local node	Blockchain node	Node software vulnerabilities (CWE-20, CWE-119, CWE-787, CWE-434, CWE-400) Outdated libraries and dependencies (CWE-937, CWE-1104, CWE-829) Weak access controls (CWE-284, CWE-287, CWE-285, CWE-798, CWE-613, CWE-522) API vulnerabilities (CWE-200, CWE-287, CWE-285, CWE-352, CWE-918, CWE-601, CWE-611) Misconfiguration flaws (CWE-16, CWE-346, CWE-703, CWE-290, CWE-295, CWE-601) Local storage vulnerabilities (CWE-922, CWE-256, CWE-798, CWE-759, CWE-311) Resource exhaustion and consumption (CWE-400, CWE-770, CWE-404, CWE-701, CWE-776, CWE-835, CWE-409)
TS7	Malicious actors targeting the interaction between the blockchain node and the distributed ledger	Blockchain platform/network and distributed ledger	Misconfigurations and logical flaws in consensus algorithms (CWE-20, CWE-354, CWE-703, CWE-285, CWE-642) Weak access controls for nodes (CWE-284, CWE-287, CWE-798, CWE-522, CWE-269) Unsecured local node to ledger communication channels (CWE-319, CWE-311, CWE-295, CWE-326, CWE-918, CWE-601)

(Continued)

Table 3.12 (Continued)

Threat scenario ID	Threat scenario description	Targeted assets	Candidate weaknesses (CWEs) for threat exploitation
TS8	Malicious actors targeting smart contracts for DeFi dApp loan and borrowing processes	Smart contracts	Smart contract vulnerabilities (*)(CWE-120, CWE-400, CWE-787, CWE-840, CWE-284, CWE-703, CWE-352, CWE-641) Logic flaws in smart contracts (CWE-20, CWE-128, CWE-129, CWE-345, CWE-682, CWE-190, CWE-191, CWE-284, CWE-285, CWE-287, CWE-288, CWE-364, CWE-703, CWE-841, CWE-664, CWE-346, CWE-708, CWE-1221, CWE-532, CWE-611, CWE-400) Outdated and vulnerable third-party frameworks/libraries (CWE-937, CWE-1104, CWE-829, CWE-502, CWE-20) Weak/insecure smart contract state and management (CWE-922, CWE-256, CWE-798, CWE-285, CWE-362, CWE-642) Flawed transaction validation logic (CWE-20, CWE-345, CWE-703, CWE-302, CWE-470)
TS9	Malicious actors target the oracle and its data feed/sources	Oracle platform and external data sources, smart contracts	Oracle to blockchain integration flaws (CWE-20, CWE-345, CWE-352, CWE-703, CWE-502, CWE-290) Insecure communication (CWE-319, CWE-311, CWE-295, CWE-326, CWE-918) Outdated and vulnerable oracle software (CWE-937, CWE-1104, CWE-829) Smart contracts and dependencies (CWE-285, CWE-400, CWE-502, CWE-471) Compromised oracle external data feeds (CWE-352, CWE-918, CWE-345, CWE-1188, CWE-922, CWE-256) Exposure of sensitive information (CWE-200, CWE-345, CWE-311, CWE-319, CWE-256, CWE-522, CWE-532, CWE-918, CWE-922, CWE-209, CWE-354)

For this exercise, we provide a mapping of the known CWEs and CVEs publicly disclosed for the technology stack components associated with the dApp. This mapping assumes that the vulnerability detection tools employed are capable of identifying the relevant CWEs and CVEs, thereby facilitating a contextualized TA.

The final mapping of threats to identified weaknesses and vulnerabilities by vulnerability assessment tests and processes executed on the dApp is presented in Table 3.13.

Note: It is important to distinguish between the intrinsic severity of a CWE and the overall risk severity calculated during the risk analysis stage (e.g. as likelihood × impact in PASTA Stage VII). A CWE's severity is an inherent rating that reflects its potential to be exploited, but it may not accurately represent the actual risk in a specific context. For example, an oracle vulnerability might be assigned a high CWE severity by default; however, if the oracle is not easily accessible to attackers, its likelihood of exploitation and therefore its overall risk might be significantly lower when evaluated using the risk analysis framework. In the best-case scenario, vulnerabilities should be rescored in context to ensure that mitigation efforts are prioritized based on a true reflection of their risk.

This threat to vulnerability mapping establishes a critical foundation for the next AM stage, where the focus shifts to understanding how specific AVs could exploit the identified weaknesses and vulnerabilities in potential attack scenarios. By linking identified vulnerabilities to the analyzed threats, this approach ensures a comprehensive understanding of the dApp's security posture and facilitates effective mitigation strategies.

Note: This mapping is provided as an example of the most common vulnerabilities typically identified by tools such as SAST, DAST, SCA, penetration testing tools, or specialized smart contract scanning tools like MythX, Slither, and Remix IDE. SCA tools generally report CVEs when vulnerabilities are detected in third-party libraries and packages. SAST and DAST tools primarily report CWEs, often accompanied by a risk rating that can be calculated using frameworks like CVSS v4.

In general, any reported vulnerability, whether identified by automated tools or manual processes, should be mapped to relevant threat scenarios, threat events, and AVs to assess exposure. When vulnerabilities in source code are fixed according to tool recommendations, evidence of remediation is typically demonstrated through subsequent scans that no longer report these vulnerabilities. In the context of threat modeling, this vulnerability mapping provides valuable insights into how the presence of a specific vulnerability could affect the likelihood of exploitation and its potential impact on the asset. This step helps quantify the risk to the dApp and informs prioritization of mitigations.

Table 3.13 Mapping of threats to identified weaknesses and vulnerabilities.

Threat scenario description	Targeted dApp asset	Targeted asset tech stack components	Identified CWEs and CVEs ranked by severity (see note)
TS1: Malicious actors targeting critical processes and assets in a dApp's mobile UI and associated components	Mobile or browser client/UI	React Native, Flutter, Ionic Enterprise; Web3 Modal (wallet integration), TailwindCSS (styling), Formik (input validation)	Critical severity: CVSS 9.8 CVE-2022-32514 CVSS 9.8 CWE-287: Improper authentication High severity: CVSS 8.8 CWE-359: Exposure of private information ("privacy violation") CVSS 7.8 CVE-2018-1000850 CVSS 7.8 CWE-1104: Use of unmaintained third-party components CVSS 7.5 CVE-2022-23520 CVSS 7.5 CWE-20: Improper input validation Medium severity: CVSS 4.7 CVE-2019-11771 CVSS 4.7 CWE-922: Insecure storage of sensitive information Low severity: CVSS 2.0 CWE-104: Excessive logging

TS2: Malicious actors targeting critical processes and assets in a dApp's interaction with digital wallets and associated components	Digital wallet (e.g. S/W and H/W) smart contracts, on-chain, off-chain components	Trust Wallet, MetaMask; Fireblocks (commercial); WalletConnect SDK (multiwallet support), BIP39 (mnemonic management)	Critical severity: CVSS 9.8 CWE-287: Improper authentication High severity: CVSS 7.5 CWE-190: Integer overflow or wraparound Medium severity: CVSS 5.9 CVE-2022-32969 CVSS 5.9 CWE-922: Insecure storage of sensitive information CVSS 6.5 CVE-2022-38380 CVSs 6.4 CWE-284: Improper access control Low severity: CVSS 2.5 CWE-540: Information exposure through debugging information
TS3: Malicious actors targeting critical processes and assets in the interaction between the dApp and the distributed digital identity system	Distributed digital identity system, smart contracts, on-chain and off-chain, bridge components	uPort, Civic; DID-JS, Veramo (decentralized identifiers and cryptographic proofs)	Critical severity: CVSS 9.8 CWE-284: Improper access control High severity: CVSS 8.8 CWE-346: Origin validation error CVSS 8.8 CWE-400: Resource exhaustion ("gas limit exploits") Medium severity: CVSS 6.3 CWE-601: Open redirect CVSS 5.8 CWE-1285: Improper validation of specified conditions Low severity: CVSS 3.5 CWE-758: Reliance on undefined behavior

(Continued)

Table 3.13 (Continued)

Threat scenario description	Targeted dApp asset	Targeted asset tech stack components	Identified CWEs and CVEs ranked by severity (see note)
TS4: Malicious actors targeting critical processes and assets in the interaction between the dApp UI and the dApp frontend	dApp frontend and components	Vue.js, AngularJS, React; Web3.js, ethers.js (blockchain interaction); Apollo Client (API data fetching); Okta (API security)	Critical severity: CVSS 9.6 XSS CVSS 9.3 CWE-352: Cross-site request forgery (CSRF) CVSS 9.1 CWE-346: Origin validation error High severity: CVSS 8.8 CWE-200: Exposure of sensitive information CVSS 7.5 CVE-2024-21490 CVSS 7.5 CWE-400: Uncontrolled resource consumption Medium severity: CVSS 5.4 CVE-2020-7676 CVSS 5.4 CWE-79: Improper neutralization of input during web page generation (XSS) Low severity: CVSS 2.0: CWE-561: Dead code
TS5: Malicious actors targeting the interaction between the dApp frontend and dApp middleware (JSON-RPC)	dApp middleware (JSON-RPC)	Infura, Alchemy; QuickNode (commercial middleware); web3-provider-engine, rpc-websockets (middleware libraries)	Critical severity: CVSS 9.8 CWE-502: Deserialization of untrusted data High severity: CVSS 7.5 CVE-2021-21369 CVSS 7.5 CWE-400: Uncontrolled resource consumption, DoS

TS6: Malicious actors targeting the interaction between the dApp middleware and blockchain local node	Blockchain node	Ethereum Node Software (Geth, OpenEthereum, Besu) Solana Node Software Solana Validator Node, Solana RPC Node Polygon Bor (block producer), Heimdall (PoS layer) API protocols JSON-RPC, WebSocket, gRPC	High severity: CVSS 7.5 CVE-2021-21369 CVSS 7.5 CVE-2020-26241 CVSS 7.5 CVE-2020-26264 CVSS 7.5 CVE-2023-42319 CVSS 7.5 CWE-400: Uncontrolled resource consumption Medium severity: CVSS 6.5 CVE-2022-23327 CVSS 6.5 CWE-20: Improper input validation CVSS 6.5 CVE-2022-23328 CVSS 6.5 CWE-400: Uncontrolled resource consumption Low severity: CVSS 2.5 CWE-540: Information exposure through debugging information

(Continued)

Table 3.13 (Continued)

Threat scenario description	Targeted dApp asset	Targeted asset tech stack components	Identified CWEs and CVEs ranked by severity (see note)
TS7: Malicious actors targeting the interaction between the blockchain node and the distributed ledger	Blockchain platform/network and distributed ledger	Ethereum Networking protocols (DevP2P, Libp2p) Solana Turbine, Libp2p Polygon DevP2P, Libp2p Consensus Protocols PoW (legacy), PoS (Ethereum 2.0) PoH, PoS, Plasma	Critical severity: CVSS 9.8 CWE-284: Improper access control CVSS 9.8 CWE-703: Improper check or handling of exceptional conditions CVSS 9.8 CWE-20: Improper input validation High severity: CVSS 7.5 CVE-2020-26241 CVSS 7.5 CWE-20: Improper input validation CVSS 7.5 CVE-2020-26264 CVSS 7.5 CWE-284: Improper access control CVSS 7.5 CVE-2021-21369 CVSS 7.5 CWE-354: Improper validation of integrity check value CVSS 7.5 CVE-2023-42319 CVSS 7.5 CWE-287: Improper authentication Medium severity: CVSS 6.5 CVE-2024-21668 CVSS 6.4 CWE-319: Cleartext transmission of sensitive information Low severity: CVSS 2.5 CWE-354: Improper validation of integrity check value

TS8: Malicious actors targeting smart contracts for DeFi dApp loan and borrowing processes	Smart contracts	Languages: Ethereum (Solidity, Vyper) Solana (Rust, C, C++) Polygon (Solidity) and dev tools Infura, Alchemy, Web3.js, Ethers.js Solana CLI, Solana Web3.js	Critical severity: CVSS 9.1 CVE-2024-54134 CVSS 9.1 CWE-284: Improper access control CVSS 9.1 CWE-502: Deserialization of untrusted data High severity: CVSS 7.5 CVE-2024-30253 CVSS 7.5 CWE-400: Uncontrolled resource consumption CVSS 7.5 CWE-770: Allocation of resources without limits, viper compiler vulnerabilities High severity: CVSS 8.2 CWE-20: Improper input validation CVSS 8.2 CWE-703: Improper handling of exceptions High severity: CVSS 8.8 CWE-829: Untrusted control sphere CWE-502: Deserialization CVSS 8.8 CWE-922: Insecure storage High severity: CVSS 7.5 CWE-20: Improper input validation CVSS 7.5 CWE-284: Improper access control CVSS 7.5 CWE-362: Concurrent execution using shared resource with improper synchronization (reentrancy)

(Continued)

Table 3.13 (Continued)

Threat scenario description	Targeted dApp asset	Targeted asset tech stack components	Identified CWEs and CVEs ranked by severity (see note)
			Medium severity: CVSS 6.5 CWE-703: Improper check or handling of exceptional conditions CVSS 6.5 CWE-345: Insufficient verification of data authenticity CVSS 6.5 CWE-641: Improper restriction of names for function identifiers CVSS 6.5 CWE-1104: Use of unmaintained third-party components CVSS 6.5 SWC-110: Business logic flaws related to medium impact Low severity: CVSS 3.5 CWE-398: Code quality issues CVSS 3.5 CWE-547: Hardcoded gas values CVSS 3.5 CWE-563: Assignment to variable without use
TS9: Malicious actors target the oracle and its data feed/sources	Oracle platform and external data sources, smart contracts	Chainlink, Chainlink-SDK (integration), API3 (SLA-backed oracles)	Severity critical: CVSS 9.1 CWE-345: Insufficient verification of data authenticity Severity high: CVSS 8.8 CWE-345: Insufficient verification of data authenticity CVSS 8.2 CWE-918: Server-side request forgery (SSRF) CVSS 7.5 CWE-311: Missing encryption of sensitive data Severity medium: CVSS 6.3 CWE-326: Inadequate encryption strength Severity low: CVSS 3.0 CWE-758: Reliance on undefined behavior

3.2.3.7 Stage VI – Attack Modeling

In the context of a DeFi dApp for lending and borrowing, AM focuses on simulating attacks and AVs to derive scenarios that validate the effectiveness of countermeasures deployed to mitigate both theoretical (i.e. theoretically possible) and realistic threats (i.e. those that have led to data breaches or incidents). These scenarios directly target the platform's lending and borrowing functionality. The process leverages insights from earlier stages of risk-based threat modeling, such as identifying threats, vulnerabilities, weaknesses, AVs, trust boundaries, and data flows. The primary objective is to assess the dApp's security measures in addressing identified vulnerabilities and design flaws, uncover areas of risk exposure, and recommend mitigation strategies to strengthen defenses.

For a lending and borrowing dApp, adversary modeling explores how adversaries might exploit vulnerabilities in critical components such as smart contracts, oracles, wallets, and frontend interfaces. It models the attacker's most likely paths using methodologies like attack trees to identify probable routes to compromise assets and attack kill chains to analyze the sequence of events leading to the exploit. For example, vulnerabilities such as improper state handling in smart contracts or manipulated oracle price feeds could be analyzed to determine their impact on lending and borrowing processes, such as unauthorized liquidations, overleveraged loans, or fund withdrawals.

To further enrich this analysis, frameworks like MITRE ATT&CK [195] can be applied, particularly to model attacker behaviors targeting supporting infrastructure, such as wallets, APIs, cloud-based oracles, and frontend layers. While not designed specifically for blockchain, MITRE ATT&CK offers a structured taxonomy of tactics and techniques (e.g. credential theft, phishing, resource hijacking) that can complement blockchain-specific threat modeling. When used alongside attack trees, the DeFi threat matrix, or vulnerability registries like the SWC registry [119], ATT&CK helps provide a more holistic view of attacker behavior across both traditional and Web3 domains.

Through the analysis of attack scenarios, it is possible to gain actionable insights into how effective existing countermeasures are in reducing the likelihood and impact of security risks. The AM stage of a risk-centric threat modeling process plays a vital role in this analysis by demonstrating how adversaries might exploit vulnerabilities, such as disrupting collateral validation, bypassing lending limits, or manipulating borrower interest rates to compromise the system. This process helps teams prioritize mitigation strategies based on real-world impact, ensuring the dApp is designed with robust security measures that preserve its functionality, integrity, and the trust of its users.

Importantly, a well-developed attack model should also account for lateral movement within the system, where an attacker may exploit a low-severity CWE as an initial foothold to pivot into higher-privilege components of the architecture. This form of privilege escalation is often underrepresented in threat models but is essential for capturing realistic multistage attack paths. The use of the cyber kill chain framework can further support this analysis by mapping the progression of an attack, from initial access to persistence, privilege escalation, and exploitation, and helps identify how early-stage, seemingly minor weaknesses can lead to high-impact outcomes. Considering such escalation scenarios is critical to identifying compounded risks and reinforcing the dApp's resilience against sophisticated adversarial behavior.

The first step of attack analysis involves examining the various types of AVs that could be used to exploit the weaknesses and vulnerabilities identified during the VA stage. This analysis considers the specific AVs associated with each threat event within the respective threat scenarios.

For instance, in the threat scenarios like "Malicious actors targeting critical processes and assets in a dApp's mobile UI and associated components," the TA conducted in Stage IV of the risk-centric threat model identified a list of threats. This includes Threat Events 1 through 15, each with associated AVs. These AVs originate from entry points located at the trust boundaries previously analyzed using DFD threat models during the TA stage in Stage IV. In Stage V, threats were mapped to corresponding weaknesses (CWEs) and vulnerabilities (CVEs), enabling the analysis of the dApp's asset exposure to the identified vulnerabilities. A detailed mapping of AVs, entry points, and their correlation with weaknesses in each threat scenario is presented in Table 3.14.

Table 3.14 Mapping of AVs, entry points, and weaknesses in each threat scenario.

Threat scenario: malicious actors targeting critical processes and assets in a dApp's mobile UI and associated components
Attack entry: mobile or browser client/UI

Weakness targeted (CWE)	Weakness description	Severity risk	AVs
CWE-287	Improper authentication	9.8 (critical)	Phishing attacks, stolen credentials, session hijacking
CWE-359	Exposure of private information	8.8 (high)	Intercepting unprotected communication or insecure data storage
CWE-1104	Use of unmaintained third-party components	7.8 (high)	Exploiting vulnerabilities in outdated libraries or frameworks
CWE-20	Improper input validation	7.5 (high)	Manipulating input data to bypass validation checks

Note: The mapping of attack entry points and AVs for all threat scenarios TS1 through TS9 are detailed in Appendix B: Mapping of Threat Scenarios to Targeted Weaknesses and Asset Impacted.

The next step is to develop threat scenarios that incorporate AVs, enabling the simulation of realistic attack scenarios.

The table that documents the AVs and critical/high-severity CWEs serves as a valuable foundation for deriving realistic attack scenarios. To do this, it is important to analyze how adversaries could exploit the listed AVs by interacting with system components like APIs, smart contracts, or middleware. These AVs should be mapped to the associated CWEs, evaluating potential exploitation impacts such as unauthorized access, data manipulation, or DoS. Identifying entry points, such as the UI, middleware, or blockchain nodes, helps determine where the attacks originate and ensures alignment with the system's architecture.

Each scenario should be contextualized by defining the targeted assets and potential impacts. For example, in threat scenario TS2: Malicious actors targeting critical processes and assets in a dApp's interaction with digital wallets and associated components, weaknesses such as CWE-284 (improper access control) could result in unauthorized wallet access and fund theft. Establishing the adversary's likely objectives, such as financial gain or service disruption, provides further clarity. Attack scenarios can then be structured into sequences of steps, starting with the AV and CWE and including intermediate actions like reconnaissance, payload injection, and postexploitation activities. Incorporating threat intelligence ensures that threat scenarios are grounded in real-world relevance by aligning them with documented incidents and established TTPs. For those new to this approach, the ATT&CK framework is especially valuable as it clearly illustrates how various TTPs interrelate, providing a structured perspective on adversary behavior.

For example, by considering the threat scenario previously analyzed by considering weaknesses and AVs for threat scenario TS4 "Malicious actors targeting critical processes and assets in the interaction between the dApp UI and the dApp frontend exploit vulnerabilities in session management, data integrity, API configurations, and business logic":

- **AV:** "Injecting malicious scripts via input fields"
- **CWE Targeted:** CWE-79 (XSS)
- **Severity:** Critical
- **Attack Entry Point:** dApp frontend through the UI web interface
- **Targeted Asset:** User credentials, session data

The attack scenario that can be deducted is the following: An attacker exploits an XSS vulnerability in the dApp frontend by injecting malicious scripts into input fields. The malicious code executes in the victim's browser, allowing the attacker to steal user session tokens, impersonate the user, and execute unauthorized transactions on the dApp.

Another example for the threat scenario TS8: Malicious actors targeting smart contracts exploit vulnerabilities in transaction execution, validation, and state management to compromise the integrity of loan approval and repayment processes:

- **AV:** Exploiting reentrancy vulnerabilities to repeatedly withdraw funds during loan processing
- **CWE Targeted:** CWE-362 (concurrent execution using shared resource with improper synchronization)
- **Severity:** High
- **Attack Entry Point:** Smart contract
- **Targeted Asset:** Loan approval and fund withdrawal processes

The attack scenario that can be deducted in this case can be documented as this: An attacker identifies a reentrancy vulnerability in the smart contract responsible for processing loans. By crafting a malicious contract, the attacker invokes the vulnerable smart contract's loan withdrawal function. During execution, the attacker's malicious contract repeatedly calls back into the withdrawal function before the original function completes, bypassing proper state updates. This allows the attacker to drain funds from the contract.

The attack scenarios for threat scenarios TS1 through TS9 are documented in the Appendix C: Mapping of Threat Scenarios to Exploitable Attack Paths. Deriving these attack scenarios is crucial for understanding threats in the context of realistic exploitation by attackers. They address critical questions, such as what could go wrong and the potential impact or losses if the identified weaknesses or vulnerabilities in the dApp were exploited by a threat agent to compromise the targeted assets.

The analysis of the attack scenarios provide a structured approach to understanding vulnerabilities by simulating how attackers might exploit weaknesses and identifying the potential impacts. This information can also guide attack-driven testing to determine the exploitability of identified vulnerabilities, validate their impact, and verify the effectiveness of remediation efforts. Attack scenarios play a crucial role in supporting attack-driven testing by providing a structured approach to evaluating vulnerabilities. They outline the specific steps attackers might use to exploit vulnerabilities, which can then be translated into attack-driven test cases to simulate potential exploits in controlled conditions. This approach helps determine whether the vulnerabilities are exploitable in their current state, providing evidence of risk and highlighting gaps in existing defenses. Furthermore, these tests can assess the extent of damage or loss an exploit could cause, such as financial theft, data breaches, or operational downtime, thus quantifying the potential business impact. Once vulnerabilities are addressed or countermeasures are implemented, the same attack-driven tests can be repeated to verify that the fixes or mitigations are effective.

The following is an explanation with examples for attack simulated tests for threat scenarios; a comprehensive list of tests can be found Appendix D: Threat Scenarios Attack Simulation Tests.

Attack Simulated Example #1: Exploit of CWE-79 (XSS)

Threat Scenario TS4: Malicious actors targeting critical processes and assets in the interaction between the dApp UI and the dApp frontend. This involves exploiting vulnerabilities in session management, data integrity, API configurations, and business logic, specifically focusing on weaknesses such as CWE-79 (XSS).

Attack Scenario: An attacker injects malicious scripts into input fields on the dApp frontend. These scripts execute in a victim's browser, allowing the attacker to steal session tokens and impersonate users.

Steps of Attack Simulation Test

1. Simulate the attack by injecting common XSS payloads (e.g. *<script>alert('XSS');</script>*) into vulnerable input fields.
2. Validate if the payload is executed in the browser and whether it provides unauthorized access or sensitive information.

Apply Countermeasures: Implement a Content Security Policy (CSP) that sanitizes the input and encodes the output for all user inputs.

Retest: After implementing CSP and input sanitization/encoding, inject the same payloads to confirm that:

- The payloads are sanitized (e.g. rendered as plain text).
- The browser blocks unauthorized script execution as per the CSP.
- Session tokens remain secure, and no unauthorized actions occur.

Attack Simulated Example #2: Exploit of CWE-362 (Reentrancy)

Threat Scenario: Malicious actors targeting smart contracts exploit vulnerabilities in transaction execution, validation, and state management. This compromises the integrity of loan approval and repayment processes by exploiting the weakness CWE-362 (reentrancy).

Attack Scenario: An attacker creates a malicious smart contract that repeatedly calls a vulnerable loan withdrawal function before the state updates, draining funds from the lending pool.

Attack Simulation Steps

1. Deploy a testing environment with the vulnerable smart contract.

2. Simulate the reentrancy attack by executing a crafted contract that repeatedly invokes the withdrawal function.
3. Observe whether the lending pool's funds are drained due to improper state handling.

Apply Countermeasure: Implement the checks-effects-interactions (CEI) pattern to update the contract state before external calls and use OpenZeppelin's ReentrancyGuard.

Retest: After implementing the countermeasures:

1. Execute the same malicious contract to ensure the state update blocks reentrant calls.
2. Verify that the lending pool's balance remains intact and the withdrawal function prevents unauthorized repetitions.

These attack-driven tests align with the objectives of prioritized attack risks and mitigation recommendations by providing a structured framework to address vulnerabilities and risks. These simulation attack tests reflect how identified vulnerabilities can be exploited by malicious actors. Each test scenario incorporates key elements such as AVs, entry points, CWEs, and targeted assets, demonstrating the realistic paths attackers might take to compromise the system. By executing these tests, organizations can verify the exploitability of vulnerabilities and their impact on assets such as middleware, blockchain nodes, and smart contracts.

Each attack-driven test focuses on critical and high-severity CWEs, ensuring that the most probable and impactful threats are addressed first. For example:

- CWE-502 (deserialization of untrusted data) in threat scenario TS5, "Malicious actors targeting the interaction between the dApp frontend and dApp middleware (JSON-RPC)" is a critical exploit because it can lead to arbitrary command execution, severely compromising middleware functionality.
- CWE-362 (reentrancy) in threat scenario TS8, "Malicious actors targeting smart contracts for DeFi dApp loan and borrowing processes" should be prioritized because if exploited it will cause a critical business impact such as to drain funds, directly impacting financial integrity of the dApp.

Attack-driven tests offer a systematic approach to identifying, testing, and remediating vulnerabilities based on their risk level and potential impact. This prioritization enables organizations to allocate resources effectively and mitigate high-priority threats, reducing the likelihood of successful attacks and minimizing their impact on critical assets.

The countermeasures outlined in the tests provide actionable risk mitigation strategies to address specific control gaps and weaknesses. Examples of these risk mitigation strategies might include:

- **Implementing input validations (e.g. whitelist, sanitization) and transactional data signatures to secure data integrity** for threat scenario TS6 "Malicious actors targeting the interaction between the dApp middleware and blockchain local node" and threat scenario TS9 "Malicious actors target the oracle and its data feed/sources."
- **Enforcing rate limiting (e.g. API rate limiting at API gateways, IP-based rate limiting) and load balancing (i.e. software, cloud ALBs) to mitigate resource exhaustion** for threat scenario TS5 "Malicious actors targeting the interaction between the dApp frontend and dApp middleware (JSON-RPC)" and TS6 "Malicious actors targeting the interaction between the dApp middleware and blockchain local node."
- **Using the CEI pattern and runtime monitoring to prevent reentrancy attacks** in TS8 "Malicious actors target smart contracts for DeFi dApp loan and borrowing processes."

Note: CEI pattern is a secure coding practice designed to reduce the risk of reentrancy by structuring smart contract functions in a secure sequence [264].

These tests also empower dApp security product owners and vulnerability risk managers to effectively manage critical risks since their corresponding mitigation strategies are prioritized for the most critical vulnerabilities and assets, such as transaction validation in middleware or integrity checks in oracle feeds. This ensures that high-priority threats, such as unauthorized data manipulation or DoS attacks, are addressed comprehensively.

Retesting after implementing countermeasures ensures that vulnerabilities remain mitigated and security measures are effective to reduce (i.e. to an acceptable residual risk) or eliminate the risk. This iterative process supports continuous improvement in security posture, ensuring that the dApp DeFi applications and systems are resilient against evolving threats.

The next step of AM is to analyze and simulate adversarial tactics. By studying adversarial tactics and simulating attacks, developers and security teams can focus on proactively identifying weaknesses in the dApp's architecture, such as improper access controls, reentrancy flaws, or API misconfigurations that have been known to be exploited in public disclosed security incidents. These simulations aim to identify exposure of dApp assets to vulnerabilities before these are exploited in real-world scenarios, allowing for timely remediation before production deployments. Analyzing adversarial tactics also helps to contextualize how attackers target specific components, such as smart contracts, oracles, middleware, or wallets. For DeFi lending and borrowing apps, understanding tactics like price manipulation, flash loan exploits, or reentrancy attacks ensures a clearer picture of the threat landscape. Simulating attacks informs the design of robust security controls, such as using cryptographic proofs for transaction integrity, implementing reentrancy guards in smart contracts, and enforcing MFA for wallets.

Simulated attacks allow for testing the effectiveness of countermeasures. For instance, testing a reentrancy guard under simulated attack conditions ensures it mitigates reentrancy vulnerabilities without impacting legitimate functionality. This iterative testing and validation builds confidence in the security of the dApp. Since DeFi lending and borrowing apps handle high-value transactions, simulating attacks that target financial mechanisms (e.g. flash loans, collateral valuations) helps mitigate risks of financial losses. Effective testing ensures that financial workflows are resilient to manipulation or fraud. Last and not the least, understanding adversarial tactics helps prepare robust incident response plans and playbooks. Simulations reveal attack pathways, helping teams define detection, containment, and mitigation strategies for real-world attacks.

Simulating Adversarial Tactics: A straightforward approach to simulating adversarial tactics involves extracting IoCs and TTPs from threat intelligence gathered from previously reported security incidents. By analyzing the attack paths followed by threat actors and identifying the weaknesses and vulnerabilities they exploited, organizations can better understand how attackers achieved their objectives. This process highlights how these attacks resulted in monetary losses for dApp business owners, providing actionable insights for strengthening security measures.

An effective approach to incorporating real-world threat scenarios involves leveraging threat events recorded in security incidents and shared through OSINT or publicly disclosed sources. These "blockchain-specific intelligence sources" enable the replication of real-world attacker behaviors, focusing on the weaknesses and vulnerabilities that have been exploited in past documented attacks. This approach highlights the potential impacts on business functionality, data integrity, and financial losses, including cryptocurrency theft. It also helps address a critical cyber risk question: if threat actors targeted your application using similar TTPs, would it withstand the attacks or experience comparable financial losses?

Table 3.15 provides an example of IoCs and TTPs from publicly disclosed incidents, highlighting the associated CWEs and their severity. These incidents, which have already occurred, serve as evidence of risk by demonstrating the likelihood of exploitation and the quantified monetary losses resulting from these security breaches (please read the note) and on how to put threat intelligence and threat modeling in the right context for execution of risk-centric threat models.

Table 3.15 IoCs and TTPs from publicly disclosed incidents.

Incident occurred date	Attack scenario	Targeted asset	IoCs	TTPs	Primary target CWEs
Lendf.me and Uniswap (April 2020)	Attackers exploited a reentrancy vulnerability that allowed them to repeatedly call a function in the smart contract before the initial execution was completed, leading to unauthorized withdrawals.	Smart contracts for lending and borrowing	• Unusual repeated transactions involving ERC-777 tokens • Drained contract balances • Interaction with imBTC	• Reentrancy attack: Exploited state changes by calling functions before completion • ERC-777 token vulnerability: Leveraged reentrant calls within the token standard	• CWE-703: Improper check or handling of exceptional conditions • CWE-285: Improper authorization • CWE-284: Improper access control • CWE-841: Improper enforcement of behavioral workflow
Aave Protocol (November 2022)	Attackers manipulated the price feed of CRV tokens using oracles to trigger liquidations. This forced collateral liquidations, creating bad debt in the protocol.	Oracles feeding asset price data	• Large-scale borrowing of CRV tokens • Collateral imbalance with disproportionate USDC deposits • Triggered liquidations	• Price manipulation: Leveraged oracles to feed false price data • Liquidation exploit: Forced collateral liquidation by manipulating asset prices	• CWE-345: Insufficient verification of data authenticity • CWE-400: Uncontrolled resource consumption • CWE-20: Improper input validation • CWE-918: Server-side request forgery (SSRF)

(Continued)

Table 3.15 (Continued)

Incident occurred date	Attack scenario	Targeted asset	IoCs	TTPs	Primary target CWEs
Onyx Protocol (November 2023)	Attackers used flash loans to manipulate asset prices and exploit the protocol's lending mechanisms. This caused manipulated collateral calculations and unauthorized withdrawals.	Price-sensitive lending logic	• Flash loan transactions in large volumes • Sudden asset price anomalies • Rapid fund movement across wallets and platforms	• Flash loan attack: Utilized flash loans to manipulate pricing and lending logic • Collateral exploit: Exploited price-sensitive operations via manipulated asset values	• CWE-345: Insufficient verification of data authenticity • CWE-400: Uncontrolled resource consumption • CWE-285: Improper authorization • CWE-502: Deserialization of untrusted data • CWE-284: Improper access control

The critical question is whether the enumerated threats (T1 through T100; refer Appendix A: Threat Modeling Matrix for details) and threat scenarios (TS1 through TS9) can be utilized to develop realistic attack-driven test cases and evaluate the associated risks and potential monetary losses tied to these vulnerabilities.

By examining the targeted assets of these incidents and the corresponding CWEs, alignment with these threat scenarios can be established, for example:

1. **Smart Contract Weaknesses:** Focus on CWE-362 (reentrancy) to test for repeated state manipulation attacks.
2. **Oracle Weaknesses:** Address CWE-345 (insufficient verification of data authenticity) and CWE-502 (deserialization of untrusted data) to validate the integrity of data feeds.
3. **Middleware and Frontend:** Ensure strong session management and API security to prevent unauthorized interactions.

By utilizing these scenarios, you can perform targeted attack-driven tests, simulate adversarial behaviors, and prioritize the remediation of high-risk vulnerabilities to prevent potential losses. This comprehensive approach ensures secure by design practices are effectively implemented across all components of a DeFi dApp.

Note: The PASTA methodology integrates threat identification with threat classification through the use of threat intelligence and structured libraries (e.g. CAPEC, MITRE ATT&CK). While the methodology accounts for this as part of its sequential stages, particularly in Stage III (Threat Analysis) and Stage VI (Attack Modeling), it is important to recognize that tables mapping real-world IoCs and TTPs as shown above can and ideally should be curated and maintained outside the immediate scope of the threat modeling exercise by a threat intelligence team. In mature security programs, such references often already exist and are continuously updated. However, in organizations where this function is absent or underdeveloped, threat modeling teams may need to work closely with threat analysts or even take on this responsibility themselves. In such cases, adopting a hybrid approach where threat modeling and threat intelligence functions collaborate in parallel can significantly streamline the process. Relying solely on Stage VI to begin building these mappings may delay the overall modeling effort, especially when retroactively analyzing past threat events. Proactive collaboration and early integration of threat intelligence greatly enhance the speed and fidelity of the risk modeling process.

Simulating Adversarial Tactics with MITRE ATT&CK: To dive in deeper in the simulation of real world attacks, we could leverage standard attack frameworks such as the MITRE ATT&CK for blockchain. The MITRE ATT&CK for blockchain is an extension of the MITRE ATT&CK framework, which is a globally recognized, structured knowledge base of adversarial TTPs used by attackers. Specifically, the blockchain domain of the framework focuses on threats and vulnerabilities unique to blockchain-based systems, including smart contracts, dApps, consensus mechanisms, and cryptocurrency platforms. The MITRE ATT&CK for Blockchain framework organizes blockchain-related attack methods into tactics and techniques that illustrate how adversaries exploit blockchain technologies, providing a detailed view of potential AVs. This can help security teams anticipate, detect, and mitigate attacks tailored to the unique characteristics of blockchain systems. We can show here how MITRE ATT&CK for blockchain [195] (please read the note) can be used to simulate and address these specific weaknesses across smart contracts, oracles, middleware, and frontend components in the context of threat scenarios we analyzed.

Example Attack Simulations with MITRE ATT&CKs Techniques include:

- **Attack Smart Contract Weakness CWE-362 (Reentrancy)**
  - **MITRE ATT&CK Technique:** Exploitation of smart contracts (T1580), which focuses on reentrancy attacks where attackers repeatedly invoke a function in a smart contract before the state is updated.
  - **Attack Simulation:** To simulate this attack, set up a controlled environment with a deliberately vulnerable smart contract, such as one implementing the ERC-777 token standard. Using MITRE ATT&CK, emulate adversarial TTPs by crafting a malicious contract designed to repeatedly call the withdrawal function of the vulnerable contract. Observe how the reentrancy vulnerability allows the malicious contract to drain funds by exploiting the failure to properly handle state updates during reentrant calls.
  - **Apply Countermeasures:** After simulating the attack, test countermeasures by implementing the CEI pattern within the smart contract and integrating protections such as OpenZeppelin's ReentrancyGuard.
  - **Retests:** Rerun the simulation to verify that the countermeasures effectively block reentrant calls and prevent unauthorized withdrawals. The expected outcome is the mitigation of repeated state manipulation attacks, ensuring the integrity of the lending and borrowing functionalities in the dApp. By addressing CWE-362 through attack simulation and countermeasure validation, the smart contract's resilience to this type of exploit is significantly improved.
- **Attack Oracle Weaknesses CWE-345 (Insufficient Verification of Data Authenticity) and CWE-502 (Deserialization of Untrusted Data)**
  - **MITRE ATT&CK Techniques:** Oracle manipulation (T1582) and data injection (T1591) techniques target vulnerabilities in oracles by tampering with data feeds or exploiting insecure deserialization processes to manipulate the dApp's decision-making and data integrity.
  - **Attack Simulation:** To simulate an attack exploiting CWE-345, set up a controlled environment with an oracle integrated into a smart contract. Use MITRE ATT&CK to emulate adversarial TTPs by feeding the oracle

falsified price data or manipulated external inputs. Observe how the absence of proper data authenticity checks allows the manipulated data to influence critical operations, such as altering collateral values or triggering unauthorized liquidations in lending and borrowing workflows. For CWE-502, simulate an attack by deserializing manipulated data payloads from an external source. This can involve crafting a malicious input that, when deserialized by the oracle or middleware, introduces unexpected or malicious behavior, such as bypassing validation rules or injecting fraudulent transactions.

  - **Apply Countermeasures:** Implementing robust cryptographic validation of oracle data (e.g. signed and verified data feeds) and secure deserialization practices, such as sanitizing and validating incoming data before processing. Rerun the simulation to ensure that the countermeasures prevent oracle manipulation and deserialization attacks, preserving the integrity of the data and the associated dApp operations.
  - **Retest:** The expected outcome is the mitigation of attacks exploiting insufficient data verification and insecure deserialization, ensuring the reliability and accuracy of oracle-supplied data and protecting the dApp's lending and borrowing functionalities. By addressing CWE-345 and CWE-502 through attack simulation and countermeasure validation, the dApp's resilience to these types of exploits is significantly enhanced.

- **Frontend Weaknesses CWE-284 (Improper Access Control) and CWE-287 (Improper Authentication)**
  - **MITRE ATT&CK Techniques:** API exploitation (T1586) and session hijacking (T1584) techniques focus on the importance of securing session management and API interactions to prevent unauthorized access and ensure robust access controls.
  - **Attack Simulation:** To simulate the attack, a session hijacking scenario can be emulated by stealing session tokens that are stored insecurely, such as in local storage, or intercepting them during unencrypted communication. Weak authentication mechanisms, such as improperly validated tokens or absent session expiration policies, could then be exploited to gain unauthorized access to the application. Similarly, API exploitation can be simulated by crafting unauthorized requests to middleware or frontend components, bypassing authentication or authorization mechanisms. Using the MITRE ATT&CK framework, adversarial tactics can be replicated to assess how attackers might manipulate session tokens or inject malicious payloads to compromise vulnerable endpoints.
  - **Apply Countermeasures:** Countermeasures for session management involve implementing secure session tokens, such as JWTs, with short expiration times and mandatory rotation policies. Tokens should be stored

securely, avoiding local storage for sensitive data, and transmitted exclusively over encrypted channels. Robust session validation techniques, such as IP binding or device verification, should also be applied. For API security, enforcing strict authentication mechanisms, including API keys, OAuth, or mTLS, helps protect middleware and frontend endpoints. Input validation and rate limiting can mitigate abuse, prevent brute force attacks, and reduce resource consumption. Tools like DAST or custom testing scripts can be used to validate the effectiveness of these security measures against the simulated attacks.

- **Retest:** The expected outcome of these simulations is the validation that unauthorized interactions with middleware and frontend components are effectively blocked, ensuring session integrity, proper access control, and secure API communication. By addressing CWE-284 and CWE-287 through attack simulation and countermeasure validation, the dApp's resilience against threats targeting session management and API vulnerabilities is significantly strengthened.

Practical applications of MITRE ATT&CK for blockchain extend to various use cases, including enhancing the security of DeFi dApps, enterprise blockchain systems, and cryptocurrency wallets and exchanges. In the context of DeFi dApps, this framework can be used to assess risks associated with lending and borrowing protocols by modeling attacks such as reentrancy, oracle manipulation, or flash loan exploits. It also supports validating transaction integrity and ensuring asset security within DeFi ecosystems.

For enterprise blockchain systems, MITRE ATT&CK enables organizations to test the resilience of consensus mechanisms and permissioned blockchain implementations against node compromise or tampering. It also provides a structured approach for validating secure communication channels between blockchain nodes, ensuring data integrity and system reliability in enterprise deployments. When applied to cryptocurrency wallets and exchanges, the framework facilitates the simulation of phishing attacks or private key theft scenarios to identify and mitigate weaknesses in wallet implementations. Additionally, it helps protect cryptocurrency exchanges from advanced persistent threats targeting user accounts or funds. By leveraging MITRE ATT&CK for blockchain, organizations can model real-world threats, identify vulnerabilities, and implement targeted countermeasures to strengthen their overall security posture.

Note: While the MITRE ATT&CK for blockchain matrix is highly relevant to DeFi and dApp security, it should not be used in isolation. Additional MITRE ATT&CK matrices, such as the mobile matrix (given the mobile interfaces commonly used in dApps) and the enterprise matrix (for covering traditional web applications,

APIs, and supporting infrastructure), should also be referenced. These provide broader coverage of TTPs that may be leveraged in hybrid environments where blockchain technologies coexist with conventional application stacks.

Attack Visualizations with Cyberattack Kill Chain: Another method that can be used for visualizing the sequential steps that an adversary can follow to carry out a cyberattack is the kill chain methodology [265]. The cyberattack kill chain provides a structured approach to understanding and analyzing attack methodologies, helping defenders identify vulnerabilities and implement countermeasures at each stage. By visualizing adversarial TTPs, the kill chain enhances proactive security measures, enabling organizations to disrupt attacks before they achieve their objectives.

This approach is particularly valuable in risk-centric threat modeling, such as PASTA, where it aids in attack scenario validation during the AM stage. The kill chain for CWE-362 (reentrancy), as observed in the context of threat scenario TS8, involves a sequence of attacker steps targeting vulnerabilities in DeFi smart contracts for lending and borrowing. The stages include reconnaissance, where attackers identify contracts with reentrancy vulnerabilities using tools like MythX or Slither.

In the weaponization stage, attackers craft malicious contracts designed to exploit these flaws programmed for invoking vulnerable functions. During delivery, the malicious contract is deployed onto the blockchain, exploiting weaknesses such as improper access control. The exploitation stage triggers the reentrancy flaw in the vulnerable contract, enabling the malicious contract to bypass state updates. Installation involves executing reentrant calls to withdraw funds from the lending pool repeatedly. In the command and control (C2) stage, attackers maintain control over the exploit, ensuring repeated invocations. Finally, the actions on objective stage sees attackers draining the contract funds, achieving unauthorized withdrawals.

This cyber kill chain of events can be visualized in Figure 3.5.

Attack Visualizations with Attack Trees: Another structured method for modeling potential attack scenarios against a system or application are "attack trees." Attack trees represent attacks as a hierarchical tree structure, starting with a root node that defines the attacker's primary goal, such as unauthorized fund withdrawal or data theft. Each branch represents different attack paths, broken down into intermediate steps or subgoals, showing how an adversary might exploit weaknesses to achieve their objective.

This visualization helps security teams analyze vulnerabilities, evaluate the likelihood and impact of different attack paths, and prioritize defenses. Attack trees are particularly effective for identifying attack surfaces in complex systems like DeFi dApps, where various components (smart contracts, oracles, APIs)

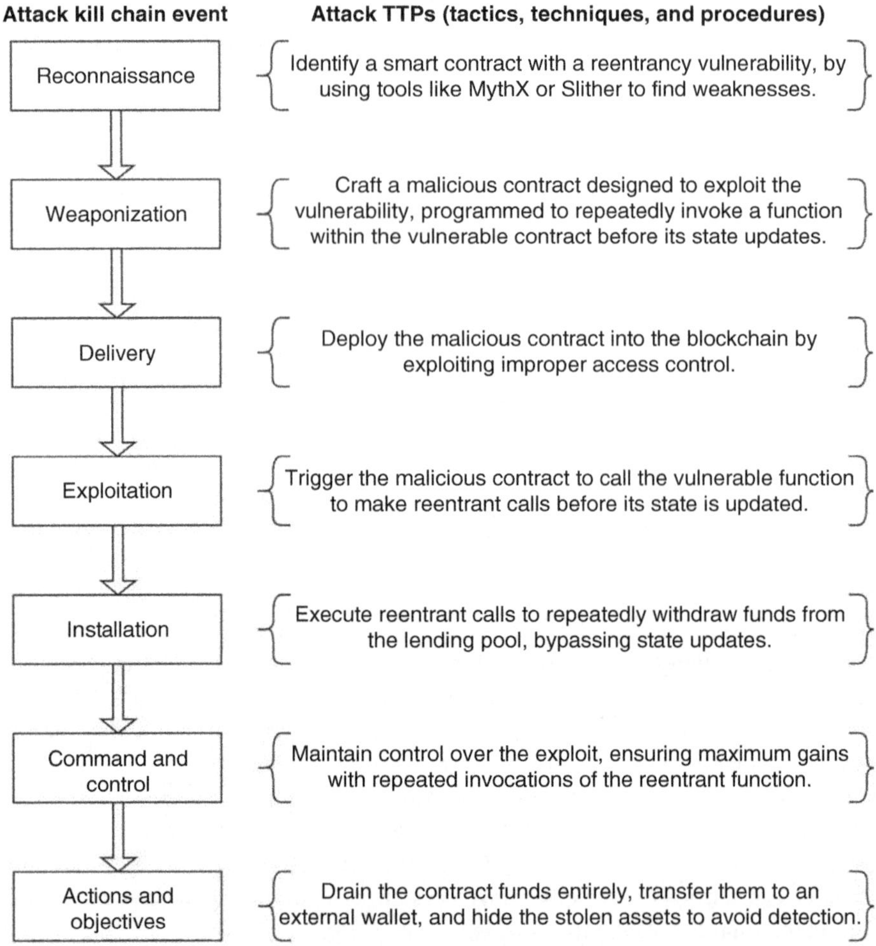

Figure 3.5 dApp attack kill chain.

may be targeted. By mapping realistic attack scenarios, organizations can better understand potential risks, design countermeasures, and enhance the system's resilience against adversarial threats.

In the case of DeFi dApp for lending and borrowing, for threat scenario TS4: dApp frontend and components, exploiting CWE-79 (XSS) and TS5: dApp middleware (JSON-RPC) exploiting (deserialization of untrusted data) with the attacker goal to compromise the dApp such as to manipulate or withdraw funds. This goal can be reached by exploiting each vulnerability separately or in combination. The various attack paths that can be followed by an attacker are shown in Figure 3.6.

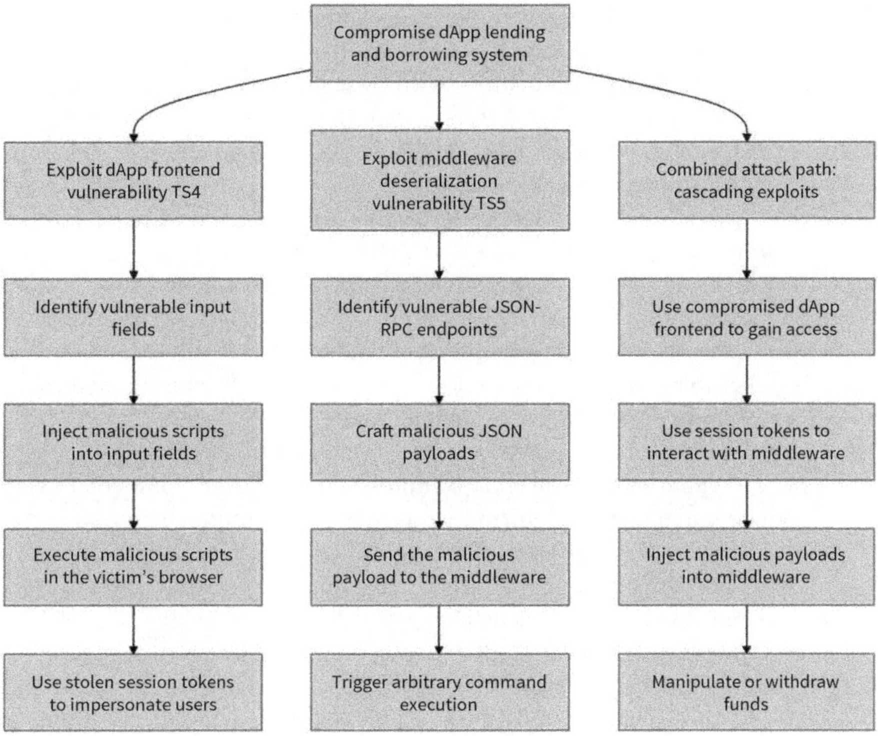

Figure 3.6 dApp attack tree.

This combined attack tree visually and logically demonstrates how attackers can exploit vulnerabilities in both the frontend (TS4) and middleware (TS5), progressing step by step to compromise the dApp's lending and borrowing functionality. It emphasizes the interconnected risks within complex systems and highlights critical points for applying security countermeasures:

1. **Left Branch: Exploit dApp Frontend Vulnerability**
   - Step 1 involves identifying vulnerable input fields using tools like Burp Suite or OWASP ZAP to detect areas lacking proper sanitization, which are susceptible to script injection.
   - Step 2 follows with injecting malicious scripts by inserting a crafted XSS payload into the identified vulnerable input field.
   - Step 3, the malicious scripts are executed in the victim's browser, allowing the attacker to steal session tokens, capture user credentials, or manipulate user sessions.

- Step 4, the attacker uses the stolen session tokens to impersonate users and execute unauthorized transactions on the dApp.

2. **Central Branch: Exploit Middleware Deserialization Vulnerability**
   - Step 1 focuses on identifying vulnerable JSON-RPC endpoints by leveraging fuzzing tools or reconnaissance techniques to locate middleware endpoints with deserialization weaknesses.
   - Step 2 involves crafting malicious JSON payloads designed to exploit the identified deserialization flaws in the middleware.
   - Step 3, the attacker sends the crafted payload to the vulnerable middleware endpoint.
   - Step 4, the payload triggers arbitrary command execution, enabling the attacker to manipulate transaction data, execute malicious commands, or compromise the stability of the middleware.
3. **Right Branch: Combined Attack Path – Cascading Exploits**
   - Step 1 begins with the attacker compromising the dApp frontend (TS4) to gain an initial foothold, obtaining user session tokens in the process.
   - Step 2, these stolen session tokens are used to interact with middleware endpoints (TS5), allowing the attacker to authenticate as a legitimate user.
   - Step 3 involves injecting malicious payloads into the middleware, exploiting deserialization vulnerabilities to execute arbitrary commands and compromise transaction integrity.
   - Step 4, the attacker chains these exploits to manipulate transactions or withdraw funds, achieving financial gain by targeting both the frontend and middleware components.

Attack trees can be a powerful tool for analyzing the potential paths an attacker might take to achieve their objectives. One critical application of attack trees is associating each path with the cost for an attacker to achieve their end goal. This analysis helps identify the path that is least costly and, therefore, the most likely to be exploited. The cost for an attacker depends on several factors that influence their decision-making:

First, complexity of execution plays a significant role. Attack paths requiring fewer resources, skills, or specialized tools are inherently less costly and more appealing to attackers. Next, the accessibility of the attack surface is crucial; entry points like the dApp frontend, which are directly exposed to external users, are typically easier to exploit compared to those requiring deeper system penetration, such as middleware endpoints. Additionally, the number of steps in the attack path impacts the cost; shorter paths with fewer sequential steps tend to reduce the overall complexity and effort required. Lastly, the risk of detection influences the cost significantly; paths that minimize interactions with monitoring systems or bypass security controls lower the risk of exposure and make these paths less costly for an attacker to pursue; see Ref. [266] for the cost-damage analysis of attack trees on how this can be done in practice.

By combining these factors with an attack tree's structure, security architects can identify high-risk paths that attackers are most likely to exploit, enabling focused mitigation strategies to strengthen defenses where they are most needed.

For the threat scenarios we have scoped for attack tree analysis TS4, TS5 and TS4 + TS5 combined as cascading exploits, we can rank the costs as following:

1. **Path A: Frontend Exploitation:** This path involves injecting malicious scripts into input fields to steal session tokens, impersonate users, and execute unauthorized transactions. It requires basic web scripting skills and targets publicly accessible components, making reconnaissance and execution straightforward. The success probability is high if input validation and CSPs are weak.
   Cost Ranking: Low
2. **Path B: Middleware Exploitation:** This path involves crafting and sending malicious JSON-RPC payloads to middleware endpoints, triggering arbitrary command execution and compromising transaction integrity. It requires knowledge of deserialization processes and API endpoints, with added complexity due to potential authentication and monitoring mechanisms. This increases the effort, time, and risk of detection.
   Cost Ranking: Moderate
3. **Path C: Combined Exploitation Path:** This combined approach uses XSS to capture middleware API tokens, followed by exploiting deserialization vulnerabilities to execute malicious commands. It requires proficiency in multiple attack techniques and involves more steps, making it more complex and increasing the risk of failure or detection.
   Cost Ranking: High

We could assume that the least costly path is Path A: Frontend Exploitation due to its simplicity and accessibility. Strengthening security measures like strict input validation, CSPs, and secure session management can effectively mitigate this risk for the vulnerabilities that could be exploited in the threat scenarios we have evaluated.

In summary, simulating attacks and AVs to derive scenarios that validate the effectiveness of countermeasures for both theoretical (potentially possible) and realistic threats (those resulting in documented breaches or incidents) is essential.

By analyzing real-world security incidents, leveraging sources of threat intelligence, adversarial tactics, and frameworks such as MITRE ATT&CK, and visualizing attack steps through kill chains and attack trees, organizations can bridge the gap between theoretical threat modeling and practical security assessments. This approach enables teams to identify exploit paths, analyze AVs, and perform targeted testing to validate a dApp's security measures. It also facilitates the development of actionable risk mitigation strategies. Without AM, risk assessments provide only estimates of likelihood and impact. With AM, organizations can move from theoretical probability to measurable risk exposure, ensuring a precise and actionable risk management strategy.

Integrating AM into the threat modeling process ensures that security teams move beyond threat likelihood and impact estimations and into validated impacts and effect of countermeasures in reducing or eliminating the risks. By incorporating real-world adversary TTPs, testing security controls, and demonstrating exploit feasibility, organizations can effectively prioritize risk mitigation efforts before a dApp is deployed into a high-risk environment. This data-driven approach not only strengthens application security but also aligns with broader risk management and compliance objectives, enabling organizations to proactively address security gaps before they are exploited.

These activities align with the next and final stage of PASTA risk-centric threat modeling: Risk Analysis and Management.

3.2.3.8 Stage VII – Risk Analysis and Management

The RAM stage of the PASTA threat modeling framework is a critical phase for ensuring the risks are controlled and managed enabling business and customers trustworthiness of the DeFi dApp.

This stage is essential for any business leader, product security owner, or executive responsible for the risk and security posture of the dApp, as it aligns security objectives with business imperatives ensuring that the dApp remains resilient and compliant and security is credible to customers and stakeholders.

At this stage, the focus is on systematically evaluating and prioritizing risks using intelligence from threat identification, vulnerability assessments, and attack simulations. By assessing the likelihood and impact of previously analyzed threat scenarios, we establish a documented linkage between threats, vulnerabilities, affected dApp components, and business assets.

This structured mapping allows for a comprehensive risk perspective, helping decision-makers understand how vulnerabilities could be exploited to compromise smart contract logic, liquidity pools, user funds, or transaction integrity.

The RAM stage is a risk-driven decision-making process aimed at:

- Analyzing risks based on identified threats, vulnerabilities, and attack simulation data
- Prioritizing risks according to their likelihood and business/technical impact
- Developing targeted mitigation strategies that integrate with security, compliance, and operational objectives

A critical outcome of this stage is the ability to deliver actionable risk analysis that informs key decision-makers/owners, whether a CISO, head of product security, or a business risk owner, on how best to mitigate potential threats before they materialize. These risk management insights serve as a foundation for implementing effective security measures, reducing financial exposure, ensuring regulatory compliance, and building customer trust in the dApp ecosystem. Even when applying a comprehensive risk analysis and management methodology, one that evaluates all risks from the perspective of threats, vulnerabilities, and their impact on assets, both technical and organizational, the ultimate goal for CISOs and head of product security, is to inform business-driven, risk-aware decision-making. However, it is equally important not to lose sight of a practical approach when determining which risks to mitigate. The decision to remediate should not rely solely on the criticality of a vulnerability as measured by severity or risk score. Additional factors such as the effort required to remediate, the ease of exploitability, and the characteristics of potential threat agents must be taken into account.

In many real-world scenarios, lower-severity vulnerabilities, especially those that are easy to discover and exploit, can become attractive targets for opportunistic attackers. These so-called low-hanging fruits may not rank highest in a theoretical risk matrix, yet they can still result in meaningful business impact if left unresolved. Furthermore, such vulnerabilities are often quick and inexpensive to fix, offering a high return on investment (ROI). Proactively addressing them can help reduce the overall attack surface, strengthen the security baseline, and create momentum within security and development teams. This pragmatic, context-driven approach to risk treatment enables more efficient resource allocation and better reflects the behavior and priorities of actual adversaries.

Ultimately, the goal for CISOs and product security heads is to ensure that the DeFi dApp is secured, resilient, and protected, not just as a technical system but as a business-critical product that directly impacts organizational success and user confidence. The RAM stage of PASTA risk-centric threat modeling acts as the final checkpoint before deployment, ensuring that security risks are not just understood but actively mitigated, preserving the integrity and long-term viability of the DeFi platform.

One of the key final outcomes of the risk-based threat assessment is the actionable implementation of security measures that secure the dApp by design, by development, by deployment, and by ongoing operations. This ensures that the dApp is protected against real-world threats by integrating proactive security controls that mitigate risks posed by threat actors targeting its attack surface.

A critical artifact produced in this stage is a comprehensive risk register, documenting identified risks across previously analyzed threat scenarios. This includes an evaluation of potential threat events, weaknesses, and vulnerabilities that could be exploited to compromise dApp assets. The risks are ranked based on severity, likelihood, and their technical and business impact, particularly in the context of DeFi lending and borrowing operations.

The risk register not only contains detailed descriptions of each risk but is also supported by evidence from attack simulations, offering an in-depth assessment of the estimated likelihood and impact of these attacks after countermeasures are applied. This allows for an evaluation of the residual risk, ensuring that mitigation strategies are effectively reducing the overall risk exposure of the dApp.

Furthermore, securing the dApp requires a holistic approach that addresses proactive risk mitigation measures across multiple layers:

- **Securing the dApp by Design:** Incorporating threat modeling outcomes to identify trust boundaries, attack surfaces, and security-critical components
- **Securing the dApp by Development:** Applying secure coding practices, secure libraries, and automated security testing to mitigate vulnerabilities before deployment
- **Securing the dApp by Deployment:** Implementing hardened configurations, runtime security monitoring, and defensive controls to minimize exposure
- **Securing the dApp by Operations:** Enforcing continuous monitoring, adaptive threat intelligence, and proactive security measures to defend against evolving attack tactics

By linking threat scenarios to specific AVs and exploited vulnerabilities, this risk assessment output serves as a strategic reference for prioritizing security investments, refining DevSecOps processes, and ensuring regulatory compliance.

Ultimately, the risk management stage empowers security leaders, product security teams, and business risk owners with a structured, risk-driven approach to securing the dApp against active and emerging threats while preserving trust, financial integrity, and operational resilience.

The initial step in dApp risk management should focus on analyzing threats in the context of mapped weaknesses and vulnerabilities identified during the VA stage. Details of these mappings are in Appendix B: Mapping of Threat Scenarios to Targeted Weaknesses and Asset Impacted. These weaknesses and vulnerabilities represent opportunities for early remediation, particularly during the design phase, as they are identified through architectural analysis and threat scenario modeling in Stage IV. By proactively addressing these risks before development, organizations can strengthen security by design in the most cost-efficient way and reduce exposure to potential exploits.

In application security risk management, mitigation efforts should begin as soon as risks are identified through architecture risk reviews and initial threat modeling exercises. This process involves analyzing the dApp's architecture, its various components, and their exposure to potential threats and vulnerabilities. Weaknesses, classified using CWEs, and vulnerabilities, identified through CVEs, are typically assigned risk severity scores using standard methodologies such as CVSS. This scoring mechanism provides a structured approach for prioritizing remediation based on severity, categorizing vulnerabilities as critical, high, medium, or low risk.

The most critical aspect of addressing these risks is ensuring organizational visibility and aligning risk management efforts with GRC frameworks. This is a fundamental responsibility for CISOs, CSOs, and heads of product security, who operate within the organization's vulnerability management policy. Effective risk mitigation strategies should be tied to defined SLAs, particularly for DeFi dApp services, to ensure timely remediation and compliance with internal security policies and external regulatory expectations.

Establishing clear SLAs for vulnerability management including enforcement, tracking, and measurement through security metrics is one of the most impactful steps an organization can take to make application security risk management actionable. By embedding secure by design principles into the risk management process, organizations can proactively mitigate vulnerabilities, enhance trust and security in DeFi applications, and ensure long-term resilience against evolving cyber threats. It signifies that the organization has adopted and formalized an SLA for vulnerability remediation, independent of the tools or processes used to identify these vulnerabilities, whether through threat modeling, security architecture reviews, vulnerability scanning, static and dynamic analysis, or penetration testing. If an organization has not yet adopted a vulnerability risk management policy, the first step should be to establish one that addresses the risks associated with CWEs and CVEs identified during VA. This policy should define remediation timelines aligned with industry standards, such as FedRAMP-like SLAs [267], ensuring vulnerabilities are addressed based on their severity. By implementing a structured vulnerability RMF, the organization can demonstrate its commitment to reducing risk exposure, improving security posture, and enhancing trust among users and stakeholders.

In addition to establishing a vulnerability risk management policy and defining SLAs for remediation, it is essential to implement a structured vulnerability risk management process. This process begins with vulnerability assessment, where identified weaknesses (CWEs and CVEs) are evaluated and assigned CVSS severity scores. In remediation planning, each vulnerability is mapped to an SLA, and remediation timelines are tracked.

Vulnerability Remediation SLAs: In vulnerability remediation planning, each vulnerability severity is remediated based upon a vulnerability remediation SLA, which should be in place at organizational level and determined based on multiple factors, such as alignment with business risk management, and product management objectives. For the SLA to be effective, organizations must first acquire the capability to identify, triage vulnerabilities to filter out false positives, and initiate remediation through tracked tickets, ensuring that fixes are planned, implemented, and tested (aka closed) within the designated SLA timeframes.

It is important to emphasize that remediation, in the context of an SLA, is only considered complete when the solution is fully deployed in the production environment and has been verified for effectiveness. This means the fix must not only be coded but also have passed all required QA or staging tests, integrated into a release cycle, deployed, and confirmed to mitigate the original vulnerability without introducing regressions. Without this E2E closure, remediation cannot be deemed fulfilled under the SLA, and the associated risk remains open from both a security and compliance perspective. This process requires close collaboration between engineering, information security, and risk management teams to ensure alignment with security and business goals. When adopting a standardized SLA, such as a CISA's *Vulnerability Remediation Requirements for Internet-Accessible Systems* [268], vulnerabilities should be addressed based on severity (please read the note):

- Critical vulnerabilities must be remediated within 15 days from identification/recording.
- High vulnerabilities must be remediated within 30 days.
- Medium vulnerabilities must be remediated within 60 days.
- Low vulnerabilities must be remediated within 180 days.

To enforce these SLAs, vulnerability remediation tracking, metrics, and measurements must be in place, ensuring timely resolution through vulnerability scanning, patch management, and continuous monitoring to prevent SLA violations. Accountability involves maintaining a vulnerability register that records CWE/CVE IDs, CVSS scores, and remediation status while ensuring transparency through regular stakeholder updates.

An example of vulnerability management SLA applied to dApp identified vulnerabilities is shown in Table 3.16.

Note: While standardized SLAs such as those defined in CISA's Vulnerability Remediation Requirements for Internet-Accessible Systems provide a valuable baseline for remediation timelines (e.g. 15 days for critical vulnerabilities), organizations must apply these thresholds with contextual flexibility. For example, if a critical vulnerability (such as a CVE in a newly introduced library) is discovered during development and has not yet been released to production, the SLA should mandate that it be remediated prior to the next release, rather than relying on the 15-day production window. Furthermore, the 30-day window for critical issues in some SLA

Table 3.16 Vulnerability remediation SLA.

Vulnerability	CWE ID	Severity	SLA	Status	Target completion date
Input validation	CWE-20	High (7.8)	30 days	In progress	2025-02-15
Reentrancy	CWE-362	Critical (9.1)	15 days	Open	2025-02-01
Sensitive logging	CWE-532	Medium (5.4)	60 days	Not started	2025-03-15
Weak encryption	CWE-326	Low (3.8)	180 days	Open	2025-07-28

models may be too long in high-risk scenarios. Attacks like Log4Shell (Log4j) demonstrated how rapidly threat actors exploit critical vulnerabilities in the wild. In such cases, businesses and product owners who ultimately own the risk must make informed risk management decisions based on operational feasibility, release cycles, and organizational capabilities. Prioritizing critical security fixes in hotfix or out-of-band releases may be necessary. Where immediate remediation is not possible, compensating controls such as custom WAF rules, runtime protection, or configuration hardening should be deployed to reduce exposure until a proper fix is implemented.

Fixing vulnerabilities in isolation without contextualizing their likelihood and impact within real-world threat scenarios can lead to inefficient mitigation efforts that do not effectively reduce risk. For instance, a medium-severity vulnerability flagged by a scanner may receive undue attention while a seemingly low-severity flaw, when correlated with a likely threat actor and AV, may actually pose a far greater business risk. To address this, Appendix I: Risk Analysis Report provides a concrete example of how the OWASP risk rating methodology can be applied to calculate risk scores for each threat scenario. These calculations consider both technical severity and business impact, including threat agent attributes and vulnerability factors, ensuring that remediation priorities are driven by realistic, context-aware assessments rather than raw severity alone.

One of the primary values of conducting a threat modeling exercise is to ensure that mitigation efforts are focused on reducing the actual risks that specific threat scenarios pose to the organization's critical assets, such as the DeFi lending dApp being developed and deployed.

Prioritizing vulnerabilities, whether identified through application vulnerability assessments via scanning tools or as security design flaws during architecture reviews, does not always require direct mapping to threat scenarios. However, integrating threat likelihood and impact analysis into the vulnerability management process provides a more comprehensive and risk-driven approach to security. This approach moves beyond remediating vulnerabilities based solely on severity scores (e.g. CVSS ratings) and instead considers how these vulnerabilities could be exploited in a real attack, leading to both technical and business consequences.

By framing vulnerabilities within relevant threat scenarios, security teams can assess which weaknesses pose the greatest risks to the dApp's security, financial integrity, and compliance obligations. A vulnerability might have a high CVSS base score, but if there is no feasible attack path or adversary motivation, the actual risk to the organization may be lower. Conversely, a medium-severity vulnerability in isolation may pose a critical risk when chained with other weaknesses in a highly probable attack scenario.

A risk-centric approach ensures that remediation efforts are prioritized effectively, focusing on threat scenarios that are most likely to be realized and that pose the most significant risk to the dApp and its users. This allows the organization to optimize security resources, reduce unnecessary remediation efforts, and ensure the dApp is designed, developed, deployed, and operated with proactive security controls that directly mitigate risks posed by real-world adversaries rather than just addressing vulnerabilities in isolation. For example, in a DeFi lending dApp, identifying a missing input validation check in a smart contract that calculates collateral-to-loan ratios, potentially enabling price manipulation via oracle exploits, should be prioritized over a frontend library with a known but nonexploitable vulnerability. While both may be flagged during code reviews or scanning, the former aligns directly with a high-impact threat scenario such as the draining of lending pools due to manipulated loan parameters. Prioritizing such vulnerabilities ensures that risk remediation efforts are tied to actual business impact, safeguarding user funds and protocol integrity in ways that are directly meaningful to the dApp's threat landscape.

For application security engineering managers, the primary focus is ensuring that the dApp application, its functionality, and its components, including internally developed smart contracts and externally integrated platforms such as blockchain backend, are secure by design. This involves identifying and addressing gaps in security controls, particularly authentication, authorization, input validation, and vulnerabilities that could expose critical dApp assets. Given that the dApp facilitates lending and borrowing of customer crypto assets, its security posture must be proactively strengthened to mitigate risks to both users and the organization.

From a product security perspective, threat modeling must be integrated into the secure product life cycle and aligned with modern agile and DevSecOps practices. Similar to other automated security tools such as SAST, SCA, DAST, and RASP, threat modeling should be incorporated into development workflows and triggered by key events such as code merges, pull requests, builds, and deployments. Security assessments should also be executed during integration, QA, and staging phases to ensure that potential risks are identified and mitigated prior to production deployment, supporting a continuous and proactive approach to product security.

The core objective is to embed threat modeling as an integral part of the SDLC, ensuring that security is not a one-time checkpoint but a continuous and iterative process. By aligning threat modeling with development workflows, actionable risk mitigation strategies can be proactively implemented, evaluating vulnerabilities within the context of real-world attack scenarios rather than in isolation.

This risk-based approach not only enhances security but also strengthens the organization's liability management. By demonstrating due diligence in securing its dApp as a critical business asset, the organization reinforces its commitment to business continuity, regulatory compliance, and customer trust critical elements for maintaining a competitive and resilient DeFi ecosystem.

For organizations operating in regulated environments such as FinTech and Financial Services, there is a top-down management focus on ensuring that any products or services offered to customers comply with security, auditing, and regulatory requirements. This is particularly relevant for a DeFi dApp designed for lending and borrowing, as it must adhere to security by design principles while aligning with financial compliance mandates.

For example, a DeFi lending platform handling cryptocurrency transactions may be subject to AML and KYC regulations, requiring robust security measures for identity verification, transaction monitoring, and fraud prevention. Similarly, a dApp facilitating cross-border transactions may need to comply with FATF guidelines, ensuring that transaction flows are secure and traceable to prevent illicit activities.

When a dApp is developed and deployed, it becomes an extension of the organization's RMF, particularly if it handles sensitive customer data such as PII, transaction records, or financial credentials. The need to comply with privacy laws such as GDPR, CCPA, and Markets in Crypto Assets (MiCA) further reinforces the necessity of implementing data protection controls, secure authentication mechanisms, and encryption to mitigate potential breaches.

Industries such as banking, payments, and financial services have long-established security risk assessment and audit practices, ensuring compliance with international frameworks like ISO 27001 that is international standard for information security management systems (ISMS) and applies across all industries, including blockchain. Helps demonstrate a systematic approach to managing sensitive information securely. SOC 2 (System and Organization Controls Type 2) that is used to evaluate how an organization manages data based on trust services criteria such as security, availability, processing integrity, confidentiality, and privacy. SOC 2 is critical for software as a service (SaaS)-based DeFi services and platforms offering B2B integrations or serving institutional clients. PCI-DSS (Payment Card Industry Data Security Standard) is aimed to protect cardholder data and applies to all entities that store, process, or transmit credit card information and is important when DeFi platforms or custodial wallets interact with fiat payment systems or offer payment card integrations. NIST SP 800-53 is a US standard but is widely adopted outside the United States too, especially for blockchain projects involving federal contracts or aligning with critical infrastructure requirements.

In Europe, compliance with the TIBER [269] framework is primarily directed at traditional financial institutions, central banks, and payment infrastructure operators. However, the methodology is increasingly being referenced in the assessment of cyber resilience across broader fintech ecosystems, including blockchain-based platforms. This trend reflects a growing regulatory focus on extending accountability and security assurance practices to DeFi applications. The CRA [270] is a proposed EU regulation aimed at ensuring that digital products and software (including connected devices, applications, and cloud-based services) are secure by design and remain secure throughout their life cycle. While not DeFi-specific, any blockchain-based product marketed in the EU, particularly those interfacing with users, financial data, or acting as service layers, must comply. CRA [271] also recognizes open-source components as part of a shared security responsibility. The NIS2 (Directive on Security of Network and Information Systems) applies to "essential" and "important" entities in sectors such as banking, financial market infrastructure, digital infrastructure, and ICT service providers, which can include blockchain node operators, wallet services, and critical DeFi infrastructure. Although noncustodial DeFi platforms might currently fall in gray areas, infrastructure providers (e.g. oracles, nodes, DeFi aggregators, custodians, off-chain service connectors) may be considered critical operators, and therefore must implement structured cybersecurity governance aligned with NIS2.

For DeFi projects, demonstrating adherence to regulatory and security best practices is not only about compliance but also about building customer trust and securing institutional adoption. As regulatory scrutiny on digital assets increases, embedding compliance-driven security measures into dApp development, deployment, and operations will be a key differentiator for long-term sustainability and market credibility.

The adoption of a risk-based approach to threat modeling can be driven by compliance requirements, as it provides evidence that the dApp is designed following security by design principles while also protecting critical data assets. This includes payment transactions, financial operations involving crypto assets, and sensitive customer data such as PII, where applicable. Ensuring compliance is not just about meeting regulatory mandates but also about demonstrating that the dApp has built-in security controls to mitigate risks associated with fraud, unauthorized access, and data breaches.

When following PASTA's risk-centric threat modeling methodology, compliance and cyber risks are evaluated as part of the assessment scope in Stage I. This ensures both cyber risks whose business impacts on the assets are evaluated and regulatory compliance are factored during early development of the dApp and providing evidence of how these requirements influence security decisions throughout the dApp SDLC.

Asset protection is a priority, particularly for products classified as high-risk due to their business impact and data sensitivity. A comprehensive risk analysis should assess the inherent risk of assets based on data classification levels whether public, confidential, or highly confidential – as well as the likelihood of threats materializing. By integrating threat modeling techniques, organizations can evaluate both the potential impact of vulnerabilities and the realistic likelihood of exploitation, ensuring a proactive security posture that aligns with regulatory requirements and business continuity objectives.

To refine our approach, it is crucial to emphasize that the core objective of risk-centric threat modeling is to proactively analyze and mitigate potential threats and risks associated with the application in scope for assessment.

In Stage I of the PASTA risk-centric threat modeling framework, we establish security and business objectives, ensuring that the assessment aligns with the organization's risk tolerance and security priorities. Once the technical scope is defined in Stage II, outlining the assessment boundaries, we move to Stage III, where we decompose the application. This involves systematically identifying the dApp's components and assets, mapping their roles within the overall architecture. This step is crucial for pinpointing potential attack surfaces, understanding how different elements interact, and recognizing vulnerabilities that threat actors could exploit in structured threat scenarios.

As we transition into the TA phase in Stage IV, we define threats as structured threat scenarios and systematically map them to vulnerabilities identified during Stage V, Vulnerability Analysis. This stage establishes a clear linkage between threats and vulnerabilities, setting the foundation for Stage VI, Attack Modeling, where threats to the dApp are further contextualized. In this phase, threat scenarios evolve from theoretical risks into simulated attack scenarios, directly targeting application components using adversarial tactics. The goal is to evaluate the dApp's exposure to real-world threats, measure the impact of potential attacks, and assess the effectiveness of existing security controls and countermeasures. Finally, in Stage VII, we analyze and manage risks by evaluating the previously identified threats, vulnerabilities, and attack scenarios. This stage focuses on prioritizing risks, assessing their potential impact, and implementing mitigation strategies to strengthen the overall security posture.

This structured approach enables an adversary-informed, attack-driven, and risk-prioritized security validation of the dApp, ensuring that identified threats are tested against realistic AVs. The ultimate objective is to be able to articulate the real cyber risks to a level of detail to analyze how a threat actor might exploit known weaknesses and vulnerabilities to compromise critical assets, allowing security teams to implement targeted risk mitigation strategies. By evaluating the actual exposure to threats and the resilience of security controls, organizations can ensure that defensive measures are designed, deployed, and operationalized effectively to protect the dApp's integrity, user assets, and business operations.

For businesses deploying dApps, proactive risk management means continuously understanding the threat exposure of the dApp and its potential impact on DeFi assets. By analyzing and documenting identified threats, vulnerabilities, and attack scenarios, organizations can make informed decisions to mitigate risks before real digital assets are compromised or exploited in fraudulent activities. By systematically conducting threat modeling, both on the dApp's baseline architecture and in response to architectural changes, DeFi organizations can ensure that risk mitigation is an ongoing process rather than a one-time assessment, strengthening security throughout the dApp's life cycle.

For practical and effective risk management, it is essential to go beyond theoretical estimations of likelihood and impact by validating risks through attack simulation and testing. Threat scenarios provide measurable insights into how easily threats can materialize as attacks and how vulnerabilities can be actively exploited to compromise the dApp and its assets. By simulating real-world attacks, security teams can validate exploitability to determine if vulnerabilities are truly actionable by adversaries; assess tangible impact by measuring the real consequences of exploitation such as fund theft, data breaches, or service disruption; and prioritize mitigation efforts by focusing on the most exploitable and high-impact threats first. This evidence-driven approach ensures that risk management is actionable, aligning security efforts with actual attack feasibility rather than relying on assumptions.

When measuring risk based on threat likelihood and impact, a fundamental risk management question arises: Given the existence of a vulnerability, how likely is it that a threat actor will exploit it, and what would be the resulting impact on critical assets? If an organization is exposed to a specific weakness or vulnerability, assessing the probability of exploitation and its potential consequences on the asset is essential for prioritizing risk mitigation efforts effectively.

While theoretical analysis of threat scenarios and vulnerability mappings can provide insight into risk exposure, security testing validates these assumptions by determining how threats materialize in practice. This aligns with the NIST definition of an attack, which states that an attack is the realization of a threat, meaning that a threat is a potential danger, while an attack occurs when the threat is executed, leading to security breaches or disruptions.

This is why risk-centric threat modeling in PASTA includes an AM stage, where the focus shifts from estimated risk assessments of likelihood and exposure to validated and tested threat scenarios. By simulating attacks, organizations can move beyond theoretical evaluations and gain tangible insights into actual exposure. This process involves correlating identified threat scenarios with vulnerabilities and assessing the feasibility of exploitation. The ultimate goal is to provide the business with a realistic understanding of how threats can impact critical assets, ensuring that mitigation efforts are aligned with real-world risks.

One of the primary reasons for adopting PASTA is to simulate attacks against threat scenarios in scope. These simulations, typically executed by red teams, emulate adversarial tactics (TTPs) based on real-world threat intelligence. Unlike traditional penetration testing, AM prioritizes attack paths by focusing on high-impact vulnerabilities tied to threat scenarios. This approach allows security teams to demonstrate real business impact by showcasing exploit feasibility in a controlled testing environment.

The results of these simulated attacks serve as a crucial validation tool, answering key risk management questions:

- How easily can a critical vulnerability be exploited?
- Is the vulnerability accessible to external attackers without authentication?
- Are there existing security controls to detect or prevent the attack?

Theoretical threat assessments assume that an attacker possesses the necessary tools, techniques, and knowledge to exploit a vulnerability. However, these assumptions require validation through AM and adversary-driven testing to ensure an accurate evaluation of threat exposure. Traditional risk assessments utilize probability models to evaluate threat likelihood (the probability of occurrence) and asset impact (potential effects on confidentiality, integrity, and availability [CIA]).

In the following, we present examples of risk frameworks that can be applied to quantify cyber risks. These frameworks are introduced in the context of Stage VII: Risk Analysis and Management of the PASTA methodology, where the goal is to prioritize mitigation efforts based on validated threat exposure and impact analysis. In this stage, the assessment of likelihood and impact for each threat scenario requires structured methodologies that support both qualitative and quantitative analysis. Frameworks such as the OWASP risk rating methodology [272], FAIR [168], OCTAVE [167], and NIST SP 800-30 [150] provide standardized models for evaluating cyber risk. Each enables teams to factor in threat agent capabilities, vulnerability characteristics, and the potential business impact of security failures.

Risk Assessments Using NIST RMF [273]: The NIST RMF, for example, as described in NIST SP 800-30 (Guide for Conducting Risk Assessments), defines risk as a function of likelihood of threat occurrence and impact on the asset.

$$Risk = Likelihood\ of\ Threat\ Occurrence \times Threat\ Impact\ on\ the\ Asset$$

The likelihood is the probability that a threat event will exploit a vulnerability and cause an adverse impact. It is determined by two key components: the threat event likelihood, that is the probability that a specific threat event will occur, and the vulnerability exploitability, that is the probability that an identified vulnerability will be successfully exploited if a threat event occurs.

Threat Likelihood: The threat likelihood factors (per NIST SP 800-30) are:

- **Threat Actor Capability:** Skill level, resources, and motivation of attackers.
- **Existence of Known Vulnerabilities:** If a vulnerability is publicly known and has available exploits, likelihood increases.
- **Exploitability:** Ease with which the vulnerability can be exploited (e.g. unauthenticated RCE versus local privilege escalation).
- **Existing Mitigations and Security Controls:** Strong security measures reduce likelihood (e.g. MFA, firewalls, EDR, SIEM detection rules).

To measure the risks qualitatively as a rating instead of scores, it is possible to assign a likelihood scale to the factor of threat likelihood such as:

- High (1.0–0.7) – Exploitable with minimal effort: known exploits exist.
- Medium (0.69–0.4) – Some effort required: no public exploit but feasible.
- Low (0.39–0.1) – Difficult to exploit: requires sophisticated skills/tools.

Threat Impact: In NIST RMF the threat impact is the consequences of the threat being realized. Impact represents the potential damage if a threat successfully exploits a vulnerability. It is measured by evaluating the impact on the CIA triad such as:

- **Confidentiality Impact:** Data exposure, privacy violations, regulatory non-compliance (e.g. PII leaks)
- **Integrity Impact:** Data corruption, unauthorized modifications, financial fraud
- **Availability Impact:** Downtime, DoS, operational disruption

To measure the impact qualitatively, it is possible to assign an impact scale to the factor of threat impact such as:

- High (1.0–0.7) – Severe damage: major financial loss, reputational damage, system compromise

- Medium (0.69–0.4) – Moderate impact: financial penalties, limited service disruption
- Low (0.39–0.1) – Minor disruption: minimal financial loss

Calculating threat risk based solely on threat likelihood and impact on the asset provides an initial assessment of potential risks without factoring in the severity of vulnerabilities (e.g. CVSS scores) that could be exploited. This preliminary evaluation can be conducted during Stage IV.

Threat Risk Assessments Using DREAD [274] (read the note): DREAD is a methodology to estimate the risk of threats typically used in conjunction with the STRIDE threat modeling framework. DREAD, as a qualitative risk assessment model, evaluates threats based on five key factors:

- Damage Potential: The extent of harm caused if exploited.
- Reproducibility: How easily the attack can be repeated.
- Exploitability: The complexity of executing the attack.
- Affected Users: The scale of impact on users or assets.
- Discoverability: The ease of identifying the vulnerability.

Each factor is scored from 0 to 10, and the DREAD Score is calculated as:

$$\text{DREAD Risk Score} = \{D + R + E + A + D\}/5$$

Note: The DREAD methodology is referenced in this section solely as a historical example to illustrate how risk factors (such as damage potential, reproducibility, exploitability, affected users, and discoverability) were once used to quantify threat risks in conjunction with the STRIDE threat modeling framework. Originally developed at Microsoft, DREAD was intended to help teams prioritize threats identified during threat modeling exercises. However, it is important to clarify that DREAD has not been maintained or recommended since around 2010. The security community has generally moved away from using DREAD due to concerns over scoring consistency and subjectivity. It is no longer considered appropriate for formal risk reporting, especially in enterprise or regulatory settings. For current best practices in cyber risk quantification and reporting, methodologies such as FAIR are preferred. FAIR provides a more structured, quantitative, and business-aligned approach to risk assessment and is increasingly adopted by CISOs and security governance teams.

In both NIST threat risk assessment and DREAD, factoring the weakness and vulnerability severity such as the risk score measured for a CVE and CWE with CVSS methodologies is not explicitly required, allowing these assessments to be conducted independently of specific vulnerability data. As a result, they can serve as preliminary threat risk analyses during Stage IV of PASTA, the TA phase, where risks are assessed based on threat scenarios before a detailed vulnerability evaluation is performed.

Risk Assessments Using the OWASP Risk Rating Methodology *(read the note)*: Besides NIST and DREAD methodologies to rate threats, methodologies like the OWASP risk rating methodology [272] provides an application security risk framework used to assess risk based on likelihood and impact, frequently applied in web and software security.

Note: The OWASP risk rating methodology is used here as a nonproprietary, community-based framework for evaluating security risks. It offers a structured approach that incorporates both technical and business impact factors, as well as threat agent characteristics and vulnerability factors, to calculate a comprehensive likelihood score. Although the OWASP risk rating methodology is not an actively maintained project, it remains a valuable reference especially for organizations seeking to assess risk beyond just technical severity metrics like CVSS. Unlike CVSS, which primarily focuses on technical attributes of vulnerabilities, OWASP's methodology brings in contextual elements such as business impact and attacker profiles, making it particularly useful in early-stage risk assessments and threat modeling exercises.

All these risk analysis methodologies offer valuable risk prioritization frameworks, particularly in the context of application threat modeling, where the focus extends beyond analyzing threats to also assessing the weaknesses and vulnerabilities that adversaries could exploit to impact critical assets.

Regarding the type of risk analysis framework used to determine the risk of threats during this stage of risk-centric threat modeling, no specific methodology is prescribed. Instead, the recommended approach is to align with the organization's existing RMFs, standards, and governance processes. Since the business ultimately owns the risks associated with its products and operations, the role of application security is to provide the necessary insights to support informed decision-making on mitigating technical risks and their business impact. The objective of this risk analysis and management stage is to ensure that threat modeling integrates seamlessly with the broader enterprise risk management strategy, offering a structured assessment of risks arising from threats, vulnerabilities, and asset exposure.

It is essential to make risk management decisions based on the best available information, which should include an analysis of threats in the context of identified vulnerabilities from various security assessments beyond threat modeling. A more comprehensive approach should incorporate CVSS scores or other vulnerability severity metrics to enhance the accuracy of risk assessments. For instance, OWASP's risk rating methodology explicitly integrates vulnerabilities into the assessment process, particularly through factors such as "ease of discovery" and "ease of exploitability." This ensures that risks are prioritized not only based on the broader threat landscape but also on the specific technical security gaps that could be exploited, leading to a more effective risk mitigation strategy.

Among the standard risk assessment methodologies used in the application security domain, the OWASP risk rating methodology provides a structured approach to evaluating security risks by assessing the likelihood of a threat exploiting a weakness and the potential impact of that exploitation and can be easily adopted in application security for its simplicity and inclusion of threat agent and business impact factors.

The OWASP risk rating methodology is best suited for application security teams and DevSecOps engineers who need a simple, structured way to assess and prioritize risks. It combines technical and business impact with threat and vulnerability factors, making it useful during threat modeling and secure development. Its ease of use helps bridge communication between developers and risk stakeholders, offering practical guidance. More complex models like FAIR [168] can be considered as alternative to OWASP especially for CISO-level risk reporting; FAIR models threat business impact in monetary terms, helping executives prioritize risk based on potential loss exposure.

This OWASP risk rating methodology considers both technical and business factors to ensure a comprehensive risk assessment. The final risk score is determined by multiplying the likelihood and impact scores with the formula:

$$\textit{Risk Score} = \textit{Likelihood} \times \textit{Impact}$$

Likelihood: Likelihood is evaluated based on multiple factors, including the skill level and motivation of potential threat actors, the attack surface exposure, and the ease of exploitability of identified vulnerabilities. Impact assessment considers both technical consequences, such as data exposure, financial loss, and system compromise, and business repercussions, including regulatory penalties and reputational damage.

Likelihood is determined by assessing both the characteristics of the threat actor and the nature of the vulnerability. Threat actors are categorized based on skill level, motivation, opportunity, and the size of the attack surface. Highly skilled attackers, such as blockchain security researchers and state-sponsored groups, pose a greater risk than lower-skilled attackers, such as casual hackers or script kiddies. Motivation plays a significant role, with financially motivated attackers being more likely to target DeFi applications due to the potential for direct financial gain. The opportunities available to an attacker, such as the presence of exposed smart contracts and API endpoints, also influence the likelihood of exploitation. Finally, the attack surface size, which considers the number of users interacting with the application, determines how widespread the impact of an attack could be.

Vulnerability factors measure how easily an attacker can exploit the weakness. Ease of discovery assesses whether the vulnerability is widely known and documented or requires advanced research and reverse engineering. Ease of exploitability considers whether exploitation can be automated using widely available tools or if it requires deep knowledge of blockchain security. Awareness and detection influence the risk, as vulnerabilities that are widely known to security researchers are more likely to be detected and mitigated before they can be exploited. The final likelihood score is derived by averaging these factors, providing a numerical representation of how probable an attack is.

Impact: Impact is assessed from both technical and business perspectives. Technical impact factors include confidentiality, integrity, availability, authentication, and authorization. A breach in confidentiality could expose sensitive user data, including private keys and wallet transactions. Integrity failures allow attackers to manipulate financial transactions or modify lending rules. Availability concerns arise when an attack disrupts lending pools or halts transactions, impacting the dApp's ability to function. Authentication and authorization weaknesses can allow unauthorized access to user wallets or financial operations.

Business Impact: Business impact factors include financial damage, reputation damage, regulatory compliance risks, and privacy violations. A successful exploit can lead to direct financial losses through the theft of crypto assets or the draining of liquidity pools. Reputation damage occurs when a DeFi protocol suffers an attack, leading to a loss of user trust and potential withdrawal of funds. Regulatory and compliance risks arise if the exploit results in violations of financial regulations, such as AML laws or KYC requirements. Privacy violations can expose borrower and lender transaction histories, leading to further compliance concerns and reputational harm.

Finally the final impact score is derived by averaging the technical and business impact factors. Multiplying the likelihood and impact scores provides the final OWASP risk score, which categorizes risks into low, medium, high, or critical severity levels.

We can apply the OWASP risk rating methodology to assess the risks of threat scenario TS1, "Malicious actors targeting dApp's mobile UI," within a DeFi lending and borrowing use case. This methodology factors in both technical and business impacts, as well as the likelihood of an attack, by considering the broader threat landscape, including the skills, motivations, and capabilities of potential attackers, along with the exposure of the attack surface. For example, the ease of exploitability is assessed based on the skills and tools required to conduct the attack. Importantly, while numerical scores are used in the background to quantify risk, the methodology assigns descriptive text to these scores, ensuring that the risks can be easily explained and understood, a key benefit of this approach.

A low skill requirement for an attack arises when authentication mechanisms have weaknesses or when sensitive data is insufficiently protected. In our dApp, known vulnerabilities such as improper authentication (CWE-287) and insecure storage (CWE-922) create high-priority AVs due to their frequent exploitation.

The widespread availability of public exploits and well-documented attack techniques further increase the likelihood of compromise, emphasizing the urgent need for robust security controls to mitigate these risks.

The technical impact factor can be assessed by combining technical consequences, such as unauthorized fund transfers, exposure of private wallet keys, and DoS attacks. For the business impact we can factor possible financial loss, reputation damage, and regulatory penalties: for example an attacker exploiting CWE-287 would lead to theft of digital assets hence classifying the business impact as critical. A threat seeking to exploit weakness CWE-359 (exposure of private information) is factored as business impact being high due to the potential for privacy violations and fraud. Threats targeting weaknesses like CWE-104 (excessive logging) are calculated as lower risk impact because of the low impact of resulting information leakage rather than direct financial or operational disruption.

We provide comprehensive risk rating calculations for threat scenario TS1 in Appendix E: Threat Scenario Weakness and Vulnerabilities Risk Ratings. The key takeaways of the risk analysis for this threat scenarios are:

1. **Authentication Vulnerabilities Pose Critical Risks,** potentially resulting in direct financial losses, regulatory scrutiny, and diminished user trust. Unlike traditional financial systems, DeFi lacks centralized account recovery, so weaknesses in authentication can lead to irreversible asset theft. Similarly, improper access control (CWE-284) may allow unauthorized users to execute privileged functions, such as modifying loan terms or withdrawing funds.
2. **Exploit Targeting Weaknesses in Protecting Sensitive Information Pose High Risks:** Exploit of CWE-359 lead to exposure of sensitive information to an unauthorized actor, where an application incorrectly handles or exposes sensitive data, allowing unintended access to PII, authentication tokens, financial records, or other confidential data. These exploits can result in identity theft, account takeovers, or financial fraud.
3. **Exploits Targeting Improper Input Validation Pose Moderate to High Risks:** For example, input validation vulnerability CVE-2022-23520 is an XSS vulnerability. OWASP ratings for this one is moderate to high since it can lead to data exposure, phishing attacks, and regulatory violations.

If we apply OWASP risk rating methodology to the threat scenario TS8, "Exploit of smart contract vulnerabilities in DeFi loan processing," we can also assess the threat risks and factor likelihood and technical and business impact. Likelihood can be determined by evaluating the threat landscape, considering attacker skill levels, motivations, attack surface exposure, and ease of exploitability. In this case, skilled blockchain cybercriminals can be targeting financial gain attacking improper access controls (CWE-284), deserialization vulnerabilities (CWE-502), and reentrancy flaws (CWE-362). One important factor to consider is that the public nature of smart contracts, their interaction with lending pools, and historical exploits in the DeFi space contribute to a high probability of compromise. Vulnerability factors we could consider are the ease of discovery, the exploitability, and existing security mitigations for disclosed vulnerabilities such as (e.g. CVE-2024-54134, CWE-922). Another factor to consider is that there are well-documented attack patterns for targeting smart contracts such as exploiting insufficient verification of authenticity and this significantly elevates the risk of exploits.

From a technical perspective, exploiting improper access controls (CWE-284) can allow attackers to gain unauthorized control over lending and borrowing transactions, leading to liquidity theft or unauthorized fund movement. Deserialization of untrusted data (CWE-502) could enable smart contract manipulation, potentially leading to unauthorized execution of high-value transactions. Last and not the least, exploits of smart contract reentrancy vulnerabilities (CWE-362) have historically led to complete depletion of DeFi liquidity pools. Inadequate encryption (CWE-922) and improper validation of conditions (CWE-1285) further increase the attack surface by exposing sensitive data and allowing malicious transactions to bypass security checks.

From a business impact perspective, the financial risks of exploiting smart contract vulnerabilities are usually very severe. Successful exploitation of these weaknesses could result in direct financial losses, lender/borrower fund drains, and smart contract insolvency. Additionally, the reputation damage to the dApp would be significant, as DeFi platforms rely heavily on user trust and security assurances. Given the growing regulatory scrutiny of DeFi applications, compliance risks also add to the business impacts as substantial violations of financial security standards could result in penalties or restrictions imposed by regulatory bodies.

Using the OWASP risk rating methodology, risk scores were calculated factoring the following (see note for clarifications):

1. **Critical Risk Ratings:** CWE-284 (improper access control) and CWE-502 (deserialization of untrusted data), both factors were high, resulting in critical risk ratings. CWE-841 (reentrancy) received a critical risk rating due to its history of enabling multimillion-dollar fund exploits.
2. **High Risk Ratings:** CWE-922 (insecure storage) and CWE-1285 (improper validation of specified conditions) were assessed as high risks, given potentially high impact business consequences.
3. **Low Risk Ratings:** CWE-398 (code quality issues) and CWE-547 (hardcoded gas values), received low risk ratings since they primarily contribute to maintainability and efficiency concerns rather than immediate security threats.

The key takeaways of OWASP risk analysis for the threat scenario TS8 Smart contract vulnerabilities in dApp for DeFi loan processing are:

1. **Smart Contract Vulnerabilities Pose Severe Risks:** Improperly enforced access controls (CWE-284) can allow unauthorized users to modify or execute privileged functions within a smart contract, potentially leading to unauthorized fund transfers, contract manipulation, or full asset depletion. If the contract owner role is not securely restricted, attackers can escalate privileges, modify critical parameters, or drain contract funds. Similarly, insecure deserialization (CWE-502) arises when a smart contract improperly processes user-supplied serialized data, potentially enabling remote code execution or data corruption. Attackers can inject malicious payloads to alter contract logic or execute arbitrary functions. Another high-risk vulnerability is reentrancy (CWE-841), which occurs when an external contract call is made before state changes are committed, enabling attackers to repeatedly invoke a function and withdraw more funds than intended.
2. **Business Impact Is Significant,** with potential for direct financial loss, regulatory scrutiny, and reputational damage. For example, exploitation of CWE-284 resulted in dForce Lendf.me attack (2020) [275], where access control flaws allowed hackers to manipulate privileged functions, ultimately leading to a $25 million theft. A notable case of incident exploiting CWE-502 was the Parliament Governance Attack (2021) [276], where an EVM-based serialization flaw was exploited to execute unauthorized governance token manipulations. For CWE-841, the DAO Hack (2016) [277] remains one of the most infamous cases of this exploit, where attackers drained $60 million in ETH, leading to an Ethereum hard fork to reverse the damage. Direct asset loss occurs when vulnerabilities allow hackers to drain funds. Indirect financial impact arises from legal fees and compensations to affected users. Noncompliance with AML and KYC regulations can lead to fines or platform shutdowns.
3. **Preventative Security Controls for Smart Contracts Need to Be High-Priority for Risk Mitigations,** focusing on improved security for access controls, implementing secure deserialization, focusing on formal verification for smart contracts implementation of reentrancy guards. Table 3.17 provides a summary of the security controls to be implemented to mitigate these risks:

Note: Please note that OWASP risk analysis calculation of critical risk differs from the CVSS critical severity rating. While CVSS focuses primarily on technical exploitability, the OWASP approach factors in likelihood, business impact, and contextual nuances specific to the use case. This holistic evaluation provides a more comprehensive perspective for prioritizing mitigation efforts in real-world operational settings.

Table 3.17 Security controls and risk mitigations.

Security control	Mitigation control	Risk mitigated
RBAC	Secure privileged functions	Improper access control (CWE-284)
Multi-Sig authentication	Prevent unauthorized contract modifications	Privilege escalation attacks
Reentrancy guards (nonreentrant)	Block recursive function calls	Reentrancy attacks (CWE-841)
Secure deserialization practices	Prevent arbitrary execution	Deserialization flaws (CWE-502)
Formal verification and audits	Prove smart contract correctness	Logic flaws, access control bugs
Timelocks for admin functions	Delay high-risk operations	Insider privilege abuse
Event monitoring and anomaly detection	Detect abnormal contract behavior	On-chain exploit detection

Finally, we provide herein the OWASP risk analysis of all threat scenarios in scope for the dApp DeFi lending and borrowing in scope for threat modeling. We provide the details of the risk calculations in Appendix L: Stakeholder Risk Communication Report. Providing detailed risk calculations improves transparency for audit and governance reporting by offering tangible evidence of how risk decisions are made. Documenting the rationale behind likelihood and impact scoring, based on threat realism, exploitability, and potential business consequences, strengthens the justification for mitigation prioritization (Table 3.18). This traceability not only supports informed risk acceptance or remediation decisions but also enables developers and security teams to connect identified risks with specific remediation paths. Ultimately, such documentation demonstrates due diligence, aligns with best practices in risk management, and enhances the credibility of the organization's security posture during audits or regulatory reviews.

This risk analysis has the purpose to be a reference for risk mitigation decision-makers to prioritize countermeasures for mitigating both technical and business risks. Given that the analyzed system is a DeFi platform for lending and borrowing, the assessment places special emphasis on financial risks, regulatory exposure, and reputational damage, in addition to standard cybersecurity threats.

This structured approach ensures that security efforts are directed toward the most pressing vulnerabilities (CWEs) within each targeted component, whether it be the mobile UI, digital wallets, smart contracts, or blockchain infrastructure.

Table 3.18 Threat scenario risk qualification assessment.

Threat scenario	Likelihood (score)	Impact (score)	Overall risk (score)	Commentary/rationale
TS1: Mobile or browser UI risks	High (9.0)	Critical (9.5)	Critical (85.5)	Due to high exploitability and severe financial consequences, improper authentication and insecure storage make this a significant threat.
TS2: Digital wallet risks	High (8.5)	Critical (9.5)	Critical (80.8)	Wallets are a prime target; improper authentication and integer overflow could lead to direct asset theft.
TS3: Distributed digital identity system risks	High (8.0)	High (8.8)	High (70.4)	Identity spoofing and resource exhaustion could lead to fraudulent activities and systemic disruptions.
TS4: dApp frontend risks	High (8.2)	High (8.9)	High (73.0)	Frontend vulnerabilities like XSS and CSRF allow for transaction manipulation, significantly impacting platform integrity.
TS5: dApp middleware (JSON-RPC) risks	High (8.5)	Critical (9.5)	Critical (80.8)	Middleware risks include deserialization attacks, enabling unauthorized transactions and system compromise.
TS6: blockchain node risks	Medium (7.2)	High (8.5)	High (61.2)	Disrupting node operations affects transaction validation, causing operational delays and financial instability.
TS7: Blockchain platform and distributed ledger risks	Medium (7.0)	Critical (9.2)	High (64.4)	Platform vulnerabilities could lead to unauthorized transactions and network disruption.
TS8: Smart contract risks	High (9.0)	Critical (9.5)	Critical (85.5)	Smart contract breaches could result in full platform liquidation and significant financial losses.
TS9: Oracle and external data feeds risks	High (8.5)	Critical (9.5)	Critical (80.8)	Oracle manipulation can lead to price exploits, draining liquidity and causing insolvency.

By quantifying threat risks, teams can make informed decisions on remediation strategies, aligning security investments with the highest-impact threats. This risk classification provides a prioritization framework for mitigating threats in DeFi lending and borrowing systems. Addressing critical risks for smart contracts, digital wallets, and middleware dApp to blockchain communication and oracles and external data feeds should be top priorities to ensure platform resilience and regulatory compliance.

Finally a useful tool for risk mitigation is mapping specific countermeasures to each threat scenario as actionable reference for risk mitigation, ensuring that security controls are prioritized based on real-world attack feasibility and business impact. The countermeasures focus on a combination of secure design principles, preventive security controls, and proactive monitoring strategies to minimize exposure to threats and enhance the security posture of the dApp throughout its development, deployment, and operational life cycle (Table 3.19).

Finally, we provide structured outputs from the risk analysis and management stage of the PASTA threat modeling framework to ensure that risks to the DeFi lending and borrowing dApp are systematically identified, assessed, mitigated, and continuously monitored. Additionally, tools like JIRA [199] are used to log and track action items from the threat model, ensuring that remediation efforts are effectively documented and followed up.

These outputs form the foundation for aligning risk management efforts with business objectives, regulatory compliance, and security best practices. Additionally, they provide a transparent accountability framework for security teams, developers, and business stakeholders to track security progress over time and demonstrate due diligence in securing the platform.

To support risk-based decision-making, the following key deliverables, referenced in the Appendices, document the threat modeling exercise in a structured and actionable manner:

- **Threat Modeling Matrix (Appendix A):** This matrix provides a structured threat modeling analysis of the dApp ecosystem, mapping threat events, AVs, threat agents, and targeted assets to their associated impacts and mitigation strategies. It serves as a critical reference for risk management, security architecture decisions, and the implementation of proactive defense measures to strengthen the platform's overall security posture.
- **Mapping of Threat Scenarios to Targeted Weaknesses and Asset Impacted (Appendix B):** This is a structured mapping of threat scenarios to vulnerabilities and impacted assets, providing a risk-driven approach to prioritizing security controls. This mapping helps security teams focus remediation efforts where they are most impactful and identify security gaps that require architectural improvements.

Table 3.19 Threat scenarios and proposed countermeasures.

Threat scenario	Countermeasures/risk mitigation
TS1: Mobile or browser UI risks	• Enforce strong authentication mechanisms (MFA, WebAuthn) to mitigate CWE-287 (improper authentication). • Encrypt and secure sensitive storage to prevent CWE-922 (insecure storage of sensitive information). • Implement anti-phishing protections to reduce credential theft risks.
TS2: Digital wallet risks	• Enforce secure key storage mechanisms and hardware-based authentication to mitigate CWE-287. • Validate integer operations in smart contracts to prevent CWE-190 (integer overflow). • Secure BIP39 mnemonics against unauthorized access to avoid large-scale wallet compromises.
TS3: Distributed digital identity system risks	• Strengthen access controls and enforce strict identity validation to prevent CWE-284 (improper access control). • Implement replay attack protections and secure transmission to mitigate CWE-346 (origin validation error). • Prevent API exhaustion to reduce impact from CWE-400 (resource exhaustion).
TS4: dApp frontend risks	• Sanitize all user input to prevent CWE-79 (XSS) and CWE-352 (CSRF). • Implement strict CSP to reduce XSS risks. • Secure API endpoints to prevent unauthorized access and misconfigurations.
TS5: dApp middleware (JSON-RPC) risks	• Enforce strict data serialization policies to prevent CWE-502 (deserialization of untrusted data). • Secure middleware authentication to mitigate unauthorized transaction execution risks. • Monitor and log API transactions for suspicious activity.
TS6: Blockchain node risks	• Implement DDoS protection and rate limiting to mitigate CWE-400 (uncontrolled resource consumption). • Ensure smart contract validation mechanisms to prevent CWE-20 (improper input validation). • Strengthen transaction validation to prevent double-spending risks.
TS7: Blockchain platform and distributed ledger risks	• Enhance consensus validation processes to prevent CWE-703 (improper handling of exceptions). • Enforce RBACs to mitigate CWE-284 (improper access control). • Ensure transaction integrity mechanisms to avoid unauthorized blockchain modifications.

(Continued)

Table 3.19 (Continued)

Threat scenario	Countermeasures/risk mitigation
TS8: Smart contract risks	• Conduct rigorous smart contract audits to prevent CWE-284 (improper access control) and CWE-502 (deserialization). • Implement reentrancy guards to mitigate CWE-362 (reentrancy). • Secure contract upgrade mechanisms to prevent unauthorized protocol changes.
TS9: Oracle and external data feeds risks	• Use multiple oracles and data sources to mitigate CWE-345 (insufficient data verification). • Secure API endpoints and restrict access to prevent CWE-918 (SSRF). • Validate incoming data before executing transactions to prevent price manipulation attacks.

- **Mapping of Threat Scenarios to Exploitable Attack Paths (Appendix C):** This mapping provides a structured framework to correlate potential threats in dApps with real-world attack scenarios. This document is designed to assist security architects, blockchain developers, and risk assessors in understanding, prioritizing, and mitigating security risks in smart contracts, DeFi, and blockchain ecosystems.
- **Threat Scenarios Attack Simulation Tests (Appendix D):** This is a documented record of attack scenarios tested during the AM phase, highlighting exploitability, security control effectiveness, and residual risk after countermeasures are applied. These results offer an evidence-based approach to risk assessment, validating which threats are most feasible in real-world attack conditions and ensuring that mitigation strategies effectively reduce risk exposure.
- **Threat Scenario Weakness and Vulnerabilities Risk Ratings (Appendix E):** Threat risk ratings play a critical role in prioritizing security efforts for dApps. This document provides a practical example of how risk scores are calculated for threat-targeted CVEs and CWEs using the OWASP risk rating methodology. These risk ratings provide a structured basis for prioritizing vulnerability mitigation prior to production deployment, ensuring the most critical issues are addressed first, by considering factors such as likelihood of threat agent, vulnerability exploitation, technical and business impact, ultimately helping to guide risk decision-makers in setting mitigation priorities.
- **Risk Mitigation Plan (Appendix F):** This is an example of an actionable risk mitigation plan following a risk mitigation strategy or mitigating the identified risks. This includes implementation timelines, defined milestones, resource allocations, and clear responsibility assignments. The plan prioritizes high-risk

vulnerabilities while ensuring that security measures are aligned with business goals, compliance obligations, and governance frameworks.

- **Threats Risk Register (Appendix G):** This is a centralized record cataloging all identified risks, ranked by severity and likelihood, linked to threat scenarios, vulnerabilities, and impacted components. This repository provides the status of the risk mitigation efforts and monitoring security posture evolution.
- **Attack Simulation Testing Report (Appendix H):** This document provides a record of attack scenarios tested during the AM phase, including identified weaknesses and the effectiveness of existing countermeasures. The findings support evidence-based risk assessments and demonstrate the validity of mitigation strategies.
- **Risk Analysis Report (Appendix I):** This is a high-level summary tailored for risk owners, developers, and business leaders, outlining critical risks, prioritized mitigation strategies, and their impact on organizational security and financial stability. This report ensures transparency, alignment, and informed decision-making across cross-functional teams, helping to bridge the gap between security and business risk management.

By integrating these artifacts into the risk management process, organizations ensure comprehensive oversight, structured risk documentation, and a proactive security strategy. These outputs reinforce accountability, provide visibility into evolving risks, and ensure continuous security improvements to safeguard the DeFi lending and borrowing ecosystem.

3.2.4 Security-Driven Tools and Techniques for dApps

To effectively conduct security testing for the dApp, it's essential to understand the development process, the technology stack in use, and the target environments for development, testing, and deployment. This knowledge helps determine the appropriate tools to integrate during the development life cycle. These tools will perform security scans as the source code is built, integrated, and tested across various environments, such as staging, QA, and production deployment.

Before integrating tools into the development pipeline to automate vulnerability scanning of the source code during development, build, and testing phases, we need to take a step back and reflect on the overall objectives we want to achieve with the deployment of these security tools, given the technology stack for the application in scope. Additionally, it's crucial to plan ahead for handling the remediation of any vulnerabilities identified by these tools. This involves understanding the impact and risk associated with each vulnerability and determining the appropriate measures needed to eliminate them effectively.

Furthermore, it's essential to have a vulnerability management standard in place that defines the remediation requirements for the vulnerabilities identified. This could be outlined as single level objectives (internally) or SLAs for remediation based on the criticality/severity of the issues. Moreover, an organization might also be contractually obligated to remediate vulnerabilities in a timely manner and to disclose reports, such as vulnerability penetration testing reports and static scans, for the software products released to customers. These products might be provisioned as integrated with APIs, offered as SaaS, or delivered as dApps to be used directly by product customers/users.

To avoid overpromising and underdelivering, it is crucial to achieve these capabilities internally. This means accurately identifying vulnerabilities and remediating them before production deployments to prevent these vulnerabilities from being discovered by customers, which could become potential liabilities if not disclosed.

With these preliminary considerations in place, it is significantly important for those responsible for the security of the dApp whether it's the head of dApp product security, security architect, or security engineer/lead. By integrating security testing into the continuous integration and deployment (CI/CD) pipeline, you ensure that security checks are performed regularly and consistently throughout the development process. This includes incorporating SCA, SAST, DAST, and Infrastructure as a Code (IaaC) scanning tools into the pipeline, ensuring that security is a continuous priority from development to deployment. Moreover, every security requirement, design assumption, or bug discovered should be accompanied by an automated test to prevent regressions and reinforce system reliability. This approach allows teams to build confidence in the security of the application without needing to rely on full-scale penetration tests for every code change, ultimately supporting more agile and secure development cycles.

By setting these standards and considering contractual obligations, you ensure that vulnerabilities are addressed promptly and transparently, minimizing potential risks to the application and its users.

A comprehensive and risk preventative approach toward security testing your dApp should at minimum routinely simulate cyberattacks (refer to step Attack Modeling of PASTA) to uncover vulnerabilities and ensure that your dApp can withstand potential threats. Automated scans of your smart contracts are crucial to identify common issues, while thorough examination of the source code helps spot potential security problems early.

Automated security testing is critical for ensuring the security, integrity, and resilience of blockchain applications, dApps, and Web3.0 ecosystems. Unlike traditional applications, dApps operate in a decentralized and immutable environment, where vulnerabilities can lead to irreversible financial losses, data breaches, and systemic failures. To mitigate these risks, automated security testing tools must be integrated into the SDLC to proactively identify, assess, and remediate vulnerabilities in key components such as smart contracts, blockchain nodes, APIs, identity systems, and distributed storage solutions. Security testing for dApps and Web3.0 applications extends beyond standard penetration testing and includes formal verification, smart contract auditing, DID validation, and blockchain platform security assessments. This book presents an automated security testing framework that integrates various security methodologies and tools to enhance the security posture of Web3.0 and blockchain-based applications.

A comprehensive security testing strategy for DeFi dApps should integrate automated tools and techniques to evaluate vulnerabilities across multiple attack surfaces, including UIs, smart contracts, blockchain networks, APIs, and identity management components. The following sections outline the essential automated security testing methodologies that enhance the resilience and trustworthiness of DeFi applications:

1. **Penetration Testing:** Penetration testing involves simulating real-world cyberattacks to identify vulnerabilities within a DeFi dApp. It focuses on attack surfaces such as web interfaces, mobile applications, APIs, and command-line interfaces (CLIs) to uncover weaknesses that adversaries could exploit. The scope should include not only functional components but also informed perspectives when threat modeling artifacts are available. A penetration tester with access to the threat model such as documented trust boundaries, attack surfaces, and known threat scenarios can produce significantly more qualitative and targeted results than traditional black-box testing. This white-box or gray-box approach increases testing accuracy, reveals higher-priority vulnerabilities, and reduces false positives. Furthermore, it enhances ROI by demonstrating how threat modeling upfront helps optimize downstream security validation efforts, thereby offsetting the cost of the modeling itself through more effective remediation and risk reduction. The scope should include:
   - **UI Testing:** Automated penetration testing tools assess web-based and mobile interfaces built with frameworks like React, Angular, Vue.js, Swift, Kotlin, and Flutter. These tests help detect XSS, Cross-Site Request Forgery (CSRF), and injection flaws.
   - **API and Backend Security:** Tools like Burp Suite, OWASP ZAP, and SonarQube can automatically scan API endpoints for vulnerabilities, including unauthorized API access, missing authentication, and data exposure risks.
   - **Automated DAST:** Continuous security assessments using automated fuzzing tools detect runtime vulnerabilities and simulate attacker behaviors to uncover potential weaknesses in the dApp's security model.

This includes the use of DAST tools (*read the note*) to evaluate the application in its running state by simulating real-world attack scenarios and identifying runtime vulnerabilities such as injection flaws, authentication bypasses, and misconfigurations.

- **Automated E2E Testing:** It is essential to verify the security and reliability of entire user workflows, ensuring that security controls function correctly across integrated components, from frontend interfaces and middleware to smart contracts and data storage layers. These testing approaches help ensure that DeFi applications remain resilient, secure, and compliant throughout the SDLC.

Note: DAST tools can be integrated into the CI/CD pipeline and configured to run automatically when the application progresses to staging or QA environments. These environments are commonly used for more realistic testing, including penetration tests, making them an ideal context for executing automated dynamic security tests in a controlled and consistent manner.

2. **Blockchain Platform Security Testing:** Since blockchain networks are the backbone of DeFi applications, automated security testing must evaluate the consensus mechanism, node integrity, and network resilience to prevent attacks on blockchain nodes and transaction validation processes. The scope should include:
   - **Automated Consensus Security Validation:** Ensuring robustness against Sybil attacks, fork manipulation, and block validation exploits.
   - **Node Security Audits:** Automated node configuration scans can detect improper permission settings, missing security patches, and network configuration issues.
   - **DDoS & Resource Consumption Protection:** Testing tools can assess gas limit exploits, spam transactions, and DoS vectors that target blockchain nodes.
3. **Smart Contract Auditing:** Smart contracts execute financial transactions, manage governance protocols, and enforce business logic in DeFi applications. Given their immutable nature, any security flaws introduced at deployment can be exploited indefinitely. Automated security testing is crucial to prevent vulnerabilities before deployment. The scope should include:
   - **Static Code Analysis:** Tools such as Slither, Mythril, and Oyente can automatically scan Solidity and Vyper-based smart contracts to detect reentrancy attacks (CWE-362), improper access controls (CWE-284), and integer overflows (CWE-190).
   - **Symbolic Execution and Fuzzing:** Automated testing tools like Manticore and Echidna simulate smart contract execution across various

input conditions, uncovering potential logic errors, edge case exploits, and security misconfigurations.
- **CI/CD Pipeline Integration:** Smart contract auditing should be automated within the CI/CD pipeline, triggering security scans before deployment to prevent introducing high-risk vulnerabilities into production environments.

4. **Formal Verification of Smart Contracts:** Formal verification [278] represents a rigorous, mathematically grounded approach to ensuring that smart contracts behave exactly as intended across all possible execution paths. Unlike traditional testing methods that examine specific scenarios, formal methods [279] aim to prove the correctness of the contract's logic under every conceivable condition. This is especially critical for mission-critical DeFi functions such as lending, staking, and cross-chain bridging, where security and predictability are paramount. The process typically involves a combination of advanced techniques and tools:
   - **Automated Theorem Proving:** Tools such as Certora [280] and Securify [281] automatically verify smart contract logic against formal specifications, helping detect logical flaws, exploitable edge cases, or unintended behaviors before deployment.
   - **Code Modeling and Simulation:** This involves translating smart contract code into mathematical models and simulating every possible state transition to identify vulnerabilities such as undefined behavior, faulty conditions, or security loopholes.
   - **Proof Generation:** Once the verification process is complete, these tools produce formal proofs that confirm the contract logic adheres to its defined security and functional requirements, offering a high level of assurance in its correctness.
5. **Identity and Access Management (IAM) Security Testing:** DeFi applications rely on decentralized authentication methods such as wallet-based authentication (e.g. MetaMask, WalletConnect) instead of traditional username/password models. Automated testing is essential to validate authentication, access control, and identity management mechanisms:
   - **Authentication and Authorization Testing:** Automated IAM testing tools evaluate RBACs, public key cryptography implementations, and wallet authentication security.
   - **Key Management and Encryption Validation:** Automated tools like OpenSSL can test TLS/SSL encryption, key strength, and key exchange security in wallet interactions.
   - **DID Protocol Testing:** Custom automation scripts validate decentralized identifiers and VCs for compliance with Web3 identity security standards.

6. **Network Security Testing:** Since DeFi dApps rely on P2P networks, automated network security testing ensures the blockchain network is resilient against partitioning attacks, DoS, and data manipulation attempts. Automated security testing should validate:
   - **P2P Network Security Analysis:** Automated tests assess network data propagation, peer authentication mechanisms, and communication channel encryption.
   - **DDoS Resilience Testing:** Tools simulate transaction spam attacks, gas exhaustion exploits, and network flooding techniques to validate mitigation measures.
   - **Blockchain Node Security Validation:** Automated scans identify weak node configurations, outdated software, and misconfigured security settings.
7. **Testing Cross-Chain and Interoperability Components:** Cross-chain functionality is essential for bridging liquidity between DeFi platforms, but it introduces security risks in APIs, cross-chain messaging, and smart contract integrations. Automated security testing should validate:
   - **Cross-Chain Bridge Security:** Automated tests ensure secure asset transfers, integrity of cross-chain proofs, and protection against double-spend attacks.
   - **API Security for Interoperability:** Postman, OWASP ZAP, and Burp Suite automate security assessments for cross-chain API authentication, authorization, and data integrity.
   - **Data Verification Mechanisms:** Automated validation of message authenticity and cryptographic proofs ensures that cross-chain transactions are resistant to replay attacks and message tampering.
8. **Performance and Scalability Testing:** DeFi applications are expected to handle high transaction volumes, which makes performance testing essential for ensuring scalability, low network latency, and efficient resource utilization. Automated performance and security testing should be conducted to meet defined SLAs, ensuring that system requirements are not only met but also verifiably tested under real-world conditions. In light of this, automated security testing should cover:
   - **Load Testing:** It simulates large-scale user interactions to assess system behavior under heavy loads.
   - **Stress Testing:** Automated stress tests evaluate the network's ability to withstand extreme conditions, such as high transaction spikes or validator failures.
   - **Scalability Testing:** Ethereum TestRPC [282], Hyperledger Caliper [283], and Chainlink benchmarking tools [284] validate how well the application scales with increased transactions and network growth.

9. **Compliance and Regulatory Testing:** DeFi platforms must comply with AML/KYC regulations, financial integrity requirements, and privacy laws. Automated compliance testing helps validate security against regulatory requirements:
   - **ISO/IEC 27001 Compliance Checks** [285]: Automated security audits verify adherence to ISMS requirements.
   - **GDPR and Data Protection Compliance:** Tools like OpenZeppelin Defender [286] can help to track smart contract compliance with GDPR policies.
   - **Financial Compliance Testing:** Automated smart contract verification [278] ensures that lending protocols, AMMs, and staking contracts meet investor protection standards (e.g. Securities and Exchange Commission [SEC], FATF, MiCA guidelines).

Security testing for DeFi applications requires automated, continuous testing methodologies to detect vulnerabilities across smart contracts, APIs, blockchain nodes, and DID systems. By integrating automated security testing into the CI/CD pipeline, DeFi developers can proactively identify and remediate vulnerabilities before deployment, ensuring that protocols remain secure, resilient, and compliant with financial regulations.

These methods, when combined, provide a comprehensive approach to securing blockchain applications, dApps, and Web3.0 platforms. For vulnerability scanning of source code as soon as being built, integrating automated security testing into the CI/CD pipeline ensures that security checks are performed regularly and consistently throughout the development process.

To summarize, we provide herein Table 3.20 as quick reference to the key components of the tech stack for building a dApp, along with the security testing scope and tools. Adopting a tech stack-driven approach to security testing is crucial because vulnerabilities can arise not only in the source code that is being developed but also in third-party libraries, software frameworks, and APIs that are integrated with the dApp. These integrated components may have their own vulnerabilities, and their remediation depends on the versions that have been integrated and the efforts of the open-source community or the commercial entities maintaining these components.

Table 3.20 provides a comprehensive security testing scope for each key component of a DeFi dApp, outlining the testing scope and methodologies applicable to each asset, along with security tools that can be integrated into automated testing workflows. This structured approach helps mitigate security risks, validate cryptographic integrity, enforce secure transaction execution, and ensure compliance with industry security best practices.

Table 3.20 dApp security testing scope.

Asset	Description	Tech stack	Testing scope and tools
UI or mobile interface	The primary interaction layer for users, enabling access to functionalities; ensures secure session management, input sanitization, and MFA	React Native, Flutter, Ionic Enterprise; Web3Modal (wallet integration), TailwindCSS (styling), Formik (input validation)	Testing scope: penetration testing, input validation, authentication testing Tools: OWASP ZAP, Burp Suite, SonarQube
Frontend application	Connects the UI to blockchain wallets and smart contracts, processing user inputs and facilitating API communication	Vue.js, AngularJS, React; Web3.js, ethers.js (blockchain interaction); Apollo Client (API data fetching); Okta (API security)	Testing scope: API security, encryption, authentication testing Tools: Postman, OWASP ZAP, Snyk
Identity solutions	Manages user authentication and PII securely using decentralized identifiers and cryptographic protections	uPort, Civic; DID-JS, Veramo (decentralized identifiers and cryptographic proofs)	Testing scope: IAM, cryptographic security Tools: OpenSSL, Auth0 Pen Test Suite
Distributed storage	Stores transaction data, loan records, and user activity on decentralized networks	IPFS, Arweave; OrbitDB (P2P databases), Textile (IPFS management), Ceramic Network, AWS Blockchain Templates	Testing scope: data integrity, access control, redundancy Tools: IPFS Security Tools, Swarm Security Tools
JSON-RPC provider middleware	Bridges the frontend and blockchain nodes, enabling smart contract interactions	Infura, Alchemy; QuickNode (commercial middleware); web3-provider-engine, rpc-websockets (middleware libraries)	Testing scope: API security, rate limiting, data serialization Tools: Postman, OWASP ZAP, API Gateway Security

Asset	Description	Tech stack	Testing scope and tools
Digital wallets	Manages private keys for signing transactions and authenticating users	Trust Wallet, MetaMask; Fireblocks (commercial); WalletConnect SDK (multiwallet support), BIP39 (mnemonic management)	Testing scope: key management, secure authentication, MFA Tools: Ledger Security Suite, Fireblocks Security Suite
Blockchain layer	Hosts smart contracts and facilitates decentralized operations	Ethereum, Solana, Polygon; Ganache (testing), Hardhat (development); Azure Blockchain (managed services)	Testing scope: node security, consensus mechanism testing Tools: Ganache, Hardhat, Blockchain Security Audit Tools
Smart contracts	Automates functionalities; requires secure coding, testing, and auditing	OpenZeppelin (secure templates), Solidity-Parser (code analysis), Foundry (testing); MythX, Slither, Trail of Bits (auditing)	Testing scope: smart contract security, reentrancy testing, formal verification Tools: MythX, Slither, Trail of Bits Security Tools
Oracles and external data sources	Provides off-chain data such as asset prices to smart contracts	Chainlink, Chainlink-SDK (integration), API3 (SLA-backed oracles)	Testing scope: data integrity, API security, external feed verification Tools: Chainlink Security Tools, API3 Auditing Suite
Indexing applications	Retrieves and indexes blockchain data for user transactions and balances	The Graph, graphql-request (querying), Dgraph (performance optimization)	Testing scope: data consistency, cryptographic validation Tools: GraphQL Security Suite, Blockchain Indexing Tools
Scaling applications	Enhances scalability by offloading computations to sidechains or Layer 2 networks	zkSync, Optimism; zkSync SDK, PlasmaJS (scaling integrations); ConsenSys Quorum (commercial scaling)	Testing scope: sidechain security, state sync testing Tools: zkSync Security Framework, Plasma Security Suite
Transaction monitoring	Tracks and analyzes on-chain activities for compliance and security	ETX; Chainalysis, Elliptic (commercial tools)	Testing scope: compliance monitoring, anomaly detection, risk scoring Tools: Chainalysis, Elliptic, Blockchain Forensic Tools

Automated security testing is essential for identifying and mitigating vulnerabilities in dApps and Web3.0 ecosystems. Given the immutable nature of blockchain transactions and the high-value assets involved in DeFi applications, security flaws must be detected and addressed before deployment to prevent financial losses, data breaches, and systemic failures. This section examined security testing methodologies, emphasizing the need for integrating automated tools within the CI/CD pipeline to maintain security throughout the SDLC.

A well-structured security testing strategy must cover multiple layers of a DeFi dApp, including the UI, smart contracts, blockchain infrastructure, APIs, and DID components. Security assessments should extend beyond source code analysis to include penetration testing, blockchain security validation, formal verification, and interoperability testing. The integration of automated security tools ensures that vulnerabilities are detected early, assessed for their impact, and remediated before deployment. By embedding security checks at each stage of development, organizations can ensure proactive risk management and maintain a strong security posture.

The evolving attack landscape in DeFi necessitates continuous security monitoring, real-time threat intelligence, and proactive mitigation to safeguard decentralized financial ecosystems. Security testing must also align with regulatory compliance requirements, including ISO 27001, GDPR, FATF guidelines, and MiCA regulations, to ensure that DeFi applications adhere to industry security standards and financial integrity mandates. By automating security testing and integrating robust security frameworks, dApps can enhance their resilience against emerging threats while ensuring the protection of user assets and maintaining trust in DeFi platforms.

The security testing framework table provided in this section offers a structured reference for assessing key DeFi dApp components, defining security testing scopes, and selecting appropriate security tools to automate testing processes. This tech stack-driven approach enables developers and security engineers to systematically identify vulnerabilities not only within proprietary code but also across third-party libraries, frameworks, and APIs. By implementing these automated security testing measures, organizations can strengthen their defenses, reduce exposure to security threats, and ensure the reliability of DeFi applications.

3.3 Auditing Blockchain Applications for Compliance

A risk mitigation strategy for a DeFi application must extend beyond technical security measures to encompass compliance with global regulatory frameworks governing financial integrity, investor protections, and security requirements. Ensuring adherence to these frameworks not only reduces regulatory risk but also strengthens the dApp's credibility among users, investors, and financial institutions.

To achieve this, we will outline how to audit a DeFi application for compliance with key regulatory frameworks that are most relevant to DeFi. This includes assessing security controls, transaction monitoring, and identity verification mechanisms against established legal and regulatory standards.

This section will demonstrate how security and compliance audits can be conducted using automated and manual assessment methodologies, ensuring that the dApp not only meets security best practices but also aligns with the broader legal requirements governing decentralized financial ecosystems. By integrating compliance verification into the security testing framework, DeFi applications can proactively address regulatory obligations, mitigate legal exposure, and build a resilient financial infrastructure that meets industry standards while maintaining decentralization and user autonomy.

To develop an effective risk mitigation strategy for a DeFi application, it is crucial to first identify the regulatory frameworks that apply to both dApps and enterprise solutions leveraging blockchain technology in regulated environments. By aligning security and operational measures with these regulations, DeFi platforms can enhance trust, ensure legitimacy, and maintain access to key financial markets. The financial and security requirements set by regulators might include the following.

The **MiCA (Regulation)** framework [287], introduced by the EU, outlines security and compliance measures for crypto asset service providers, including DeFi protocols. A core requirement under MiCA is the implementation of robust wallet security practices, ensuring that private keys are properly protected through encryption, multi-sig authentication, and institutional-grade custody solutions. Additionally, MiCA enforces stringent security measures for smart contract implementations, mandating protections against unauthorized contract modifications, transaction integrity failures, and exploitable vulnerabilities. By aligning with MiCA, a DeFi lending platform can adhere to European financial security standards, reduce exposure to compliance risks, and enhance investor confidence in on-chain lending and borrowing transactions.

The **FATF Guidelines** [288] focus on AML and KYC regulations, requiring crypto platforms including DeFi lending applications to establish financial transparency mechanisms. The **FATF Travel Rule**, a key component of these guidelines, mandates that platforms prevent illicit financial activities such as money laundering, terrorist financing, and fraudulent lending schemes. To comply, DeFi platforms must integrate secure identity verification mechanisms, real-time transaction monitoring, and risk-based due diligence. Failure to meet FATF standards can result in regulatory blacklisting, restricting a platform's ability to engage with banking institutions and limiting its operational reach in compliant jurisdictions.

Adopting AML/KYC controls not only ensures compliance but also strengthens the legitimacy of DeFi lending services by deterring malicious actors from exploiting the platform.

In the United States, **SEC** [289] **and CFTC crypto regulations** [290] impose strict compliance requirements on DeFi platforms that facilitate lending, borrowing, or derivatives trading. The SEC has intensified enforcement actions against DeFi protocols that fail to disclose risks, misrepresent financial returns, or engage in unregistered securities offerings. DeFi lending platforms must integrate investor protection measures, including transparent risk disclosures, compliance reviews, and mechanisms that ensure compliance with securities laws. Similarly, the CFTC regulates derivatives and lending-related financial products, requiring DeFi platforms to implement security controls that prevent market manipulation, ensure fair pricing, and maintain operational transparency. Platforms that fail to meet these obligations risk enforcement penalties, loss of market access, and reputational damage. By incorporating compliance measures aligned with SEC and CFTC mandates, DeFi lending applications can mitigate regulatory risks and appeal to institutional investors looking for secure, compliant DeFi solutions.

Ensuring alignment with these regulatory frameworks requires continuous security assessments, transaction monitoring, and governance measures that integrate compliance into the platform's security strategy. The next step involves defining an audit framework that evaluates compliance with these global regulations, demonstrating how DeFi applications can implement automated security checks, risk reporting mechanisms, and policy enforcement tools to meet legal and financial obligations.

Aligning risk mitigation strategies with global regulatory frameworks is essential for ensuring the sustainability, security, and legitimacy of DeFi lending and borrowing platforms. By incorporating compliance requirements into security and operational measures, DeFi applications can effectively navigate regulatory challenges, mitigate legal risks, and foster trust among users, investors, and financial institutions.

A regulatory-aligned risk mitigation strategy offers several key benefits for DeFi platforms:

1. **Reducing legal and compliance risks** by proactively addressing regulatory obligations, thereby preventing enforcement actions, financial penalties, or platform shutdowns due to noncompliance with frameworks like MiCA, FATF, SEC, and CFTC.
2. **Enhancing financial transparency** by ensuring that on-chain transactions, liquidity pools, and lending mechanisms operate within legal boundaries, fostering confidence among lenders, borrowers, and institutional investors. This includes compliance with AML/KYC requirements,

auditability of financial activities, and adherence to investor disclosure mandates.

3. **Improving security posture** by meeting regulatory standards for wallet protection, smart contract security, and transaction integrity, reducing the likelihood of financial fraud, contract exploits, and operational failures. Implementing security controls that align with MiCA's wallet management policies or SEC's investor protection guidelines strengthens the resilience of DeFi protocols.
4. **Ensuring long-term sustainability** in regulated financial markets by designing DeFi platforms that can scale while remaining compliant with evolving legal frameworks. Compliance with FATF's Travel Rule, for example, enables DeFi lending protocols to maintain banking relationships and continue operations in key jurisdictions.

Regulatory frameworks are not just legal constraints – they provide a structured foundation for securing DeFi lending protocols and ensuring resilience against financial crimes, investor fraud, and systemic risks. A compliance-aligned risk mitigation strategy enables DeFi platforms to operate with predictability and legitimacy, making them more sustainable in the long term.

As global regulators continue to refine policies governing DeFi, DeFi platforms must establish continuous monitoring, risk assessment, and compliance reporting mechanisms that integrate seamlessly into their security architecture. The next step involves developing an audit framework that assesses how DeFi applications meet compliance standards, ensuring proactive regulatory adherence while maintaining decentralized innovation.

Ensuring compliance with security standards and regulatory frameworks requires a structured approach to documentation, auditability, and security controls in DeFi applications, dApps, and blockchain-based platforms. Regulatory frameworks mandate transparency in financial transactions, identity verification, and smart contract security, while security standards like ISO 27001, SOC 2, and NIST provide a structured approach to security governance and risk management.

To demonstrate compliance, organizations must provide auditors with verifiable evidence of security controls, operational procedures, and risk mitigation efforts. This requires a well-defined strategy that includes engaging relevant stakeholders, utilizing appropriate security tools, and maintaining an audit-ready repository of security artifacts.

Key Stakeholders for Compliance Audits: Compliance with regulatory frameworks and security standards requires a cross-functional engagement of security, risk, compliance, and operational teams. The following stakeholders should be actively involved in preparing for audits:

- **CISO:** Oversees security governance, risk management, and compliance strategy for blockchain and DeFi applications
- **Head of Blockchain Product Security/Smart Contract Security Team:** Ensures that security controls for smart contracts, cryptographic security, and key management meet regulatory and industry standards
- **GRC Team:** Manages regulatory compliance reporting, audit documentation, and security policies
- **Legal and Regulatory Affairs Team:** Ensures that DeFi protocols comply with jurisdictional requirements, including financial regulations, AML/KYC standards, and privacy laws (e.g. GDPR)
- **Blockchain Developers/DevOps:** Implement security tools, enforce secure software development practices, and ensure CI/CD pipeline security
- **External Auditors and Third-Party Security Firms:** Conduct independent security assessments, penetration testing, and regulatory compliance evaluations to validate adherence to security frameworks
- **Data Protection Officer (DPO) (if applicable):** Ensures compliance with data protection regulations (e.g. GDPR, CCPA) in blockchain applications involving user data

Categories of Compliance Evidence for Auditors: To pass compliance audits for DeFi, dApps, and blockchain applications, organizations must maintain structured security documentation and evidence repositories that auditors can review. The following categories of evidence are critical:

- **Risk Register and Threat Models:** Provides a detailed record of identified threats, vulnerabilities, and their associated risks, ensuring a structured approach to risk management
- **Regulatory Mapping and Compliance Documentation:** Aligns DeFi security controls with MiCA, FATF, SEC, CFTC, ISO 27001, and SOC 2 standards to demonstrate regulatory adherence
- **Security Policies and Procedures:** Includes smart contract security guidelines, key management policies, and secure coding practices for blockchain applications
- **Smart Contract and Blockchain Security Audits:** These are reports that provide evidence of smart contract audits such as:
  - Verifiable third-party audits from firms evaluating vulnerabilities in smart contracts
  - Evidence that smart contract logic adheres to mathematically proven security guarantees
  - Results from blockchain security testing demonstrating resilience against real-world AVs

Transaction Security and Financial Compliance: These are reports that provide evidence/proof of secure transaction processing and compliance such as:

- **Evidence of AML and KYC** enforcement, including integration with identity verification providers.
- **On-Chain Transaction Monitoring:** Reports from ETX, CipherTrace, and Chainalysis tracking on-chain activity for suspicious transactions.
- **Oracle Security and Data Integrity Reports:** Proof that oracles (e.g. Chainlink, API3) follow security best practices, preventing price manipulation and data integrity attacks.

Infrastructure Security and Node Management: These are reports that provide evidence/proof network security and secure nodes such as:

- **Blockchain Node Security Documentation:** Evidence of secure node configurations, access controls, and logging mechanisms
- **DDoS Protection and Network Resilience Reports:** Testing results demonstrating resistance to Sybil attacks, DoS, and transaction flooding using network security tools like Wireshark, Metasploit, and Snort

Secure Development and CI/CD Security: These are reports that provide evidence/proof that security tools are integrated in the CI/CD pipeline to enable secure software development life cycle (SSDLC) such as:

- **SAST/DAST Reports:** Results from security tools such as SonarQube, Checkmarx, and OWASP ZAP, validating application security
- **Software Bill of Materials (SBOM):** Provides a detailed inventory of third-party libraries, dependencies, and open-source components to ensure supply chain security compliance

To streamline compliance audits and evidence collection, organizations should integrate automated security testing tools that generate real-time reports for auditors. Ensuring the security, compliance, and resilience of DeFi applications requires the integration of specialized security tools that address various aspects of blockchain infrastructure, smart contracts, identity verification, financial compliance, and secure software development. Security automation is critical in mitigating risks, maintaining regulatory adherence, and providing auditors with verifiable evidence of compliance. The following sections outline key security tools used to enhance blockchain security, financial transparency, and secure software development practices in DeFi ecosystems.

Blockchain and Smart Contract Security Tools: Smart contracts serve as the backbone of DeFi applications, executing financial transactions, governance mechanisms, and lending operations. Given their immutable nature, security vulnerabilities in smart contracts can lead to irreversible financial losses. Several tools provide automated analysis, vulnerability detection, and security enforcement to strengthen smart contract integrity:

- **Slither** [189] is a static analysis tool for Solidity smart contracts, enabling developers to detect vulnerabilities such as reentrancy attacks, integer overflows, and improper access controls before deployment.
- **Mythril** [188] performs advanced vulnerability detection for Ethereum smart contracts, identifying security flaws through symbolic execution and control-flow analysis.
- **Certora** [280] facilitates formal verification of smart contracts, mathematically proving the correctness of contract logic and preventing exploits before deployment.
- **OpenZeppelin Defender** [192] provides continuous security monitoring for deployed smart contracts, offering automated incident response and contract governance controls.

By integrating these tools into the CI/CD pipeline, developers can automate security testing throughout the smart contract development life cycle, ensuring compliance with industry best practices. Section 3.2.4 of this chapter covers a variety of tools such as Slither, Mythril, Certora, and OpenZeppelin Defender and others that help identify, analyze, and remediate smart contract vulnerabilities before and after deployment.

Financial Compliance Tools: Automated tools help enforce compliance by monitoring transactions and detecting suspicious activity:

- **Chainalysis** [291] provides blockchain transaction monitoring and AML compliance solutions, identifying illicit activities and ensuring financial transparency.
- **Elliptic** [292] specializes in blockchain forensics, analyzing wallet transactions to detect financial fraud and illicit activities in dApps.

These tools assist DeFi platforms in meeting regulatory obligations by integrating automated compliance checks, transaction risk scoring, and real-time monitoring, ensuring adherence to FATF guidelines, SEC/CFTC regulations, and MiCA compliance requirements.

3.4 Conclusions

This chapter has emphasized the critical importance of security in blockchain applications, highlighting the unique risks and vulnerabilities associated with dApps and blockchain-based platforms. Given the irreversible nature of blockchain transactions and the financial implications of security breaches, proactive security strategies must be implemented from the early stages of development.

We explored lessons learned from high-profile security incidents in the blockchain space, showcasing the real-world consequences of unpatched vulnerabilities, flawed smart contract logic, and weaknesses in consensus mechanisms. A key takeaway is that security failures in blockchain applications often stem from predictable AVs – such as reentrancy, private key mismanagement, and insecure API integrations – underscoring the necessity of secure software design.

To mitigate these risks, we outlined strategies for identifying design flaws and vulnerabilities early in the SDLC. This includes the integration of threat modeling to systematically assess potential attack surfaces and security gaps, helping developers identify key assets, threats, and countermeasures. We also examined practical examples of security testing tools that can be integrated into automated CI/CD pipelines to enforce security best practices continuously.

Beyond technical security measures, auditing blockchain applications for compliance with security standards and regulatory frameworks was discussed as a crucial step in ensuring the legitimacy and long-term viability of blockchain projects. Compliance with global regulations helps organizations navigate legal risks while maintaining the trust of users, investors, and financial institutions.

Call to Action for dApp Product Owners and Software Developers: dApp product owners and software developers working on blockchain applications must prioritize security at every phase of the SDLC. As blockchain ecosystems continue to evolve, security cannot be an afterthought; it must be embedded into the development process. The following security best practices should be followed:

1. **Adopt secure coding practices for smart contracts** to prevent vulnerabilities such as reentrancy, integer overflows, and unauthorized contract modifications (see Section 2.4.10.3: Best Practices for Smart Contracts Security for more details).
2. **Leverage threat modeling techniques** to anticipate potential AVs and proactively design robust security countermeasures.
3. **Implement automated security testing tools** such as Slither, Mythril, and Certora for smart contract security analysis, open-source tools like Burp Suite, OWASP ZAP, and commercial tools like Checkmarx and Synk.
4. **Conduct regular security audits and penetration testing** to validate the effectiveness of security controls before deployment.
5. **Maintain regulatory compliance** by integrating AML/KYC checks, transaction monitoring, and financial transparency measures in DeFi applications.
6. **Foster a security-first culture** within blockchain development teams by prioritizing risk assessment and integrating security into continuous deployment workflows.

Future Directions in Blockchain Security Research and Development: As blockchain technology continues to evolve, new challenges and security threats will emerge. Future research and development in blockchain security should focus on several key areas:

- **Advancements in Formal Verification:** Increasing the adoption of formal verification methods to mathematically prove the correctness of smart contract logic and blockchain protocols
- **AI and Machine Learning for Threat Detection:** Leveraging AI-driven analytics to detect fraudulent transactions, anomalous activity, and smart contract exploits in real time
- **ZKPs for Privacy:** Enhancing user privacy in blockchain transactions through advanced cryptographic techniques such as ZKPs while maintaining compliance with financial regulations
- **Decentralized IAM:** Strengthening authentication mechanisms through DID solutions to reduce reliance on traditional KYC/AML procedures
- **Secure Cross-Chain Interoperability:** Addressing vulnerabilities in cross-chain bridges to prevent exploits such as double-spending and replay attacks
- **Quantum-Resistant Cryptography:** Researching cryptographic protocols that can withstand emerging threats posed by quantum computing

As blockchain adoption grows across industries, security will remain a fundamental concern that dictates the long-term success and trustworthiness of dApps. The continued collaboration between blockchain developers, security researchers, regulators, and auditors will be crucial in strengthening blockchain security standards and fostering a more resilient and secure digital financial ecosystem.

4

Securing Blockchain Applications: Practical Examples

For the things we have to learn before we can do them, we learn by doing them.
—Aristotle

4.1 Introduction

This chapter introduces practical examples to help you get started on your blockchain security practitioner journey. You'll focus on two key aspects: smart contract auditing and building a decentralized application (dApp) on Amazon Web Services (AWS) by leveraging AWS-managed blockchain [48], with the explicit goal of preparing both for threat modeling and security code reviews.

The code vulnerability auditing example targets specific code areas where blockchain vulnerabilities commonly occur. Previous chapters outlined security risks and best practices – now, we shift to real-world code examples that may contain flaws. You'll examine blockchain node software, wallet software, and dApps, with a special emphasis on smart contract code. Each component presents unique challenges but requires detailed code reviews to detect bugs, design flaws, and misconfigurations.

Auditing this code is essential for ensuring system integrity, preventing unauthorized access, and reducing the risk of financial or reputational damage. We'll connect known vulnerabilities from earlier chapters to actual code samples, demonstrating how to apply security principles in practice. The goal is to equip developers, security engineers, and architects with practical techniques to audit and strengthen blockchain implementations.

In the dApp-building example, you'll walk through creating a dApp that integrates blockchain technology with AWS services, providing a secure and scalable infrastructure. By combining smart contracts with AWS Lambda, Application

Blockchain Application Security: How to Design Secure and Attack Resilient Blockchain Applications, First Edition. Marco Morana, Harpreet Singh and Francesco Piccoli.
© 2026 John Wiley & Sons, Inc. Published 2026 by John Wiley & Sons, Inc.

Programming Interface (API) Gateway, and InterPlanetary File System (IPFS), this project shows how to automate cloud processes while enabling decentralized data storage. This dApp will serve as a foundation for applying threat modeling techniques and conducting code vulnerability audits.

By combining these practical exercises, you'll gain the tools to identify vulnerabilities, build secure dApps, and integrate these components into a broader threat modeling and security review strategy.

4.2 dApp Creation Example

This guide provides a step-by-step approach to building a dApp that integrates blockchain technology with AWS services to create a scalable and secure infrastructure. By combining smart contracts with AWS Lambda, API Gateway, and IPFS, you'll learn how to automate cloud processes while enabling decentralized data storage.

4.2.1 Architecture

The architecture of this dApp consists of the following components (Figure 4.1):

- Blockchain: Smart contracts deployed on Ethereum to handle decentralized logic.
- AWS Lambda: Automates backend processes triggered by blockchain events.
- Amazon API Gateway: Serves as the API interface between blockchain and AWS services.
- IPFS: Provides decentralized data storage, ensuring immutability and accessibility.
- Amazon S3: Used for additional off-chain data storage.

4.2.2 Project Components

The core of the dApp is built using Solidity smart contracts. This project implements an Ethereum Request for Comments (ERC)-20 token contract that allows users to mint and transfer tokens.

4.2.2.1 Token.sol (ERC-20 Token Contract)

```
// SPDX-License-Identifier: MIT
pragma solidity ^0.8.0;

import "@openzeppelin/contracts/token/ERC20/ERC20.sol";

contract MyToken is ERC20 {
  constructor(uint256 initialSupply) ERC20("MyToken", "MTK") {
     _mint(msg.sender, initialSupply);
  }
}
```

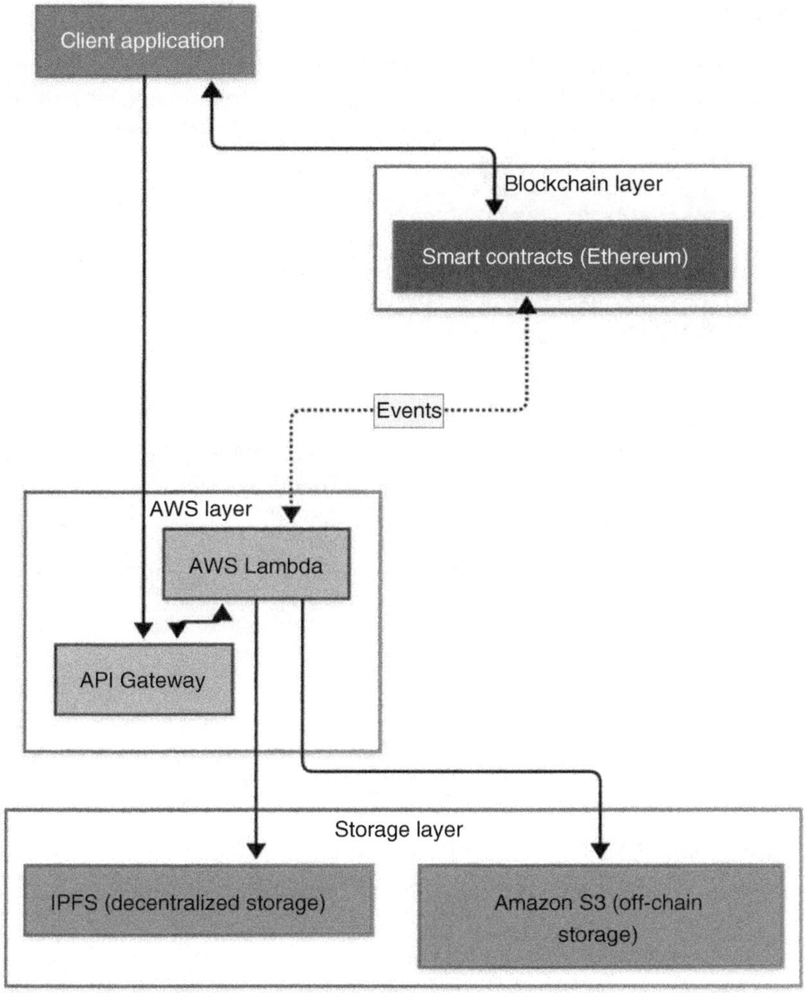

Figure 4.1 dApp architecture.

4.2.2.2 Smart Contract Deployment

1. Open Remix.
2. Copy and paste the `Token.sol` contract.
3. Compile the contract using the Solidity Compiler.
4. Deploy the contract to a testnet (e.g. Sepolia or Ropsten) using MetaMask.
5. Interact with the contract using functions like `mint()` and `transfer()`.

4.2.3 AWS Integration

To scale the dApp efficiently, AWS services handle backend logic and event-based automation. AWS Lambda Setup (Node.js) Navigate to AWS Lambda and create a new function. Select "Author from Scratch" and choose Node.js as the runtime. Paste the following code to listen for blockchain events:

```
exports.handler = async  (event)  => {
  try {
    const transactionDetails = JSON.parse(event.body);
    console.log("Transaction details received:",
transactionDetails);

    return {
       statusCode:  200,
       body:  JSON.stringify({ message:  'Lambda executed
successfully' })
        };
    } catch  (error)  {
       console.error("Error processing transaction:",  error);
       return {
         statusCode:  500,
         body:  JSON.stringify({ message:  'Internal
Server Error' })
        };
    }
};
```

4.2.3.1 API Gateway Setup

To configure the API Gateway setup for your dApp, follow these steps:

4.2.3.2 Create a New API in Amazon API Gateway

1. Open the AWS Management Console and navigate to API Gateway.
2. Click "Create API" and select "REST API" (not private).
3. Choose "New API" and provide a name for your API (e.g. "Web3dAppAPI").
4. Select "Regional" for the Endpoint Type.
5. Click "Create API" to finish the initial setup.

4.2.3.3 Link the API to AWS Lambda Function

1. In the left sidebar of your newly created API, click on "Resources."
2. Select the root resource ("/") and click "Actions" > "Create Resource."
3. Name the resource (e.g. "transactions") and click "Create Resource."

4. With the new resource selected, click "Actions" > "Create Method."
5. Choose "POST" from the dropdown and click the checkmark.
6. In the setup pane:
   - Select "Lambda Function" for Integration type.
   - Check "Use Lambda Proxy integration."

```
const Web3 = require('web3');
const ABI =  [/* Paste your ERC20 ABI here */];
const CONTRACT_ADDRESS = 'YOUR_CONTRACT_ADDRESS';

exports.handler = async  (event)  => {
  const web3 = new Web3('https://goerli.infura.io/v3/
YOUR_INFURA_PROJECT_ID');
  const contract = new web3.eth.Contract(ABI,  CONTRACT_ADDRESS);

  try {
      const userAddress = event.address;
      const balance = await contract.methods.balanceOf
(userAddress).call();
      return {
        statusCode:  200,
        body:  JSON.stringify({ balance }),
       };
   } catch  (error)  {
      console.error(error);
      return {
         statusCode:  500,
         body:  JSON.stringify({ error:  'Failed to
fetch balance' }),
       };
   }
};
```

4.2.3.4 Define API Methods

1. With the POST method selected, click "Method Request."
2. Under "Settings":
   - Set "Authorization" to "NONE" (or any Appropriate option).
   - Set "API Key Required" to "true" if you want to use API keys.
3. Expand "Request Body" and add a model:
   - Click "Add model."
   - Set "Content type" to "application/json."
   - Select "Empty" for the model.

4. Click on "Integration Request" to customize how the request is sent to Lambda:
   - Leave most settings as default for Lambda Proxy integration.
   - You can add mappings here if needed, but it's often unnecessary with proxy integration.
5. To handle cross-origin resource sharing (CORS; if your frontend is on a different domain):
   - Go back to the resource level.
   - Click "Actions" > "Enable CORS."
   - Accept the default settings or customize as needed.
   - Click "Enable CORS and replace existing CORS headers."
6. Deploy the API:
   - Click "Actions" > "Deploy API."
   - Choose "New Stage" and name it (e.g. "dev" or "prod").
   - Click "Deploy."
7. Note the "Invoke URL" provided after deployment. This is the endpoint your frontend or smart contracts will use to interact with the API.

4.2.3.5 Additional Configuration

1. To secure your API, consider adding:
   - API keys: In API Gateway, go to "API Keys" to create and manage keys.
   - Usage plans: Set up usage plans to control and monitor API usage.
   - AWS Identity and Access Management (IAM) roles: For more granular access control.
2. Set up logging and monitoring:
   - In your API's settings, find "CloudWatch Settings."
   - Enable CloudWatch Logs and set the log level.
3. Configure stage variables if needed:
   - In the stage editor, go to the "Stage Variables" tab.
   - Add key-value pairs that can be used in your API configuration.

By following these steps, you'll have a fully configured API Gateway that can handle requests from your dApp's frontend or smart contracts, forwarding them to your AWS Lambda function for processing. Remember to update your Lambda function to properly handle the incoming API Gateway events and return appropriate responses.

4.2.4 Create a Frontend

The following example details step-by-step instructions to develop frontend using React framework.

4.2.4.1 Create React App

```
npx create-react-app dApp-frontend
cd dApp-frontend
npm install web3 axios
```

4.2.4.2 Create Frontend Code

```
import React, { useState } from 'react';
import Web3 from 'web3';
import axios from 'axios';

const App = () => {
  const [account, setAccount] = useState('');
  const [balance, setBalance] = useState('');

  const connectWallet = async () => {
    if (window.ethereum) {
       const web3 = new Web3(window.ethereum);
       await window.ethereum.
request({ method: 'eth_requestAccounts' });
          const accounts = await web3.eth.getAccounts();
          setAccount(accounts[0]);
       } else {
           alert('Please install MetaMask!');
       }
   };

   const fetchBalance = async () => {
     try {
      const response = await axios.post('https://your-api-gateway-
url.amazonaws.com/default/BlockchainHandler', {
               address: account,
           });
           setBalance(response.data.balance);
       } catch (error) {
           console.error(error);
           alert('Failed to fetch balance');
       }
   };

   return (
       <div>
          <h1>ERC20 dApp</h1>
          <button onClick={connectWallet}>Connect Wallet</button>
          {account && <p>Connected Account: {account}</p>}
```

```
          <button onClick={fetchBalance} disabled=
{!account}>Fetch Balance</button>
          {balance && <p>Token Balance:  {balance}</p>}
       </div>
   );
};

export default App;
```

Now your frontend of the application is ready. Congratulations! You have created the first dApp. Now you want to learn secure practices while developing dApps.

4.2.5 Security Review

This is the most important part of your implementation. No matter how well you write the code and deploy your services and applications, security vulnerabilities will remain and it is very critical to ensure those threats, risks, and vulnerabilities. The following section details few of the secure practices based on known vulnerabilities.

4.2.5.1 Smart Contract Vulnerabilities

Issue: Smart contracts, such as the ERC-20 token contract, may contain bugs or vulnerabilities like reentrancy attacks, unchecked arithmetic, or improper access control.

Remediation:

- Conduct tests using tools like MythX or OpenZeppelin Defender.
- Use libraries like OpenZeppelin for implementing secure, pretested ERC-20 contracts.
- Test contracts extensively on testnets before deploying to mainnets.

4.2.5.2 AWS Lambda Security

Issue: The Lambda function processes blockchain events but lacks input validation and could be vulnerable to injection attacks or malformed data.

Remediation:

- Validate and sanitize all incoming data before processing.
- Use IAM roles with the principle of least privilege to restrict Lambda's access to other AWS resources.
- Enable AWS Secrets Manager for securely handling sensitive credentials (e.g. API keys).
- Scan your Lambda function code for secrets and common vulnerabilities using tools like AWS Inspector, Semgrep, or TruffleHog. This helps detect

misconfigurations, insecure coding practices, or hardcoded credentials early in the development life cycle, reducing the risk of exposure or compromise.

- Regularly scan third-party dependencies used by the Lambda function using tools such as AWS Inspector or Snyk. Supply chain attacks present serious risks; attackers may exploit known vulnerabilities or introduce malicious packages capable of data exfiltration, credential theft, or even reverse shell access.
- Keep your Lambda runtime and packages up-to-date, and configure alerting and monitoring for any unusual behavior or excessive privilege usage to improve detection and response.

4.2.5.3 API Gateway Misconfigurations

Issue: The API Gateway is potentially exposing endpoints without enforcing proper authentication or rate limiting, increasing the risk of unauthorized access and Denial of Service (DoS) attacks. This exposure is further compounded by the absence of an AWS Web Application Firewall (WAF) in front of the API Gateway, which would otherwise help protect against common web and API-based attack vectors.

Remediation:

- Deploy AWS WAF or similar in front of the API Gateway with managed rule sets to block Open Web Application Security Project (OWASP) Top 10 threats like cross-site scripting (XSS), SQL injection (SQLi), etc.
- Enforce OAuth 2.0, JSON Web Tokens (JWTs), or API keys for all API endpoints.
- Enable throttling and rate limiting in API Gateway to mitigate abuse.
- Use burst limits and quotas to control API usage patterns and prevent abuse.
- Apply strict input validation using API schema definitions (OpenAPI/Swagger).
- Reject unexpected input to mitigate injection attacks.
- Implement authentication mechanisms such as AWS Cognito for user management or API keys for access control.
- Use HTTPS endpoints to encrypt data in transit.
- Enable caching, logging, and anomaly detection features.
- Use private endpoints for internal services.
- Use AWS Shield Standard for basic distributed DoS (DDoS) protection.
- Enable CloudWatch logs for API Gateway.
- Use AWS GuardDuty, WAF logs, and Athena to detect malicious behavior.

4.2.5.4 Data Storage Risks

Issue: IPFS ensures decentralized storage but does not inherently encrypt files, while Amazon S3 may store sensitive off-chain data unencrypted.

Remediation:

- Encrypt files before uploading them to IPFS using cryptographic libraries.
- Enable server-side encryption (e.g. AES-256) for Amazon S3 buckets and enforce bucket policies to restrict access.

4.2.5.5 Blockchain Event Handling

Issue: Blockchain events processed by AWS Lambda via API Gateway could result in missed events due to failures in the pipeline (e.g. network issues).

Remediation:

- Use Amazon SQS as a buffer between API Gateway and Lambda to ensure reliable event processing.
- Implement dead-letter queues (DLQs) for failed event processing.

4.2.5.6 Cross-Origin Resource Sharing (CORS)

Issue: Improperly configured CORS settings in API Gateway could expose the dApp to XSS attacks.

Remediation:

- Restrict allowed origins, methods, and headers in CORS policies to trusted sources only.

4.2.5.7 Frontend Integration Risks

Issue: The frontend interacts with smart contracts and APIs via MetaMask but may expose sensitive information through poorly secured JavaScript code.

Remediation:

- Avoid hardcoding sensitive data (e.g. private keys or API URLs) in the frontend code.
- Use environment variables or secure configuration files during deployment.
- Ensure that communication between the server and the frontend is encrypted.
- Minify and obfuscate the frontend source code during deployment in order to remove developer comments and other details that can make it harder for an attacker to collect all the required information about the application.

4.2.6 Conclusion

This project showcases how Web3 dApps can integrate with AWS to create scalable and automated decentralized systems. By leveraging smart contracts, AWS Lambda, and IPFS, you can develop a secure, decentralized infrastructure with cloud-based automation.

4.3 Code Auditing Examples

4.3.1 Introduction

This chapter focuses on specific code areas where blockchain vulnerabilities commonly occur. Previous chapters outlined security risks and best practices; here, we shift to concrete examples of code that may contain flaws. We will examine blockchain node software, wallet software, and dApps, with a special emphasis on smart contract code. Each category presents different challenges, but all require careful reviews to detect bugs, design weaknesses, or misconfigurations.

Auditing this code is essential for preserving system integrity, preventing unauthorized access, and reducing the risk of financial or reputational damage. Throughout this chapter, we will connect known vulnerabilities from earlier chapters to real-world code samples. By doing so, we establish how to apply security principles in practice. The goal is to equip developers, security engineers, and other stakeholders with techniques to systematically audit and strengthen their blockchain implementations.

4.3.2 Rationale for Secure Coding Practices

The $50 million hack of the DAO in 2016 [132] serves as a stark reminder of the critical importance of secure coding practices and post coding security assurances in the world of blockchain and smart contracts. This incident not only shook the Ethereum community but also highlighted the potential vulnerabilities that can exist in decentralized systems.

The DAO Hack: A watershed moment, the DAO, short for Decentralized Autonomous Organization, was an ambitious project aimed at creating a decentralized venture capital fund on the Ethereum blockchain. However, a small coding error in its smart contract led to catastrophic consequences. An attacker exploited a vulnerability in the code, allowing them to drain millions of dollars worth of Ether from the organization. This hack wasn't just a financial loss; it was a wake-up call for the entire blockchain industry. It underscored the fact that in the world of immutable, self-executing smart contracts, even a minor coding flaw can have far-reaching implications.

The Imperative of Secure Coding: The DAO hack emphasizes several crucial aspects of secure coding:

- **Rigorous Testing:** Thorough testing, including penetration testing and code audits, is essential before deploying any smart contract or blockchain application. The DAO's code had been audited, but the critical vulnerability was missed, highlighting the need for even more stringent review processes.

- **Formal Verification:** Implementing formal verification techniques can help ensure that smart contracts behave exactly as intended under all possible scenarios. This mathematical approach to code verification could have potentially caught the recursive call vulnerability that led to the DAO's downfall.
- **Simplicity in Design:** Complex systems are more prone to errors and harder to secure. The DAO's structure was intricate, making it difficult to foresee all potential attack vectors. Striving for simplicity in code design can significantly reduce the risk of vulnerabilities.

Building a Culture of Security: Beyond technical practices, the incident highlights the need for a fundamental shift in how we approach software development in high-stakes environments:

- **Security-First Mindset:** Developers must adopt a security-first approach, considering potential vulnerabilities at every stage of the development process. This mindset should be ingrained in the culture of organizations dealing with blockchain and other critical systems.
- **Continuous Learning:** The rapidly evolving nature of blockchain technology and cyber threats necessitates ongoing education and skill development for developers. Staying updated with the latest security best practices is crucial.
- **Transparency and Collaboration:** Open-source development and collaborative security efforts can help identify and address vulnerabilities more effectively. The collective knowledge of the community can be a powerful tool in enhancing security.

The Path Forward: The DAO hack was a turning point that led to significant improvements in smart contract security. It spurred the development of new tools, frameworks, and best practices for secure blockchain development. However, as the technology continues to evolve and find new applications, the need for vigilance remains paramount.

By prioritizing secure coding practices, fostering a culture of security, and learning from past incidents, we can build more robust and trustworthy decentralized systems. The DAO hack serves as a powerful reminder that in the world of blockchain and smart contracts, code is law – and secure code is the foundation of trust and innovation.

4.3.3 Auditing Smart Contract Code

Smart contracts encapsulate the core logic of many blockchain applications. They dictate how assets are transferred, how user interactions are processed, and how data is stored on-chain. If improperly written or audited, these contracts can expose the application to significant security risks. This section addresses common vulnerabilities, discusses practical auditing tools and approaches, and outlines best practices for remediation and secure deployment.

4.3.3.1 Common Smart Contract Vulnerabilities: Reentrancy

Explanation: Reentrancy occurs when a contract calls an external untrusted function (e.g. sending Ether to a user), and that external function reenters the original contract to modify its state in unexpected ways. Attackers exploit this by performing repeated calls before the contract state is updated, often draining funds or causing double-spend conditions.

Vulnerable Example (Solidity):

```
mapping(address => uint256) public balances;

function deposit() public payable {
  balances[msg.sender] += msg.value;
}

// Vulnerable withdraw function
function withdraw() public {
    require(balances[msg.sender] > 0, "No balance");
    (bool sent, ) = msg.sender.call{value: balances[msg.sender]}
("");
    require(sent, "Failed to send Ether");

    // State update happens after the external call
    balances[msg.sender] = 0;
}
```

Explanation of Vulnerability: Because `balances[msg.sender]` is updated after sending Ether, a malicious contract can trigger the `withdraw()` function repeatedly before the balance is reset.

Improved Example (Solidity):

```
function withdraw() public {
  uint256 amount = balances[msg.sender];
  require(amount > 0, "No balance");

  // Update state first
  balances[msg.sender] = 0;

  (bool sent, ) = msg.sender.call{value: amount}("");
  require(sent, "Failed to send Ether");
}
```

Why It Works: By setting the balance to zero before the external call, any subsequent reentrant call finds the balance already depleted, preventing further withdrawals.

4.3.3.2 Integer Overflows and Underflows

Explanation: Older versions of Solidity allowed unsigned integers to overflow or underflow, meaning they would wrap around when exceeding their maximum or falling below zero. Attackers could exploit these vulnerabilities to manipulate balances, token supplies, or other numeric state variables.

To mitigate this risk, it is essential to always include guard checks before updating or returning any numeric values. Implementing these checks, either manually or by using safe arithmetic libraries like SafeMath, or by leveraging built-in overflow protections in newer Solidity versions, ensures that the contract can detect and prevent arithmetic anomalies, thereby protecting against attacks that exploit numerical wraparound vulnerabilities.

Vulnerable Example (Solidity):

```
function multiply(uint256 a, uint256 b) public pure returns
(uint256) {
  return a * b; // Potential for overflow

}
```

If `a` and `b` are large enough, the multiplication can overflow, leading to incorrect values.

Improved Example with Checked Operations:

```
import "@openzeppelin/contracts/utils/math/SafeMath.sol";

function multiplySafe(uint256 a, uint256 b) public pure
returns (uint256) {
  return SafeMath.mul(a, b); // Reverts on overflow
}
```

Why It Works: Libraries like OpenZeppelin's `SafeMath` [293] (or Solidity's built-in checked arithmetic [294] in versions >= 0.8.0) revert the transaction when an overflow or underflow occurs.

4.3.3.3 DoS of Smart Contracts

Explanation: A DoS can be triggered if a contract function consumes excessive gas, enters a locked state, or relies on unbounded loops or external calls. Attackers can exploit these conditions to stall critical operations, preventing other users from executing them.

Example (Solidity):

```
// Potential DoS if the array grows large
address[] public stakeholders;

function distributeFunds() public {
  // If stakeholders array is huge, this might exceed the
block gas limit
  for  (uint i = 0;  i < stakeholders.length;  i++) {
        payable(stakeholders[i]).transfer(1 ether);
    }
}
```

Why It Happens: As the array size grows, the loop can consume more gas than is available. An attacker might add many entries to `stakeholders`, effectively making this function fail.

Mitigation Strategies:

- Implement gas-efficient data structures or break large loops into smaller transactions.
- Use off-chain processes or oracles to handle lengthy operations where feasible.

4.3.3.4 Access Control Failures

Explanation: Access control failures happen when a contract's permissions are not correctly defined or enforced. Common scenarios include missing `onlyOwner` modifiers or misconfiguration of roles in upgradeable or proxy contracts.

Example (Solidity):

```
address public owner;

modifier onlyOwner() {
  require(msg.sender == owner,  "Not owner");
   _;
}

function setOwner(address _newOwner)  public {
  // Access control missing:  anyone can change the owner
  owner = _newOwner;
}
```

Why It Happens: Developers may forget to add the onlyOwner modifier or may implement custom access checks incorrectly.

Fix:

```
function setOwner(address _newOwner)  public onlyOwner {
  owner = _newOwner;
}
```

Why It Works: The modifier ensures only the current owner can transfer ownership, preventing unauthorized role changes.

4.3.3.5 Logic Flaws and Business Logic Errors

Explanation: Even if code is syntactically correct, the contract may not reflect the intended financial or operational rules. Mistakes in function sequencing, conditional checks, or external calls can lead to unintended outcomes that attackers can exploit.

Example (Token Vesting):

```
// Simplified vesting example
uint256 public totalTokens;
uint256 public releasedTokens;
uint256 public vestingStart;
uint256 public vestingEnd;

function releaseTokens() public {
  require(block.timestamp > vestingEnd,  "Vesting not ended");
  uint256 amount = totalTokens - releasedTokens;

  // Business logic error:  no checks on partial vesting or cliff
  _transfer(msg.sender,  amount);
  releasedTokens += amount;
}
```

Potential Problem: The example assumes tokens should only be released after vestingEnd in one lump sum, ignoring partial vesting schedules. Attackers or authorized users might release tokens prematurely if the logic is not strict, or users might be blocked from legitimate partial withdrawals.

Mitigation: Carefully design and review the contract's intended behavior, ideally with domain experts, to ensure the logic follows correct business rules.

4.3.4 Audit Processes and Tools for Smart Contracts

4.3.4.1 Manual Code Review

Manual analysis is essential and has been the de facto standard for understanding contract logic and verifying that it matches the intended design. It involves:

- **Line-by-Line Review:** Checking for hidden assumptions, verifying that error handling and access controls are properly in place
- **Contextual Understanding:** Confirming the contract's logic aligns with the application's broader requirements, such as business rules or compliance mandates

Example Checklist Items

- Are state-changing operations properly guarded by checks or modifiers?
- Is external code called only after updating internal state variables to avoid reentrancy?
- Are event logs used consistently to track critical operations?

4.3.4.2 Automated Static Analysis Tools

Automated tools can detect common vulnerabilities, code smells, or suspicious patterns:

- **Slither** [295]**:** Identifies reentrancy, unused return values, and code complexity issues
- **Mythril** [296]**:** Performs symbolic execution to discover logic flaws or overflow problems
- **Almanax** [297]**:** AI-powered security analyzer that can scan code for vulnerabilities and generate recommendations for best practices

Implementation Issues:

- Ensure these tools are integrated in your development continuous integration/continuous deployment (CI/CD) pipeline or used before deployment.
- Note that automated tools may produce false positives or fail to detect complex logic bugs. While AI tools have shown great promise in finding vulnerabilities, manual validation is still critical.

4.3.4.3 Unit and Integration Testing

Test frameworks (e.g. Truffle [298], Hardhat [299], Foundry [300]) can simulate contract interactions and check whether functions operate as intended:

- **Unit Tests:** Target specific functions. Confirm that expected and edge-case conditions trigger correct behavior.
- **Integration Tests:** Combine multiple contracts and test real-world scenarios, such as multicontract calls or interactions with external oracles.

Example (Testing Reentrancy Mitigation):

```
// Pseudocode using a Truffle test
contract("MyContract", (accounts) => {
```

```
  it("should prevent reentrancy attacks", async  () => {
    const myContract = await MyContract.deployed();
    await myContract.deposit({ from:  accounts[0],  value:  10 });
    try {
      await myContract.
reentrantWithdraw({ from:  attackerContract });
    } catch  (error)  {
      assert(error.message.includes("revert"),
"Reentrancy not prevented");
    }
  });
});
```

4.3.4.4 Formal Verification

Formal methods [279] mathematically model a contract's execution and verify that specific properties always hold (e.g. "No user can withdraw more than their balance"):

- **Process**
  - Translate contract or parts of its logic into a formal specification language.
  - Use theorem provers or model checkers to exhaustively verify correctness.
- **Benefits**
  - Reduces ambiguity by treating the code as a logical system.
  - Proves correctness for critical functions.
- **Limitations**
  - Time-consuming and requires specialized expertise.
  - Doesn't replace a broader security audit; used to supplement it.

Certora Prover [301], for example, is a popular tool that can formally verify ERC-20 token properties like balance updates in transfers.

4.3.5 Best Practices in Smart Contract Audits

4.3.5.1 Security by Design

- **Defensive Programming**
  - Use checks before external calls, validate all inputs, and handle unexpected errors gracefully.
- **Consistent Architecture**
  - Apply consistent patterns for access control, error handling, and logging.

Implementation Details:

- Keep contract modules small and clearly scoped.
- Use recognized libraries for token standards [302] or mathematical operations.

4.3.5.2 Remediation and Secure Redeployment

- **Address Audit Findings Promptly**
  - Fix high-severity bugs first, then retest.
  - Keep a structured change log to track modifications.
- **Retesting and Phased Deployment**
  - Rerun the test suite after patches.
  - Deploy to testnets for trial runs and community feedback.
  - Perform a final production deployment only after verifying stability.

Key Points:

- Use versioning (e.g. contract version numbers) for clarity in case multiple releases exist.
- Engage third-party auditors and the community to assess updates.

Summary: Auditing smart contracts is a multilayered process involving manual review; automated scanning with tools like Slither, Mythril, and Almanax; thorough testing; and formal verification. By integrating security by design principles and prioritizing remediation, developers can reduce exploit risks and maintain trust in their blockchain applications.

4.3.6 Auditing Blockchain Node Software

Nodes underpin a blockchain network by maintaining the ledger, validating transactions, and participating in consensus. Because they function as the backbone of a distributed ledger, vulnerabilities in node software can lead to major security breaches, consensus manipulation, or service disruption. In this section, we focus on common node-related vulnerabilities, discuss approaches to auditing node implementations, and provide code-oriented examples that highlight potential pitfalls and solutions.

4.3.6.1 Types of Blockchain Nodes

1. **Full Nodes**
   - Store a complete copy of the blockchain ledger.
   - Validate all blocks and transactions.
   - Typically provide higher security at the cost of higher resource consumption.
2. **Light Nodes**
   - Store partial blockchain data (e.g. block headers only).
   - Rely on full nodes for transaction validation or state verification.
   - Offer lower resource consumption but depend more on trusted peers.
3. **Mining/Validator Nodes**
   - Participate directly in consensus, creating or validating new blocks.

- Require additional security because they hold private keys for block signing or hashing.
- Prime targets for attacks aimed at network manipulation or reward theft.

Each node type faces unique risks. Full nodes handle large amounts of data, while light nodes must trust external sources. Mining/validator nodes involve critical private keys, making them attractive to attackers.

4.3.6.2 Typical Vulnerabilities in Node Implementations

4.3.6.2.1 Consensus Algorithm Weaknesses

Description: Implementation errors in consensus algorithms – be it Proof of Work (PoW), Proof of Stake (PoS), or others – may allow double-spending, block withholding, or chain reorganizations if attackers can exploit logic flaws.

- **Example Flaw:** An off-by-one error in PoS validator selection may allow malicious nodes to gain undue influence over block creation.
- **Implementation Issue:**

```
// Hypothetical PoS validator selection  (Go)
// Example snippet where an off-by-one can allow an extra validator
if validatorIndex < len(validators)  {
    selectedValidator :=  validators[validatorIndex]
    // ...
} else {
   // Edge condition not properly handled
   // Potentially skipped or incorrectly assigned
}
```

Mitigation:

- Thoroughly test edge conditions in validator selection or block reward logic.
- Use formal verification or advanced static analysis (including AI-based tools) on consensus-critical portions of the code.

4.3.6.2.2 Networking Stack and P2P Protocol Issues

Description: Nodes rely on peer-to-peer (P2P) communication to exchange blocks and transactions. Bugs in the networking layer can lead to DoS, partitioning attacks (where legitimate nodes are isolated), or code injection vulnerabilities.

- **Example Flaw:** Unbounded queue size in a network message buffer, allowing an attacker to flood the node with bogus requests

- **Implementation Issue:**

```
// Hypothetical Rust snippet for a message handler
let mut buffer: Vec<Message> = Vec::new();

// Potentially no size limit:
fn handle_incoming_message(msg: Message) {
   buffer.push(msg);
    // Attackers can send massive amounts of data, eventually
causing an out-of-memory error
}
```

Mitigation:

- Enforce rate limiting and message size restrictions.
- Define maximum allowable connections or resource thresholds.
- Include comprehensive error handling to drop malformed or suspicious messages.

Keep in mind that not every mitigation suggestion is a quick fix for a node operator; some require changes at the protocol level or are already built into the protocol itself (e.g. gas limits in Ethereum).

4.3.6.2.3 Resource Exhaustion Attacks

Description: Nodes can be overwhelmed through high computational or memory-intensive tasks, especially in networks with large transaction volumes. Attackers might exploit transactions with high verification costs or spam the network with useless data.

Example Flaw:

- Sending an inordinate number of valid-but-complex transactions (e.g. large multisignature inputs or sophisticated smart contract calls) to drain node CPU/memory
- Flooding the network with tiny, spammy transactions to bloat mempools

Implementation Issue (C++ Pseudocode):

```
bool verifyTransaction(const Transaction& tx) {
  // Potentially expensive cryptographic checks inside a loop
  for (auto &input : tx.inputs) {
       if (!verifySignature(input, tx)) {
            return false;
        }
    }
    return true;
}
```

- When many such transactions are processed in quick succession, the node's CPU usage can spike or remain pegged, delaying validation of legitimate transactions.

Mitigation:

- Implement rate limiting or backpressure mechanisms to reject or delay processing when resource usage is high.
- Impose fees or gas limits for computationally intensive transactions, ensuring that complex operations are economically disincentivized.
- Monitor node resources (CPU, memory, disk) and automate protective actions (e.g. temporary ban of abusive peers) when thresholds are exceeded.

4.3.6.2.4 Configuration and Key Management Errors

Description: Node operators often misconfigure security settings or store private keys in unsafe locations. Such errors expose validator/miner keys to unauthorized access or allow attackers to compromise the node's Remote Procedure Call (RPC) endpoints.

Example Flaw:

- Leaving default credentials (e.g. admin:admin) active on an RPC interface
- Storing unencrypted private validator keys on a publicly accessible server

Implementation Issue (Config File Snippet):

```
[rpc]
enable = true
auth_user = admin
auth_password = admin   ;  Weak or default password

[keys]
private_key_path = "/home/user/validator_key.pem"   ;
Plaintext storage
```

- Attackers who gain access to this configuration can connect remotely and control the node or simply copy the private key.

Mitigation:

- Enforce strong credentials and TLS/HTTPS for all remote procedure call endpoints.
- Encrypt validator or miner keys in secure hardware modules (SHMs) or at least protect them with file-level encryption and strict permissions.
- Periodically rotate keys, and maintain separate access policies for different functional roles (e.g. read-only versus validator permissions).

4.3.6.3 Approaches to Node Software Auditing

4.3.6.3.1 Source Code Review

Manually inspecting node software (commonly written in Go, Rust, or C++) allows auditors to:

- **Identify Dead Code or Hidden Backdoors:** Look for suspicious conditionals, outdated references, or debugging shortcuts left by developers.
- **Validate Security Libraries:** Check if cryptographic libraries are used correctly, verifying ephemeral keys and random number generation.
- **Confirm Proper Error Handling:** Ensure that unhandled exceptions or unexpected states do not silently compromise node stability.

Example Checklist Items:

1. Are all network inputs sanitized?
2. Do cryptographic operations follow recommended best practices?
3. Are sensitive error messages (stack traces, debug data) leaking in production builds?

4.3.6.3.2 Penetration Testing

Penetration testing involves simulating malicious behaviors from the perspective of an external attacker.

- **P2P Flooding Attacks:** Test how the node reacts to large volumes of connection attempts or messages.
- **Replay or Fork Attacks:** Attempt to replay blocks or transactions to detect if node logic fails to handle duplicates.
- **Over-Resource Consumption:** Launch scenarios that aim to overwhelm CPU, memory, or disk usage to see if the node remains stable.

Implementation Tip: Run node software under a controlled environment (e.g. Docker containers or VMs) while using specialized tools or frameworks to flood the node with traffic. Monitor logs, CPU usage, and memory consumption for anomalies.

4.3.6.3.3 CI/CD Checks

Integrate security checks and testing in the build pipeline for each release:

- **Static Analysis Tools**
  - **AI-Based:** Scan node code to spot suspicious patterns, code quality issues, or known vulnerabilities.

    - **Traditional Linters and Code Scanners:** Look for memory leaks, concurrency bugs, or known common vulnerabilities and exposures (CVEs) in third-party libraries.
- **Automated Tests and Regression Suites**
    - Run functional tests on commits to detect any break in node behavior.
    - Employ fuzzing frameworks that repeatedly execute random transactions or messages against the node to unearth corner-case errors.

Implementation Detail:

```
# Example GitHub Actions or CI/CD snippet
name: Node Software Build and Test

on: [push, pull_request]

jobs:
  build-and-test:
   runs-on: ubuntu-latest
   steps:
     - name: Checkout
       uses: actions/checkout@v2

     - name: Install Dependencies
       run: sudo apt-get install ...

     - name: Build
       run: make build  # or cargo build for Rust

     - name: Static Analysis
       run: almanax scan --path=./src

     - name: Unit Tests
       run: make test   # or cargo test for Rust

     - name: Fuzz Tests

       run: ./fuzz.sh
```

This approach ensures code is continuously vetted for security flaws and performance regressions before deployment.

4.3.7 Auditing Wallet Software

Wallets facilitate key management, transaction signing, and user interaction with the blockchain. Because they directly handle private keys or seed phrases, they are prime targets for attackers. A compromise at the wallet level can result in direct asset theft or unauthorized transaction approvals. In this section, we review different wallet types, common vulnerabilities, and auditing practices, complete with illustrative code snippets to highlight issues and solutions.

4.3.7.1 Types of Wallets

1. **Software Wallets (Desktop, Mobile, Web)**
   - Run on general-purpose operating systems (Windows, macOS, iOS, Android, browser environments).
   - Keys are stored locally in software-managed keystores.
   - Easier to distribute and update but rely on the security posture of the underlying device/OS.
2. **Hardware Wallets**
   - Store private keys in a dedicated, tamper-resistant chip.
   - Generally more secure against software exploits.
   - Can still be vulnerable to supply chain or physical tampering attacks.

Another important distinction in wallets is **custodial versus noncustodial**:

- **Custodial:** Private keys held by a third-party service provider. Users rely on the provider's security.
- **Noncustodial:** Users control their private keys directly. More responsibility but also more freedom.

Each wallet type has unique considerations. Hardware wallets tend to mitigate certain software risks but introduce hardware reliability and supply chain risks. Software wallets must deal with endpoint security challenges and potential malicious code injection. Custodial solutions pivot trust to a service provider's security practices.

Cold, Warm, and Hot Wallets: Wallets are often categorized based on their connectivity and operational context. Each category presents different risk profiles and auditing approaches.

Cold Wallets

- Typically remain offline, making them less susceptible to remote exploits.
- Can include hardware wallets, paper wallets, or air-gapped machines.
- Physical security is a focus: attacks may involve physical tampering, supply chain breaches, or key extraction attempts.
- Audits emphasize secure key generation, offline transaction signing procedures, and tamper-evident hardware design.

Warm Wallets

- Often kept mostly offline but brought online for scheduled use, such as batch processing of transactions.
- Frequently found in institutional environments that need a balance between security and operational efficiency.
- Auditing examines access controls during the transition from offline to online, as well as protocols for authorizing and verifying transactions.

Hot Wallets

- Hot wallets stay online to facilitate frequent transactions. This connectivity exposes them to a wider attack surface, including remote intrusion.
- Risk areas include malware, network-based exploits, and compromised client applications.
- Audits focus on secure coding, segregation of duties, least privilege principles, and continuous monitoring of access and network traffic.
- Security measures often incorporate automated alerts and rapid incident response plans to handle potential breaches.

4.3.7.2 Wallet-Specific Vulnerabilities

4.3.7.2.1 Private Key Exposure

Description: Unencrypted or improperly secured keys allow attackers to read or exfiltrate them. This risk is especially high if private keys are stored in plaintext on disk or in browser local storage.

Vulnerable Example (Web-Based Wallet JavaScript):

```
// Storing private key in localStorage in plain text
localStorage.setItem('privateKey',  '0xABC123...');
```

Problem: Any script with access to localStorage (including malicious browser extensions) can extract this private key.

Mitigation:

- Encrypt private keys with a passphrase before storing.
- Use secure storage APIs.
- Segregate sensitive logic to minimize the risk of script injection.

4.3.7.2.2 User Interface Manipulation (Phishing or Spoofing)

Description: Attackers can trick users with deceptive prompts or URLs. This is particularly dangerous for web wallets or browser extensions that prompt transaction approvals.

Example Attack Vector:

1. A malicious site injects HTML or JavaScript that mimics the wallet's transaction confirmation pop-up.
2. The user approves a transfer to the attacker's address, believing it is a legitimate action from their wallet.

Mitigation:

- Implement strong domain and TLS checks.
- Use cryptographic proofs (e.g. verifying wallet extension signatures) to ensure authenticity.
- Educate users to confirm transaction details (e.g. recipient address, gas fees) carefully.
- Conduct transaction simulations to ensure funds are being moved as expected.

4.3.7.2.3 Transaction Handling Errors

Description: Wallets may incorrectly parse or validate transaction data, leading to misdisplayed amounts or addresses. Attackers can exploit these errors by sending specially crafted inputs.

Mitigation:

- Validate input types and formats (e.g. addresses must be valid checksummed hex).
- Display the final parsed transaction data to the user before requesting confirmation.
- Reject malformed or out-of-range values.

4.3.7.2.4 Third-Party Library Issues

Description: Wallet software often relies on cryptographic libraries or third-party frameworks. Compromises or vulnerabilities in these libraries can propagate into the wallet.

Example:

- An NPM or PyPI package update introduces malicious code that exfiltrates private keys.

Mitigation:

- Pin specific versions of critical libraries.
- Audit third-party dependencies regularly (e.g. using npm audit, cargo audit, or pip-audit).
- Integrate AI-based scanners like Almanax to detect suspicious patterns in dependencies.

4.3.7.3 Wallet Security Audits and Testing

4.3.7.3.1 Code Review for Cryptographic Routines

Manual review confirms that cryptographic primitives are used properly:

- **Check Key Generation:** Ensure random number generation is truly random (e.g. using crypto.getRandomValues in JavaScript [303] or OS-level randomness in native code).
- **Verify Signature Schemes:** Confirm usage of secure and standardized algorithms (e.g. secp256k1 for Ethereum [304]).

Example (Verifying Key Generation in Node.js):

```
const crypto = require('crypto');

// Should use a secure random function for key generation

const privateKey = crypto.randomBytes(32);
```

Any deviation (like using predictable seeds) compromises wallet security.

4.3.7.3.2 UI/UX Security Testing

The user interface must accurately represent on-chain data. Security testers focus on:

- **Detecting Phishing-Like Prompts:** Make sure transaction prompts cannot be easily cloned or replaced.
- **Error Handling:** Confirm the wallet does not crash or reveal sensitive data on unexpected inputs.

Implementation Tip:

- Perform code injection tests against the wallet's frontend.
- Use JavaScript-based frameworks to run end-to-end tests that detect suspicious Document Object Model (DOM) modifications.

4.3.7.3.3 Secure Build and Deployment

Ensuring reproducible builds and verifying the integrity of wallet binaries or web bundles is critical:

- **Signed Releases:** Sign your executables or web deployments with a developer key to confirm authenticity.
- **Version Control for Build Scripts:** Keep track of build dependencies and configuration (e.g. using Git submodules, lock files).

- **CI with Security Scans**
  - Integrate AI-powered tools for code scanning to spot new vulnerabilities before release.
  - Add linting, unit tests, and integration tests for each commit.

4.3.7.3.4 Compliance with Regulatory and Industry Standards

Wallets may need to meet certain legal or industry requirements, such as General Data Protection Regulation (GDPR) [50] or Payment Card Industry Data Security Standard (PCI DSS) [305] (if facilitating payments):

- **Privacy Protections:** Ensure user data is not inadvertently collected or transmitted without consent.
- **Auditable Logs:** Maintain a secure audit trail of critical actions (e.g. key import, transaction signing).

4.3.8 Auditing dApps

dApps often integrate on-chain smart contracts, off-chain services, and diverse frontend clients. This combination introduces various security and privacy challenges. Auditors must examine how these parts interact and ensure each component properly validates and sanitizes data. This section outlines key dApp architecture elements, frequent vulnerabilities, and recommended auditing strategies.

4.3.8.1 dApp Architecture Components

1. **Smart Contract**
   - Core on-chain logic, often written in Solidity or Vyper (for Ethereum-based projects) or other languages like Rust for blockchain platforms like Solana and Stellar.
   - Responsible for transaction processing, asset management, and business rules.
   - Security flaws here can lead to direct financial losses or protocol manipulation.
2. **Frontend Web or Mobile Interface**
   - The user-facing layer that interacts with smart contracts and possibly other backend services.
   - If compromised, attackers can perform phishing attacks or mislead users to sign malicious transactions.
   - Often built with JavaScript frameworks (React, Vue.js, etc.) or mobile frameworks (React Native, Swift, Kotlin).

3. **Off-Chain Services or Oracles**
   - Services that gather external data (e.g. real-world prices, event outcomes) and pass it into the blockchain.
   - If oracles or off-chain backends are unreliable or manipulated, the entire dApp logic can be subverted.
   - May involve custom APIs, data feeds, and bridging protocols that require thorough security checks.

By mapping out each component, auditors gain a holistic view of how data travels between the frontend, smart contract layer, and external systems.

4.3.8.2 Common dApp Vulnerabilities

4.3.8.2.1 Frontend Vulnerabilities

Description: The client-side interface interacts with users and the blockchain. Attackers can exploit it via malicious scripts, code injection, or impersonation of the legitimate domain.

Example Flaw (JavaScript Injection):

```
<input type="text" id="inputAddress" />
<script>
  // Potentially unsanitized user input for an address
  function initiateTransfer() {
    const recipient = document.getElementById
('inputAddress').value;
    // If we directly build a transaction call without
validating 'recipient',
    // an attacker could inject malicious code or redirect funds.
    someSmartContract.transfer(recipient,  amount);
  }
</script>
```

Mitigation:

- Validate all user inputs (addresses, amounts).
- Use strict Content Security Policies (CSP).
- Secure the domain with HTTPS/TLS (read the note) and be wary of XSS vectors.

Note: It is important to ensure that secure and up-to-date versions of TLS protocols are used to prevent vulnerabilities such as downgrade attacks. Older or misconfigured TLS versions can expose applications to man-in-the-middle risks and weaken encrypted communications. Teams are encouraged to test their TLS configurations using tools like testssl.sh, an open-source script that scans for protocol support, cipher strength, and known TLS/SSL vulnerabilities.

4.3.8.2.2 Smart Contract Integration Flaws

Description: Even if individual smart contracts are secure, their integration points can introduce risks. For example, the frontend or additional contracts might call functions in an incorrect sequence or pass in malformed arguments.

Mitigation:

Strictly match contract Application Binary Interface (ABI) in the frontend.

- Perform end-to-end testing to confirm correct function calls.
- Enforce type checks and data validation before contract invocation.

4.3.8.2.3 Data Privacy and Confidentiality Gaps

Description: Many dApps store sensitive user or business data on-chain or in logs. Unless carefully designed, anyone can access this data since blockchains are generally public. Off-chain storage can also leak confidential information if misconfigured.

Example Flaw (Inadvertent Log Exposure):

```
event UserEmailStored(address indexed user,  string email);
// This email now appears in the public blockchain logs,
accessible to anyone.
```

Mitigation:

- Store only necessary data on-chain, and consider zero-knowledge proofs or encryption when feasible.
- Use private channels or permissioned ledgers if confidentiality is critical.
- Mask personally identifiable information (PII) or use hashed references instead of storing plain text.

4.3.8.3 dApp Auditing and Testing

4.3.8.3.1 End-to-End Testing

Description: End-to-end tests simulate real user workflows – connecting the frontend client, smart contracts, and any off-chain services. This helps identify breaks in the user flow or inconsistent data handling.

Implementation Tip:

```
# Using a JavaScript-based test suite  (e.g.  Cypress)
npx cypress run

# These tests might:
# 1)  Launch local dev environment
```

```
# 2)  Interact with a testnet
# 3)  Validate actual transaction flows
```

Why It Matters:

- Ensures that changes in one component (e.g. frontend UI) don't break the rest
- Confirms accurate contract interactions under realistic user conditions

4.3.8.3.2 Penetration Testing and Ethical Hacking

Description: Pen testers act as external attackers, trying to exploit vulnerabilities in both the web interface and smart contract layer.

Areas of Focus:

- **Frontend Attacks:** XSS, phishing, or domain impersonation
- **Smart Contract Exploits:** Checking if the dApp's contract calls can be triggered in unexpected sequences
- **Oracle Manipulation:** Attempting to feed false data from oracles

4.3.8.3.3 Security Scans in CI/CD

Description: CI/CD pipelines can integrate security checks for code changes across the entire dApp.

- **Static Analysis for Frontend**
  - Linting JavaScript/TypeScript code for potential injection vectors or unsafe DOM usage.
  - Dependency checks for outdated or vulnerable libraries.
- **Smart Contract Scanning**
  - Tools like **Slither**, **Mythril**, or **Almanax** to detect known vulnerability patterns before each deployment.
- **Integration Testing**
  - Validate that newly introduced UI components or contract interactions do not break existing logic or introduce new flaws.

Integrating these scans into daily or per-commit workflows helps maintain consistent security hygiene.

4.3.9 Consolidating Findings and Reporting

Security audits generate detailed insights about vulnerabilities across smart contracts, node implementations, wallet software, and dApps. However, the value of these findings depends on how effectively they are analyzed and communicated to developers and stakeholders. This section describes the recommended framework for reporting discovered issues, outlines the coordination process with development teams, and explains best practices for responsible disclosure.

4.3.9.1 Security Reporting Framework

- **Issue Summaries**
  - Provide a concise description of each vulnerability, including its **type** (e.g. reentrancy, access control flaw, misconfiguration) and **affected components** (e.g. smart contract function, node config file, frontend UI).
  - Attach any relevant code snippets or logs that illustrate how the vulnerability was discovered.
- **Severity Ratings**
  - Assign a clear **risk level** (generally "informational," "low," "medium," "high," "critical") based on potential impact, exploitability, and likelihood of occurrence (read the note).
  - Include a rationale: if an issue is deemed critical, explain how it could lead to loss of funds, consensus failures, or large-scale user compromise.
- **Recommended Fixes**
  - Provide both **short-term remediations** (e.g. disabling a feature, raising fees, or limiting access) and **long-term solutions** (e.g. refactoring contract logic, reimplementing an algorithm, upgrading libraries).
  - Reference any established best practices, such as using the **OpenZeppelin** library for ERC standards or recommended node configurations from official documentation.
- **Supporting Evidence and Tools Used**
  - Mention specific techniques (manual review, automated static analysis, AI-based tools, etc.) to validate the thoroughness of the audit.
  - Where possible, attach logs, proof of concepts, or screenshots of the detected vulnerabilities to help developers reproduce the issues.

Note: If any team members have experience with the CVSS (Common Vulnerability Scoring System), it can be very helpful when assigning severity levels to discovered issues. CVSS provides a standardized way to evaluate impact, exploitability, and likelihood, which aligns well with risk categorization (e.g. informational, low, medium, high, critical). Since CVSS is also widely used by penetration testing teams, adopting it internally can streamline communication and make discussions around vulnerability triage and mitigation more consistent and effective.

4.3.9.2 Coordination with Development Teams (see the note)

1. **Prioritizing Fixes**
   - Align on the most critical issues first, especially those with high exploitability or major business impact.
   - Maintain a **remediation schedule** or sprint backlog that balances security needs with development resources.
2. **Patch Testing and Retesting**
   - After fixes, rerun automated tools and manual checks to ensure no regressions or secondary flaws emerge.
   - Employ CI pipelines that automate regression tests, vulnerability scans, and code reviews.
3. **Documentation and Knowledge Transfer**
   - Prepare internal "lessons learned" docs to guide future development and prevent recurring vulnerabilities.
   - Involve stakeholders beyond the development team – project and product managers, DevOps, QA, legal/regulatory (if relevant) – for a broader security culture.

Note: It is highly useful to define an internal SLA (service level agreement) for fixing security issues and to clearly communicate this with the development team. Establishing such SLAs helps set expectations for resolution timelines based on vulnerability severity levels and ensures consistent prioritization. In addition, documenting a step-by-step workflow, from the moment an issue is discovered internally or reported by external researchers, helps streamline remediation efforts, promotes accountability, and reinforces collaboration between security and engineering teams.

4.3.9.3 Disclosure Best Practices

1. **Responsible Disclosure Timelines**
   - Coordinate with the affected project or team on a **private disclosure**. Give them a reasonable time frame (often 30–90 days) to fix the issue before public announcement.
   - For critical zero-day exploits, immediate partial disclosure may be necessary if there is an active threat to users.
2. **Security Advisory Publication**
   - Once patches are released or the window expires, publish a **security advisory** outlining the discovered vulnerabilities, steps for remediation, and any recommended version upgrades.
   - Encourage users, node operators, and integrators to apply updates promptly.

3. **Bug Bounties and Community Programs**
   - In many blockchain ecosystems, bug bounty platforms or dedicated security funds exist to reward researchers who responsibly report vulnerabilities.
   - Encourage community-driven security by offering incentives for timely disclosures.

4.3.10 Conclusion

This chapter illustrated how blockchain vulnerabilities can manifest in different codebases – smart contracts, node software, wallets, and dApps more broadly. Each section provided examples of common security flaws and included implementation pitfalls and recommended auditing approaches. Key takeaways include:

1. **Smart Contracts**
   - Reentrancy, integer overflows, access control, and logic flaws remain major attack vectors.
   - Manual reviews, automated tools, and formal verification offer multiple layers of defense.
2. **Node Software**
   - Consensus-level bugs, P2P protocol issues, and improper key management can undermine network integrity.
   - Operators can mitigate resource exhaustion or misconfiguration attacks with policy adjustments, secure credential handling, and continuous monitoring.
3. **Wallet Software**
   - Sensitive data (private keys, user details) needs strong encryption and safe storage.
   - UI consistency, transaction validation, and robust dependency management minimize the risk of phishing, spoofing, and supply chain attacks.
4. **dApps**
   - Integration points among contracts, user interfaces, and off-chain services are frequent sources of error.
   - End-to-end tests, penetration testing, and multilayer security checks (frontend, backend, contract layer) help detect both technical and logical flaws.
5. **Consolidating Findings and Reporting**
   - Effective audits require organized reporting and timely coordination with development teams.
   - Responsible disclosure and patch testing ensure vulnerabilities are addressed without causing undue risk to users.

Auditing is not just about identifying weaknesses – it also involves guiding teams toward remediation and architectural improvements. Employing best practices at each level of the blockchain stack helps maintain robust, trustworthy services. Promising work in the industry involves deeper automated scanning, usage of more sophisticated AI models, wider adoption of formal methods, and improved standards for secure deployment.

A

Threat Modeling Matrix

Blockchain Application Security: How to Design Secure and Attack Resilient Blockchain Applications, First Edition. Marco Morana, Harpreet Singh and Francesco Piccoli.
© 2026 John Wiley & Sons, Inc. Published 2026 by John Wiley & Sons, Inc.

Threat Scenario TS1 Malicious actors targeting critical processes and assets in a dApp's mobile UI and associated components.

Threat ID	Threat event	Threat description	Threat agents	Attack vectors	Assets targeted	Threat impacts	Countermeasures and risk mitigation strategies
T1	Exploitation of authentication and session management flaws	Weak authentication mechanisms or stolen session tokens allow unauthorized access or actions.	Malicious users, external attackers	Phishing attacks, stolen credentials, session hijacking	User accounts, authentication tokens	Unauthorized access to accounts or execution of fraudulent transactions	Implement strong authentication (e.g. MFA), secure session management, and token integrity checks.
T2	Data tampering at the mobile UI boundary	Malicious actors alter data inputs or manipulate transaction payloads between the user and mobile UI.	Malicious users, external attackers	Input manipulation, malicious payload injection	Transaction payloads, input data	Corruption of data, compromised transactions, and loss of trust	Validate inputs, encrypt data in transit, and use secure communication protocols.
T3	Information disclosure through insecure communication	Sensitive data is exposed due to unencrypted communication or weak encryption protocols.	External attackers, advanced persistent threats	Eavesdropping on communication, exploiting weak encryption	Data in transit between user and mobile UI	Data breaches and exposure of sensitive user information	Use strong encryption protocols (e.g. TLS) for data in transit and secure session management.

T4	Operational disruptions	Excessive API calls, denial-of-service attacks, or compromised third-party services disrupt operations.	Automated threats	Flooding APIs, exploiting third-party dependencies	Application resources, backend services	Service downtime, loss of availability, and operational disruptions	Implement controls to monitor API usage, enforce rate limiting, and validate third-party service reliability.
T5	Privilege escalation risks	Insufficient or absent role-based access controls lead to elevated privileges for attackers.	Malicious users, insiders	Exploiting weak RBAC settings, privilege escalation scripts	Access control policies, sensitive functionalities	Unauthorized actions or access to sensitive functionalities or data	Implement robust RBAC policies and enforce the principle of least privilege.
T6	API vulnerabilities and misconfigurations	Weak or unprotected APIs are exploited to perform unauthorized actions or access sensitive data.	External attackers, automated threats	Exploiting open or improperly configured APIs	API endpoints, sensitive user data	Data leaks, system compromise, or unauthorized actions through exposed APIs	Harden APIs with authentication, input validation, and rate limiting.
T7	Insecure data storage	Sensitive data, such as private keys or credentials, is stored insecurely in local storage.	Insiders, malicious users, external attackers	Accessing local storage without encryption, exploiting insecure storage paths	Local storage, sensitive data (e.g. private keys)	Exposure of sensitive user data and potential data breaches	Use secure storage solutions like Keychain or Keystore and encrypt sensitive data.

(Continued)

Threat Scenario TS1 (Continued)

Threat ID	Threat event	Threat description	Threat agents	Attack vectors	Assets targeted	Threat impacts	Countermeasures and risk mitigation strategies
T8	Exploitation of outdated dependencies	Libraries or frameworks used in the app are outdated, introducing known vulnerabilities.	Supply chain attackers, external attackers	Exploiting known vulnerabilities in outdated libraries or frameworks	Application libraries, software dependencies	Compromise of app integrity or exploitation of known vulnerabilities	Regularly update libraries and frameworks to patch known vulnerabilities.
T9	Code tampering and reverse engineering	Attackers reverse engineer the app to extract sensitive information or alter its behavior.	External attackers, advanced persistent threats	Reverse engineering using debugging tools, injecting malicious code	Application logic, sensitive algorithms	Compromise of app logic, enabling attackers to bypass security controls or launch further attacks	Apply code obfuscation and tamper detection to protect app logic and sensitive information.
T10	Sensitive data logging	Debug logs inadvertently expose sensitive information, such as private keys or session tokens.	Insiders, malicious users	Accessing logs in debug mode, reading sensitive data from logs	Debug logs, sensitive credentials	Unintended exposure of sensitive data and increased risk of exploitation	Sanitize logs to exclude sensitive data and disable logging in production environments.
T11	Unauthorized loan creation	Exploiting weak input validation or authentication mechanisms to create fraudulent loan requests.	Malicious users, external attackers	Sending unauthorized loan requests, bypassing input validation	Loan request records, user authentication mechanisms	Financial losses due to fraudulent loans and reputational damage	Implement robust input validation and multifactor authentication for loan requests.

T12	Manipulation of collateral valuation	Using compromised oracle feeds to alter collateral values, enabling fraudulent borrowing.	Insiders, external attackers	Manipulating oracle data feeds, sending spoofed data to oracles	Collateral valuation data, oracle feeds	Incorrect collateral valuation leading to financial imbalances and platform instability	Secure oracle feeds with cryptographic proofs and regular audits of data sources.
T13	Replay attacks on repayment transactions	Exploiting nonce or timestamp vulnerabilities to replay repayment transactions multiple times.	Automated threats, external attackers	Replaying valid transaction payloads using network sniffing tools	Repayment records, transaction logs	Multiple repayments falsely credited, causing financial inaccuracies and trust issues	Use nonce-based mechanisms and timestamp validation to prevent replay attacks.
T14	Exploitation of smart contract vulnerabilities	Manipulating loan or repayment logic in smart contracts to bypass business rules or exploit assets.	External attackers, advanced persistent threats	Exploiting logic flaws in smart contracts via custom transactions	Smart contracts, business logic rules	Smart contract exploitation leading to significant financial losses and system vulnerabilities	Conduct rigorous audits of smart contracts and apply runtime monitoring for anomalies.
T15	Data integrity issues	Injecting or tampering with user-submitted data to create inconsistencies in loan or collateral records.	Malicious users, insiders	Sending malformed or tampered user-submitted data for injection attacks	User-submitted data, loan and collateral records	Data inconsistencies undermining trust in the platform and leading to operational failures	Validate all user-submitted data and maintain integrity checks for loan and collateral records.

Threat Scenario TS2 Malicious actors targeting critical processes and assets in a dApp's interaction with digital wallets and associated components exploit vulnerabilities in authentication, transaction integrity, API security, and external services like oracles.

Threat instance ID	Threat event	Threat description	Threat actors	Attack vectors	Assets targeted	Threat impact	Countermeasures and risk mitigation strategies
T16	Spoofing user identity	Attackers impersonate legitimate users to gain unauthorized access to wallets or transactions.	Malicious users, external attackers	Phishing, credential theft, or session hijacking	User credentials, session tokens, digital wallets	Unauthorized transactions, theft of funds, and loss of user trust	Implement robust identity verification (e.g. MFA) and session validation.
T17	Tampering with transaction payloads	Malicious actors alter transaction data during communication between the dApp UI and the digital wallet.	External attackers, automated threats	Payload modification, injection attacks	Transaction payloads, data integrity	Corruption of transaction data, leading to financial inaccuracies and system mistrust	Use end-to-end encryption and validate transaction payloads for integrity.
T18	Replay attacks on signed transactions	Reusing valid signed transactions to perform unauthorized or duplicate operations.	External attackers, automated threats	Replaying intercepted signed transactions	Signed transaction records, financial integrity	Duplicate or unauthorized transactions causing financial losses and operational disruptions	Implement nonce and timestamp checks to prevent replay attacks.

T19	Exploitation of insecure API endpoints	Exploiting vulnerable API endpoints to bypass security and execute unauthorized actions.	Malicious users, external attackers	Exploiting API misconfigurations or insufficient authentication	API endpoints, user accounts, and transaction systems	Data breaches, unauthorized transactions, and loss of system integrity	Harden APIs with strong authentication, rate limiting, and access controls.
T20	Exploitation of hardware wallet vulnerabilities	Physical or side-channel attacks on hardware wallets to extract sensitive data, such as private keys.	Advanced persistent threats, external attackers	Physical tampering, side-channel attacks on hardware wallets	Private keys, sensitive wallet data	Loss of private keys, leading to unauthorized transactions and asset theft	Use tamper-resistant hardware wallets and implement physical attack defenses.
T21	Compromised software wallet security	Exploiting insecure software wallet implementations to gain unauthorized access to keys or transactions.	Malicious users, external attackers	Exploiting weak encryption or access control in software wallets	Software wallet implementations, user funds	Unauthorized access to user funds or manipulation of transaction logic	Secure software wallets with encryption and access controls; audit implementation regularly.
T22	Outdated wallet dependencies	Using outdated or vulnerable software dependencies to compromise wallet security.	Supply chain attackers, external attackers	Leveraging known vulnerabilities in outdated dependencies	Wallet dependencies, application integrity	Exploitation of known vulnerabilities resulting in system compromise or data theft	Regularly update wallet dependencies and libraries to patch vulnerabilities.

(Continued)

Threat Scenario TS2 (Continued)

Threat instance ID	Threat event	Threat description	Threat actors	Attack vectors	Assets targeted	Threat impact	Countermeasures and risk mitigation strategies
T23	Insufficient authentication for loan signing	Insufficient authentication mechanisms enable unauthorized users to sign loan-related transactions.	Malicious users, external attackers	Exploiting weak or absent authentication mechanisms	Loan approval systems, authentication mechanisms	Fraudulent loan approvals and significant financial losses	Enforce strong authentication mechanisms for all loan-related transactions.
T24	Manipulated oracle data for collateral valuation	Manipulating oracle-fed data to influence collateral valuation and loan approval processes.	Insiders, external attackers	Sending spoofed or manipulated data to oracles	Collateral data, oracle feeds, loan systems	Incorrect collateral valuations disrupting loan approvals and financial operations	Secure oracle feeds with cryptographic proofs and conduct regular audits.
T25	Unauthorized access to digital wallet APIs	Unauthorized access to wallet APIs allows attackers to modify or execute financial transactions.	External attackers, malicious users	Exploiting weak or unprotected wallet API endpoints	Wallet APIs, user accounts, financial transactions	Unauthorized actions compromising user accounts and financial integrity	Harden wallet APIs with strong access controls, monitoring, and secure coding practices.

T26	Information disclosure during data transmission	Sensitive user or transaction data is exposed during transmission between components.	Advanced persistent threats, external attackers	Sniffing unencrypted communications during data transmission	User data, transaction records, communication channels	Exposure of sensitive data leading to potential regulatory violations and user distrust	Encrypt data in transit using strong protocols (e.g. TLS) and secure session management.
T27	Denial-of-service attacks on the trust boundary	Flooding the trust boundary with requests to disrupt wallet connectivity and dApp functionality.	Automated threats, competitors	Overloading trust boundary endpoints with excessive requests	Wallet connectivity, dApp functionality	Service disruptions impacting lending and borrowing processes and user experience	Implement rate limiting, monitoring, and CAPTCHA to prevent denial-of-service attacks.

Threat Scenario TS3 Malicious actors targeting critical processes and assets in the interaction between the dApp and the distributed digital identity system (uPort) exploit vulnerabilities in identity validation, API security, data transmission, and trust configurations.

Threat ID	Threat event	Threat description	Threat agents	Attack vectors	Assets targeted	Threat impact	Countermeasures and risk mitigation strategies
T28	Spoofing user identity	Attackers impersonate users by forging or hijacking distributed identity credentials.	Malicious users, external attackers	Credential forgery, phishing, session hijacking	User identity credentials, session tokens	Unauthorized access to the system, fraudulent loans, and compromise of user accounts	Use cryptographic proofs for identity validation and multifactor authentication (MFA).
T29	Manipulation of identity payloads	Malicious actors tamper with identity data transmitted between the dApp UI and uPort.	Malicious users, external attackers	Tampering with transmitted payloads, injecting malicious data	Identity payloads, user records	Corruption of identity records, resulting in denial of legitimate access or fraud	Apply end-to-end encryption and validate identity payloads for integrity.
T30	Replay attacks on identity validation	Reusing previously valid identity tokens or requests to gain unauthorized access.	External attackers, automated threats	Replaying previously valid tokens or requests	Identity tokens, validation records	Unauthorized user impersonation, enabling fraudulent activities	Implement nonce-based mechanisms and timestamp validation to prevent replay attacks.

T31	Exploitation of insecure uPort APIs	Exploiting vulnerabilities in uPort APIs to manipulate or extract sensitive identity data.	External attackers, malicious users	Exploiting insecure API endpoints, input manipulation	uPort APIs, sensitive identity data	Unauthorized access to identity data and potential identity theft	Harden APIs with strong authentication, access control, and input validation.
T32	Insecure storage of identity credentials	uPort credentials stored insecurely, exposing them to theft or unauthorized access.	Malicious users, insiders	Accessing insecure storage, credential theft	Stored credentials, user identity information	Exposure of sensitive user credentials leading to impersonation and misuse of identities	Use secure storage mechanisms (e.g. encrypted storage) for all credentials.
T33	Outdated uPort dependencies	Using outdated libraries or frameworks in the uPort system introduces exploitable vulnerabilities.	Supply chain attackers, external attackers	Exploiting known vulnerabilities in outdated libraries	uPort dependencies, libraries, frameworks	Exploitation of known vulnerabilities, leading to data breaches or system compromise	Regularly update libraries and frameworks to patch known vulnerabilities.
T34	Insufficient cryptographic validation	Weak or absent cryptographic checks for uPort credentials during identity verification.	Malicious users, external attackers	Forged or invalid cryptographic signatures	Identity verification processes, cryptographic data	Acceptance of forged credentials, enabling fraudulent or unauthorized access	Enforce robust cryptographic checks, including signature validation for all identity credentials.

(Continued)

Threat Scenario TS3 (Continued)

Threat ID	Threat event	Threat description	Threat agents	Attack vectors	Assets targeted	Threat impact	Countermeasures and risk mitigation strategies
T35	Denial-of-service attacks on identity validation	Overwhelming the uPort system with identity validation requests, disrupting service availability.	Automated threats, competitors	Flooding identity validation systems with excessive requests	Identity validation systems, availability of services	Inability to validate identities, leading to denial of service for legitimate users	Implement rate limiting, CAPTCHA, and monitoring to detect and mitigate denial-of-service attacks.
T36	Information disclosure during identity transmission	Sensitive identity data exposed during communication between the dApp UI and uPort.	External attackers, advanced persistent threats	Intercepting unencrypted data, exploiting weak encryption	Personal identity data, communication channels	Unauthorized access to personal information, resulting in privacy breaches and regulatory violations	Use strong encryption protocols (e.g. TLS) for all data transmissions.
T37	Fraudulent creation of uPort identities	Attackers create fake distributed identities to exploit lending and borrowing systems.	Malicious users, external attackers	Creating fake identities, exploiting lax identity checks	Identity creation mechanisms, user accounts	Fraudulent loan applications and potential financial losses	Enforce strict identity creation policies and use trusted third-party verifications where necessary.

T38	Misconfigured uPort trust settings	Improper configuration of trust relationships between the dApp and uPort allows unauthorized interactions.	Insiders, misconfigured systems	Exploiting misconfigured trust relationships	Trust configurations, identity transaction records	Unauthorized identity transactions and potential data leakage	Audit and validate all trust configurations and enforce least privilege principles.
T39	Collusion or compromise of uPort nodes	Malicious nodes within the uPort network manipulate or falsify identity validation data.	Compromised nodes, malicious users	Manipulating consensus or injecting falsified data via malicious nodes	uPort network nodes, consensus integrity	Acceptance of invalid identities, resulting in fraud or misuse of the system	Use decentralized consensus mechanisms and cryptographic proofs to ensure node trustworthiness.

Threat Scenario TS4 Malicious actors targeting critical processes and assets in the interaction between the dApp UI and the dApp frontend exploit vulnerabilities in session management, data integrity, API configurations, and business logic.

Threat ID	Threat event	Threat description	Threat agents	Attack vectors	Assets targeted	Threat impacts	Countermeasures and risk mitigation strategies
T40	Spoofing user session	Threat actors hijack user sessions to impersonate legitimate users in the dApp.	Malicious users, external attackers	Session hijacking, stolen tokens	User sessions, transaction data	Unauthorized transactions, financial fraud, and potential identity theft	Implement session management with cryptographic session tokens and secure session validation.
T41	Injection attacks on data flows	Malicious actors inject payloads into data flows between the dApp UI and dApp frontend.	Malicious users, external attackers	Payload manipulation, injection attacks	Transaction data flows, data integrity	Corrupted transactions, unauthorized commands, or data breaches	Validate and sanitize all user inputs and use secure API communication.
T42	API misconfiguration	Misconfigured APIs in the dApp frontend allow unauthorized access or data manipulation.	External attackers, misconfigured systems	Exploiting misconfigured API endpoints	API endpoints, system configurations	Unauthorized actions, data exposure, and compromise of lending or borrowing logic	Regularly audit and enforce secure API configurations with strong authentication and rate limiting.

T43	Replay attacks on transaction requests	Replaying valid requests to repeat unauthorized operations like borrowing or loan repayment.	External attackers, automated threats	Replaying intercepted requests	Transaction requests, system records	Financial discrepancies and operational instability in lending or borrowing processes	Use nonce-based mechanisms and timestamp validation for transaction requests.
T44	Cross-site scripting (XSS)	Malicious scripts injected into the dApp frontend exploit user browsers or steal sensitive data.	Malicious users, external attackers	Injecting malicious scripts via input fields	User credentials, session data	User credential theft, session hijacking, or unauthorized actions performed on behalf of users	Implement strict content security policies (CSP) and sanitize user-generated content.
T45	Cross- site request forgery (CSRF)	Attackers execute unauthorized actions on the dApp frontend by exploiting user authentication.	Malicious users, external attackers	Forged requests exploiting user sessions	Authenticated sessions, user accounts	Unauthorized transactions or changes to loan terms, leading to financial losses	Use anti-CSRF tokens and verify origin headers for all sensitive requests.
T46	Information disclosure during data transmission	Sensitive user or transaction data is exposed during communication between the dApp UI and frontend.	Advanced persistent threats, external attackers	Intercepting unencrypted data transmissions	Data in transit, user privacy	Privacy breaches, data theft, and potential regulatory penalties	Use end-to-end encryption with strong protocols like TLS for all data in transit.

(Continued)

Threat Scenario TS4 (Continued)

Threat ID	Threat event	Threat description	Threat agents	Attack vectors	Assets targeted	Threat impacts	Countermeasures and risk mitigation strategies
T47	Outdated dApp frontend dependencies	Use of outdated or vulnerable libraries/frameworks in the dApp frontend introduces security risks.	Supply chain attackers, external attackers	Exploiting known vulnerabilities in outdated libraries	dApp frontend libraries, frameworks	Exploitation of known vulnerabilities leading to system compromise or data breaches	Regularly update dependencies and apply security patches promptly.
T48	Lack of access control enforcement	Weak or absent access controls on the dApp frontend allow unauthorized users to manipulate transactions.	Malicious users, external attackers	Weak RBAC or lack of access control	Access controls, transaction records	Unauthorized modifications to loan data or business logic, causing financial losses	Implement robust role-based access control (RBAC) and the principle of least privilege.
T49	Denial-of-service (DoS) attacks	Flooding the dApp frontend with requests to disrupt service availability and transaction processing.	Automated threats, competitors	Flooding API endpoints with excessive requests	API endpoints, system availability	Service downtime, operational disruptions, and user dissatisfaction	Apply rate limiting, IP blacklisting, and load balancing to mitigate DoS attacks.

T50	Fraudulent transaction injection	Attackers inject fraudulent transactions into the data flow from the UI to the dApp frontend.	Malicious users, external attackers	Injecting unauthorized transactions via insecure paths	Transaction data, lending records	Approval of illegitimate loans, financial discrepancies, and damage to system integrity	Validate all transaction data and enforce cryptographic signatures before processing.
T51	Business logic vulnerabilities in the frontend	Flaws in the implementation of lending and borrowing workflows allow bypassing rules or controls.	Malicious users, external attackers	Exploiting logic flaws in business rules	Business logic, lending/ borrowing workflows	Exploitation of system rules, resulting in financial inaccuracies and operational risks	Conduct regular security testing for business logic flaws and enforce robust validation checks.

Threat Scenario TS5 Malicious actors targeting the interaction between the dApp frontend and dApp middleware (JSON-RPC) exploit vulnerabilities in authentication, data integrity, API configurations, and middleware logic.

Threat ID	Threat event	Threat description	Threat agent	Attack vectors	Assets targeted	Threat impact	Countermeasures and risk mitigation strategies
T52	Spoofing middleware identity	Attackers impersonate the middleware to intercept or manipulate transactions.	Malicious users, external attackers	Forging middleware credentials, impersonating trusted middleware	Middleware credentials, transaction data	Unauthorized access, transaction manipulation, and compromise of system integrity	Use mutual TLS or API keys for authentication between the frontend and middleware.
T53	Tampering with JSON-RPC requests	Threat actors alter transaction payloads during transmission to the middleware.	Malicious users, external attackers	Altering transaction payloads during transmission	Transaction payloads, data integrity	Corrupted or fraudulent transactions, leading to financial losses or operational inconsistencies	Validate payloads, encrypt all communications, and use digital signatures for transaction integrity.
T54	Replay attacks on JSON-RPC requests	Reusing previously valid requests to execute unauthorized or duplicate operations.	External attackers, automated threats	Replaying valid JSON-RPC requests intercepted during transmission	Transaction records, validation processes	Financial discrepancies and potential over approval of loans or repayments	Implement nonce and timestamp validation for all JSON-RPC requests.

T55	Exploitation of insecure API endpoints	Vulnerabilities in middleware APIs allow unauthorized access or data manipulation.	External attackers, malicious users	Exploiting improperly secured API endpoints	Middleware APIs, sensitive data	Unauthorized actions, data leakage, and compromise of business logic	Harden APIs with strong authentication, access controls, and input validation.
T56	Information disclosure during communication	Sensitive data transmitted between the frontend and middleware is exposed.	Advanced persistent threats, external attackers	Intercepting unencrypted communication channels	Data in transit, middleware integrity	Privacy breaches, regulatory violations, and user distrust	Encrypt all communications using strong protocols like TLS and secure session management.
T57	Outdated middleware dependencies	Middleware relies on outdated or vulnerable libraries, introducing security risks.	Supply chain attackers, external attackers	Leveraging vulnerabilities in outdated libraries	Middleware libraries, system dependencies	Exploitation of known vulnerabilities, leading to system compromise or data breaches	Regularly update middleware dependencies and apply security patches.
T58	Denial-of-service (DoS) attacks on middleware	Attackers flood the middleware with excessive requests to disrupt its availability.	Automated threats, competitors	Flooding middleware with excessive traffic or requests	Middleware resources, service availability	Service unavailability, disruption of lending and borrowing operations, and user dissatisfaction	Apply rate limiting, IP whitelisting, and load balancing to mitigate DoS attacks.

(Continued)

Threat Scenario TS5 (Continued)

Threat ID	Threat event	Threat description	Threat agent	Attack vectors	Assets targeted	Threat impact	Countermeasures and risk mitigation strategies
T59	Lack of access control enforcement	Weak or absent access controls on the middleware allow unauthorized actions or data access.	Malicious users, external attackers	Exploiting weak or absent access controls	Middleware access controls, transaction systems	Unauthorized manipulation of transactions or sensitive information, leading to financial losses	Implement role-based access control (RBAC) and enforce the principle of least privilege.
T60	Fraudulent transaction injection	Attackers inject unauthorized transactions directly into the middleware via insecure endpoints.	External attackers, malicious users	Injecting fraudulent transactions via insecure endpoints	Transaction validation processes, data records	Approval of illegitimate loans, financial discrepancies, and potential system compromise	Enforce input validation and cryptographic transaction signing for all requests.

T61	Exploitation of middleware configuration flaws	Improperly configured middleware settings expose vulnerabilities that attackers can exploit.	Malicious users, misconfigured systems	Exploiting misconfigured middleware settings	Middleware configurations, system settings	System compromise, data leakage, and operational disruptions	Audit middleware configurations and implement secure by default settings.
T62	Business logic flaws in middleware	Logical errors in transaction processing within the middleware allow bypassing DeFi rules.	Malicious users, external attackers	Exploiting logic errors in transaction processing	Middleware logic, DeFi rules	Financial inaccuracies, unauthorized loan approvals, and operational risks	Conduct regular security assessments of middleware business logic and enforce robust validation checks.

Threat Scenario TS6 Malicious actors targeting the interaction between the dApp middleware and blockchain node exploit vulnerabilities in authentication, data transmission, API security, and local storage integrity.

Threat ID	Threat event	Threat description	Threat agents	Attack vectors	Assets targeted	Threat impacts	Countermeasures and risk mitigation strategies
T63	Spoofing blockchain node identity	Attackers impersonate blockchain nodes to intercept or manipulate transactions.	Malicious users, external attackers	Forging blockchain node credentials, impersonating trusted nodes	Blockchain node credentials, transaction data	Unauthorized access, transaction manipulation, and compromise of blockchain integrity	Use mutual TLS or API keys for authentication between middleware and blockchain nodes.
T64	Tampering with blockchain transaction payloads	Threat actors alter transaction payloads during transmission from middleware to blockchain.	Malicious users, external attackers	Modifying transaction payloads during transmission	Transaction payloads, data integrity	Corrupted or fraudulent transactions, leading to financial losses or operational inconsistencies	Validate transaction payloads, encrypt all communications, and use digital signatures for integrity.
T65	Replay attacks on blockchain transactions	Reusing valid transaction payloads to execute duplicate or unauthorized operations on the blockchain.	Automated threats, external attackers	Replaying valid blockchain transactions intercepted during communication	Blockchain transaction records, validation processes	Financial discrepancies and unauthorized transactions processed on the blockchain	Implement nonce and timestamp validation for all blockchain transactions.

T66	Exploitation of insecure blockchain node APIs	Exploiting vulnerable blockchain node APIs to access or manipulate sensitive data.	Malicious users, external attackers	Exploiting improperly secured blockchain APIs	Blockchain APIs, sensitive data	Unauthorized actions, data leakage, and compromise of blockchain business logic	Harden blockchain APIs with strong authentication, access controls, and input validation.
T67	Information disclosure during middleware–blockchain communication	Sensitive data transmitted between middleware and blockchain nodes is exposed.	Advanced persistent threats, external attackers	Intercepting unencrypted middleware–blockchain communication	Data in transit, middleware–blockchain integrity	Privacy breaches, regulatory violations, and user distrust	Encrypt all communications using strong protocols like TLS and secure session management.
T68	Outdated blockchain node dependencies	Blockchain node relies on outdated libraries, introducing exploitable vulnerabilities.	Supply chain attackers, external attackers	Leveraging vulnerabilities in outdated blockchain node libraries	Blockchain node libraries, system dependencies	Exploitation of known vulnerabilities, leading to system compromise or data breaches	Regularly update blockchain node dependencies and apply security patches.
T69	Denial-of-service (DoS) attacks on blockchain node	Attackers flood the blockchain node with requests to disrupt transaction processing.	Automated threats, competitors	Flooding blockchain nodes with excessive requests	Blockchain node resources, service availability	Service unavailability, disruption of blockchain-based operations, and user dissatisfaction	Apply rate limiting, IP whitelisting, and load balancing to mitigate DoS attacks.

(Continued)

Threat Scenario TS6 (Continued)

Threat ID	Threat event	Threat description	Threat agents	Attack vectors	Assets targeted	Threat impacts	Countermeasures and risk mitigation strategies
T70	Lack of access control enforcement on blockchain node	Weak or absent access controls on the blockchain node allow unauthorized actions or data access.	Malicious users, external attackers	Exploiting weak or absent access controls on blockchain nodes	Blockchain node access controls, transaction systems	Unauthorized manipulation of blockchain transactions or sensitive data, leading to financial losses	Implement role-based access control (RBAC) and enforce the principle of least privilege.
T71	Compromise of local blockchain node storage	Attackers gain unauthorized access to local blockchain storage to manipulate transaction history or extract sensitive data.	Advanced persistent threats, malicious users	Accessing local blockchain storage using compromised credentials	Local blockchain storage, transaction history	Manipulated transaction history or stolen sensitive data from local storage	Encrypt local blockchain storage and implement secure access controls.
T72	Business logic flaws in smart contract execution	Logical errors in smart contract execution allow attackers to bypass DeFi lending and borrowing rules.	Malicious users, external attackers	Exploiting logic flaws in smart contracts	Smart contracts, business logic rules	Financial inaccuracies, unauthorized loan approvals, and operational risks	Conduct regular security assessments of smart contracts and enforce robust validation checks.

Threat Scenario TS7 Malicious actors targeting the interaction between the blockchain node and the distributed ledger exploit vulnerabilities in transaction validation, consensus mechanisms, and block storage integrity.

Threat ID	Threat event	Threat description	Threat agents	Attack vectors	Assets targeted	Threat impacts	Countermeasures and risk mitigation strategies
T73	Spoofing distributed ledger nodes	Attackers impersonate distributed ledger nodes to alter transaction validation or block creation.	Malicious users, external attackers	Forging ledger node credentials, impersonating trusted nodes	Ledger node credentials, transaction data	Unauthorized transactions, manipulation of ledger state, and compromise of ledger integrity	Implement mutual TLS or blockchain-specific authentication protocols for node communication.
T74	Tampering with transaction data before block creation	Threat actors modify transaction data during validation, leading to corrupted block storage.	Malicious users, external attackers	Modifying transaction data during validation	Transaction validation processes, block data	Corrupted block storage, financial losses, and operational inconsistencies	Validate transaction data integrity using cryptographic signatures before block storage.
T75	Replay attacks during transaction validation	Reusing valid transactions during validation to execute duplicate operations.	Automated threats, external attackers	Replaying valid transactions intercepted during validation	Transaction records, consensus operations	Duplicate or unauthorized transactions affecting platform trust and stability	Enforce nonce and timestamp validation to prevent replay attacks during validation.

(Continued)

Threat Scenario TS7 (Continued)

Threat ID	Threat event	Threat description	Threat agents	Attack vectors	Assets targeted	Threat impacts	Countermeasures and risk mitigation strategies
T76	Exploitation of consensus mechanism vulnerabilities	Exploiting flaws in the consensus mechanism to manipulate transaction validation or block ordering.	Malicious users, advanced persistent threats	Exploiting design flaws in consensus mechanisms	Consensus algorithms, transaction integrity	Manipulated validation processes, allowing malicious transactions or block reordering	Conduct rigorous testing and auditing of consensus mechanisms.
T77	Information disclosure in transaction validation process	Sensitive transaction data exposed during validation or consensus operations.	Advanced persistent threats, external attackers	Intercepting unencrypted data during validation or consensus processes	Transaction data in validation, user privacy	Privacy breaches, regulatory violations, and exposure of sensitive transaction data	Encrypt all data in transit using strong protocols like TLS and secure validation processes.
T78	Outdated blockchain platform dependencies	Blockchain platform relies on outdated or vulnerable software components.	Supply chain attackers, external attackers	Leveraging vulnerabilities in outdated blockchain platform components	Blockchain platform software, dependencies	Exploitation of vulnerabilities leading to data breaches or system compromise	Regularly update blockchain platform dependencies and apply security patches.

T79	Denial-of-service (DoS) attacks on ledger nodes	Flooding ledger nodes with excessive requests to disrupt consensus or validation.	Automated threats, competitors	Flooding ledger nodes with excessive traffic	Ledger node resources, consensus availability	Disruption of consensus operations, impacting transaction validation and platform trust	Apply rate limiting, node whitelisting, and load balancing to mitigate DoS attacks.
T80	Exploitation of blockchain platform misconfigurations	Improperly configured blockchain platform settings exposing vulnerabilities.	Malicious users, external attackers	Exploiting misconfigured platform settings	Blockchain platform configurations, operational settings	System compromise, data leakage, and operational disruptions due to misconfigurations	Audit platform configurations and enforce secure by default settings.
T81	Logical errors in consensus affecting transaction ordering	Errors in consensus logic allow attackers to bypass validation rules or reorder transactions.	Malicious users, advanced persistent threats	Exploiting logical inconsistencies in consensus rules	Consensus logic, transaction ordering	Unintended transaction ordering or approval, undermining system reliability	Regularly review and test consensus logic to identify and fix logical errors.

Threat Scenario TS8 Malicious actors targeting smart contracts exploit vulnerabilities in transaction execution, validation, and state management to compromise the integrity of loan approval and repayment processes.

Threat ID	Threat event	Threat description	Threat agents	Attack vectors	Assets targeted	Threat impacts	Countermeasures and risk mitigation strategies
T82	Exploitation of smart contract vulnerabilities	Attackers exploit flaws in smart contracts to manipulate loan approval or repayment logic.	Malicious users, external attackers	Injecting malicious payloads into smart contract logic	Smart contracts, loan approval/ rejection rules	Unauthorized loan approvals, financial losses, and system instability	Conduct rigorous audits and runtime monitoring of smart contracts.
T83	Tampering with loan approval terms during execution	Threat actors modify loan terms, such as interest rates or collateral requirements, during execution.	Malicious users, competitors	Modifying loan approval parameters via crafted transactions	Loan terms, borrower agreements	Corruption of loan agreements, leading to financial inaccuracies and disputes	Validate and verify loan terms cryptographically before execution.
T84	Replay attacks on transaction approval	Reusing valid loan approval transactions to authorize duplicate or fraudulent operations.	Automated threats, external attackers	Replaying previously valid loan approval transactions	Loan transaction records, approval logs	Duplicate loan transactions undermining system trust and stability	Enforce nonce and timestamp validation for transaction approval.

T85	Gas limit exploitation in contract execution	Exploiting gas limits to interrupt loan approval execution, causing incomplete processing.	External attackers, malicious users	Forcing execution failure through gas limit manipulation	Gas consumption metrics, execution integrity	Incomplete loan processing causing operational delays or failures	Apply appropriate gas limits and prioritize critical executions.
T86	Sensitive data exposure during contract execution	Sensitive data, such as loan terms or borrower details, is exposed during contract execution.	Advanced persistent threats, external attackers	Intercepting unencrypted data during smart contract processing	Sensitive loan data, borrower details	Privacy breaches and potential regulatory violations	Encrypt all sensitive data during contract processing.
T87	Use of outdated or vulnerable smart contract frameworks	Outdated frameworks or libraries in smart contracts create exploitable vulnerabilities.	Supply chain attackers, external attackers	Exploiting known flaws in outdated contract frameworks	Smart contract frameworks, dependent libraries	Data breaches or manipulation due to unpatched vulnerabilities	Regularly update frameworks and libraries with security patches.
T88	Denial-of-service (DoS) attacks targeting loan execution contracts	Attackers overwhelm smart contracts with requests to disrupt loan processing operations.	Competitors, automated threats	Sending excessive requests to overload smart contract operations	Contract execution availability, processing capacity	Disruption of loan execution processes, user dissatisfaction	Monitor contract activity and apply rate-limiting measures.

(Continued)

Threat Scenario TS8 (Continued)

Threat ID	Threat event	Threat description	Threat agents	Attack vectors	Assets targeted	Threat impacts	Countermeasures and risk mitigation strategies
T89	Logic flaws in loan and repayment terms	Errors in contract logic allow bypassing key validation checks for loan or repayment terms.	Malicious users, external attackers	Exploiting untested or flawed validation logic in contracts	Loan validation logic, repayment rules	Financial risks from bypassed validation checks, operational failures	Thoroughly test validation logic to prevent bypassing of checks.
T90	Compromise of deployment keys for loan contracts	Unauthorized access to deployment keys allows attackers to modify or deploy malicious contracts.	Advanced persistent threats, malicious users	Using stolen deployment keys to alter smart contract code	Deployment keys, contract integrity	Compromised contracts resulting in malicious modifications	Secure deployment keys with multifactor authentication and encryption.
T91	Manipulation of distributed ledger updates for loan status	Manipulating ledger updates to falsify loan status or repayment records.	Malicious users, external attackers	Injecting falsified updates into the distributed ledger	Distributed ledger entries, loan status records	Falsified loan status records undermining platform reliability	Validate ledger updates with cryptographic proofs and state consistency checks.

Threat Scenario TS9 Threat actors exploit vulnerabilities in the oracle platform, its data sources, and the interaction between oracles and the blockchain platform.

Threat ID	Threat event	Threat description	Threat agents	Attack vectors	Assets targeted	Threat impacts	Countermeasures and risk mitigation strategies
T92	Exploitation of oracle vulnerabilities including smart contract vulnerabilities	Attackers exploit vulnerabilities in the oracle to manipulate data feeds for the blockchain.	Malicious users, external attackers	Exploiting flaws in oracle implementation or data feed logic	Oracle services, data feeds	Manipulated data feeds leading to incorrect blockchain decisions	Conduct regular audits of oracle implementation and smart contracts.
T93	Tampering with data feed from external sources	Threat actors modify data being transmitted from external sources to the oracle.	Malicious users, advanced persistent threats	Modifying data in transit between external sources and the oracle	External data sources, feed integrity	Corrupted data feeds resulting in financial inaccuracies or fraud	Encrypt communication between external sources and the oracle.
T94	Compromise of external API or database feeding the oracle	Exploiting vulnerabilities in external APIs or databases to manipulate data fed into the oracle.	Advanced persistent threats, supply chain attackers	Compromising APIs or databases providing data to the oracle	APIs, external databases feeding the oracle	Tampered external data causing incorrect or misleading blockchain outputs	Harden APIs and secure databases providing data to the oracle.

(Continued)

Threat Scenario TS9 (Continued)

Threat ID	Threat event	Threat description	Threat agents	Attack vectors	Assets targeted	Threat impacts	Countermeasures and risk mitigation strategies
T95	Data exposure during oracle communication with blockchain	Sensitive data transmitted between the oracle and blockchain is exposed.	Advanced persistent threats, external attackers	Intercepting unencrypted communication between oracle and blockchain	Communication integrity between oracle and blockchain	Privacy breaches and potential regulatory violations	Encrypt all communication between the oracle and blockchain.
T96	Outdated oracle software or dependencies	Oracle relies on outdated libraries or software, introducing vulnerabilities.	Supply chain attackers, external attackers	Leveraging vulnerabilities in outdated oracle software	Oracle libraries, dependent frameworks	Exploitation of vulnerabilities causing system compromise or data manipulation	Regularly update oracle software and dependencies with security patches.
T97	Manipulation of oracle-signed transactions	Threat actors alter oracle-signed transactions to manipulate their content or authenticity.	Malicious users, advanced persistent threats	Altering signed transactions created by the oracle	Signed oracle transactions, blockchain integrity	Altered signed transactions impacting transaction authenticity and trust	Validate and verify the authenticity of oracle-signed transactions.

T98	Denial-of-service (DoS) attacks on oracle services	Flooding the oracle service with excessive requests to disrupt its availability.	Automated threats, competitors	Overloading the oracle with excessive data or requests	Oracle service availability, processing capacity	Service disruptions affecting accuracy (i.e. up-to-date of market data)	Apply rate limiting and load balancing to protect oracle services.
T99	Compromise of oracle node credentials	Unauthorized access to credentials used to operate oracle nodes.	Advanced persistent threats, malicious users	Using stolen credentials to operate or manipulate oracle nodes	Oracle node credentials, operational security	Unauthorized operation or manipulation of oracle services	Secure oracle node credentials with multifactor authentication and encryption.
T100	Injection of malicious data through oracle integration	Injecting falsified or malicious data into the oracle's input from external sources.	Malicious users, external attackers	Feeding falsified data into the oracle from external sources	Oracle inputs, data accuracy	Falsified data affecting loan approvals and financial calculations	Validate all external data sources feeding the oracle with cryptographic proofs.

B

Mapping of Threat Scenarios to Targeted Weaknesses and Asset Impacted

Blockchain Application Security: How to Design Secure and Attack Resilient Blockchain Applications, First Edition. Marco Morana, Harpreet Singh and Francesco Piccoli.
© 2026 John Wiley & Sons, Inc. Published 2026 by John Wiley & Sons, Inc.

Threat Scenario TS1 Malicious actors targeting critical processes and assets in a dApp's mobile UI and associated components.

Weakness targeted (CWE)	Weakness description	Severity risk	Asset impacted	Attack vectors
CWE-287	Improper authentication	9.8 (critical)	Mobile or browser client/UI	Phishing attacks, stolen credentials, session hijacking
CWE-359	Exposure of private information	8.8 (high)	Mobile or browser client/UI	Intercepting unprotected communication or insecure data storage
CWE-1104	Use of unmaintained third-party components	7.8 (high)	Mobile or browser client/UI	Exploiting vulnerabilities in outdated libraries or frameworks
CWE-20	Improper input validation	7.5 (high)	Mobile or browser client/UI	Manipulating input data to bypass validation checks
CWE-922	Insecure storage of sensitive information	4.7 (medium)	Mobile or browser client/UI	Accessing sensitive data in local storage without encryption
CWE-104	Excessive logging	2.0 (low)	Mobile or browser client/UI	Accessing sensitive information from improperly sanitized debug logs
Threat Scenario TS2: Malicious actors targeting critical processes and assets in a dApp's interaction with digital wallets and associated components exploit vulnerabilities in authentication, transaction integrity, API security, and external services like oracles.				
CWE-287	Improper authentication	9.8 (critical)	Digital wallet/smart contracts	Phishing, credential theft, session hijacking
CWE-284	Improper access control	6.5 (medium)	Digital wallet/smart contracts	Payload modification, injection attacks

CWE-284	Improper access control	6.5 (medium)	Digital wallet/smart contracts	Replaying intercepted signed transactions
CWE-284	Improper access control	6.5 (medium)	Digital wallet/smart contracts	Exploiting API misconfigurations, insufficient authentication
CWE-922	Insecure storage of sensitive information	5.9 (medium)	Digital wallet/smart contracts	Physical tampering, side-channel attacks
CWE-540	Information exposure through debugging information	2.5 (low)	Digital wallet/smart contracts	Accessing debug logs containing sensitive information
Threat Scenario TS3: Malicious actors targeting critical processes and assets in the interaction between the dApp and the distributed digital identity system (uPort) exploit vulnerabilities in identity validation, API security, data transmission, and trust configurations.				
CWE-502	Deserialization of untrusted data	9.8 (critical)	Distributed digital identity system/trust boundary	Credential forgery, phishing, session hijacking
CWE-284	Improper access control	9.8 (critical)	Distributed digital identity system/trust boundary	Tampering with transmitted payloads, injecting malicious data
CWE-346	Origin validation error	8.8 (high)	Distributed digital identity system/trust boundary	Replaying previously valid tokens or requests
CWE-400	Resource exhaustion ("gas limit exploits")	8.8 (high)	Distributed digital identity system/trust boundary	Exploiting insecure API endpoints, input manipulation
CWE-601	Open redirect	6.3 (medium)	Distributed digital identity system/trust boundary	Intercepting unencrypted data, exploiting weak encryption

(Continued)

Threat Scenario TS3 (Continued)

Weakness targeted (CWE)	Weakness description	Severity risk	Asset impacted	Attack vectors
CWE-1285	Improper validation of specified conditions	5.8 (medium)	Distributed digital identity system/trust boundary	Creating fake identities, exploiting lax identity checks
CWE-758	Reliance on undefined behavior	3.5 (low)	Distributed digital identity system/trust boundary	Flooding identity validation systems with excessive requests
Threat Scenario TS4: Malicious actors targeting critical processes and assets in the interaction between the dApp UI and the dApp frontend exploit vulnerabilities in session management, data integrity, API configurations, and business logic.				
CWE-352	Cross-site request forgery (CSRF)	9.3 (critical)	dApp frontend	Session hijacking, stolen tokens
CWE-79	Cross-site scripting (XSS)	9.6 (critical)	dApp frontend	Payload manipulation, injection attacks
CWE-346	Origin validation error	9.1 (critical)	dApp frontend	Replaying intercepted requests
CWE-200	Exposure of sensitive information	8.8 (high)	dApp frontend	Exploiting misconfigured API endpoints
CWE-400	Uncontrolled resource consumption	7.5 (high)	dApp frontend	Exploiting known vulnerabilities in outdated libraries
CWE-79	Improper neutralization of input during web page generation	5.4 (medium)	dApp frontend	Injecting unauthorized transactions via insecure paths
CWE-561	Dead code	2.0 (low)	dApp frontend	Exploiting logic flaws in business rules

Threat Scenario TS5: Malicious actors targeting the interaction between the dApp frontend and dApp middleware (JSON-RPC) exploit vulnerabilities in authentication, data integrity, API configurations, and middleware logic.				
CWE-502	Deserialization of untrusted data	9.8 (critical)	dApp middleware (JSON-RPC)	Forging middleware credentials, impersonating trusted middleware
CWE-502	Deserialization of untrusted data	9.8 (critical)	dApp middleware (JSON-RPC)	Altering transaction payloads during transmission
CWE-400	Uncontrolled resource consumption	7.5 (high)	dApp middleware (JSON-RPC)	Flooding middleware with excessive traffic or requests
CWE-400	Uncontrolled resource consumption	7.5 (high)	dApp middleware (JSON-RPC)	Exploiting improperly secured API endpoints
CWE-400	Uncontrolled resource consumption	7.5 (high)	dApp middleware (JSON-RPC)	Intercepting unencrypted communication channels
CWE-400	Uncontrolled resource consumption	7.5 (high)	dApp middleware (JSON-RPC)	Leveraging vulnerabilities in outdated libraries
CWE-400	Uncontrolled resource consumption	7.5 (high)	dApp middleware (JSON-RPC)	Exploiting weak or absent access controls
Threat Scenario TS6: Malicious actors targeting the interaction between the dApp middleware and blockchain node exploit vulnerabilities in authentication, data transmission, API security, and local storage integrity.				
CWE-400	Uncontrolled resource consumption	7.5 (high)	Blockchain node	Forging blockchain node credentials, impersonating trusted nodes
CWE-400	Uncontrolled resource consumption	7.5 (high)	Blockchain node	Replaying valid blockchain transactions intercepted during communication

(Continued)

Threat Scenario TS6 (Continued)

Weakness targeted (CWE)	Weakness description	Severity risk	Asset impacted	Attack vectors
CWE-400	Uncontrolled resource consumption	7.5 (high)	Blockchain node	Exploiting improperly secured blockchain APIs
CWE-400	Uncontrolled resource consumption	7.5 (high)	Blockchain node	Leveraging vulnerabilities in outdated blockchain node libraries
CWE-400	Uncontrolled resource consumption	7.5 (high)	Blockchain node	Flooding blockchain nodes with excessive requests
CWE-20	Improper input validation	6.5 (medium)	Blockchain node	Modifying transaction payloads during transmission
CWE-20	Improper input validation	6.5 (medium)	Blockchain node	Accessing local blockchain storage using compromised credentials
CWE-20	Improper input validation	6.5 (medium)	Blockchain node	Exploiting logic flaws in smart contracts
CWE-540	Information exposure through debugging information	2.5 (low)	Blockchain node	Intercepting unencrypted middleware–blockchain communication
Threat Scenario TS7: Malicious actors targeting the interaction between the blockchain node and the distributed ledger exploit vulnerabilities in transaction validation, consensus mechanisms, and block storage integrity.				
CWE-284	Improper access control	9.8 (critical)	Blockchain platform/ distributed ledger	Forging ledger node credentials, impersonating trusted nodes

CWE-703	Improper check or handling of exceptional conditions	9.8 (critical)	Blockchain platform/ distributed ledger	Modifying transaction data during validation
CWE-20	Improper input validation	9.8 (critical)	Blockchain platform/ distributed ledger	Replaying valid transactions intercepted during validation
CWE-20	Improper input validation	9.8 (critical)	Blockchain platform/ distributed ledger	Exploiting misconfigured platform settings
CWE-703	Improper check or handling of exceptional conditions	9.8 (critical)	Blockchain platform/ distributed ledger	Exploiting logical inconsistencies in consensus rules
CWE-354	Improper validation of integrity check value	7.5 (high)	Blockchain platform/ distributed ledger	Exploiting design flaws in consensus mechanisms
CWE-284	Improper access control	7.5 (high)	Blockchain platform/ distributed ledger	Leveraging vulnerabilities in outdated blockchain platform components
CWE-287	Improper authentication	7.5 (high)	Blockchain platform/ distributed ledger	Flooding ledger nodes with excessive traffic
CWE-319	Cleartext transmission of sensitive information	6.5 (medium)	Blockchain platform/ distributed ledger	Intercepting unencrypted data during validation or consensus processes
Threat Scenario TS8: Malicious actors targeting smart contracts exploit vulnerabilities in transaction execution, validation, and state management to compromise the integrity of loan approval and repayment processes.				
CWE-284	Improper access control	9.1 (critical)	Smart contracts	Forging ledger node credentials, impersonating trusted nodes

(Continued)

Threat Scenario TS8 (Continued)

Weakness targeted (CWE)	Weakness description	Severity risk	Asset impacted	Attack vectors
CWE-502	Deserialization of untrusted data	9.1 (critical)	Smart contracts	Injecting malicious serialized objects into smart contracts
CWE-400	Uncontrolled resource consumption	7.5 (high)	Smart contracts	Flooding smart contract methods with excessive transactions or data
CWE-20	Improper input validation	8.2 (high)	Smart contracts	Exploiting input validation flaws in contract compilation or execution
CWE-829	Untrusted control sphere	8.8 (high)	Smart contracts	Leveraging backdoored NPM packages integrated into smart contracts
CWE-362	Concurrent execution using shared resource with improper synchronization	7.5 (high)	Smart contracts	Exploiting reentrancy vulnerabilities in contract logic
CWE-703	Improper check or handling of exceptional conditions	6.5 (medium)	Smart contracts	Triggering unhandled exceptions to disrupt contract operations
CWE-1104	Use of unmaintained third-party components	6.5 (medium)	Smart contracts	Leveraging outdated third-party libraries integrated into smart contracts
CWE-547	Hardcoded gas values	3.5 (low)	Smart contracts	Exploiting inflexible gas values in contract execution
CWE-345	Insufficient verification of data authenticity	9.1 (critical)	Oracle platform/external data sources	Injecting malicious payloads into smart contract logic

Threat Scenario TS9: Threat actors exploit vulnerabilities in the oracle platform, its data sources, and the interaction between oracles and the blockchain platform.

CWE-345	Insufficient verification of data authenticity	9.1 (critical)	Oracle platform/external data sources	Using stolen deployment keys to alter smart contract code
CWE-918	Server-side request forgery (SSRF)	8.2 (high)	Oracle platform/external data sources	Sending malicious requests to manipulate external data feeds
CWE-311	Missing encryption of sensitive data	7.5 (high)	Oracle platform/external data sources	Intercepting unencrypted data during processing
CWE-345	Insufficient verification of data authenticity	8.8 (high)	Oracle platform/external data sources	Modifying loan approval parameters via crafted transactions
CWE-400	Uncontrolled resource consumption	7.5 (high)	Oracle platform/external data sources	Overwhelming loan execution contracts with excessive requests
CWE-326	Inadequate encryption strength	6.3 (medium)	Oracle platform/external data sources	Replaying valid transactions to authorize fraudulent operations
CWE-758	Reliance on undefined behavior	3.0 (low)	Oracle platform/external data sources	Leveraging undefined behavior in smart contract operations

C

Mapping of Threat Scenarios to Exploitable Attack Paths

Threat scenario (TS)	Attack vector	CWE	Entry point	Derived attack scenario
TS1: mobile or browser client/UI	Exploiting improper authentication via stolen session tokens	CWE-287 (improper authentication)	Mobile or browser client	An attacker uses stolen session tokens obtained through phishing to impersonate legitimate users, allowing unauthorized access and fraudulent transactions.
TS2: digital wallet/ smart contracts	Reusing valid transactions to perform replay attacks	CWE-326 (inadequate encryption strength)	Digital wallet API	The attacker intercepts and replays previously signed wallet transactions due to weak encryption, leading to duplicate payments or withdrawals.
TS3: distributed digital identity system/trust boundary	Manipulating identity data transmitted between the dApp and identity system	CWE-345 (insufficient verification of data authenticity)	Identity validation API	The attacker tampers with transmitted identity payloads, corrupting user data or enabling fraudulent access by impersonating a legitimate user.

(Continued)

Blockchain Application Security: How to Design Secure and Attack Resilient Blockchain Applications, First Edition. Marco Morana, Harpreet Singh and Francesco Piccoli.
© 2026 John Wiley & Sons, Inc. Published 2026 by John Wiley & Sons, Inc.

(Continued)

Threat scenario (TS)	Attack vector	CWE	Entry point	Derived attack scenario
TS4: dApp frontend and components	Injecting malicious scripts into input fields	CWE-79 (cross-site scripting)	dApp frontend	Malicious scripts injected into input fields execute in the victim's browser, allowing the attacker to steal session tokens and execute unauthorized transactions.
TS5: dApp middleware (JSON-RPC)	Exploiting deserialization vulnerabilities to inject malicious payloads	CWE-502 (deserialization of untrusted data)	Middleware endpoint	An attacker sends crafted JSON-RPC payloads containing malicious data, leading to the execution of arbitrary commands in the middleware.
TS6: blockchain node	Flooding blockchain nodes with excessive requests	CWE-400 (uncontrolled resource consumption)	Node API	The attacker initiates a denial-of-service (DoS) attack by flooding blockchain nodes with excessive requests, disrupting transaction processing.
TS7: blockchain platform/ distributed ledger	Exploiting improperly configured platform settings	CWE-20 (improper input validation)	Blockchain platform configuration	The attacker exploits misconfigured platform settings to bypass validation checks, enabling unauthorized transactions or reordering blocks in the ledger.
TS8: smart contracts	Exploiting reentrancy vulnerabilities in loan processing	CWE-362 (concurrent execution using shared resource with improper synchronization)	Smart contract logic	The attacker uses a malicious contract to repeatedly call a loan withdrawal function, bypassing state updates and draining funds from the lending pool.
TS9: oracle platform/ external data sources	Manipulating oracle-fed data to influence collateral valuation	CWE-345 (insufficient verification of data authenticity)	Oracle data feed	The attacker injects falsified data into the oracle, manipulating collateral values to enable fraudulent borrowing, destabilizing the lending platform.

D

Threat Scenarios Attack Simulation Tests

Blockchain Application Security: How to Design Secure and Attack Resilient Blockchain Applications, First Edition. Marco Morana, Harpreet Singh and Francesco Piccoli.
© 2026 John Wiley & Sons, Inc. Published 2026 by John Wiley & Sons, Inc.

Threat scenario (TS)	CWE	Attack scenario	Attack-simulation test	Controls to be applied	Retest verification after controls applied
TS1: mobile or browser client/UI	CWE-287 (improper authentication)	An attacker uses stolen session tokens obtained through phishing to impersonate legitimate users, allowing unauthorized access and fraudulent transactions.	Simulate the attack by replaying a valid session token obtained from a phishing simulation.	Implement multifactor authentication (MFA), secure session token management, and periodic token rotation.	Replay the same session token and verify that it is invalidated by token rotation.
TS2: digital wallet/smart contracts	CWE-326 (inadequate encryption Strength)	The attacker intercepts and replays previously signed wallet transactions due to weak encryption, leading to duplicate payments or withdrawals.	Intercept a signed transaction during its transmission and replay it to check for duplicate payments.	Use stronger encryption for wallet transactions and enforce nonce-based transaction validation.	Replay intercepted transactions to ensure they are rejected due to nonce mismatches.
TS3: distributed digital identity system	CWE-345 (insufficient verification of data authenticity)	The attacker tampers with transmitted identity payloads, corrupting user data or enabling fraudulent access by impersonating a legitimate user.	Modify identity payloads in transit using a proxy tool like Burp Suite and check if altered payloads are accepted.	Enforce digital signatures and cryptographic proofs for identity payloads.	Repeat the tampering attempt and verify that altered identity payloads are rejected.
TS4: dApp frontend and components	CWE-79 (cross-site scripting)	Malicious scripts injected into input fields execute in the victim's browser, allowing the attacker to steal session tokens and execute unauthorized transactions.	Insert XSS payloads into input fields and observe execution behavior in the frontend.	Implement input sanitization and content security policies (CSP).	Reapply the same XSS payloads and confirm they are sanitized and blocked.

TS5: dApp middleware (JSON-RPC)	CWE-502 (deserialization of untrusted data)	An attacker sends crafted JSON-RPC payloads containing malicious data, leading to the execution of arbitrary commands in the middleware.	Create and send a malicious JSON-RPC payload to test if it executes arbitrary commands.	Validate all serialized inputs before deserialization and implement sandboxing.	Send the malicious payload again and verify it is rejected.
TS6: blockchain node	CWE-400 (uncontrolled resource consumption)	The attacker initiates a denial-of-service (DoS) attack by flooding blockchain nodes with excessive requests, disrupting transaction processing.	Use stress-testing tools like LOIC to flood node endpoints and observe system response.	Implement rate limiting, load balancing, and IP whitelisting for node endpoints.	Repeat the flood test and verify that rate limiting blocks excessive requests.
TS7: blockchain platform/ distributed ledger	CWE-20 (improper input validation)	The attacker exploits misconfigured platform settings to bypass validation checks, enabling unauthorized transactions or reordering blocks in the ledger.	Inject invalid data or bypass input constraints when submitting blockchain transactions.	Apply strict validation rules at all input layers and audit platform configurations.	Repeat invalid transaction submissions and confirm rejection at all input points.

(Continued)

(Continued)

Threat scenario (TS)	CWE	Attack scenario	Attack-simulation test	Controls to be applied	Retest verification after controls applied
TS8: smart contracts	CWE-362 (reentrancy)	The attacker uses a malicious contract to repeatedly call a loan withdrawal function, bypassing state updates and draining funds from the lending pool.	Deploy a testing environment with a vulnerable smart contract and simulate a reentrancy attack.	Implement the checks-effects-interactions pattern and use OpenZeppelin's ReentrancyGuard.	Execute the malicious contract to confirm reentrant calls are blocked.
TS9: oracle platform/ external data sources	CWE-345 (insufficient verification of data authenticity)	The attacker injects falsified data into the oracle, manipulating collateral values to enable fraudulent borrowing, destabilizing the lending platform.	Inject falsified data into oracle feeds and observe changes in collateral valuation.	Secure oracle feeds with cryptographic proofs and implement periodic audits of data sources.	Reinject falsified data and confirm it is rejected due to authenticity checks.

E

Threat Scenario Weakness and Vulnerabilities Risk Ratings

E.1 Risk Assessment for Threat Scenario TS1

The OWASP risk rating methodology was applied to assess the risk severity of CVEs and CWEs specifically targeted by threat agents in threat scenario TS1. This scenario involves malicious actors exploiting a DeFi dApp's mobile user interface by targeting vulnerabilities related to authentication, data exposure, and input validation. The risk evaluation focused on how these known weaknesses, such as CWE-287 (improper authentication) and CWE-359 (exposure of private information), could be exploited by skilled attackers and financially motivated cybercriminals.

The likelihood of exploitation was determined by assessing the threat landscape, attacker capabilities, and the ease of discovering and exploiting the vulnerabilities. In parallel, the impact was evaluated from both a technical and business perspective, considering consequences such as unauthorized access, financial loss, and regulatory exposure. This structured approach enables precise prioritization of vulnerabilities within TS1, helping guide mitigation efforts before production deployment.

E.2 Risk Severity Evaluation of CWEs and CVEs in Threat Scenario TS1 Using the OWASP Risk Rating Methodology

The OWASP risk rating methodology assesses risk by combining the likelihood of a threat successfully exploiting a weakness with the potential impact of that exploitation. Likelihood is evaluated based on multiple factors, including attacker skill level, motivation, attack surface exposure, and ease of exploitability. In the context of TS1, the presence of skilled blockchain security researchers and financially

Blockchain Application Security: How to Design Secure and Attack Resilient Blockchain Applications, First Edition. Marco Morana, Harpreet Singh and Francesco Piccoli.
© 2026 John Wiley & Sons, Inc. Published 2026 by John Wiley & Sons, Inc.

Threat targeted CVEs/CWEs	Weakness description	Weakness severity risk	Threat risk factor likelihood	Threat impact risk factor (technical and business factors)	Overall threat targeted weakness risk score (risk rating)
CVE-2022-32514	Improper authentication	9.8 (critical)	9.5 (high)	9.5 (critical)	90.25 (● critical)
CWE-287	Improper authentication	9.8 (critical)	9.5 (high)	9.5 (critical)	90.25 (● critical)
CWE-359	Exposure of private information	8.8 (high)	8.5 (high)	8.5 (high)	72.25 (● high)
CVE-2018-1000850	Use of unmaintained third-party components	7.8 (high)	7.5 (medium)	7.5 (medium)	56.25 (● high)
CWE-1104	Use of unmaintained third-party components	7.8 (high)	7.5 (medium)	7.5 (medium)	56.25 (● high)
CVE-2022-23520	Improper input validation	7.5 (high)	7.0 (medium)	7.0 (medium)	49.00 (● high)
CWE-20	Improper input validation	7.5 (high)	7.0 (medium)	7.0 (medium)	49.00 (● high)
CVE-2019-11771	Insecure storage of sensitive information	4.7 (medium)	4.5 (low)	5.0 (medium)	22.50 (● medium)
CWE-922	Insecure storage of sensitive information	4.7 (medium)	4.5 (low)	5.0 (medium)	22.50 (● medium)
CWE-104	Excessive logging	2.0 (low)	3.0 (low)	3.5 (low)	10.50 (● low)

motivated cybercriminals significantly raises the probability of successful exploitation, particularly in cases involving improper authentication (CWE-287).

The final OWASP risk score is calculated as Likelihood × Impact, producing risk severity ratings that guide remediation priorities. In TS1, CWE-287 received a critical risk score of 90.25, while CWE-359 (exposure of private information) scored 72.25 (high). In contrast, medium-risk issues such as CWE-922 (insecure storage) received a lower score of 22.50, reflecting reduced exploitability or limited business impact.

Impact assessments factored in both technical consequences (e.g. unauthorized fund transfers, wallet key exposure) and business risks (e.g. financial losses, reputational damage, and compliance violations). CWE-287 stood out as a critical risk due to its direct role in authentication failures, while CWE-359 was rated high for its potential to cause privacy breaches and enable fraud. Lower-impact issues, such as CWE-104 (excessive logging), received lower scores due to their minimal disruption to platform operations.

These risk ratings provide a structured basis for prioritizing vulnerability mitigation prior to production deployment, ensuring the most impactful issues are addressed first.

E.3 OWASP Risk Rating Calculation for CVE-2022-32514

This example illustrates how the OWASP risk rating methodology is applied to evaluate the overall risk severity of a known vulnerability, CVE-2022-32514, which maps to CWE-287: improper authentication within the context of TS1.

The calculation demonstrates a structured, evidence-based approach to scoring risk by analyzing both the likelihood of exploitation and the potential technical and business impact. Factors such as attacker skill level, exploitability, and real-world consequences (e.g. unauthorized access, financial loss, and privacy breaches) are considered to produce a quantifiable risk score.

This methodology supports consistency across risk assessments and allows for prioritization of CVEs and CWEs that represent the highest threat to the platform. In this case, the final risk score of 90.25 (critical) reflects the high severity of authentication bypass vulnerabilities in a DeFi dApp environment, especially where financial and user data are at stake.

This example serves as a reference for scoring other weaknesses in TS1, ensuring transparency and repeatability in the risk assessment:

Weakness: Improper Authentication (CWE-287)

Risk Scenario: Bypass of authentication controls in a DeFi dApp leading to unauthorized access and critical financial and data impact.

E.4 Step 1: Likelihood Score

Likelihood is the average of two subcategories:

- Threat Agent Factors
- Vulnerability Factors

Each subcategory has four factors, scored on a scale of 0–9.

E.4.1 Threat Agent Factors

Factor	Score	Rationale
Skill level	8	Moderate skill required; exploits are known and reusable.
Motive	10	Strong financial motivation in DeFi (direct asset access).
Opportunity	10	Publicly exposed interfaces; ease of access to attack surface.
Size	10	Large pool of threat actors (e.g. bots, cybercriminals).

Threat Agent Subtotal: (8 + 10 + 10 + 10)/ 4 = 9.5.

E.4.2 Vulnerability Factors

Factor	Score	Rationale
Ease of discovery	9	Well-known weakness (CWE-287) with clear detection patterns.
Ease of exploit	9	Simple to exploit with phishing or token interception.
Awareness	9	Widely documented, regularly found in audits.
Intrusion detection	9	Often difficult to distinguish from legitimate use.

Vulnerability Subtotal: (9 + 9 + 9 + 9)/4 = 9.0.
Final Likelihood Score: (9.5 + 9.0)/2 = 9.25.

E.5 Step 2: Impact Score

Impact is the average of two subcategories:

- Technical Impact Factors
- Business Impact Factors

E.5.1 Technical Impact Factors

Factor	Score	Rationale
Loss of confidentiality	9	Authentication bypass can lead to exposure of wallet data or PII.
Loss of integrity	9	Unauthorized actions may be taken in the dApp.
Loss of availability	8	Session misuse or abuse could degrade or disrupt service.
Loss of accountability	9	Logs become unreliable if auth is bypassed.

Technical Impact Subtotal: $(9 + 9 + 8 + 9)/4 = 8.75$.

E.5.2 Business Impact Factors

Factor	Score	Rationale
Financial damage	10	High-value assets could be stolen directly.
Reputation damage	9.5	Loss of user trust in security of DeFi platform.
Noncompliance	9	Violations of KYC/AML and data privacy requirements.
Privacy violation	10	Sensitive user data may be leaked or misused.

Business Impact Subtotal: $(10 + 9.5 + 9 + 10)/4 = 9.625$.
Final Impact Score: $(8.75 + 9.625)/2 = 9.1875 \rightarrow$ Rounded to 9.5.

E.6 Step 3: Final Risk Score

Final Risk Score = Likelihood × Impact = $9.25 \times 9.5 = 90.25 \rightarrow$ ● Critical.

F

Risks Mitigation Plan

F.1 Objective

The risk mitigation plan aims to systematically address the identified risks in the decentralized finance (DeFi) lending and borrowing dApp by implementing a combination of technical controls, procedural measures, and security best practices. The plan is designed to prioritize high-risk vulnerabilities, align security measures with organizational objectives and compliance requirements, and provide a clear roadmap with timelines, milestones, resource allocations, and responsibilities.

F.2 Risk Mitigation Strategy

Each threat scenario is mitigated through targeted risk reduction measures. To ensure accountability, each risk mitigation action must have a designated owner responsible for its implementation and execution. Ownership of these actions is essential, as it establishes clear accountability and ensures progress is measured against project timelines.

Additionally, within the project timelines, a vulnerability risk remediation SLA (service level agreement) is integrated as a mandatory requirement, defining the time frame within which identified risks must be addressed to maintain security and compliance standards.

Blockchain Application Security: How to Design Secure and Attack Resilient Blockchain Applications, First Edition. Marco Morana, Harpreet Singh and Francesco Piccoli.
© 2026 John Wiley & Sons, Inc. Published 2026 by John Wiley & Sons, Inc.

Threat scenario	Key risks (OWASP assessment)	Mitigation measures	Owner(s)	Priority	Project timeline (within SLA compliance)
TS1 – mobile/UI security risks	Improper authentication, exposure of private data, unmaintained components	Implement MFA, encrypt sensitive data, regularly audit and update third-party libraries	Security engineering team, product owner	Critical	Q1 (30 days)
TS2 – digital wallet and smart contract risks	Authentication bypass, integer overflow, insecure storage	Implement hardware wallet integration, enforce strict smart contract access controls, conduct formal verification of smart contracts	Blockchain security team, DevOps	Critical	Q1 (30 days)
TS3 – distributed identity system risks	Improper access control, origin validation errors, gas limit exploits	Adopt decentralized identity best practices, use token-binding techniques, optimize gas usage in smart contract functions	Identity management team, blockchain security team	High	Q1–Q2 (60 days)
TS4 – dApp frontend security risks	XSS, CSRF, exposure of sensitive information	Implement content security policy (CSP) and subresource integrity (SRI), use anti-CSRF tokens, secure API authentication	Web security team, AppSec engineers	High	Q1–Q2 (60 days)
TS5 – middleware/API security risks	Deserialization of untrusted data, API DoS attacks	Enforce strong API authentication, validate and sanitize all user inputs, implement rate-limiting and WAF	API security team, infrastructure engineers	Critical	Q1 (30 days)

TS6 – blockchain node security risks	Uncontrolled resource consumption, improper input validation	Enforce RBAC for blockchain node access, apply node-level monitoring and anomaly detection, implement secure node configurations	Blockchain infrastructure team	High	Q1–Q2 (60 days)
TS7 – distributed ledger security risks	Access control flaws, improper validation of integrity check values	Implement BFT consensus validation mechanisms, ensure multi-signature governance, perform periodic blockchain integrity audits	Blockchain governance team	High	Q1–Q2 (60 days)
TS8 – smart contract vulnerabilities	Reentrancy, improper access control, untrusted control sphere	Implement reentrancy guards, adopt formal verification, utilize multitier contract logic separation	Smart contract security team	Critical	Q1 (30 days)
TS9 – oracle and external data risks	Insufficient verification of data authenticity, SSRF, missing encryption	Use decentralized oracles, validate and sanitize all external API data, apply encryption and authentication for oracle transmissions	Oracle security team, smart contract engineers	Critical	Q1 (30 days)

G

Threats Risk Register

This document provides a consolidated record of key risks identified for the DeFi lending and borrowing dApp, categorized by severity and likelihood, and mapped to specific threat scenarios, vulnerabilities (CWEs), and impacted components within the platform. It functions as a central reference point for tracking the organization's progress toward the risk mitigation objectives outlined in the security strategy.

Each risk is assigned a status to reflect its current treatment:

- In Progress: Active mitigation efforts are underway.
- Mitigated: Appropriate controls have been implemented and verified.
- Accepted: The risk has been acknowledged, with no immediate action planned.

Given the critical nature of many threat scenarios, all risks listed in this register are either actively being addressed or have already been successfully mitigated.

This register supports informed decision-making by security, engineering, and business stakeholders and ensures visibility and accountability across the risk management lifecycle.

Blockchain Application Security: How to Design Secure and Attack Resilient Blockchain Applications, First Edition. Marco Morana, Harpreet Singh and Francesco Piccoli.
© 2026 John Wiley & Sons, Inc. Published 2026 by John Wiley & Sons, Inc.

Threat scenario description	Impacted asset	Associated CWEs/CVEs	Overall threat scenario risk rating	Risk mitigation status
TS1 – Malicious actors targeting dApp's mobile UI	Mobile UI	CWE-287 (improper authentication), CWE-359 (exposure of private information)	Critical	Mitigated
TS2 – Threats targeting digital wallets	Digital wallet	CWE-287 (improper authentication), CWE-190 (integer overflow)	Critical	Mitigated
TS3 – Threats to distributed identity systems	Distributed identity system	CWE-284 (improper access control), CWE-346 (origin validation error)	High	In progress
TS4 – Exploitation of dApp frontend components	dApp frontend	CWE-352 (CSRF), CWE-79 (XSS)	High	In progress
TS5 – Attacks on dApp middleware (JSON-RPC)	dApp middleware	CWE-502 (deserialization of untrusted data), CWE-400 (uncontrolled resource consumption)	Critical	Mitigated
TS6 – Exploitation of blockchain nodes	Blockchain node	CWE-400 (uncontrolled resource consumption), CWE-20 (improper input validation)	High	In progress
TS7 – Threats to blockchain platform and consensus integrity	Blockchain platform	CWE-284 (improper access control), CWE-703 (exception handling issues)	High	In progress
TS8 – Smart contract vulnerabilities in DeFi loan processing	Smart contracts	CWE-284 (improper access control), CWE-502 (deserialization of untrusted data), CWE-841 (reentrancy)	Critical	Mitigated
TS9 – Oracle manipulation attacks affecting financial transactions	Oracle platform	CWE-345 (insufficient data authenticity verification), CWE-918 (SSRF)	Critical	In progress

H

Attack Simulation Testing Report

H.1 Executive Summary

This document presents the results of targeted attack simulation testing conducted during the threat modeling phase of the DeFi dApp security assessment. Each scenario is mapped to known software weaknesses (CWEs) and tested against the current or proposed countermeasures. The objective was to validate how real-world adversaries could exploit these vulnerabilities and to assess the effectiveness of mitigation strategies in practice.

The findings provide evidence-based support for the overall risk scoring methodology and help prioritize areas requiring further control enhancements. Scenarios tested include high-impact threats such as smart contract reentrancy, oracle manipulation, authentication bypass, and middleware-level deserialization attacks, all of which are critical to the integrity and resilience of the platform.

The results confirm that while several high-risk areas are well-addressed through industry best practices, others, particularly in the user interface and external data exposure layers, would benefit from stronger or layered defenses. This analysis supports informed decision-making for ongoing risk reduction, secure development, and incident prevention efforts across the platform.

Blockchain Application Security: How to Design Secure and Attack Resilient Blockchain Applications, First Edition. Marco Morana, Harpreet Singh and Francesco Piccoli.
© 2026 John Wiley & Sons, Inc. Published 2026 by John Wiley & Sons, Inc.

Attack scenario tested	Exploited weakness (CWE)	Testing results	Countermeasures evaluated	Countermeasures effectiveness
Phishing attack on wallet authentication	CWE-287: improper authentication	Successful account takeover	• Enable multifactor authentication (MFA) • Implement phishing-resistant authentication • Deploy anti-phishing protections	Moderate
Replay attack exploiting improper access control	CWE-284: improper access control	Unauthorized access to sensitive transactions	• Implement timestamp and expiry checks • Use cryptographic message signing with context binding • Enforce nonce-based transaction validation	High
Smart contract reentrancy attack	CWE-362: reentrancy	Successful fund drain from contract	• Implement checks-effects-interactions pattern • Use OpenZeppelin's ReentrancyGuard • Limit gas available for external call	High
Cross-site scripting (XSS) in frontend UI	CWE-79: XSS	Execution of malicious scripts	• Enforce content security policy (CSP) • Use secure JavaScript frameworks with input sanitization/output encoding	Moderate
Deserialization of untrusted data in middleware	CWE-502: deserialization	Remote code execution vulnerability	• Validate and sanitize all input data • Use safe deserialization libraries • Isolate and sandbox deserialization operations	High

Gas limit exploit on blockchain node	CWE-400: resource exhaustion	Denial-of-service condition on node	• Set smart contract execution gas limits • Implement gas usage monitoring and rate limits • Optimize smart contract code efficiency	High
Bypassing smart contract authentication	CWE-287: improper authentication	Unauthorized transaction approvals	• Enforce role-based access control (RBAC) • Use cryptographic signatures for authentication (verifying signatures before executing privileged actions) • Implement multi-signature authorization	High
Oracle price manipulation attack	CWE-345: insufficient verification of data authenticity	Market distortion and profit extraction	• Use multiple independent oracle sources • Implement median and time-weighted price feeds • Cap maximum price deviation and implement circuit breakers	High
Server-side request forgery (SSRF) on external data feeds	CWE-918: SSRF	Exfiltration of sensitive information	• Restrict outbound requests to allowlisted domains • Disable unnecessary URL fetching and metadata services • Validate and sanitize user-supplied URLs	Moderate

H.2 Understanding the "Countermeasures Effectiveness" Rating

The effectiveness rating reflects how well the evaluated countermeasures prevented, mitigated, or responded to the simulated attack. Ratings are categorized as follows:

- High: The implemented or proposed countermeasures significantly reduce or fully mitigate the attack vector, making exploitation either highly unlikely or effectively unfeasible in most scenarios.
- Moderate: The countermeasures reduce the impact or likelihood of the attack but may not fully prevent exploitation under all conditions or may require additional layers of defense.
- Low: The countermeasures offer limited mitigation, leaving significant residual risk. Additional controls are required to achieve acceptable risk levels.

I

Risk Analysis Report

I.1 Executive Summary

This report presents a consolidated view of the key risks identified through the threat modeling and risk assessment of the DeFi lending and borrowing dApp. It is intended to support decision-makers, including risk owners, business leaders, and security stakeholders, in evaluating the platform's security posture, understanding the potential business and technical impact of identified threats, and prioritizing mitigation strategies. The goal is to enable informed, risk-based decisions that safeguard the platform's critical assets, maintain user trust, and ensure regulatory compliance in an evolving threat landscape.

I.2 Critical Threat Scenarios and Risk Ratings

The following table outlines the most critical threat scenarios identified during the threat modeling and risk assessment of the DeFi lending and borrowing platform. Each scenario is evaluated based on the likelihood of exploitation, the impact on security and business operations, and the resulting overall risk score. These risk scores help prioritize mitigation efforts by highlighting vulnerabilities that could lead to financial loss, platform instability, regulatory exposure, or loss of user trust. The scenarios span across the platform's key components, ranging from user-facing interfaces and wallets to smart contracts, oracles, and infrastructure.

High-severity issues, such as those related to smart contracts, UI security, middleware, and oracles, demand immediate attention due to their critical exploitability and potential to cause irreversible damage (e.g. asset theft, unauthorized transactions, or protocol manipulation). Medium-risk areas, like blockchain node

Blockchain Application Security: How to Design Secure and Attack Resilient Blockchain Applications, First Edition. Marco Morana, Harpreet Singh and Francesco Piccoli.
© 2026 John Wiley & Sons, Inc. Published 2026 by John Wiley & Sons, Inc.

and platform risks, while less exploitable, still pose serious threats that could disrupt core functionality and consensus integrity.

This overview is intended to inform risk mitigation planning and resource allocation, enabling security, engineering, and leadership teams to respond with proportionate controls, development guardrails, and continuous monitoring strategies.

Threat scenario	Likelihood (score)	Impact (score)	Overall risk (score; seethe note)	Commentary/rationale
TS1: mobile or browser UI risks	High (9.0)	Critical (9.5)	Critical (85.5)	Due to high exploitability and severe financial consequences, improper authentication and insecure storage make this a significant threat.
TS2: digital wallet risks	High (8.5)	Critical (9.5)	Critical (80.8)	Wallets are a prime target; improper authentication and integer overflow could lead to direct asset theft.
TS3: distributed digital identity system risks	High (8.0)	High (8.8)	High (70.4)	Identity spoofing and resource exhaustion could lead to fraudulent activities and systemic disruptions.
TS4: dApp frontend risks	High (8.2)	High (8.9)	High (73.0)	Frontend vulnerabilities like XSS and CSRF allow for transaction manipulation, significantly impacting platform integrity.
TS5: dApp middleware (JSON-RPC) risks	High (8.5)	Critical (9.5)	Critical (80.8)	Middleware risks include deserialization attacks, enabling unauthorized transactions and system compromise.
TS6: blockchain node risks	Medium (7.2)	High (8.5)	High (61.2)	Disrupting node operations affects transaction validation, causing operational delays and financial instability.
TS7: blockchain platform and distributed ledger risks	Medium (7.0)	Critical (9.2)	High (64.4)	Platform vulnerabilities could lead to unauthorized transactions and network disruption.

TS8: smart contract risks	High (9.0)	Critical (9.5)	Critical (85.5)	Smart contract breaches could result in full platform liquidation and significant financial losses.
TS9: oracle and external data feeds risks	High (8.5)	Critical (9.5)	Critical (80.8)	Oracle manipulation can lead to price exploits, draining liquidity and causing insolvency.

I.3 Risk Mitigation Strategy

To ensure the security and resilience of the lending and borrowing dApp, a comprehensive risk mitigation strategy has been defined based on the analysis of critical threat scenarios, their likelihood, and potential impact. This strategy addresses vulnerabilities across the entire application stack, including smart contracts, oracles, wallets, frontend interfaces, and supporting infrastructure by implementing targeted controls and proactive safeguards.

The strategy emphasizes securing the most high-risk components, such as mobile/browser UIs, digital wallets, smart contracts, and middleware layers (e.g. JSON-RPC endpoints). For each risk area, specific technical countermeasures have been proposed to reduce exploitability and minimize potential financial and operational damage.

Mitigations include:

1. Strengthening authentication and access controls, particularly for user interfaces and wallet interactions
2. Securing smart contracts through formal verification, use of trusted libraries, and clearly defined access control mechanisms
3. Protecting external data feeds and oracles from manipulation via multisource validation and on-chain data integrity checks
4. Hardening middleware and APIs with strict input validation, method whitelisting, and deserialization protections
5. Ensuring infrastructure resilience through regular patching, node redundancy, encrypted communications, and monitoring
6. Addressing identity system risks with decentralized identity protocols, resource abuse protection, and zero-knowledge proofs where applicable

Additionally, the strategy incorporates the adoption of a risk-centric threat modeling process as an integral part of the dApp's secure development life cycle. This approach goes beyond checklist-based assessments by leveraging structured methodologies such as PASTA (Process for Attack Simulation and Threat Analysis), attack trees, and cyber kill chains to simulate realistic adversary behaviors and assess how threats may propagate through the system. By modeling likely attack paths and understanding how an adversary could exploit vulnerabilities across different components, such as smart contracts, oracles, frontend interfaces, and backend infrastructure, it is possible to identify high-impact attack scenarios early, prioritize mitigations, and embed security considerations directly into design and implementation decisions.

I.4 Strategic Alignment of Risk Mitigation with Business Objectives

Given the platform's central role in DeFi lending and borrowing, risk mitigation strategies must be carefully aligned with the platform's business objectives, regulatory obligations, and user expectations for security, transparency, and reliability. This alignment ensures that security is not treated as a siloed function but as a core enabler of sustainable growth and platform integrity.

Proactively addressing security risks, through robust threat modeling, layered defenses, and secure development practices, serves multiple critical functions. It strengthens user trust, which is essential in a decentralized financial ecosystem where confidence in the protocol determines user adoption and liquidity. It also minimizes financial and legal exposure, particularly in jurisdictions with evolving compliance expectations around digital assets, consumer protection, and data privacy.

Furthermore, by embedding security into the platform's design and operations, the strategy ensures that the dApp can remain resilient against emerging threats, including protocol-level exploits, smart contract vulnerabilities, and supply chain attacks. This forward-looking posture not only safeguards current operations but also creates a foundation for scaling securely and adapting to new market opportunities and regulatory landscapes.

I.5 Next Steps and Strategic Recommendations

This report provides a high-level overview of the key risks facing the DeFi lending and borrowing platform, along with mitigation strategies aligned to business, technical, and regulatory priorities. To effectively reduce exposure and strengthen platform resilience, stakeholders are encouraged to collaborate closely with the security team to prioritize risk management initiatives. By aligning security efforts with development roadmaps, compliance obligations, and operational goals, the organization can ensure that the platform remains secure, trustworthy, and aligned with industry best practices in a rapidly evolving threat landscape.

I.6 Addendum: Risk Scoring Methodology and Calculation Example

To support transparency and consistency in the risk assessment process, this addendum details the risk scoring methodology applied to each identified threat scenario in the report. The OWASP risk rating methodology was used as the standardized framework to calculate the overall risk score, combining structured evaluations of likelihood and impact. Each score is based on a weighted analysis of subfactors within threat agent, vulnerability, technical impact, and business impact categories.

This section includes a full breakdown of the scoring process using TS1: mobile or browser UI risks as a representative example. The methodology demonstrated here has been consistently applied across all threat scenarios (TS1–TS9) to ensure comparability and alignment with industry-accepted best practices. The OWASP risk rating methodology is used to calculate the overall risk score for each threat scenario. For example, the critical score of 85.5 risk rating assigned to TS1: mobile or browser UI risks is derived by averaging key factors related to threat likelihood and impact.

This same structured approach is applied consistently across all other threat scenarios, TS1 through TS9, to ensure risk calculations use the OWASP standard risk rating methodology and are comparable.

We provide herein the step-by-step calculations that lead to TS1 risk rating of critical.

I.7 Step 1: Likelihood Score Calculations

The OWASP risk rating methodology calculates likelihood based on the average of two subcategories:

1. Threat agent (4) factors
2. Vulnerability (4) factors

Each subcategory typically includes four individual factors, each scored from 0 to 9. The final likelihood is the average of all eight values.

I.7.1 Threat Agent Factor Scores

Factor	Score	Rationale
Skill level	8	Skilled attackers capable of phishing, token theft, etc.
Motive	9	High financial incentive (access to funds)

(Continued)

(Continued)

Factor	Score	Rationale
Opportunity	9	Easy access via public-facing UI (browser, mobile)
Size	8	Large number of attackers targeting Web3 interfaces

Threat Agent Subtotal: (8 + 9 + 9 + 8)/4 = 8.5.

I.7.2 Vulnerability Factor Scores

Factor	Score	Rationale
Ease of discovery	9	CWE-287 (improper authentication): This weakness is frequently discovered during basic static analysis or black-box testing using common tools (e.g. OWASP ZAP, Burp Suite). Its patterns are easy to fingerprint in poorly secured mobile/web apps. CWE-922 (insecure storage): Easily detectable in mobile apps by analyzing local storage mechanisms (e.g. Android's SharedPreferences or iOS Keychain misuse).
Ease of exploit	9	CWE-287: No need for sophisticated exploits – attackers can bypass weak authentication or steal tokens via phishing or intercepted traffic. CWE-922: Attackers with physical or root access (e.g. via jailbroken devices) can easily extract unencrypted keys or sensitive session data.
Awareness	8	Both CWEs are highly documented, with hundreds of CVEs published. They are routinely exploited in real-world breaches (e.g. insecure session storage in mobile wallets or dApps).
Intrusion detection	7	Attacks exploiting these weaknesses are often stealthy. Improper auth bypasses (CWE-287) may appear as legitimate logins; local data theft (CWE-922) leaves no trace in centralized logs.

Vulnerability Subtotal: (9 + 9 + 8 + 7)/4 = 8.25.
Final Likelihood Score: (8.5 + 8.25)/2 = 8.375 → Rounded to 9.0 (High).

I.8 Step 2: Impact Score Calculations

The OWASP risk rating methodology calculates impact based on the average of two subcategories:

1. Technical impact (4) factors
2. Business impact (4) factors

Each subcategory typically includes four individual criteria, each scored from 0 to 9. The final impact is the average of all eight values.

I.8.1 Technical Impact Factor Scores

Factor	Score	Rationale
Loss of confidentiality	9	CWE-922: Leaked private keys or tokens can lead to irreversible fund loss and identity exposure.
Loss of integrity	8	CWE-287: Enables unauthorized access and transaction manipulation.
Loss of availability	6	CWE-287: Attackers could abuse session hijacking to lock users out or initiate draining transactions.
Loss of accountability	9	CWE-287: Authentication failures obscure audit trails, complicating attribution.

Technical Impact Subtotal: $(9 + 8 + 6 + 9)/4 = 8.0$.

I.8.2 Business Impact Factor Scores

Factor	Score	Rationale
Financial damage	9.5	Compromised UI could result in drained wallets or stolen funds.
Reputation damage	9.5	Trust is easily lost with security incidents in DeFi.
Noncompliance	9.0	Exposure of PII or account takeover could violate KYC/AML requirements.
Privacy violation	10	Critical if unencrypted user credentials or wallet data are leaked.

Business Impact Subtotal: $(9.5 + 9.5 + 9 + 10)/4 = 9.5$.
Final Impact Score: $(8.0 + 9.5)/2 = 8.75 \rightarrow$ Rounded to 9.5 (Critical).

I.9 Step 3: Overall Threat Scenario Risk Score Calculation

Risk Score = Likelihood × Impact = $9.0 \times 9.5 = 85.5$.
Overall Risk Rating: 85.5 → Critical.

I.10 Understanding the Risk Score Ratings

The overall risk scores presented in this report are calculated using the OWASP risk rating methodology, combining likelihood and impact values. The resulting score is interpreted using the following risk rating bands:

- Critical (81–100): Immediate action required. The threat poses severe financial, operational, or reputational damage. Exploitation is likely and impact is high.
- High (61–80): Significant risk requiring prioritized mitigation. Exploitation could lead to serious consequences but may be somewhat constrained in scope.
- Medium (41–60): Moderate risk. Should be addressed as part of planned security improvements.
- Low (0–40): Minimal impact or low likelihood. May not require immediate remediation but should be monitored.

This classification helps stakeholders understand the severity of each threat scenario and prioritize remediation accordingly.

References

1 **Nakamoto, S. (2008).** *Bitcoin: A Peer-to-Peer Electronic Cash System*. Retrieved from https://bitcoin.org/bitcoin.pdf

2 **International Organization for Standardization**. **(2022).** *ISO/IEC 27002: 2022 Information Security, Cybersecurity and Privacy Protection — Information Security Controls*. Retrieved from https://www.iso.org/standard/82208.html

3 **Bitcoin.** Retrieved from https://bitcoin.org

4 **Ethereum.** Retrieved from https://ethereum.org

5 **Litecoin.** Retrieved from https://litecoin.org

6 **Ripple.** Retrieved from https://ripple.com

7 **Hyperledger Fabric.** Retrieved from https://www.hyperledger.org/use/fabric

8 **R3. Corda.** Retrieved from https://www.corda.net

9 **ConsenSys.** *Quorum*. Retrieved from https://consensys.net/quorum

10 **R3.** *Corda Enterprise*. Retrieved from https://www.r3.com/corda-enterprise

11 **Dragonchain.** Retrieved from https://dragonchain.com

12 **Polkadot.** Retrieved from https://polkadot.network

13 **Ripple**. Retrieved from https://ripple.com

14 **CoinDesk.** (n.d.). *ICOs Explained*. Retrieved from https://www.coindesk.com/learn/icos-explained

15 **Ethereum.** *Ethereum Upgrades*. Retrieved from https://ethereum.org/en/upgrades/

16 **Lightning Network.** Retrieved from https://lightning.network

17 **Ethereum.** *Danksharding*. Retrieved from https://ethereum.org/en/roadmap/danksharding/

18 **Back, A., Corallo, M., Dashjr, L., Friedenbach, M., Maxwell, G., Miller, A., Poelstra, A., Timón, J., & Wuille, P. (2014).** *Enabling Blockchain Innovations with Pegged Sidechains*. Retrieved from https://blockstream.com/sidechains.pdf

Blockchain Application Security: How to Design Secure and Attack Resilient Blockchain Applications, First Edition. Marco Morana, Harpreet Singh and Francesco Piccoli.
© 2026 John Wiley & Sons, Inc. Published 2026 by John Wiley & Sons, Inc.

19 **Poon, J., & Dryja, T. (2016).** *The Bitcoin Lightning Network: Scalable Off-Chain Instant Payments.* Retrieved from https://lightning.network/lightning-network-paper.pdf

20 **XRPL.** *Consensus Structure.* Retrieved from https://xrpl.org/docs/concepts/consensus-protocol/consensus-structure

21 **Cosmos.** Retrieved from https://cosmos.network/

22 **Chainlink.** Retrieved from https://chain.link/

23 **Interledger.** Retrieved from https://interledger.org/

24 **Global Blockchain Business Council.** *Token Taxonomy Framework.* Retrieved from https://gbbcouncil.org/interwork-alliance/token-taxonomy-framework/

25 **Stellar.** Retrieved from https://stellar.org/

26 **Axoni.** Retrieved from https://axoni.com/

27 **ConsenSys.** Retrieved from https://consensys.io/

28 **Digital Asset.** Retrieved from https://www.digitalasset.com/

29 **Chain.** Retrieved from https://chain.com/

30 **Brave New Coin. (2020).** *ETC 51% Attack: What Happened and How It Was Stopped.* Retrieved from https://bravenewcoin.com/insights/etc-51-attack-what-happened-and-how-it-was-stopped

31 **Investopedia.** *Altcoins.* Retrieved from https://www.investopedia.com

32 **Investopedia.** *Stablecoins.* Retrieved from https://www.investopedia.com

33 **Chainalysis. (2025).** *Crypto Hacking: Stolen Funds 2025.* Retrieved from https://www.chainalysis.com/blog/crypto-hacking-stolen-funds-2025/

34 **Szabo, N. (1996).** *Smart Contracts: Building Blocks for Digital Markets.* Retrieved from https://www.fon.hum.uva.nl/rob/Courses/InformationInSpeech/CDROM/Literature/LOTwinterschool2006/szabo.best.vwh.net/smart_contracts_2.html

35 **Gemini.** *The DAO Hack and MakerDAO.* Retrieved from https://www.gemini.com/cryptopedia/the-dao-hack-makerdao#section-origins-of-the-dao

36 **Investopedia.** *Decentralized Finance (DeFi).* Retrieved from https://www.investopedia.com/decentralized-finance-defi-5113835

37 **Uniswap Foundation.** Retrieved from https://www.uniswapfoundation.org/

38 **Get Monero.** *Ring Signatures.* Retrieved from https://www.getmonero.org/resources/moneropedia/ringsignatures.html

39 **Bytecoin.** Retrieved from https://www.bytecoin.world/

40 **PIVX.** Retrieved from https://pivx.org/

41 **Dash.** Retrieved from https://www.dash.org/

42 **Komodo.** Retrieved from https://komodoplatform.com/en/

43 **Chainalysis.** *Introduction to Zero-Knowledge Proofs (ZKPs).* Retrieved from https://www.chainalysis.com/blog/introduction-to-zero-knowledge-proofs-zkps/

44 **Sovrin.** Retrieved from https://sovrin.one/

45 **uPort.** Retrieved from https://www.uport.me/

46 **Hyperledger Indy.** Retrieved from https://lf-hyperledger.atlassian.net/wiki/spaces/indy/overview
47 **Civic.** Retrieved from https://www.civic.com/
48 **Amazon Web Services.** *AWS Managed Blockchain.* Retrieved from https://aws.amazon.com/managed-blockchain/
49 **IBM.** *Blockchain Platform: Hyperledger Fabric.* Retrieved from https://www.ibm.com/products/blockchain-platform-hyperledger-fabric
50 **General Data Protection Regulation (GDPR).** Retrieved from https://gdpr.eu/
51 **California Office of the Attorney General.** *California Consumer Privacy Act (CCPA).* Retrieved from https://oag.ca.gov/privacy/ccpa
52 **Investopedia.** *Initial Coin Offerings (ICOs).* Retrieved from https://www.investopedia.com/initial-coin-offerings-4689763
53 **Coinbase.** *What Is a Security Token Offering (STO)?* Retrieved from https://www.coinbase.com/learn/crypto-glossary/what-is-a-security-token-offering-sto
54 **Government of Canada.** *Personal Information Protection and Electronic Documents Act (PIPEDA).* Retrieved from https://laws-lois.justice.gc.ca/ENG/ACTS/P-8.6/index.html
55 **Government of Singapore.** *Personal Data Protection Act 2012.* Retrieved from https://sso.agc.gov.sg/Act/PDPA2012
56 **Chainalysis.** (n.d.). *What Is AML and KYC for Crypto?* Retrieved from https://www.chainalysis.com/blog/what-is-aml-and-kyc-for-crypto/
57 **Digital Asset.** (2024). *Gartner Hype Cycle for Web3 and Blockchain.* Retrieved from https://www.digitalasset.com/gartner-hype-cycle-for-web3-and-blockchain-2024
58 **U.S. Department of Homeland Security (DHS), Science & Technology Directorate.** *Blockchain and Distributed Ledger Technology: Evaluating Its Suitability for Development Initiatives.* U.S. Department of Homeland Security, Washington, DC. Available online.
59 **National Institute of Standards and Technology (NIST).** *Blockchain Technology Overview.* NIST Special Publication 800-282, Gaithersburg, MD. Available online.
60 **Ethereum Foundation.** *ERC-20 Token Standard Definition.* Ethereum Improvement Proposal (EIP-20). Available online.
61 **Ethereum Foundation.** *ERC-721: Non-Fungible Token Standard.* Ethereum Improvement Proposal (EIP-721). Available online.
62 **Saltzer, J. H., & Schroeder, M. D.** *The Protection of Information in Computer Systems.* Proceedings of the IEEE, 63(9), 1278–1308, 1975. This paper introduces the principles of least privilege, fail-safe defaults, and separation of duties. Available online.
63 **NIST Cybersecurity Framework.** *Framework for Improving Critical Infrastructure Cybersecurity.* National Institute of Standards and Technology, 2018. This framework emphasizes continual monitoring, secure communication, and defense in depth. Available online.

64 **OWASP Foundation.** *OWASP Security Principles.* This reference provides a practical guide to applying security principles such as least privilege, secure defaults, and auditability in software development. Available online.

65 **ISO/IEC 27002: 2022.** *Information Security, Cybersecurity, and Privacy Protection — Information Security Controls.* International Organization for Standardization. This standard outlines the importance of monitoring, separation of duties, and secure communication. Available online.

66 **UcedaVelez, T., & Morana, M. M.** *Risk Centric Threat Modeling: Process for Attack Simulation and Threat Analysis (PASTA).* Wiley, 2015. Available online.

67 **Amazon Web Services, Inc. (2023).** *Amazon Managed Blockchain.* Retrieved from https://aws.amazon.com/managed-blockchain/

68 **IBM Corporation.** *IBM Blockchain Platform API Reference.* Retrieved from https://www.ibm.com/blockchain/platform

69 **Oracle Corporation.** *Oracle Blockchain Platform REST API Reference.* Retrieved from https://docs.oracle.com/blockchain/

70 **SAP SE.** *SAP Cloud Platform Blockchain Service API Reference.* Retrieved from https://api.sap.com/blockchain

71 **Infura.** *Infura Documentation.* Retrieved from https://infura.io/docs

72 **Alchemy.** *Alchemy API Documentation.* Retrieved from https://alchemy.com/docs

73 **Chainlink.** *Chainlink Documentation.* Retrieved from https://docs.chain.link

74 **Coinbase.** *Coinbase API Documentation.* Retrieved from https://developers.coinbase.com/api/v2

75 **Binance.** *Binance API Documentation.* Retrieved from https://binance-docs.github.io/apidocs/spot/en/

76 **Lamport, L. (2002)** *Specifying Systems: The TLA+ Language and Tools for Hardware and Software Engineers.* Addison-Wesley.

77 **Bertot, Y., & Castéran, P. (2004)** *Interactive Theorem Proving and Program Development: Coq'Art: The Calculus of Inductive Constructions.* Springer.

78 **Blockchain Solutions Ltd.** *ByzSim: Byzantine Fault Tolerance Simulation Framework.* Retrieved from https://github.com/BlockchainSolutions/byzsim

79 **Truffle Security Inc.** *TruffleHog: Secrets Management Scanner.* Retrieved from https://github.com/TruffleSecurity/truffleHog

80 **Kingsbury, K.** *Jepsen: Testing Framework for Distributed Systems.* Retrieved from https://jepsen.io

81 **Coin Sciences Ltd.** *MultiChain Streams: Blockchain Data Stream Analysis Tool.* Retrieved from https://www.multichain.com/streams/

82 **Ledger SAS.** *Ledger Wallets.* Retrieved from https://www.ledger.com

83 **SatoshiLabs s.r.o.** *Trezor Hardware Wallets.* Retrieved from https://trezor.io

84 **ConsenSys Software Inc. (2023).** *MetaMask: A Crypto Wallet & Gateway to Blockchain Apps.* Retrieved from https://metamask.io

85 **Ledger SAS.** *Ledger Nano S Product Page.* Retrieved from https://www.ledger.com/products/ledger-nano-s

86 **Ledger SAS.** *Ledger Nano X Product Page.* Retrieved from https://www.ledger.com/products/ledger-nano-x

87 **OpenZeppelin.** *OpenZeppelin: Secure Smart Contract Library for Blockchain Applications.* Retrieved from https://openzeppelin.com

88 **Ethereum Foundation.** *ERC-20 Token Standard.* Retrieved from https://eips.ethereum.org/EIPS/eip-20

89 **Ethereum Foundation.** *ERC-721 Non-Fungible Token Standard.* Retrieved from https://eips.ethereum.org/EIPS/eip-721

90 **Binance.** *BEP-20 Token Standard.* Retrieved from https://docs.binance.org/smart-chain/developer/BEP20.html

91 **TRON Foundation.** *TRC-20 Token Standard.* Retrieved from https://developers.tron.network/docs/trc20

92 **Neo Foundation.** *NEP-5 Token Standard.* Retrieved from https://docs.neo.org/docs/en-us/basic/contract/nep5.htm

93 **Solana Foundation.** *SPL Token Standard.* Retrieved from https://spl.solana.com/token

94 **Harmony.** *HRC-20 Token Standard.* Retrieved from https://docs.harmony.one/home/developers/api

95 **Vogelsteller, F., & Buterin, V. (2017).** *EIP-20: ERC-20 Token Standard.* Ethereum Improvement Proposals. Retrieved from https://eips.ethereum.org/EIPS/eip-20

96 **TokenSwap Development Team.** *TokenSwap: Decentralized Exchange for ERC-20 Tokens.* Retrieved from https://tokenswap.com/documentation

97 **CryptoLend Development Team.** *CryptoLend: Decentralized Lending Platform.* Retrieved from https://cryptolend.com/whitepaper

98 **OpenEX.** *Decentralized Exchanges and KYC: A Path to Trust and Compliance.* Retrieved from https://mirror.xyz/openexorg.eth/EmLairK_1INSXSmLhOpva5X2bKcceGVZcP-xZ7I3KX8

99 **Allen, C. (2016).** *The Path to Self-Sovereign Identity.* Retrieved from http://www.lifewithalacrity.com/2016/04/the-path-to-self-soverereign-identity.html

100 **ConsenSys.** *uPort: A Platform for Self-Sovereign Identity.* Retrieved from https://www.uport.me

101 **Sovrin Foundation.** *Sovrin: Secure and Sovereign Digital Identities.* Retrieved from https://sovrin.org

102 **Decentralized Identity Foundation.** *Understanding Decentralized IDs (DIDs).* Retrieved from https://identity.foundation

103 **OpenID Foundation.** *OpenID Connect.* Retrieved from https://openid.net/connect/

104 **Alchemy University.** *JavaScript Crash Course*. Alchemy University. Available at Alchemy University Courses

105 **Coursera.** *Solidity Smart Contract Security*. Available at https://www.coursera.org

106 **Udemy.** *Advanced Techniques in Solidity Security*. Available at https://www.udemy.com

107 **edX.** *Mastering Smart Contract Security with Solidity*. Available at https://www.edx.org

108 **Blockchain Security Professionals Association.** *Certified Blockchain Security Professional (CBSP) Program*. Further details can be accessed at Blockchain Security Professionals Association

109 **OWASP.** *Smart Contract Security Verification Standard*. OWASP. Available online at OWASP SCSVS.

110 **AuditFirst.** *Essential Security Practices for Decentralized Applications*. More information can be found at AuditFirst Security Blog.

111 **DEV Community.** *Secure dApp Development*. DEV Community Press. Detailed guidance is available on DEV.to.

112 **Ethereum.** *Solidity Style Guide*. Ethereum Foundation. More information can be found at the Solidity documentation.

113 **Ethlint.** *Ethlint: Solidity Linter for Security and Style Guidelines*. Available at Ethlint on GitHub.

114 **ConsenSys.** *ConsenSys Solidity Coding Recommendations*. ConsenSys Media.

115 **ConsenSys Software Inc.** *MythX: Security Analysis Platform for Ethereum Smart Contracts*. Available at MythX and ConsenSys Diligence.

116 **Truffle Suite.** *Truffle: A Development Environment, Testing Framework and Asset Pipeline for Blockchains*. Available at Truffle Suite.

117 **Nomic Labs.** *Hardhat: Ethereum Development Environment for Professionals*. Available at Hardhat.

118 **OpenZeppelin.** *OpenZeppelin Smart Contract Library*. Available at OpenZeppelin Contracts.

119 **SWC Registry.** *Smart Contract Weakness Classification*. Visit the SWC Registry for more information.

120 **ConsenSys Diligence.** *Solidity Security Blog*. Check out the Solidity Security Blog for comprehensive insights.

121 **Trail of Bits.** *Slither: Static Analysis for Solidity*. Slither can be accessed at GitHub – Slither.

122 **Mythril.** *Mythril: Security Analysis Tool for Ethereum Smart Contracts*. More details are available at GitHub – Mythril.

123 **Securify.** *Securify: Ethereum Smart Contract Security Scanner*. Visit Securify for further details.

124 **Truffle Suite.** *Ganache: Your Personal Blockchain for Ethereum Development*. Available at Truffle Suite Ganache.

125 OpenJS Foundation. *Mocha: Simple, Flexible, Fun JavaScript Test Framework.* Available at Mocha.

126 Chai. *Chai: BDD / TDD Assertion Library.* Available at Chai.

127 OpenZeppelin. *SafeMath: Library for Safe Mathematical Operations.* Available at GitHub – OpenZeppelin.

128 Brownie. *Brownie: A Python-Based Development Framework for Ethereum.* Available at Brownie Documentation.

129 Embark Labs. *Embark: Framework for Decentralized Application (dApp) Development.* Available at GitHub – embarklabs/embark.

130 Remix Team. *Remix: Ethereum IDE.* Available at Remix.

131 Vyper Team. *Vyper Documentation.* Available at Vyper Documentation.

132 Finley, K. (2016). *A $50 Million Hack Just Showed That the DAO Was All Too Human.* Wired. Available at Wired.com.

133 Norvill, R., Zohar, A., Teutsch, J., & Luu, L. (2019). *SoK: Transparent Dishonesty: Front-Running Attacks on Blockchain.* arXiv. Available at arXiv.org.

134 Apriorit. (2019). *Fomo3D Round Conclusion Vulnerability.* Apriorit. https://www.apriorit.com/dev-blog/556-fomo3d-vulnerability

135 NVD. CVE-2018-10299. https://nvd.nist.gov/vuln/detail/CVE-2018-10299

136 DeversiFi Team. (2020). *$23.7 Million Dollar Ethereum Transaction Fee: Post-Mortem.* Rhino.fi Blog. Available at https://rhino.fi/blog/23-7-million-dollar-ethereum-transaction-fee-post-mortem/

137 Quantstamp. (2020). *10 Quick and Dirty Facts about the bZx Hacks* https://quantstamp.com/blog/10-quick-and-dirty-facts-about-the-bzx-hacks

138 Gate.io. (2024). *MEV Overview: Hidden Value and Risks in Blockchain Networks.* Available at https://www.gate.io/learn/articles/mev-overview-hidden-value-and-risks-in-blockchain-networks/4922

139 CertiK. (2021). *The Poly Network Hack: First Cross-Chain Bridge Attack in 2021.* Available at CertiK.

140 Investopedia. (2022). *Binance Hit By $570 Million Blockchain Bridge Hack.* Available at Investopedia.

141 TechCrunch. (2021). *Hackers abuse Nomad exploit to drain almost $200M in crypto.* Available at https://techcrunch.com/2022/08/02/nomad-chaotic-exploit-crypto/

142 Kessler, S. (2022, April 17). *Attacker Drains $182M from Beanstalk Stablecoin Protocol.* CoinDesk. Available at https://www.coindesk.com/tech/2022/04/17/attacker-drains-182m-from-beanstalk-stablecoin-protocol

143 Cointelegraph. (2021). *SushiSwap Approval Bug Leads to $3.3M Exploit.* Available at https://cointelegraph.com/news/sushiswap-approval-bug-leads-to-3-3-million-exploit

144 **Fang, V., & Werlau, P. (2025, February 28).** *ByBit Billion Dollar Hack – Part 1: Smart Contracts Forensics Timeline*. AnChain.AI. Retrieved from https://www.anchain.ai/blog/bybit

145 **Smith, J. (2023).** *Analysis of the BtcTurk Hot Wallet Hack*. https://www.halborn.com/blog/post/explained-the-btc-turk-hack-june-2024

146 **OpenZeppelin. (2017).** *On the Parity Wallet Multisig Hack*. OpenZeppelin Blog. Available at https://blog.openzeppelin.com/on-the-parity-wallet-multisig-hack-405a8c12e8f7

147 **Mist Wallet. (2021).** *Security Alert: Critical Vulnerability in Mist Browser*. Available at Mist Wallet Official.

148 **Myagmar, S., Wijayarathna, G., Wichers, D., & Manico, J. (2023).** *OWASP SAMM v2.1: Software Assurance Maturity Model*. OWASP Foundation. Available at https://owaspsamm.org

149 **McGraw, G., Migues, S., & West, J. (2021).** *BSIMM12: The Building Security in Maturity Model*. Synopsys Software Integrity Group. Available at https://www.bsimm.com

150 **Stoneburner, G., Goguen, A., & Feringa, A. (2002).** *Risk Management Guide for Information Technology Systems* (NIST Special Publication 800-30). National Institute of Standards and Technology. https://doi.org/10.6028/NIST.SP.800-30

151 **Joint Task Force Transformation Initiative. (2013).** *Security and Privacy Controls for Federal Information Systems and Organizations* (NIST Special Publication 800-53, Rev. 4). National Institute of Standards and Technology. https://doi.org/10.6028/NIST.SP.800-53r4

152 **Scarfone, K., & Barrett, M. (2016).** *Guide to Data-Centric System Threat Modeling (Draft NIST Special Publication 800-154)*. National Institute of Standards and Technology (NIST). Available at https://csrc.nist.gov/files/pubs/sp/800/154/ipd/docs/sp800_154_draft.pdf

153 **OWASP Foundation. (2023).** *OWASP Application Security Verification Standard (ASVS) Version 4.0.3*. Open Web Application Security Project (OWASP). Retrieved from https://owasp.org/www-project-application-security-verification-standard

154 **Center for Internet Security (CIS). (2021).** *CIS Critical Security Controls Version 8*. Center for Internet Security. Retrieved from https://www.cisecurity.org/controls/v8

155 **Howard, M., & Lipner, S. (2006).** *The Security Development Lifecycle: A Process for Developing Demonstrably More Secure Software*. Microsoft Press. Amazon Link.

156 **UcedaVelez, T. M., & Morana, M. M. (2015).** *Risk Centric Threat Modeling: Process for Attack Simulation and Threat Analysis*. Wiley. Wiley Link.

157 **Dimova, M., Heyman, T., Scandariato, R., & Joosen, W. (2014).** *LINDDUN: A Privacy Threat Analysis Framework*. Official Website.

158 ThreatModeler. (2021). *The VAST Methodology*. Official Link.

159 Cuellar, A., Cowan, C., Freed, M., & Torrey, J. (2008). *Trike: A Unified Methodology for Security and Privacy Threat Modeling*. https://dymaxion.org/trike/Trike_Paper_Introduction.pdf

160 Schoenfield, B. S. E. (2015). *Securing Systems: Applied Security Architecture and Threat Models*. Boca Raton, FL: CRC Press. ISBN: 978-1-4987-0255-7. Available at https://www.routledge.com/Securing-Systems-Applied-Security-Architecture-and-Threat-Models/Schoenfield/p/book/9781498702557

161 IT Governance Publishing. (2020). *EU General Data Protection Regulation (GDPR) – An Implementation and Compliance Guide* (4th ed.). https://www.itgovernanceusa.com/shop/product/eu-general-data-protection-regulation-gdpr-an-implementation-and-compliance-guide-fourth-edition

162 Brill, J., Harris, M., & Rotenberg, M. (2021). *The California Consumer Privacy Act (CCPA): An Implementation Guide*. IT Governance Publishing. https://www.amazon.com/California-Consumer-Privacy-CCPA-implementation/dp/1787781321

163 Scharfman, J. *Cryptocurrency Compliance and Operations: Digital Assets, Blockchain, and DeFi*, Cham, Switzerland: Springer, 2021. ISBN: 978-3030879996. https://www.amazon.com/Cryptocurrency-Compliance-and-Operations_-Digital-Assets_-Blockchain-and-DeFi/dp/3030879992

164 European Commission. (2022). *Proposal for a regulation of the European Parliament and of the Council on horizontal cybersecurity requirements for products with digital elements and amending Regulation (EU) 2019/1020 (Cyber Resilience Act)* (COM(2022) 454 final). Brussels: European Commission. Retrieved from https://eur-lex.europa.eu/legal-content/EN/TXT/?uri=CELEX%3A52022PC0454

165 National Institute of Standards and Technology (NIST). (2012). *Guide for Conducting Risk Assessments (NIST Special Publication 800-30, Revision 1)*. U.S. Department of Commerce. https://csrc.nist.gov/publications/detail/sp/800-30/rev-1/final

166 National Institute of Standards and Technology (NIST). (2018). *Framework for Improving Critical Infrastructure Cybersecurity, Version 1.1 (NIST Cybersecurity Framework)*. U.S. Department of Commerce. https://www.nist.gov/cyberframework

167 Alberts, C., & Dorofee, A. (2003). *Managing Information Security Risks: The OCTAVE Approach*. Addison-Wesley. https://resources.sei.cmu.edu/asset_files/Book/2003_009_001_14051.pdf

168 Freund, J., & Jones, J. (2014). *Measuring and Managing Information Risk: A FAIR Approach*. Elsevier. ISBN: 978-0124202313. https://www.fairinstitute.org/book-measuring-and-managing-information-risk

169 **National Institute of Standards and Technology (NIST). (2010).** *Contingency Planning Guide for Federal Information Systems (NIST Special Publication 800-34 Rev. 1)*. Gaithersburg, MD: U.S. Department of Commerce. Business Impact Analysis (BIA) Template. Available at https://csrc.nist.gov/files/pubs/sp/800/34/r1/upd1/final/docs/sp800-34-rev1_bia_template.docx

170 **ISO 22301: 2019 and Business Continuity Management.** *Understand How to Plan, Implement, and Enhance a Business Continuity Management System*. IT Governance Publishing. ISBN: 978-1787782996. https://www.amazon.com/ISO-22301-continuity-management-Understand/dp/1787782999

171 **MetricStream. (2023).** *Top Governance, Risk, and Compliance (GRC) Tools in 2023*. MetricStream Blog. Available at https://www.metricstream.com/blog/top-governance-risk-compliance-grc-tools.html

172 **Fouzia A., Kawther S., & Liping Z.** Architectural Patterns for Blockchain Systems and Application Design. Published in *Applied Sciences*, 2023. https://www.mdpi.com/2076-3417/13/20/11533

173 **Amazon Web Services (AWS). (2023).** *Web3 Decentralized Applications on AWS: Reference Architecture for Blockchain Integration*. AWS Architecture Center. Available at https://docs.aws.amazon.com/architecture-diagrams/latest/web3-decentralized-applications-on-aws/web3-decentralized-applications-on-aws.html

174 **Draw.io.** *Online Diagram Software*. JGraph Ltd. https://www.drawio.com/

175 **Lucid Software Inc.** (n.d.). *Lucidchart: Intelligent Diagramming Application*. Retrieved from https://www.lucidchart.com/

176 **OWASP Foundation.** (n.d.). *OWASP Threat Dragon: Open-Source Threat Modeling Tool for Security Analysis*. Retrieved from https://owasp.org/www-project-threat-dragon/

177 **Microsoft Corporation.** (n.d.). *Microsoft Threat Modeling Tool: A Security-Focused Design Analysis Tool*. Retrieved from https://learn.microsoft.com/en-us/azure/security/develop/threat-modeling-tool

178 **Duo Security.** (n.d.). *CloudMapper: Open-Source AWS Security and Network Visualization Tool*. Retrieved from https://github.com/duo-labs/cloudmapper

179 **IriusRisk.** (n.d.). *IriusRisk: Automated Threat Modeling and Security Risk Assessment Platform*. Retrieved from https://www.iriusrisk.com/

180 **Threagile.** (n.d.). *Threagile: Agile Threat Modeling for Security Risk Analysis*. Retrieved from https://threagile.io/

181 **Avocado Systems.** (n.d.). *Avocado Systems: Real-Time Application Flow Monitoring for Runtime Security*. Retrieved from https://www.avocadosys.com/

182 **ThreatModeler.** (n.d.). *The Ultimate Guide to Threat Modeling*. ThreatModeler Inc. Retrieved from https://www.threatmodeler.com/the-ultimate-guide-to-threat-modeling

183 **FIRST.org, Inc. (2019).** *Common Vulnerability Scoring System (CVSS) version 3.1: Specification Document.* Retrieved from https://www.first.org/cvss/v3.1/specification-document

184 **Bhat, M., et al. (2023, February 24).** *How to Select DevSecOps Tools for Secure Software Delivery.* Gartner. Retrieved from https://www.gartner.com/en/documents/4131199

185 **MITRE.** (n.d.). *Common Vulnerabilities and Exposures (CVE).* Retrieved from https://cve.mitre.org/

186 **MITRE. (2024).** *Common Weakness Enumeration (CWE) Version 4.16.* Retrieved from https://cwe.mitre.org/data/index.html

187 **DeepFactor.** (n.d.). *Security Scanning Tools Defined: SAST*, IaC, SCA, DAST, IAST, RASP, *Container Runtime Security, and Runtime SCA.* DeepFactor. Retrieved from https://www.deepfactor.io/security-scanning-tools-defined-sast-iac-sca-dast-iast-rasp-container-runtime-security-and-runtime-sca/

188 **MythX.** (n.d.). *MythX: Smart Contract Security Service for Ethereum.* Retrieved from https://mythx.io

189 **Slither: Trail of Bits.** (n.d.). *Slither: Static analyzer for Solidity and Vyper.* Retrieved from https://github.com/crytic/slither

190 **Remix IDE: Remix Project.** (n.d.). *Remix - Ethereum IDE documentation.* Retrieved from https://remix-ide.readthedocs.io/

191 **OWASP Foundation.** (n.d.). *OWASP Zed Attack Proxy (ZAP).* Retrieved from https://owasp.org/www-project-zap/

192 **OpenZeppelin.** (n.d.). *Defender.* Retrieved from https://openzeppelin.com/defender/

193 **npm.** *npm Audit.* Retrieved from https://docs.npmjs.com/cli/v7/commands/npm-audit

194 **Snyk Ltd.** (n.d.). *Snyk.* Retrieved from https://snyk.io/

195 **Abdullah, M., Saleh, A., & Parvez, M. (2023).** Blockchain and cryptocurrency security from a new layered perspective and a novel MITRE ATT&CK-based approach for understanding cyberattacks and mitigating their impacts. *International Journal of Engineering Trends and Technology*, 72(4), 101–112. Retrieved from https://ijettjournal.org/Volume-72/Issue-4/IJETT-V72I4P101.pdf

196 **Kordy, B., Kordy, P., Mauw, S., & Schweitzer, P. (2013).** *ADTool: Security Analysis with Attack–Defense Trees.* Springer. Retrieved from https://satoss.uni.lu/members/piotr/adtool/

197 **Gartner.** (n.d.). *Breach and Attack Simulation (BAS) Tools, Market Reviews and Ratings.* Gartner. Retrieved from https://www.gartner.com/reviews/market/breach-and-attack-simulation-bas-tools

198 **Archer Integrated Risk Management.** (n.d.). *Build a GRC Framework.* Archer IRM. Retrieved from https://www.archerirm.com/solution/build-a-grc-framework

199 **Li, P. (2013).** *JIRA 5.2 Essentials.* Packt Publishing. Retrieved from https://www.amazon.com/JIRA-5-2-Essentials-Patrick-Li/dp/1782179992

200 **Antonopoulos, A. M., & Wood, G. (2018).** *Mastering Ethereum: Building Smart Contracts and dApps.* O'Reilly Media. Retrieved from https://www.oreilly.com/library/view/mastering-ethereum/9781491971932/

201 **Binance Chain Development Team. (2020).** *BNB Smart Chain Whitepaper.* Retrieved from https://github.com/bnb-chain/whitepaper/blob/master/WHITEPAPER.md

202 **Andres, S. (2022).** *Solana Development with Rust and Anchor.* Independently published. Retrieved from https://www.amazon.com/Solana-Development-Rust-Anchor-Sebastian/dp/B0BHVJQMM4

203 **Adams, H., Zinsmeister, N., & Robinson, D. (2020).** *Uniswap v2 Core.* Retrieved from https://uniswap.org/whitepaper.pdf

204 **Stani, K. (2020).** *Aave Protocol Whitepaper.* Retrieved from https://github.com/aave/protocol-v2/blob/master/aave-v2-whitepaper.pdf

205 **Maker Foundation. (2020).** *The Maker Protocol: MakerDAO's Multi-Collateral Dai (MCD) System.* Retrieved from https://makerdao.com/whitepaper/

206 **Pourpouneh, M., Nielsen, K., & Ross, O. (2020).** *Automated Market Makers.* University of Copenhagen, Department of Food and Resource Economics. Retrieved from https://okonomi.foi.dk/workingpapers/WPpdf/WP2020/IFRO_WP_2020_08.pdf

207 **Ponciano, J. (2021, August 10).** *More Than $600 Million Stolen in Ethereum and Other Cryptocurrencies—Marking One of Crypto's Biggest Hacks Ever.* Forbes. Retrieved from https://www.forbes.com/sites/jonathanponciano/2021/08/10/more-than-600-million-stolen-in-ethereum-and-other-cryptocurrencies-marking-one-of-cryptos-biggest-hacks-ever/

208 **Elliptic. (2022, February 3).** *$325 Million Stolen from Wormhole DeFi Service.* Elliptic. Retrieved from https://www.elliptic.co/blog/325-million-stolen-from-wormhole-defi-service

209 **Howell, S. T., Niessner, M., & Yermack, D. (2023, May 22).** *Anatomy of a Run: The Terra Luna Crash.* Harvard Law School Forum on Corporate Governance. Retrieved from https://corpgov.law.harvard.edu/2023/05/22/anatomy-of-a-run-the-terra-luna-crash/

210 **Ponciano, J. (2022, March 29).** *Second Biggest Crypto Hack Ever: $600 Million in Ether Stolen from NFT Gaming Blockchain.* Forbes. Retrieved from https://www.forbes.com/sites/jonathanponciano/2022/03/29/second-biggest-crypto-hack-ever-600-million-in-ethereum-stolen-from-nft-gaming-blockchain/

211 **Government of Singapore. (2012).** *Personal Data Protection Act 2012 (PDPA).* Singapore Statutes Online. Retrieved from https://sso.agc.gov.sg/Act/PDPA2012

212 **Chainalysis Team. (2023, March 7).** *Oracle Manipulation Attacks are Rising, Creating a Unique Concern for DeFi.* Retrieved from https://www.chainalysis.com/blog/oracle-manipulation-attacks-rising/

213 **Coinbase. (2020).** *Around the Block #3: Analysis on the bZx Attack, DeFi Vulnerabilities and the State of Debit Cards in Crypto.* Retrieved from https://www.coinbase.com/learn/market-updates/around-the-block-issue-3

214 **Immunefi. (2021, October 27).** *Hack Analysis: Cream Finance Oct 2021.* Retrieved from https://medium.com/immunefi/hack-analysis-cream-finance-oct-2021-fc222d913fc5

215 **Chainalysis Team. (2020, September 28).** *The KuCoin Hack: What We Know So Far and How the Hackers Are Using DeFi Protocols to Launder Stolen Funds.* Chainalysis. Retrieved from https://www.chainalysis.com/blog/kucoin-hack-2020-defi-uniswap/

216 **Commodity Futures Trading Commission.** (n.d.). *Digital Assets.* Retrieved February 20, 2025, from https://www.cftc.gov/digitalassets/index.htm

217 **Carlisle, D. (2024).** *The Crypto Launderers: Crime and Cryptocurrencies from the Dark Web to the Blockchain.* Wiley. https://www.amazon.com/Crypto-Launderers-Crime-Cryptocurrencies-Beyond/dp/1394203195

218 **Kelman Law. (2019, May 15).** *Money Transmitter Licensing for U.S. Crypto Companies.* Retrieved from https://kelman.law/money-transmitter-licensing-for-u-s-crypto-companies/

219 **OpenZeppelin.** (n.d.). *Defender Documentation.* Retrieved February 20, 2025, from https://docs.openzeppelin.com/defender/

220 **Chainalysis, Inc.** (n.d.). *Digital Asset Solutions for Regulators.* Retrieved February 20, 2025, from https://www.chainalysis.com/regulators/

221 **Eisenman, B. (2016).** *Learning React Native: Building Native Mobile Apps with JavaScript.* O'Reilly Media. https://www.oreilly.com/library/view/learning-react-native/9781491929049/

222 **Windmill, E., & Rischpater, R. (2020).** *Flutter in Action.* Manning Publications. https://www.manning.com/books/flutter-in-action

223 **Kumar, N. (2023).** *Mastering Blockchain Development: Building Decentralized Applications with Ethereum and Web3.* https://www.amazon.com/Mastering-Web3-Crypto-Investors-Builders/dp/B0CY7B2P98

224 **Khulbe, V. (2021, June 14).** *Tailwind CSS from Zero to Hero – Extracting Components and Setting up for Production.* daily.dev. Retrieved from https://daily.dev/blog/tailwind-css-from-zero-to-hero-extracting-components-and-setting-up-for-production-part-4-of-4

225 **Knight, F. (2020, March 20).** *Building Forms with Formik and React – Part 2.* DEV Community. Retrieved from https://dev.to/farleyknight/building-forms-with-formik-and-react-part-2-27a8

226 **Naik, N. (2020).** *uPort Open-Source Identity Management System: An Assessment of Self-Sovereign Identity and User-Centric Data Platform Built on Blockchain.* IEEE. https://ieeexplore.ieee.org/document/9272223

227 **Civic Technologies, Inc.** (n.d.). *Civic: Digital Identity Solutions for Web3.* Retrieved February 20, 2025, from https://www.gemini.com/cryptopedia/civic-identity-cvc-crypto-civic-crypto-cvc-token

228 **Benet, J. (2014).** *IPFS – Content Addressed, Versioned, P2P File System.* Retrieved from https://ipfs.io/ipfs/QmR7GSQM93Cx5eAg6a6MRyLJSa1JkUE1J1zpYoCFm2LkZg

229 **Williams, S. (2019).** *Arweave: A Protocol for Economically Sustainable Information Permanence.* Retrieved from https://www.arweave.org/yellow-paper.pdf

230 **Paul, H., & Holm, J. (2019).** *OrbitDB: Serverless, Distributed, Peer-to-Peer Database.* Retrieved from https://github.com/orbitdb/orbit-db

231 **Textile Team.** (n.d.). *Textile Documentation.* Retrieved from https://docs.textile.io/

232 **Ceramic Network Team.** (n.d.). *Ceramic Network Documentation.* Retrieved from https://developers.ceramic.network/

233 **Amazon Web Services.** (n.d.). *Amazon Managed Blockchain Documentation.* Retrieved from https://docs.aws.amazon.com/managed-blockchain/index.html

234 **Infura Team. (2018).** *Ethereum RPCs, Methods and Calls.* Retrieved from https://www.infura.io/blog/ethereum-rpcs-methods

235 **Alchemy Team.** (n.d.). *How to Read Data with JSON-RPC.* Retrieved from https://docs.alchemy.com/docs/how-to-read-data-with-json-rpc

236 **QuickNode Team.** (n.d.). *Core API – The Solution for Blockchain Apps.* Retrieved February 20, 2025, from https://www.quicknode.com/core-api

237 **MetaMask. (2024).** *web3-provider-engine: A JavaScript Library for Composing Ethereum Provider Objects Using Middleware Modules.* GitHub. Retrieved from https://github.com/MetaMask/web3-provider-engine

238 **MetaMask. (2024).** *Add WebSocket Subprovider.* GitHub. Retrieved from https://github.com/MetaMask/web3-provider-engine/issues/189

239 **Trust Wallet Team. (2024).** *Trust Wallet in 2024: A Year of Setting New Standards for Web3 Wallets.* Retrieved from https://trustwallet.com/blog/announcements/trust-wallet-in-2024-a-year-of-setting-new-standards-for-web3-wallets

240 **Fireblocks Team.** (n.d.). *Fireblocks Platform Overview and Demo.* Retrieved from https://www.fireblocks.com/webinars/fireblocks-platform-overview-and-demo-replay/

241 **Palaniachari, M. (2021).** *Building Enterprise Blockchain Solutions on AWS: A Developer's Guide to Build, Deploy, and Manage Apps Using Ethereum, Hyperledger Fabric, and AWS Blockchain.* BPB Publications. https://www.amazon.com/Building-Enterprise-Blockchain-Solutions-AWS/dp/9390684439

242 **Zellic Team. (2024).** *Solidity-Parser: A Python Library for Solidity Source Code Analysis.* Retrieved from https://github.com/Zellic/solidity-parser

243 **Foundry Project Contributors. (2024).** *Foundry Book: Writing Tests.* Retrieved from https://book.getfoundry.sh/forge/writing-tests

244 **Chainlink Labs.** (n.d.). *Security and Operation Best Practices.* Retrieved from https://docs.chain.link/chainlink-nodes/resources/best-security-practices

245 **API3 Foundation. (2020).** *API3: Decentralized APIs for Web 3.0.* Retrieved from https://old-docs.api3.org/api3-whitepaper-v1.0.3.pdf

246 **The Graph Team.** (n.d.). *Indexing Overview.* Retrieved from https://thegraph.com/docs/en/indexing/overview/

247 **Prisma.** (n.d.). *graphql-request: A Minimal GraphQL Client for Node and Browsers.* Retrieved from https://github.com/prisma-labs/graphql-request

248 **Dgraph Labs.** (n.d.). *Query Optimization in Graph Databases.* Retrieved from https://dgraph.io/blog/post/query-optimization/

249 **zkSync Documentation Team.** (n.d.). *zkSync SDK Documentation.* Retrieved from https://sdk.zksync.io/

250 **OP Labs Team. (2025).** *OP Stack Specification.* Retrieved from https://specs.optimism.io/

251 **Nelson, M. (2021).** *What Is ConsenSys Quorum*? Retrieved from https://consensys.io/blog/what-is-consensys-quorum

252 **POA Network.** (n.d.). *Blockscout: Open Source Ethereum Explorer.* Retrieved from https://blockscout.com/

253 **Wu, J., Lin, D., & Zheng, Z. (2024).** *Blockchain Transaction Data Analytics: Complex Network Approaches.* Springer. https://link.springer.com/book/10.1007/978-981-97-4430-5

254 **National Institute of Standards and Technology.** *Security Requirements for Cryptographic Modules.* FIPS PUB 140-2, U.S. Department of Commerce, 2001. https://csrc.nist.gov/publications/detail/fips/140/2/final.

255 **Ledger SAS.** *Ledger Nano X: Secure Hardware Wallet for Digital Assets.* Ledger SAS, 2023. https://www.ledger.com/products/ledger-nano-x

256 **Ledger SAS.** *Ledger Nano S Plus: Advanced Hardware Wallet for Cryptocurrency Management.* Ledger SAS, 2023. https://www.ledger.com/products/ledger-nano-s-plus

257 **Yubico.** *YubiHSM 2: Hardware Security Module for Cryptographic Protection.* Yubico, 2023. https://www.yubico.com/product/yubihsm-2/

258 **PeckShield Inc.** *Uniswap/Lendf.Me Hacks: Root Cause and Loss Analysis.* PeckShield, 2020. https://peckshield.medium.com/uniswap-lendf-me-hacks-root-cause-and-loss-analysis-50f3263dcc09

259 **Warmuz, J., Chaudhary, A., & Pinna, D.** *Toxic Liquidation Spirals.* arXiv, 2022. https://arxiv.org/abs/2212.07306

260 **CertiK.** *Post Mortem: Onyx Protocol.* CertiK, 2024. https://www.certik.com/resources/blog/post-mortem-onyx-protocol

261 **IntelOwl Project.** *IntelOwl: Open Source Threat Intelligence Platform.* IntelOwl, 2023. https://github.com/intelowlproject/IntelOwl

262 **Wang, B., et al.** *BLOCKEYE: Hunting for DeFi Attacks on Blockchain.* arXiv, 2021. https://arxiv.org/abs/2103.02873

263 **Signorini, M., et al.** *BAD: Blockchain Anomaly Detection.* arXiv, 2018. https://arxiv.org/abs/1807.03833

264 **Solidity Documentation Authors.** *Security Considerations.* In *Solidity Documentation*, Ethereum Foundation, 2023. https://docs.soliditylang.org/en/latest/security-considerations.html

265 **Hutchins, E. M., Cloppert, M. J., & Amin, R. M.** *Intelligence-Driven Computer Network Defense Informed by Analysis of Adversary Campaigns and Intrusion Kill Chains.* Lockheed Martin Corporation, 2011. https://www.lockheedmartin.com/content/dam/lockheed-martin/rms/documents/cyber/LM-White-Paper-Intel-Driven-Defense.pdf

266 **Lopuhaä-Zwakenberg, M., & Stoelinga, M. (2023).** *Cost-Damage Analysis of Attack Trees.* Retrieved from https://arxiv.org/abs/2304.05812

267 **Federal Risk and Authorization Management Program (FedRAMP).** *FedRAMP Vulnerability Scanning Requirements.* Version 3.0, February 2023. U.S. General Services Administration. Available at https://www.fedramp.gov/assets/resources/documents/CSP_Vulnerability_Scanning_Requirements.pdf

268 **Cybersecurity and Infrastructure Security Agency (CISA).** *Binding Operational Directive 19-02: Vulnerability Remediation Requirements for Internet-Accessible Systems.* CISA, 2019. https://www.cisa.gov/news-events/directives/bod-19-02-vulnerability-remediation-requirements-internet-accessible-systems

269 **European Central Bank. (2018).** *TIBER-EU: Framework for Threat Intelligence-Based Ethical Red Teaming.* Frankfurt am Main: European Central Bank. Retrieved from https://www.ecb.europa.eu/pub/pdf/other/ecb.tiber_eu_framework.en.pdf

270 **European Commission. (2022).** *Proposal for a Regulation on Horizontal Cybersecurity Requirements for Products with Digital Elements (Cyber Resilience Act).* Brussels: European Commission. https://eur-lex.europa.eu/legal-content/EN/TXT/?uri=CELEX:52022PC0454

271 European Union. (2022). Directive (EU) 2022/2555 of the European Parliament and of the Council of 14 December 2022 on measures for a high common level of cybersecurity across the Union (NIS2 Directive). *Official Journal of the European Union*, L 333, 80–152. https://eur-lex.europa.eu/eli/dir/2022/2555/oj

272 Williams, J. (2010). *OWASP Risk Rating Methodology*. Open Web Application Security Project (OWASP). https://owasp.org/www-community/OWASP_Risk_Rating_Methodology

273 National Institute of Standards and Technology (NIST). *Guide for Conducting Risk Assessments (NIST Special Publication 800-30, Revision 1)*. Gaithersburg, MD: U.S. Department of Commerce, September 2012. Available at https://csrc.nist.gov/publications/detail/sp/800-30/rev-1/final

274 Howard, M., & LeBlanc, D. (2003). *Writing Secure Code* (2nd ed.). Microsoft Press. https://www.microsoftpressstore.com/store/writing-secure-code-9780735617223

275 Riley, D. (2020, April 19). *$25M in cryptocurrency stolen in hack of Lendf.me and Uniswap*. SiliconANGLE. Retrieved from https://siliconangle.com/2020/04/19/25m-cryptocurrency-stolen-hack-lendf-uniswap/

276 Immunefi. (2022, April 18). *Hack Analysis: Beanstalk Governance Attack, April 2022*. Retrieved from https://medium.com/immunefi/hack-analysis-beanstalk-governance-attack-april-2022-f42788fc821e

277 Siegel, D. (2016). *Understanding the DAO Hack for Journalists*. Retrieved from https://www.coindesk.com/understanding-dao-hack-journalists

278 Palina T., Li, Y., Lin, S.-W., Liu, Y., & Li, Z. (2020). *A Survey of Smart Contract Formal Specification and Verification*. arXiv preprint arXiv:2008.02712 https://arxiv.org/abs/2008.02712?utm_source=chatgpt.com

279 Ethereum Foundation. (n.d.). Formal Verification. Ethereum Developer Documentation. Retrieved February 25, 2025, from https://ethereum.org/en/developers/docs/smart-contracts/formal-verification/

280 Certora. *Certora Technology White Paper*. Certora, 2022 https://www.certora.com/blog/white-paper?utm_source=chatgpt.com

281 Tsankov, P., et al. "Securify: Practical Security Analysis of Smart Contracts." *Proceedings of the 2018 ACM SIGSAC Conference on Computer and Communications Security*, 2018, pp. 67–82. https://dl.acm.org/doi/10.1145/3243734.3243780?utm_source=chatgpt.com

282 Truffle Suite. *Ganache Documentation*. Truffle Suite, 2022. https://archive.trufflesuite.com/docs/ganache/?utm_source=chatgpt.com

283 Hyperledger Caliper. *Hyperledger Caliper Documentation*. The Linux Foundation, 2023. https://lf-hyperledger.atlassian.net/wiki/spaces/caliper/overview?utm_source=chatgpt.com

284 **SmartContractKit.** *Chainlink Testing Framework*. GitHub, 2025. https://github.com/smartcontractkit/chainlink-testing-framework?utm_source=chatgpt.com

285 **Kosutic, D.** *Secure & Simple: A Small-Business Guide to Implementing ISO 27001 on Your Own*. Advisera Expert Solutions Ltd, 2016. ISBN: 978-953-57452-5-9. https://advisera.com/books/secure-and-simple-a-small-business-guide-to-implementing-iso-27001-on-your-own/?utm_source=chatgpt.com

286 **OpenZeppelin.** *Defender Documentation*. OpenZeppelin, 2025. https://docs.openzeppelin.com/defender/?utm_source=chatgpt.com

287 **European Parliament and Council of the European Union.** Regulation (EU) 2023/1114 of 31 May 2023 on Markets in Crypto-Assets. *Official Journal of the European Union*, L 150, 9 June 2023, pp. 40–205. https://en.wikipedia.org/wiki/Markets_in_Crypto-Assets?utm_source=chatgpt.com

288 **Financial Action Task Force.** *International Standards on Combating Money Laundering and the Financing of Terrorism & Proliferation: The FATF Recommendations*. FATF, 2012.https://www.fatf-gafi.org/content/dam/fatf-gafi/recommendations/FATF%20Standards%20-%2040%20Recommendations%20rc.pdf?utm_source=chatgpt.com

289 **U.S. Securities and Exchange Commission.** *Crypto Assets*. SEC.gov, 2025. https://www.sec.gov/securities-topics/crypto-assets?utm_source=chatgpt.com

290 **Commodity Futures Trading Commission.** *The CFTC's Role in Monitoring Virtual Currencies*. CFTC Monitoring Report, 2020. https://www.cftc.gov/digital assets/index.htm?utm_source=chatgpt.com

291 **Chainalysis Inc.** *Crypto Compliance Solutions*. Chainalysis, 2025. Available at https://www.chainalysis.com/solution/crypto-compliance/

292 **Rodriguez, J. J.** *In-Depth Look at Chainalysis, Elliptic, and CipherTrace: Forensic Science in Blockchain Analysis*. The Truth About Forensic Science, 2025. Available at https://thetruthaboutforensicscience.com/in-depth-look-at-chainalysis-elliptic-and-ciphertrace-forensic-science-in-blockchain-analysis/

293 **OpenZeppelin.** (n.d.). Math. OpenZeppelin Contracts 2.x API Documentation. Retrieved February 25, 2025, from https://docs.openzeppelin.com/contracts/2.x/api/math

294 **Solidity Team.** (n.d.). Solidity v0.8.0 Breaking Changes. Solidity 0.8.29 Documentation. Retrieved February 25, 2025, from https://docs.soliditylang.org/en/latest/080-breaking-changes.html

295 **Feist, J., & De Certeau, F.** (n.d.). Slither: A Solidity Static Analysis Framework. GitHub. Retrieved February 25, 2025, from https://github.com/crytic/slither

296 **Consensys.** (n.d.). Mythril: Security Analysis Tool for EVM Bytecode. GitHub. Retrieved February 25, 2025, from https://github.com/ConsenSys/mythril

297 **Almanax.** (n.d.). AI-Powered Security Analyzer. Retrieved February 25, 2025, from https://www.almanax.ai/

298 **Truffle Suite.** (n.d.). Test Your Contracts. Retrieved February 25, 2025, from https://trufflesuite.com/docs/truffle/how-to/debug-test/test-your-contracts/

299 **Nomic Foundation.** (n.d.). Testing Contracts. Hardhat Documentation. Retrieved February 25, 2025, from https://hardhat.org/tutorial/testing-contracts

300 **Foundry.** (n.d.). Foundry Book: Testing. Retrieved February 25, 2025, from https://book.getfoundry.sh/

301 **Certora.** (n.d.). Running the Certora Prover. Certora Documentation. Retrieved February 25, 2025, from https://docs.certora.com/en/latest/docs/user-guide/running.html

302 **Ethereum Foundation.** (n.d.). Token Standards. Retrieved February 25, 2025, from https://ethereum.org/en/developers/docs/standards/tokens/

303 **MDN Web Docs.** (n.d.). Crypto.getRandomValues(). Retrieved February 25, 2025, from https://developer.mozilla.org/en-US/docs/Web/API/Crypto/getRandomValues

304 **Saleem, F. (2023, August 15).** Ethereum's Elliptic Curve Digital Signature Algorithm (ECDSA). Medium. Retrieved February 25, 2025, from https://medium.com/@fitsaleem/ethereums-elliptic-curve-digital-signature-algorithm-ecdsa-88e1659f4879

305 **PCI Security Standards Council.** (n.d.). Payment Card Industry Data Security Standard (PCI DSS) Quick Reference Guide. Retrieved February 25, 2025, from https://www.pcisecuritystandards.org/documents/PCI_DSS-QRG-v3_2_1.pdf

About the Authors

Marco Morana is the Field CISO at Avocado Systems Inc., where he leads the deployment of runtime threat modeling and product security solutions for enterprise clients where he leads the deployment of runtime threat modeling and product security solutions for enterprise clients. Previously, he held senior security leadership roles at JPMorgan Chase and Citibank, where he directed global architecture programs and led S-SDLC adoption across cloud and application platforms. At Citibank, Marco conducted architecture risk assessments for blockchain pilots including the first digital asset trade with Nasdaq via Chain.com and authored the bank's first security standards for blockchain. He is the coauthor of the PASTA threat modeling methodology and OWASP project leader focused on secure-by-design for blockchain and AI systems.

Francesco Piccoli is the cofounder and CEO of Almanax, a cybersecurity firm specializing in AI-powered solutions. He was previously the Head of Product at AnChain.AI, where he built security and compliance tools for the US SEC, IRS, and Salesforce. His team was part of $100M+ crypto hacks investigations. Prior to AnChain.AI, he led research initiatives in anomaly detection and autonomous driving. Francesco holds a Master of Engineering from UC Berkeley and lives in New York.

Blockchain Application Security: How to Design Secure and Attack Resilient Blockchain Applications, First Edition. Marco Morana, Harpreet Singh and Francesco Piccoli.
© 2026 John Wiley & Sons, Inc. Published 2026 by John Wiley & Sons, Inc.

Harpreet Singh is a seasoned engineering leader with 19 years of experience driving innovation in AI/LLM, cybersecurity, and large-scale distributed systems. He is known for a strategic approach to architecture, a focus on customer trust and safety, and a collaborative leadership style. He has successfully launched new products, integrated emerging technologies, and optimized security practices across global organizations.

Index

Note: Page numbers in *italics* and **bold** refers to figures and tables respectively.

Blockchain Application Security: How to Design Secure and Attack Resilient Blockchain Applications, First Edition. Marco Morana, Harpreet Singh and Francesco Piccoli.
© 2026 John Wiley & Sons, Inc. Published 2026 by John Wiley & Sons, Inc.

d

e